AF712371

Semirings and their Applications

Dear MyCopy Customer,

This Springer book is a monochrome print version of the eBook to which your library gives you access via SpringerLink. It is available to you at a subsidized price since your library subscribes to at least one Springer eBook subject collection.

Please note that MyCopy books are only offered to library patrons with access to at least one Springer eBook subject collection. MyCopy books are strictly for individual use only.

You may cite this book by referencing the bibliographic data and/or the DOI (Digital Object Identifier) found in the front matter. This book is an exact but monochrome copy of the print version of the eBook on SpringerLink.

Semirings and their Applications

by

Jonathan S. Golan
University of Haifa,
Haifa, Israël

Springer-Science+Business Media, B.V.

A C.I.P. Catalogue record for this book is available from the Library of Congress.

DOI 10.1007/978-94-015-9333-5

Printed on acid-free paper

This book is an updated and considerably expended
version of the book "The Theory of Semirings,
with Applications to Mathematics and Theoretical science",
published by Longinov in 1992.

All Rights Reserved
© 1999 Springer Science+Business Media Dordrecht
Originally published by Kluwer Academic Publishers in 1999.
MyCopy version of the original edition 1999
No part of the material protected by this copyright notice may be reproduced or
utilized in any form or by any means, electronic or mechanical,
including photocopying, recording or by any information storage and
retrieval system, without written permission from the copyright owner.
www.springer.com/mycopy

In memory of my mother,
Prof. Naomi Golan

יקרה היא מפנינים וכל חפציה לא ישוו בה

CONTENTS

PREFACE

There is no branch of mathematics, however abstract, which may not some day be applied to phenomena of the real world.

—Nikolai Ivanovich Lobatchevsky

This book is an extensively-revised and expanded version of "**The Theory of Semirings, with Applications in Mathematics and Theoretical Computer Science**" [Golan, 1992], first published by Longman. When that book went out of print, it became clear — in light of the significant advances in semiring theory over the past years and its new important applications in such areas as idempotent analysis and the theory of discrete-event dynamical systems — that a second edition incorporating minor changes would not be sufficient and that a major revision of the book was in order. Therefore, though the structure of the first edition was preserved, the text was extensively rewritten and substantially expanded.

In particular, references to many interesting and applications of semiring theory, developed in the past few years, had to be added. Unfortunately, I find that it is best not to go into these applications in detail, for that would entail long digressions into various domains of pure and applied mathematics which would only detract from the unity of the volume and increase its length considerably. However, I have tried to provide an extensive collection of examples to arouse the reader's interest in applications, as well as sufficient citations to allow the interested reader to locate them. For the reader's convenience, an index to these citations is given at the end of the book.

Thanks are due to the many people who, in the past six years, have offered suggestions and criticisms of the preceding volume. Foremost among them is Dr. Susan LaGrassa, who was kind enough to send me a detailed list of errors — typographical and mathematical — which she found in it. I have tried to correct them all. During the 1997/8 academic year I conducted a seminar on semirings while on sabbatical at the University of Idaho. Many thanks are due to the participants of that seminar, and in particular to Prof. Erol Barbut and Prof. Willy Brandal, for their incisive comments. Prof. Dan Butnariu of the University of Haifa was also very instrumental in introducing me to applications of semirings in the theory of fuzzy sets, as where departmental guests Prof. Ivan Chajda, Prof. E. P. Klement and Prof. Radko Mesiar, while Dr. Larry Manevitz of the Department of Computer Science at the University of Haifa was always ready to help me understand applications in artificial intelligence and other areas of computer science. I also owe a large debt to my two former Ph.D. students, Dr. Wang Huaxiong and Dr. Wu Fuming, who listened patiently to my various p-baked ideas as they formed and contributed many original insights on semiring theory, which have been incorporated in this edition.

• • • •

Semirings abound in the mathematical world around us. Indeed, the first mathematical structure we encounter – the set of natural numbers – is a semiring. Other semirings arise naturally in such diverse areas of mathematics as combinatorics, functional analysis, topology, graph theory, Euclidean geometry, probability theory, commutative and noncommutative ring theory, optimization theory, discrete event dynamical systems, automata theory, formal language theory and the mathematical modeling of quantum physics and parallel computation systems. From an algebraic point of view, semirings provide the most natural common generalization of the theories of rings and bounded distributive lattices, and the techniques used in analysing them are taken from both areas.

Historically, semirings first appear implicitly in [Dedekind, 1894] and later in [Macaulay, 1916], [Krull, 1924], [Noether,1927], and [Lorenzen, 1939] in connection with the study of ideals of a ring. They also appear in [Hilbert, 1899] and [Huntington, 1902] in connection with the axiomatization of the natural numbers and nonnegative rational numbers. Semirings per se were first considered explicitly in [Vandiver, 1934], also in connection with the axiomatization of the arithmetic of the natural numbers. His approach was later developed in a series of expository articles culminating in [Vandiver & Weaver, 1956]. Over the years, semirings have been studied by various researchers either in their own right, in an attempt to broaden techniques coming from semigroup theory or ring theory, or in connection with applications. However, despite such categorical pronouncements as "... the above shows that *the ring is not the fundamental system for associative algebra of double composition*" (italics in the original) found in [Vandiver, 1939], semirings never became popular and the interest in them among algebraists gradually petered out, although it never died completely. The only attempt to present the algebraic theory of semirings as an integral part of modern algebra seems to be in [Rédei, 1967] and [Almeida Costa, 1974]. Nonetheless, semirings – and semimodules over them – have become an important tool in applied mathematics and theoretical computer science and appear, under various names, with consistent and increasing frequency in the literature of those subjects. Were there more communication between theoretical algebraists and these utilizers of algebra, it is likely that the former would find in the work of the latter sufficiently many "naturally arising" problems to revive and revitalize research in semiring theory in its own right, while the latter would find at their disposal a supply of theoretical results which they can use.

Since the results on semirings are scattered through the mathematical literature and are for most part inaccessible, they are not easily available to those who have to use them. A further problem is that the terminology used by different authors is not standard many authors use the term "semiring" for what we call here a "hemiring" and vice versa. Others, translating directly from the German, use the term "halfring". Some do not require that a semiring have a multiplicative identity or even an additive zero. On the other hand, some insist that multiplication, as well as addition, be commutative. In [Gondran & Minoux, 1984], the term "dioïd" (i.e., double monoid) is used in place of "semiring" and the term "semiring" is used in a stronger sense, while [Shier, 1973] prefers the term "binoid" for a commutative semiring in which addition is idempotent; others use "dioïd" for that purpose. A categorical definition of "semiring" (namely as a semiadditive category having one object) is given in [Manes, 1976]. To add to the confusion, some sources, e.g. [Sturm, 1986], use the term "semiring" to mean something else entirely. The reader must therefore be extremely wary.

• • • •

The notation used throughout the book will be explained as it is introduced. In addition, we will use the following standard notation:

$\mathbb{B} = \{0, 1\}$,
$\mathbb{P}$ = the set of all positive integers,
$\mathbb{N}$ = the set of all nonnegative integers,
$\mathbb{Z}$ = the set of all integers,
$\mathbb{Q}$ = the set of all rational numbers,
$\mathbb{Q}^+$ = the set of all nonnegative rational numbers
$\mathbb{I}$ = the unit interval on the real line,
$\mathbb{R}$ = the set of all real numbers,
$\mathbb{R}^+$ = the set of all nonnegative real numbers,
$\mathbb{C}$ = the set of all complex numbers.

If n is a positive integer then:

$\mathcal{S}_n$ is the group of all permutations of $\{1, \ldots, n\}$,
$\mathcal{A}_n$ is the group of all even permutations of $\{1, \ldots, n\}$.

If A and B are sets then:

$sub(A)$ is the family of all subsets of A,
$fsub(A)$ is the family of all finite subsets of A,
B^A is the set of all functions from A to B.

1. HEMIRINGS AND SEMIRINGS: DEFINITIONS AND EXAMPLES

A **semigroup** $(M, *)$ consists of a nonempty set M on which an associative operation $*$ is defined. If M is a semigroup in which there exists an element e satisfying $m * e = m = e * m$ for all $m \in M$, then M is called a **monoid** having **identity** element e. This element can easily seen to be unique, and is usually denoted by 1_M. Note that a semigroup $(M, *)$ which is not a monoid can be canonically embedded in a monoid $M' = M \cup \{e\}$ where e is some element not in M, and where the operation $*$ is extended to an operation on M' by defining $e * m' = m' = m' * e$ for all $m' \in M'$. An element m of M **idempotent** if and only if $m * m = m$. A semigroup $(M, *)$ is **commutative** if and only if $m * m' = m' * m$ for all $m, m' \in M$.

A monoid $(M, *)$ is **partially-ordered** if and only if there exists a partial order relation $\leq$ defined on M satisfying the condition that $m \leq m'$ implies that $m*m'' \leq m' * m''$ and $m'' * m \leq m'' * m'$ for all elements m, m', and m'' in M. Basic information on partially-ordered monoids can be found in [Fuchs, 1963].

A **hemiring** [resp. **semiring**] is a nonempty set R on which operations of addition and multiplication have been defined such that the following conditions are satisfied:

(1) $(R, +)$ is a commutative monoid with identity element 0;
(2) $(R, \cdot)$ is a semigroup [resp. monoid with identity element 1_R];
(3) Multiplication distributes over addition from either side;
(4) $0r = 0 = r0$ for all $r \in R$.

As a rule, we will write 1 instead of 1_R when there is no likelihood of confusion. Note that if $1 = 0$ then $r = r1 = r0 = 0$ for each element r of R and so $R = \{0\}$. In order to avoid this trivial case, we will assume that all semirings under consideration are nontrivial, i.e. that

(5) $1 \neq 0$.

Note that 0 is clearly the only element of R satisfying (4): if z is an element of R satisfying $zr = z = rz$ for all r in R then $0 = 0z = z$. The corresponding condition for addition, namely that 1 satisfies the condition that $1 + r = r$ for all $r \in R$, will be discussed later.

Conditions (4) and (5) insure that the operations of addition and multiplication are not the same. Even given that they are not, these conditions do not follow from the others.

(1.1) EXAMPLE. Consider the set $\mathbb{N}$ together with the operation $\oplus$ defined by letting $a \oplus b$ be the least common multiple of a and b, and the usual multiplication operation $\cdot$. Then conditions (1) - (3) are satisfied, while (4) and (5) are not since 1 is the identity element with respect to both operations.

Another example is the **algebra of digraphs**, developed in an unpublished paper by Anthony P. Stone. A **digraph** is a pair (V, E) consisting of a finite set V of **vertices** and a subset E of $V \times V$ of **edges**. Let R be the set of all digraphs and define addition and multiplication on R by setting $(V, E)+(V', E') = (V \cup V', E \cup E')$ and $(V, E) \cdot (V', E') = (V \cup V', E \cup E' \cup \{V \times V'\})$. Then conditions (1) - (3) are satisfied, while (4) and (5) are not since $(\varnothing, \varnothing)$ is the identity element with respect to both operations.

If R is the family of all subsets of a nonempty set X, define operations of addition and multiplication on R by setting $a + b = a \cap b$ and $ab = (a \cup b) \setminus (a \cap b)$. Then $(R, +, \cdot)$ satisfies conditions (1), (2), (3), and (5) (with additive identity X and multiplicative identity $\varnothing$) but does not satisfy (4).

In order to construct efficient computer programs for recognizing semirings, it is sometimes helpful to reduce the number of axioms which need to be checked to as small a number as possible. Several such reductions have been obtained, of which the following result is typical.

(1.2) PROPOSITION. *A set R containing two distinct elements 0 and 1 and on which operations + and $\cdot$ are defined is a commutative semiring if and only if the following conditions are satisfied for all $a, b, c, d, e \in R$:*

(1) $a + 0 = 0 + a = a$;
(2) $a1 = a$;
(3) $0a = 0$;
(4) $[(ae + b) + c]d = db + [a(ed) + cd]$.

PROOF. Surely any commutative semiring satisfies conditions (1) - (4). Conversely, assume that these four conditions are satisfied. If $b, d \in R$ then $bd = [(00 + b) + 0]d = db + [0(0d) + 0d] = db$ and so multiplication is commutative. If $a, b \in R$ then $a + b = [(a1 + b) + 0]1 = 1b + [a(11) + 01] = b + a$ and so addition is commutative. If $a, e, d \in R$ then $(ae)d = [(ae + 0) + 0]d = a0 + [a(ed) + 0d] = a(ed)$ and so multiplication is associative. If $a, b, c \in R$ then

$$(a + b) + c = (b + a) + c = [(b1 + a) + c]1 = 1a + [b(11) + c1] = a + (b + c)$$

and so addition is associative. Finally, if $a, b, d \in R$ then

$$(a + b)d = [(a1 + b) + 0]d = db + [a(1d) + 0d] = db + ad = ad + bd$$

and so multiplication distributes over addition. Thus R is a commutative semiring. □

In this work we will be interested primarily in semirings and will refer to hemirings only tangentially, as necessary. This approach is justified by the fact that if $(R, +, \cdot)$ is a hemiring then we can canonically embed it in a semiring in the following manner: let $S = R \times \mathbb{N}$ and define operations of addition and multiplication on S by setting $(r,n)+(r',n') = (r+r', n+n')$ and $(r,n)\cdot(r',n') = (nr'+n'r+rr', nn')$ for all $(r,n), (r'n') \in S$. Then $(S, +, \cdot)$ can be easily verified to be a semiring with multiplicative identity $(0,1)$, called the **Dorroh extension** of R **by** $\mathbb{N}$.

A subset S of a semiring R is a **subhemiring** of R if it contains 0 and is closed under the operations of addition and multiplication in R. If it also contains 1, it is a **subsemiring**. Thus, for example, if R is a semiring then

$$P(R) = \{0\} \cup \{r + 1 \mid r \in R\}$$

is a subsemiring of R. If R is a hemiring and S is a subhemiring of R which is a semiring having multiplicative identity e then then set $R \times S$, on which we define operations of addition and multiplication by $(r,s) + (r',s') = (r + r', s + s')$ and $(r,s) \cdot (r',s') = (rs' + sr' + rr', ss')$, is a semiring with multiplicative identity $(0, e)$, called the **Dorroh extension** of R **by** S.

The **center** of a hemiring R is $C(R) = \{r \in R \mid rr' = r'r \text{ for all } r' \in R\}$. This set is nonempty since it contains 0, and it is easily seen to be a subhemiring of R. If R is a semiring then $1 \in C(R)$ and $C(R)$ is a subsemiring of R. The hemiring R is **commutative** if and only if $C(R) = R$.

An element r of a hemiring R is **additively idempotent** if and only if $r + r = r$ for all r in R. The set $I^+(R)$ of all additively-idempotent elements of R is nonempty since it contains 0. The hemiring R is **additively idempotent** if and only if $I^+(R) = R$. Baccelli et al. [1992] and Gunawardena [1996] use the term "dioid" as a synonym for "additively-idempotent semiring". Note that if R is an additively-idempotent semiring then $\{0, 1\}$ is a subsemiring of R. Moreover, a necessary and sufficient condition for a semiring R to be additively idempotent is that $1 + 1 = 1$. Indeed, this condition is clearly necessary while, if it holds, then for each $r \in R$ we have $r = r(1 + 1) = r + r$, proving that R is additively idempotent. The computational complexity of determining whether two formulae over an additively idempotent semiring are equivalent is discussed in [Hunt, 1983]. For the complexity of related problems refer also to [Bloniarz, Hunt & Rosenkrantz, 1984]. Additively-idempotent semirings also arise naturally in the consideration of command algebras for computers; refer to [Hesselink, 1990]. Additively-idempotent semirings having three or four elements have been completely classified by Shubin [1992].

An element a of a hemiring R is **multiplicatively idempotent** if and only if $a^2 = a$. We will denote the set of all multiplicatively idempotent elements of R by $I^\times(R)$. This set is nonempty since $0 \in I^\times(R)$. If R is a commutative semiring then $I^\times(R)$ is a submonoid of $(R, \cdot)$. The hemiring R is **multiplicatively idempotent** if and only if $I^\times(R) = R$. If $0 \neq e \in I^\times(R)$ then $eRe = \{ere \mid r \in R\}$ is a subhemiring of R which is a semiring, though not a subsemiring of R unless $e = 1$.

An element a of a hemiring R is **multiplicatively regular** if and only if there exists an element b of R satisfying $aba = a$. Such an element b is called a **generalized inverse** of a. If b is a generalized inverse of a and $b' = bab$, then $ab'a = a$ and $b'ab' = b'$. An element satisfying these two conditions is a **Thierrin-Vagner**

inverse of a and we have thus seen that an element of R is multiplicatively regular if and only if it has a Thierrin-Vagner inverse. If a is multiplicatively regular then ab is multiplicatively idempotent. Conversely, every multiplicatively-idempotent element of R is surely multiplicatively regular. A hemiring R is **multiplicatively regular** if and only if each element of R is multiplicatively regular.

Set $I(R) = I^+(R) \cap I^\times(R)$. Elements of $I(R)$ are **idempotent**. Note that if $a \in I(R)$ then $\{0, a\}$ is a semiring contained in R, though it is not a subsemiring unless $a = 1$. The hemiring R is **idempotent** if it is both additively and multiplicatively idempotent, i.e. if and only if $R = I(R)$.

A hemiring R is **zerosumfree** if and only if $r + r' = 0$ implies that $r = r' = 0$. This condition states that the monoid $(R, +)$ is as far as possible from being a group: no nonzero element has an inverse. Note that a ring cannot be zerosumfree as a semiring. Indeed, if R is a ring then $-1 + 1 = 0$ in R, while both -1 and 1 are necessarily nonzero. Note that every additively-idempotent hemiring is zerosumfree. Indeed, if R is additively idempotent and if $r + r' = 0$ then

$$r = r + 0 = r + (r + r') = (r + r) + r' = r + r' = 0$$

and similarly $r' = 0$.

If R is a zerosumfree semiring then $R' = \{0\} \cup \{r \in R \mid rb \neq 0 \text{ for all } 0 \neq b \in R\}$ is a subsemiring of R. In order for a semiring R to be zerosumfree it suffices that there exist one element $t \in R$ satisfying $t = t + 1$. Indeed, if such an element exists and if $r + r' = 0$ then

$$\begin{aligned} 0 &= (r + r')t = rt + r't = r(1 + t) + r'(1 + t) \\ &= r(1 + t) + r'(1 + 1 + t) = (r + r') + r' + (r + r')t = r' \end{aligned}$$

and so $r = r + r' = 0$ as well.

A nonzero element a of a hemiring R is a **left zero divisor** if and only if there exists a nonzero element b of R satisfying $ab = 0$. It is a **right zero divisor** if and only if there exists a nonzero element b of R satisfying $ba = 0$. It is a **zero divisor** if and only if it is either a left and a right zero divisor. A hemiring R having no zero divisors is **entire**. In [Kuntzmann, 1972], entire zerosumfree semirings are called **information algebras**.

An element a of a hemiring R is **infinite** if and only if $a + r = a$ for all $r \in R$. Such an element is necessarily unique since if a and a' are infinite elements of R then $a = a + a' = a' + a = a'$. Note that 0 can never be infinite since $0 + 1 = 1 \neq 0$. A semiring R is **simple** if and only if 1 is infinite, that is to say if and only if $a + 1 = 1$ for all elements a of R. Equivalently, R is simple if and only if $P(R) = \{0, 1\}$. Commutative simple semirings are studied in [Cao, 1984] and [Cao, Kim & Roush, 1984] under the name of **inclines**. (But note that in [Kim & Roush, 1995] the commutativity condition has been removed.) If R is simple then, in particular, $1 + 1 = 1$ which suffices to show that R is additively idempotent. Conversely, if R is additively idempotent then $\{a \in R \mid a + 1 = 1\}$ is a subsemiring of R and so R is simple precisely when this subsemiring is all of R. Note that if R is a simple semiring and if $1 \neq a \in R$ then $ab \neq 1$ for all $b \in R$. Indeed, if $ab = 1$ then $a = a1 = a(b + 1) = ab + a = 1 + a = 1$.

The condition that 1 be infinite is the dual of the condition that $0r = 0 = r0$ for all $r \in R$, which we have assumed among the defining axioms of a semiring. The opposite of the notion of simplicity is that of an antisimple semiring: a semiring R is **antisimple** if and only if $R = P(R)$. Any ring is antisimple as a semiring.

The existence of simple semirings having more than one nonzero element (see below) shows that an infinite element a of a semiring R does not necessarily satisfy $ar = a$ for all $0 \neq r \in R$. An infinite element a of R having the property that $ra = a = ar$ for all $0 \neq r \in R$ is **strongly infinite**.

A semiring R is **semitopological** if and only if it has the additional structure of a topological space such that the functions $R \times R \to R$ defined by $(r, r') \mapsto r + r'$ and $(r, r') \mapsto rr'$ are continuous. If the underlying topological space is Hausdorff, then the semiring is **topological**. Any semiring is topological with respect to the discrete topology.

Rings are clearly semirings, but there are many other interesting examples of semirings. We conclude this chapter by assembling several such examples from various branches of mathematics and its applications.

(1.3) EXAMPLE. The set $\mathbb{N}$ of nonnegative integers with the usual operations of addition and multiplication of integers is a commutative, zerosumfree, entire semiring which is not additively idempotent. The same is true for the set $\mathbb{Q}^+$ of all nonnegative rational numbers, for the set $\mathbb{R}^+$ of all nonnegative real numbers, and, in general, for $S^+ = S \cap \mathbb{R}^+$, where S is any subring of $\mathbb{R}$. Given a fixed infinite cardinal number c, it is also true for the set of all cardinal numbers $d \leq c$. The semiring $\mathbb{N}$ is also antisimple. These semirings are among the first mathematical structures we encounter. Clearly $\mathbb{N}$ is a subsemiring of $\mathbb{Q}^+$ and $\mathbb{Q}^+$ is a subsemiring of $\mathbb{R}^+$. Note that $\{0, 1, 2, 3\} \cup \{q \in \mathbb{Q} \mid q \geq 4\}$ is an example of a subsemiring of $\mathbb{R}^+$ which is not of the form S^+ for some subring S of $\mathbb{R}$.

If S is one of the semirings $\mathbb{N}$, $\mathbb{Q}^+$, or $\mathbb{R}^+$ and if r is an element of S satisfying $r \geq 1$ then $R = \{a \in S \mid a > r\} \cup \{0\}$ is a subhemiring of S which is never a semiring. If $2 > r \geq 1$ then $\{a \in \mathbb{R} \mid a > r\} \cup \{0, 1\}$ is a subsemiring of $\mathbb{R}^+$.

This example can also be extended in the following manner: following the terminology of [Brunfiel, 1979], we say that a commutative semiring S is **partially ordered** if there exists a subset P of S satisfying the following conditions

(1) $P \cap (-P) = \{0\}$;
(2) $P + P \subseteq P$;
(3) $P \cdot P \subseteq P$;
(4) $s^2 \in P$ for all $s \in S$.

In this case, it is clear that $(P, +, \cdot)$ is a commutative zerosumfree semiring. Also refer to [Craven, 1991].

In the most general setting, Lawvere [1964] defined the notion of a **natural number object** N in an arbitrary topos. On such an object one can define operations of addition and multiplication in such a way as to turn N into a commutative semiring. This construction has important applications in the generalization of the theory of recursive functions. See [Coste-Roy, Coste & Mahé, 1980]. Natural number objects have since been defined in even more general contexts, such as that of arbitrary cartesian categories. See [Román, 1989].

Let S be a semiring containing $\mathbb{R}^+$ as a subsemiring. An **invariant metric** on S is a function $d: S \times S \to \mathbb{R}^+$ satisfying the following conditions:

(1) $d(s, s') = 0$ if and only if $s = s'$;
(2) $d(s, s') = d(s', s)$ for all $s, s' \in S$;
(3) $d(as, s') = ad(s, s')$ for all $s, s' \in S$ and all $a \in \mathbb{R}^+$;
(4) $d(s, s'') \leq d(s, s') + d(s', s'')$ for all $s, s', s'' \in S$;
(5) $d(s + s'', s' + s'') = d(s, s')$ for all $s, s', s'' \in S$.

Any such invariant metric defines on S the structure of a topological semiring. See [Bourne, 1961/2a].

This situation arises often in analysis. For example, let X be a compact space and let $C^+(X)$ be the semiring of all continuous functions from X to $\mathbb{R}^+$ with pointwise addition and multiplication. If X is a Tychonoff space, this is essentially the Stone-Čech compactification of X [Acharyya, Chattopadhyay & Ray, 1993]. Define the invariant metric d on $C^+(X)$ by setting $d(f, g) = sup\{|f(x) - g(x)| \mid x \in X\}$. Similarly, if W^+ is the cancellative semiring of all convergent series of the form $f(t) = \sum_{n=0}^{\infty} a_n e^{int}$ $(a_n \in \mathbb{R}^+)$ with the usual addition and multiplication, then we can define an invariant metric d on W^+ as follows: if $f(t) = \sum a_n e^{int}$ and $g(t) = \sum b_n e^{int}$ then $d(f, g) = \sum |a_n - b_n|$.

(1.4) EXAMPLE. Let R be a ring. Dedekind was the first to observe that the set $ideal(R)$ consisting of R and all of its ideals, with the usual operations of addition and multiplication of ideals, is an additively-idempotent (and hence zerosumfree) semiring which need not be commutative or entire. We will later see that the same is true for the family of all ideals of a semiring.

Now let R be a commutative ring and let A be the set of all elements of R which are not zero divisors. Let $S = A^{-1}R$ be the total ring of quotients of R. A **fractional ideal** K of R is an R-submodule of S satisfying the condition that $aK \subseteq R$ for some $a \in A$. The set $fract(R)$ of all fractional ideals of R is closed under taking intersections, sums, and products. Moreover, $(fract(R), +, \cdot)$ is a commutative additively-idempotent (and hence zerosumfree) semiring with additive identity (0) and multiplicative identity R. The family of all finitely-generated fractional ideals of R is a subsemiring of this semiring.

A commutative integral domain R is a **Prüfer domain** if and only if every finitely-generated fractional ideal of R has a multiplicative inverse in $fract(R)$. This condition is equivalent to the condition that, in $ideal(R)$, intersection distributes over addition, i.e. that $(ideal(R), +, \cap)$ is a semiring. See Theorem 6.6 of [Larsen & McCarthy, 1971] or [Gilmer, 1972] for a proof of this fact. Moreover, the set of all finitely-generated ideals of R is a subsemiring of this semiring. The study of rings with the property that $(ideal(R), +, \cap)$ is a distributive lattice goes back to [Blair, 1953]. See [Tuganbaev, 1998] for a comprehensive bibliography of the many works in this area. Noetherian Prüfer domains are called **Dedekind domains**. These are precisely the commutative integral domains having the property that each ideal can be written as a unique product of prime ideals. The multiplicative theory of ideals of a ring is certainly one of the major sources of inspiration and problems in semiring theory.

(1.5) EXAMPLE. Another major source of inspiration for the theory of semirings is lattice theory. If $(R, \vee, \wedge)$ is a bounded distributive lattice having unique minimal element 0 and unique maximal element 1, then it is a commutative, idempotent simple semiring. Indeed, these properties uniquely characterize bounded distributive lattices: if R is a commutative, idempotent, simple semiring then $(R, +, \cdot)$ is a bounded distributive lattice having unique minimal element 0 and unique maximal element 1. Another well-known characterization of bounded distributive lattices is the following: $(R, \vee, \wedge)$ is a bounded distributive lattice having unique minimal element 0 and unique maximal element 1 if and only if it is a commutative idempotent semiring and $a \wedge (a \vee b) = a = a \vee (a \wedge b)$ for all $a, b \in R$. Henriksen [1958/9] gives another characterization of the bounded distributive lattices in the family of semirings by showing that a commutative semiring R is a bounded distributive lattice if and only if the following conditions are satisfied for every element a of R:

(1) $(1 + a)^2 = 1 + a \Rightarrow 1 + a = 1$;

(2) There exists a natural number $n(a) > 1$ such that $a^{n(a)} = a$.

Yet another such condition is given by Głazek [1968]: a semiring R is a bounded distributive lattice if and only if $ab + cd = (a + c)(a + d)(b + cd)$ for all $a, b, c, d \in R$. The theory of modules having the property that their lattice of submodules, with the operations of addition and intersection, is summarized in [Tuganbaev, 1998].

Since the dual lattice of a distributive lattice is again distributive, we see that $(R, \wedge, \vee)$ is also a commutative, simple semiring.

As a particular case, we note that every frame is a semiring. A **frame** (alias **complete brouwerian lattice**, alias **locale**, alias **local lattice**, alias **Heyting algebra**, alias **pointless topology**) is a complete lattice in which meets distribute over arbitrary joins. The study of frames is rooted in the topological work of Marshall Stone [1937] as extended in [Benabou, 1959], [Ehresmann, 1957], and [Papert, 1959, 1964] and later further extended by Dowker, Isbell, and Strauss and in the logical studies of Skolem. For further details concerning frames, see [Birkhoff, 1973], [Johnstone, 1982], or [Rasiowa & Sikorski, 1963].

If a and b are elements of a frame $(L, \vee, \wedge)$ then the **pseudocomplement of b relative to a**, denoted $(a : b)$, is the unique largest element c of L satisfying $b \wedge c \leq a$. The **pseudocomplement** of an element a of L is $(0 : a)$. If $(0 : a) = 0$, then a is **dense** in L. The dense elements of a frame are precisely those elements which are not zero divisors.

The simplest example of a frame is $\mathbb{B} = \{0, 1\}$. Note that the algebraic structure of $\mathbb{B}$ is not the same as that of the field $\mathbb{Z}/(2)$ since $1+1 = 1$ in $\mathbb{B}$, whereas $1+1 = 0$ in $\mathbb{Z}/(2)$. The variety generated by the semirings $\mathbb{B}$ and $\mathbb{Z}/(2)$ is characterized in [Guzmán, 1992]. The semiring $\mathbb{B}$ is called the **boolean semiring**; it has many applications in automata theory (see [Eilenberg, 1974]) and in switching theory, where it is often known as the **switching algebra**. For generalization of switching theory over other finite semirings, see [Berman & Mukaidono, 1984] and [Muzio & Wesselkamper, 1986]. A well-known topology on $\mathbb{B}$ is the **Sierpinski topology**, the open sets of which are $\varnothing$, $\{1\}$, and $\mathbb{B}$. Given this topology, the semiring $\mathbb{B}$ is semitopological but not topological. If R is an additively-idempotent hemiring, then it makes sense to define the **Dorroh extension** of R by $\mathbb{B}$ in exactly the same way that the Dorroh extension of R by $\mathbb{N}$ was previously defined.

The lattice of all ideals of a distributive lattice is a frame; similarly, the lattice of all closed ideals of a commutative C^*-algebra is also well-known to be isomorphic to the lattice of all open sets of a locally compact Hausdorff space and so is a frame. Hence they are canonically semirings.

Another important instance of this construction is the following: if X is any topological space then the family of all closed subsets of X is a bounded distributive lattice and hence a semiring, with addition taken to be intersection and multiplication taken to be union. (For a consideration of the problem of precisely which distributive lattices are isomorphic to lattices of closed sets, refer to [Papert, 1959].) Similarly, the family of all open subsets of X is a semiring, with addition taken to be intersection and multiplication taken to be union. If R is the semiring of all closed subsets of a topological space X then a **basis** for R is a subsemiring S of R having the property that every element of R is the intersection of elements of S. Any bounded distributive lattice is isomorphic to a basis of the semiring of closed subsets of a compact topological space. For a proof, see [Sancho de Salas, 1987].

Similarly, if R is a totally-ordered set with unique minimal element 0 and unique maximal element 1 then (R, max, min) is a distributive lattice and hence a semiring. Cechlárová and Plávka [1996] call such semirings **bottleneck algebras**, to emphasize their connection with bottleneck problems in combinatorial optimization. In particular, we have a natural semiring structure on $\mathbb{I}$. Any subset of $\mathbb{I}$ containing 0 and 1 is a subsemiring of this semiring. Similarly, $(\mathbb{N} \cup \{\infty\}, max, min)$ is a zerosumfree commutative simple semiring. Note that if R is a semiring of this sort and if S is any subset of R containing 0 and 1 then S is a subsemiring of R. Also, note that if R is a bounded distributive lattice and $0 \neq r \in R$ then r is idempotent so $rRr = [0, r] = \{r' \in R \mid r' \leq r\}$ is a subhemiring of R which is a semiring in its own right, having multiplicative identity r.

If R is a bounded distributive lattice, then the ideal topology on R turns R into a semitopological semiring. Since this topology need not be Hausdorff, the semiring is not necessarily topological. Conditions for R to be a topological semiring are discussed in [Murty, 1974].

(1.6) EXAMPLE. An element $a \neq 1_M$ of a monoid $(M, \cdot)$ is **absorbing** if and only if $ab = a = ba$ for all $b \in M$. If M has an absorbing element, it is clearly unique. From the definition of a semiring R, we note that the monoid $(R, \cdot)$ has an absorbing element 0. Conversely, let $(M, \cdot)$ be a multiplicative monoid having an absorbing element 0. Define addition on M by setting $a + b = 0$ for all $a, b \in M$. Then $(M, +)$ is an abelian semigroup, $(M, \cdot)$ is a monoid, and multiplication distributes over addition from either side. Let u be an element not in M and set $R = M \cup \{u\}$, and define operations on S as follows:

(1) If $a, b \in M$ then $a + b$ and ab are as in M;
(2) If $a \in S$ then $a + u = u + a = a$ and $au = ua = u$.

Then $(S, +, \cdot)$ is a semiring with additive identity u, which is both zerosumfree and entire. See [LaGrassa, 1995].

Thus the theory of multiplicative monoids with absorbing elements can be subsumed in the theory of semirings. Such monoids arise in various contexts, such as the modeling of knapsack problems in combinatorics, which in turn have important applications in the construction of public-key cyphers.

(1.7) EXAMPLE. Let $R-fil$ be the set of all topologizing filters of left ideals of a ring R (which, in a natural way, correspond to the linear topologies on R), and let $\cdot$ denote the Gabriel product of such filters. Then $(R-fil, \cap, \cdot)$ is a simple additively-idempotent semiring which is not, in general, commutative. The structure of $R-fil$ and its use in ring theory have been considered in detail in [Golan, 1987].

(1.8) EXAMPLE. There is no lack of finite semirings. For example, for each positive integer n consider the set $X_n = \{-\infty, 0, 1, ..., n\}$ in which $-\infty$ is assumed to satisfying the conditions that $-\infty \leq i$ and $-\infty + i = -\infty$ for all $i \in X_n$. Define operations of addition and multiplication on X_n by $i + h = max\{i, h\}$ and $ih = min\{i + h, n\}$. This gives X_n the structure of a commutative zerosumfree semiring, first studied in [Smith, 1966].

Another important family of finite semirings is considered in [Alarcón & Anderson, 1994a]. Let $n > 1$ be an integer and let $0 \leq i \leq n - 1$. Set $B(n, i) = \{0, 1, \ldots, n-1\}$ and define an operation $\oplus$ on $B(n, i)$ as follows: if $a, b \in B(n, i)$ then $a \oplus b = a + b$ if $a + b \leq n - 1$ and, otherwise, $a \oplus b$ is the unique element c of $B(n, i)$ satisfying $c \equiv a + b \pmod{n - i}$. Define the an operation $\odot$ on $B(n, i)$ similarly. Thus, $B(n, 0)$ is a ring isomorphic to $\mathbb{Z}/(n)$ and if $i > 0$ then $B(n, i) \setminus \{0\}$ is the cyclic semigroup generated by 1 of period $n-i$ and index i. Clearly $\mathbb{B} = B(2, 1)$.

(1.9) EXAMPLE. In many categorical situations we have "sums" and "products" satisfying semiring-like conditions. Rather than enter into such abstracta in detail which we have no intention of pursuing, we present here some special cases which suffice to illustrate the general situation.

Let $\mathcal{A}$ be the family of isomorphism classes of additive abelian groups and denote by $[G]$ the isomorphism class of a group G. Then $\mathcal{A}$ is a commutative semiring under the operations of addition and multiplication defined by $[G] + [H] = [G \oplus H]$ and $[G][H] = [G \otimes H]$. The multiplicative identity of $\mathcal{A}$ is $[\mathbb{Z}]$. This semiring is considered in [Feigelstock, 1980]. More generally, if R is an arbitrary ring and if $R - Mod - R$ is the family of isomorphism classes of (R, R)-bimodules then we can define operations of addition and multiplication on $R - Mod - R$ by setting $[M] + [N] = [M \oplus N]$ and $[M][N] = [M \otimes_R N]$.

Let $\mathcal{C}$ be the family of isomorphism classes of countable Boolean algebras and denote by $[R]$ the isomorphism class of a Boolean algebra R. Then $\mathcal{C}$ is a commutative semiring under the operations of addition and multiplication defined by $[R] + [S] = [R \oplus S]$ and $[R][S] = [R * S]$, where $\oplus$ denotes the direct sum of boolean algebras and $*$ denotes the free product. The additive identity of $\mathcal{C}$ is the class of one-element algebras and its multiplicative identity is the class of two-element algebras. This semiring is considered by Dobbertin [1982] and R. Pierce [1983, 1989].

If $\mathcal{D}$ is the family of homeomorphism classes of compact zero-dimensional metric spaces of finite type and if $[X]$ denotes the homeomorphism class of a space X, then $\mathcal{D}$ is a countably-infinite commutative semiring under the operations of addition and multiplication defined by $[X] + [Y] = [X + Y]$ and $[X][Y] = [X \times Y]$, where $X + Y$ is the disjoint union of X and Y and $X \times Y$ is the cartesian product of X and Y. The additive identity of $\mathcal{D}$ is $[\varnothing]$ and the multiplicative identity is $[\{x\}]$. This semiring is studied in [Pierce, 1972], where its additive structure is completely described.

Similarly, if G is a finite group then a **finite** G**-set** S is a finite set together with a left action of G on it. The family of all G-isomorphism classes of finite G-sets forms a commutative semiring $A^+(G)$ in which addition is defined by disjoint union and multiplication by the cartesian product with diagonal action. This construction is given in [tom Dieck, 1979].

(1.10) EXAMPLE. If $(M, *)$ is a semigroup, then the family $R = sub(M)$ of all subsets of M is a hemiring, with operations of addition and multiplication given by $A + B = A \cup B$ and $AB = \{a * b \mid a \in A,\ b \in B\}$. The additive identity is $\varnothing$. If M is a monoid, $sub(M)$ is a semiring with multiplicative identity $\{1_M\}$. We will make much use of semirings of this form later.

This example is weakened in [Kuntzmann, 1972], where the semigroup $(M, *)$ is replaced by a partial semigroup, i.e. a nonempty set M with an operation $*$ defined on subset of $M \times M$ but subject to the condition that if a, b, and c are elements of M for which $a * (b * c)$ and $(a * b) * c$ are both defined, then these two elements must be equal. Partial semirings and other partial algebras are finding more and more applications in the theory of abstract data types in theoretical computer science. Refer to [Manes & Arbib, 1986] and [Reichel, 1987].

If R is a semiring then we can also define the structure of a semiring on $sub(R)$ by setting $A + B = \{a + b \mid a \in A; b \in B\}$ and $AB = \{ab \mid a \in A; b \in B\}$.

(1.11) EXAMPLE. The following example presents one of the most important applications of semiring theory. If A is a nonempty set then the **free monoid** A^* is the set of all finite strings $a_1 a_2 \ldots a_n$ of elements of A (including the empty string, which is denoted by $\square$). Two strings $a_1 a_2 \ldots a_n$ and $b_1 b_2 \ldots b_m$ are equal in A^* if and only if $n = m$ and $a_i = b_i$ for all $1 \leq i \leq m$. We define on A^* the operation of concatenation:

$$a_1 a_2 \ldots a_n \cdot b_1 b_2 \ldots b_k = a_1 a_2 \ldots a_n b_1 b_2 \ldots b_k.$$

The set A^* is a monoid under this operation, the identity element of which is $\square$. The elements of A are often called **symbols** or **letters** and the elements of A^* are called **words** on these symbols. If $w = a_1 a_2 \ldots a_n$ is a word in $A^* \setminus \{\square\}$ then the natural number n is the **length** of w and is denoted by $|w|$; by convention, $|\square| = 0$. Clearly $|ww'| = |w| + |w'|$ for all words w and w' in A^*. For each $a \in A$ and $w \in A^*$ we denote by $|w|_a$ the number of occurrences of a in the word w. The function p which assigns to each $w \in A^*$ the function $p(w)\colon A \to \mathbb{N}$ defined by $p(w)\colon a \mapsto |w|_a$ is called the **Parikh mapping**.

Subsets of A^* are **(formal) languages** on the **alphabet** A. If $B \neq \square$ is a submonoid of A^* and $B' = B \setminus \{\square\}$ then $C = B' \setminus (B')^2$ is a minimal set of generators of B, called the **base** of B. A base of a free submonoid of A^* is called a **(variable length) code** over the alphabet A. Thus, for example, a nonempty subset C of $A^* \setminus \{\square\}$ satisfying the condition that $uw \notin C$ for all $u \in C$ and $\square \neq w \in A^*$ is a code. Codes of this type are known as **prefix codes**. Codes have their origin in Shannon's early work on information transmission, though they were given an algebraic formulation only much later. For an introduction to formal language theory, see [Rozenberg & Salomaa, 1980]; for codes and their uses, refer to [Berstel & Perrin, 1985], [Lallement, 1979] or [Shyr, 1979].

If the set A is finite, say $A = \{a_1, \ldots, a_k\}$, then we have a bijective function $\varphi: A^* \to \mathbb{N}$ defined by $\varphi(\square) = 0$ and $\varphi(a_{i_0}a_{i_1}\ldots a_{i_n}) = \sum_{h=0}^{n} i_h k^h$.

If m is an element of a semigroup $(M, *)$ then the set of **left divisors** of m is $LD(m) = \{m' \in M \mid m = m' * m'' \text{ for some } m'' \in M\}$ and the set of **right divisors** of m is $RD(m) = \{m'' \in M \mid m = m' * m'' \text{ for some } m' \in M\}$. These sets may be empty for an arbitrary semigroup M but that cannot be the case if M is a monoid. Indeed, if M is a monoid then $\{1, m\} \subseteq LD(m) \cap RD(m)$ for all $m \in M$. A monoid $(M, *)$ is **finitary** if and only if any element of M can be written as $m' * m''$ for only finitely-many choices of m' and m'' in M. Note that if $w \in A^*$ then $LD(w)$ and $RD(w)$ are always finite and that, as a result, free monoids are finitary. More generally, we can consider monoids of the form $M = A_1^* \times \ldots \times A_n^*$, where the A_i are nonempty sets and multiplication on M is defined componentwise:

$$(w_1, \ldots, w_n) \cdot (y_1, \ldots, y_n) = (w_1 y_1, \ldots, w_n y_n).$$

Such monoids are also finitary.

It is possible to extend the definition of A^* to include words of infinite length. Let A^∞ consist of A^* together with all countably-infinite sequences of elements of A. Define the operation $\cdot$ on A^∞ by setting ww' to be the concatenation of w and w' if $w \in A^*$ and $ww' = w$ if w is a countably-infinite sequence of elements of A. Then $(A^\infty, \cdot)$ is again a monoid the identity element of which is $\square$. Subsets of A^∞ are **∞-languages** on A. These constructions were first defined in [Park, 1981] and are important in the modeling of concurrent computational systems.

Let A be a nonempty set. As in Example 1.10, we can define operations of addition and multiplication on $sub(A^*)$ as follows: $L + L' = L \cup L'$, while $LL' = \{ww' \mid w \in L \text{ and } w' \in L'\}$. Then $(sub(A^*), +, \cdot)$ is an additively-idempotent entire semiring in which the additive identity is $\varnothing$ and the multiplicative identity is $\{\square\}$. This semiring originated in Kleene's algebraic formulation of the theory of machines [Kleene, 1956], and is at the heart of algebraic automata theory. Refer to [Conway, 1971], [Gécseg & Peák, 1972], [Lallement, 1979], and [Salomaa & Soittola, 1978]. It is easy to verify that the set of all elements of $sub(A^*)$ which are not prefix codes is a subsemiring of this semiring. The smallest subsemiring of $sub(A^*)$ containing all singletons and closed under intersections is called the semiring of all **starfree languages** on A.

Other variants on the above semiring are possible. Let A be a finite set and define an operation $*$ on A^* as follows: if $u = a_1 a_2 \ldots a_n$ and $v = b_1 b_2 \ldots b_m$ are elements of A^* then $u * v = a_1 a_2 \ldots a_n b_2 \ldots b_m$ if $a_n = b_1$ and $u * v = \square$ otherwise. If U and V are nonempty subsets of A^* set $U * V = \{u * v \mid u \in U \text{ and } v \in V\}$. If $U = \varnothing$ or $V = \varnothing$ set $U * V = \varnothing$. This is called the **Latin product** of U and V. Then we can define a semiring structure on $sub(A^*)$ by taking addition to be union as before and multiplication to be the Latin product. This semiring is of important use in enumeration problems in graph theory; see [Gondran & Minoux, 1984a]. In another variant, we define the **fusion product** on $sub(A^*)$ by setting $U * V = \{uxv \mid ux \in U; xv \in V\}$. See [Ying, 1991] for details.

Another product which can be defined on $sub(A^*)$ is the **shuffle product** $\otimes$. If $u, v \in A^*$ and $a, b \in A$, we inductively define

$$1 \otimes u = u \otimes 1 = \{u\}$$

$$(au \otimes bv) = a(u \otimes bv) + b(au \otimes v)$$

and then, for $L, L' \subseteq A^*$, we set

$$L \otimes L' = \cup\{u \otimes v \mid u \in L; v \in L'\}.$$

It is easy to check that $(sub(A^*), \cup, \otimes)$ is a semiring, which has important applications in models of concurrent computation. See [Bloom, Sabadini & Walters, 1996] or [Golan, Mateescu & Vaida, 1996].

(1.12) EXAMPLE. If R is a nonempty set then a function $\circledast$ from $R \times R$ to $sub(R)$ can be extended to an operation on $sub(R)$ by setting $A \circledast B = \cup\{a \circledast b \mid a \in A, b \in B\}$. In imitation of the definition of a hyperring - as given, for example, in [Ciampi & Rota, 1987] - we can define a **hypersemiring** to be a nonempty set R together with functions $+$ and $\cdot$ from $R \times R$ to $sub(R)$ satisfying the following conditions:

(1) Addition is associative and commutative;
(2) Multiplication is associative and distributes over addition from either side;
(3) There exists an element 0 of R such that, for all $r \in R$ we have $0 + r = \{r\}$ and $0 \cdot r = \{0\} = r \cdot 0$.
(4) There exists an element 1 of R such that, for all $r \in R$ we have $1 \cdot r = \{r\} = r \cdot 1$.
(5) $1 \neq 0$.

Note that if R is a hypersemiring then $sub(R)$ is a semiring with respect to the operations of addition and multiplication extended as above. The additive identity of $sub(R)$ is $\{0\}$ and the multiplicative identity of $sub(R)$ is $\{1\}$. A related notion is also discussed in [Nakano, 1967]; refer also to [Mockor, 1977].

As an example of such a construction, let G be a lattice-ordered group and let z be an element not in G. Set $R = G \cup \{z\}$ and extend the operation of G to R by setting $rz = zr = z$ for all $r \in R$. Also assume that $r \wedge z = r$ and $r \vee z = z$ for all $r \in R$. Define functions $+$ and $\cdot$ from $R \times R$ to $sub(R)$ as follows:

(1) If $g, g' \in G$, set $g + g' = \{g'' \in G \mid g \wedge g' = g'' \wedge g' = g \wedge g''\}$ and extend this to a function from $R \times R$ to $sub(R)$ by setting $z \in r + r'$ if and only if $r = r'$ for all $r, r' \in R$;
(2) If $r, r' \in R$ then $r \cdot r' = \{rr'\}$.

Then $(R, +, \cdot)$ is a hypersemiring; for a proof, see Example 3.4 of [Mockor, 1983].

(1.13) EXAMPLE. The additive structure of a semiring does not necessarily determine its multiplicative structure. Thus, for example, if $(R, +, \cdot)$ is any noncommutative semiring then we can define another semiring $(R, +, \circ)$, having the same additive structure, called the **opposite semiring** of $(R, +, \cdot)$, by setting $a \circ b = b \cdot a$ for all a and b in R. A more graphic set of examples is the following. A **triangular norm (t-norm)** on $\mathbb{I}$ is defined to be an operation $\sqcap$ on $\mathbb{I}$ satisfying the following conditions:

(1) $(\mathbb{I}, \sqcap)$ is a commutative monoid with identity element 1;
(2) $a \leq b$ in $\mathbb{I}$ implies that $a \sqcap c \leq b \sqcap c$ for all $c \in \mathbb{I}$.

From these conditions it follows that $0 \sqcap a = 0$ for all $a \in \mathbb{I}$. Triangular norms were first introduced by Menger [1942] and have proven useful in the theory of probabilistic metric spaces and in multivalued logic. Measures based on triangular norms are used in mathematical statistics [Dvoretzky, Wald & Wolfowitz, 1952] and [Schweizer & Sklar, 1961], capacity theory [Frank, 1979], probability theory [Schmidt, 1982], game theory [Aumann & Shapley, 1979] and [Butnariu & Klement, 1993], and pattern recognition [Sugeno, 1979]. For the use of triangular norms in defining propositional fuzzy logics, refer to [Butnariu et al., 1995]. It is straightforward to see that if $\sqcap$ is a triangular norm on $\mathbb{I}$ then $(\mathbb{I}, max, \sqcap)$ is a semiring. But there are infinitely-many triangular norms definable on $\mathbb{I}$! For example, for each $s \in \mathbb{R}^+ \cup \{\infty\}$ we can define the **fundamental triangular norm** $\sqcap_s$ on $\mathbb{I}$ as follows:

(1) $a \sqcap_0 b = min\{a, b\}$;
(2) $a \sqcap_1 b = ab$;
(3) $a \sqcap_\infty b = max\{0, a + b - 1\}$;
(4) $a \sqcap_s b = log_s\{[(s-1) + (s^a - 1)(s^b - 1)]/(s-1)\}$ for all $s \in \mathbb{R}^+ \setminus \{0, 1\}$.

Thus we have an infinite family of multiplication operations which turn the commutative monoid $(\mathbb{I}, max)$ into a commutative semiring. Other infinite families of triangular norms on $\mathbb{I}$ can also be found in [Schweizer & Sklar, 1963] and [Weber, 1983]. If $R_s = (\mathbb{I}, max, \sqcap_s)$ for each $s \in R^+ \cup \{\infty\}$ then $I^\times(R_s) = \{0, 1\}$ for all $s \neq 0$, while $I^\times(R_0) = R_0$. For various ways of looking at triangular norms, refer to [Mesiar & Pap, 1998].

Dually, a **triangular conorm** (**t-conorm**) on $\mathbb{I}$ is an operation $\sqcup$ on $\mathbb{I}$ satisfying the following conditions:

(1) $(\mathbb{I}, \sqcup)$ is a commutative monoid with identity element 0;
(2) $a \leq b$ in $\mathbb{I}$ implies that $a \sqcup c \leq b \sqcup c$ for all $c \in \mathbb{I}$.

From these conditions, it follows that $1 \sqcup a = 1$ for all $a \in \mathbb{I}$ and so $(\mathbb{I}, min, \sqcup)$ is a semiring with additive identity 1 and multiplicative identity 0. Every triangular norm $\sqcap$ defines a corresponding triangular conorm $\sqcup$ by $a \sqcup b = 1 - [(1-a) \sqcap (1-b)]$ and every triangular conorm on $\mathbb{I}$ is definable in this manner. Thus we also have an infinite family of multiplication operations which turn the commutative monoid $(\mathbb{I}, min)$ into a commutative semiring.

If $s \in \mathbb{R}^+ \cup \{\infty\}$ and $\mathbb{I}$ is topologized with its usual topology, then $(\mathbb{I}, max, \sqcap_s)$ and $(\mathbb{I}, min, \sqcup_s)$ are both topological semirings. Indeed, if $*$ is a continuous function from $\mathbb{I} \times \mathbb{I}$ to $\mathbb{I}$ such that $(\mathbb{I}, *)$ is a monoid with identity element 1 such that $a * 0 = 0 = 0 * a$ for all $a \in \mathbb{I}$ then $(\mathbb{I}, max, *)$ must be a commutative topological semiring. Similarly, if $(\mathbb{I}, *)$ is a monoid with identity element 0 such that $a * 1 = 1 = 1 * a$ for all $a \in \mathbb{I}$ then $(\mathbb{I}, min, *)$ must be a commutative topological semiring. Refer to [Frank, 1979]. For further examples of t-norms and t-conorms and their computer-generated pictorial representations, refer to [Mizumoto, 1989]. For representation of t-norms as well as similar operations which turn $(\mathbb{R}^+ \cup \{\infty\}, max)$ into a topological semiring, see [Ling, 1966].

(1.14) EXAMPLE. [Bourne, 1951; Heatherly, 1974] If $(M, +)$ is a commutative monoid with identity element 0 then the set $End(M)$ of all endomorphisms of M is a semiring under the operations of pointwise addition and composition of functions. Thus, for example, if $M = (\mathbb{R} \cup \{\infty\}, min)$ then $End(M)$ is the set of all nondecreasing functions on M. This semiring is of use in certain types of path problems in graph theory, see [Gondran & Minoux, 1984a]. If M is idempotent then, for each $m \in M$, the function $\alpha_m \colon M \to M$ defined by $\alpha_m \colon x \mapsto m$ for all $x \in M$ is a member of $End(M)$ and $\{\alpha_m \mid m \in M\}$ is a subhemiring of $End(M)$ which is not a subsemiring.

We note that $End_0(M) = \{\alpha \in End(M) \mid \alpha(0) = 0\}$ is a subsemiring of $End(M)$. If $\theta \colon M \to End_0(M)$ is a function satisfying

(1) $\theta(m+m') = \theta(m) + \theta(m')$;
(2) $\theta(m)(m') = \theta(m) \circ \theta(m')$; and
(3) $\theta(0) \colon m \mapsto 0$.

for all m and m' in M then θ defines a multiplication operation on M by setting $m \cdot m' = \theta(m)(m')$ such that $(M, +, \cdot)$ is a semiring. Indeed, all possible semiring structures on $(M, +)$ arise in this manner.

(1.15) EXAMPLE. Let R be a semiring, let $0 \neq b \in R$, and let D be a finite subset of R containing 0. Then R has **base** b and set of **digits** D if every $r \in R$ has a unique representation in the form $d_0 + d_1 b + \cdots + d_n b^n$, where the d_i belong to D. For example, for $R = \mathbb{N}$ take $2 \leq b \in \mathbb{N}$ and $D = \{0, \ldots, b-1\}$. For the use of such semirings in generating fractals, refer to [Allouche et al., 1995].

(1.16) EXAMPLE. If R is an infinite commutative integral domain and if $f(X,Y)$ and $g(X,Y)$ are polynomials over R, we can define operations $\oplus$ and $\odot$ on R by setting $a \oplus b = f(a,b)$ and $a \odot b = g(a,b)$. Necessary and sufficient conditions for $(R, \oplus, \odot)$ to be a hemiring or a semiring in this situation have been studied in [Petrich, 1965].

(1.17) EXAMPLE. In Example 1.13 we saw that the additive structure of a semiring does not necessarily determine its multiplicative structure. The multiplicative structure of a semiring does not determine the additive structure either. For instance, on the multiplicative monoid $(\mathbb{R}^+, \cdot)$ we can define two semiring structures by taking addition to be ordinary addition of numbers or taking addition to be maximum. Pearson [1966, 1968a] has classified all operations $\oplus$ on subintervals of $\mathbb{R} \cup \{\infty\}$ which, together with ordinary multiplication, turn them into topological semirings.

Let $R = \mathbb{N} \times \mathbb{N} \times \mathbb{N}$. Define addition on R componentwise and define multiplication on R by setting

$$(a, a', a'')(b, b', b'') = (a''b + ab'', a''b' + a'b'' + ab + ab' + a'b + a'b', a''b'').$$

Then R is an entire semiring. Another semiring structure on R is obtained by defining multiplication componentwise. We have already seen that $(\mathbb{N}, +, \cdot)$ has a semiring structure. If $h, k \in \mathbb{N}$, let $h \sqcup k$ be the greatest common divisor of h and k in $\mathbb{N}$. Then it is straightforward to verify that $(\mathbb{N}, \sqcup, \cdot)$ is also a semiring. More generally, a **standard thread** is a topological semigroup on a closed subinterval $I = [a, b]$ of $\mathbb{R}$, together with an operation $*$ defined on I such that $(I, *)$ is a monoid with identity element b and the element a satisfies $a * x = x * a = a$ for all $x \in I$. The problem of finding all possible operations $+$ on a standard thread over which $*$ distributes is solved in [Mak & Sigmon, 1988]. In particular, this solution yields all semiring structures on I with given multiplication $*$. Refer also to [Cao, Kim & Roush, 1984]. This situation can also occur for finite semirings. If $R = \{0, u, 1\}$ is ordered by $0 < u < 1$ then (R, max, min) is a semiring. However, R can be given another semiring structure $(R, \oplus, min)$, where $a \oplus b$ is defined to be equal to $max\{a, b\}$ unless $a = b = 1$, while $1 \oplus 1 = u$. This semiring has applications in the design of computer circuitry; see [Hu, 1975].

(1.18) EXAMPLE. On a ring R, define the operation $\circ$ by $a \circ b = a + b - ab$. This operation has been well studied and plays an important role in the definition of the Jacobson radical. Let S be a nonempty set of commuting elements of R containing 0 and closed under both $\circ$ and multiplication, which satisfies the **self-distributivity condition** with respect to multiplication, namely that $a^2bc = abc$ for all $a, b, c \in S$. Then $(S, \circ, \cdot)$ is a hemiring. Thus, for example, we could take S to be the set $I^{\times}(R) \cap C(R)$ of central idempotents of R. For a discussion of this construction in a more general setting, see [Birkenmeier, 1989]. Note that if 1_R belongs to S then S is in fact a simple semiring, since $a \circ 1_R = a$ for all $a \in S$.

(1.19) EXAMPLE. Let R be an additively-idempotent hemiring and define a new operation $\circ$ on R by $a \circ b = a + b + ab$. Then $(R, +, \circ)$ satisfies conditions (1)-(3) of a semiring but is not a semiring since its additive and multiplicative identities coincide.

An interesting interpretation of this operation also arises in the context of the **join geometries** studied in [Prenowitz & Jantosciak, 1979]. Let E be a nonempty convex subset of euclidean space and let $R = sub(E)$. For $a, b \in E$, let ab be the element of R defined as follows:

(1) $aa = \{a\}$;

(2) If $a \neq b$ then ab is the open line segment connecting a and b.

As in Example 1.10, extend the definition of product to an operation on R by setting $AB = \cup\{ab \mid a \in A, b \in B$ for all $A, B \in R\}$. Note that $AB = \varnothing$ if and only if $A = \varnothing$ or $B = \varnothing$. It is straightforward to verify that $(R, \cup, \cdot)$ is a commutative, additively-idempotent entire hemiring with additive identity $\varnothing$. Moreover, $A \in I^{\times}(R)$ if and only if A is a convex subset of E. If a and b are distinct elements of E then $a, b \notin ab$. The closed line segment connecting a and b is precisely $a \circ b = \{a\} \cup \{b\} \cup ab$.

We note that the hemiring R can be embedded in a semiring as follows: let z be a point not in E and let $S = sub(E \cup \{z\})$. Extend the definition of product by setting $az = za = a$ for all $a \in E \cup \{z\}$ and, as before, set

$$AB = \cup\{ab \mid a \in A, b \in B \text{ for all } A, B \in S\}.$$

Then $(S, \cup, \cdot)$ is a semiring with multiplicative identity $\{z\}$. Refer also to [Lyndon, 1961].

(1.20) EXAMPLE. [Martelli, 1974, 1976] Let X be a set and let $P = sub(X)$. A subset U of P is a **clutter** if and only if $A \neq B \in U$ implies that $A \not\subseteq B$. Any finite set U contains a clutter $U^{\#}$ obtained by deleting from U all supersets of sets in U. On the family R of all finite clutters of elements of P we can define addition and multiplication by setting $U+V = (\{A\cup B \mid A\in U, B\in V\})^{\#}$ and $U\cdot V = (U\cup V)^{\#}$. Under these operations, R is a commutative additively-idempotent semiring with additive identity $\{\emptyset\}$ and multiplicative identity $\emptyset$.

(1.21) EXAMPLE. A **prenucleus** on a frame $(L, \vee, \wedge)$ is a function $z: L \to L$ satisfying the following conditions:

(1) $c \leq z(c)$ for all $c \in L$;
(2) If $c \leq c'$ in L then $z(c) \leq z(c')$;
(3) $z(c \wedge c') = z(c) \wedge z(c')$ for all $c, c' \in L$.

Let $\mathbf{PN}(L)$ be the set of all prenuclei on L and define the operation $\wedge$ on $\mathbf{PN}(L)$ componentwise: $(z \wedge z'): c \mapsto z(c) \wedge z'(c)$ for all $c \in L$. Then $\mathbf{PN}(L)$, together with addition given by $\wedge$ and multiplication given by composition of functions, is a zerosumfree simple semiring which has important applications in the analysis of frames. For structural consideration of $\mathbf{PN}(L)$, where $L = R - tors$ is the frame of all (hereditary) torsion theories on a module category $R - mod$, see [Golan & Simmons, 1988].

(1.22) EXAMPLE. Let $R = \mathbb{R} \cup \{\infty\}$. Then $(R, min, +)$ is an additively-idempotent commutative semiring in which addition is the operation of taking minimum and multiplication is ordinary addition. As we shall see later, this semiring is important in solving the shortest-path problem in optimization. If $S = \mathbb{R}^+ \cup \{\infty\}$ then $(S, min, +)$ is a simple subsemiring of $(R, min, +)$ with infinite element 0. For uses of this semiring in optimization theory, see [Gondran & Minoux, 1984a]; for its uses in analysis, refer to [Maslov & Sambourskiĭ, 1992] and [Kolokol'tsov & Maslov, 1997]. It can also replace the semiring $(\mathbb{R}^+, +, \cdot)$ to obtain a new type of probability theory, first studied by Maslov [1987] and later by Akian [1995a, 1995b] and her collaborators [Akian, Quadrat & Viot, 1998]. This semiring has also found applications in multicriteria optimization, optimal control, and the theory of semantic domains [Sünderhauf, 1997]. Indeed, computation in this semiring is so important that Lam and Tong [1996] have proposed a hardware implementation in analog processing circuits.

The semiring S has a subsemiring $(\mathbb{N} \cup \{\infty\}, min, +)$, known as the **tropical semiring**, which has important applications in the theory of formal languages and automata theory, including the capture of the nondeterministic complexity of a finite automaton. Refer to [Mascle, 1986], [Simon, 1988], and [Pin, 1998]. The one-point compactification of $\mathbb{N} \cup \{\infty\}$, endowed with the discrete topology, is the set $\mathbb{N} \cup \{\omega, \infty\}$. This set can be totally-ordered by setting $i < \omega < \infty$ for all $i \in \mathbb{N}$ and has the topological structure defined by taking as open sets all subsets of $\mathbb{N} \cup \{\infty\}$ and all sets of the form $([\mathbb{N} \cup \{\infty\}] \setminus A) \cup \{\omega\}$, where A is a finite subset of $\mathbb{N} \cup \{\infty\}$. This set too can be turned into a semiring with operations $\oplus$ and $\otimes$ defined by:

(1) $a \oplus b = min\{a, b\}$;
(2) $a \otimes b = a + b$ if $a, b \in \mathbb{N}$ and $max\{a, b\}$ otherwise.

See [Leung, 1988] and [Simon, 1994] for details and applications; for applications to synchronized elementary net systems, see [Andre, 1989].

In a manner similar to the above, we see that $(\mathbb{R} \cup \{-\infty\}, max, +)$ is an additively-idempotent commutative semiring. This semiring is called the **schedule algebra** or, sometimes, the **max-plus algebra**. Cuninghame-Green [1979] illustrates how it can be used in the analysis of the behavior of industrial processes. Also refer to [Cheng, 1987], [Cruon & Hervé, 1965], [Cuninghame-Green, 1962, 1976, 1991] and [Cuninghame-Green & Borawitz, 1984]. Much of this work is based on ideas presented informally in [Giffler, 1963, 1968]. A more formal presentation of Giffler's schedule algebra is given in [Wongseelashote, 1976]. For the use of this semiring in finding critical paths in graphs, refer to [Carré, 1979]. For its use in discrete-event dynamical systems refer to [Baccelli et al., 1992], [Gaubert, 1996a], [Gaubert & Max Plus, 1997], [Gunawardana, 1994], [Olsder, 1991, 1992]. For applications to control theory, see [Mairesse, 1985] and for applications to automata theory, see [Krob, 1998]. Also refer to [Max-Plus Working Group, 1995].

Among the subsemirings of the schedule algebra which we will discuss later are $(\mathbb{Z} \cup \{-\infty\}, max, +)$ and the antisimple semiring $(\mathbb{N} \cup \{-\infty\}, max, +)$. Another important subsemiring of this semiring is $(\mathbb{R}^{+} \cup \{-\infty\}, max, +)$. This semiring has important applications in the categorical approach to the theory of metric spaces, which was first developed in [Lawvere, 1973]. In fact, we can restrict ourself to $(F \cup \{-\infty\}, max, +)$, where F is any submonoid of $(\mathbb{R}, +)$.

(1.23) EXAMPLE. If $(R, +, \cdot)$ is a semiring and X is a set together with a bijective function $\delta\colon X \to R$ then the semiring structure on R induces a semiring structure $(X, \oplus, \odot)$ on X with the operations defined by $x \oplus y = \delta^{-1}(\delta(x) + \delta(y))$ and $x \odot y = \delta^{-1}(\delta(x) \cdot \delta(y))$. Such constructions can often lead to interesting examples such as the following one, mentioned in [Mullin, 1975]: Let R be the semiring of all functions from $\mathbb{N}$ to itself with the operations of componentwise addition and multiplication. Define a function δ from $\mathbb{R}^{+}$ to R which sends each nonnegative real number into its representation as a continued fraction. Then $\delta(r)(i) = 0$ for only finitely-many $i \in \mathbb{N}$ if and only if r is irrational. Since the family of all $f \in R$ satisfying the property that $f(i) = 0$ for only finitely-many $i \in \mathbb{N}$ is closed under taking componentwise sums and products, we see that we have an induced semiring structure on $\{0\} \cup [\mathbb{R}^{+} \setminus \mathbb{Q}^{+}]$.

(1.24) EXAMPLE. Finally, we mention another example arising from theoretical computer science. Bergstra and Klop [1983, 1984, 1986, 1989] have constructed an **algebra of communicating processes (ACP)** to formalize the actions in a distributive computation environment. Such an algebra consists of a finite set R of **atomic actions**, among which is a distinguished action δ "deadlock"), on which we have operations $+$ ("choice") and $|$ ("communication merge") satisfying the conditions that $(R, +, |)$ is a commutative additively-idempotent hemiring with additive identity δ. In addition, there is another operation $\cdot$ ("sequential composition") defined on R such that $(R, \cdot)$ is a semigroup satisfying $a \cdot \delta = \delta$ for all $a \in R$ and such that $\cdot$ distributes over $+$ from the right but not necessarily from the left. Refer also to [Baeten, Bergstra & Klop, 1988]. For other related algebras of communicating processes which form semirings, see [Cherkasova, 1988] and [Hennessy, 1988].

(1.25) EXAMPLE. If $(R, +, \cdot)$ is a semiring then one can define operations $\oplus$ and $\odot$ on $sub(R)$ by setting $A \oplus B = \{a + b \mid a \in A, b \in B\}$ and $A \odot B = \{ab \mid a \in A, b \in B\}$. Then $(sub(R), \oplus, \odot)$ need not be a semiring. To see this, consider the following example, due to [Litvinov, Maslov & Sobolevskiĭ, 1998]: let $R = (\mathbb{R} \times \mathbb{R}) \cup \{-\infty\}$ and define operations of R as follows:

(1) $(a, b) + (a', b') = (max\{a, a'\}, max\{b, b'\})$ for all $a, a', b, b' \in \mathbb{R}$;
(2) $(a, b)(a', b') = (a + a', b + b')$ for all $a, a', b, b' \in \mathbb{R}$;
(3) $(-\infty) + r = r + (-\infty) = r$ for all $r \in R$;
(4) $(-\infty)r = -\infty = r(-\infty)$ for all $r \in R$.

Then $(R, +, \cdot)$ is a semiring but $(sub(R), \oplus, \odot)$ is not since, if $r = \{(0, 1), (1, 0)\}$, $r' = \{(1, 0)\}$, and $r'' = \{(0, 1)\}$, then $r \odot (r' \oplus r'') \neq (r \odot r') \oplus (r \odot r'')$.

2. SETS AND RELATIONS WITH VALUES IN A SEMIRING

The **direct product** $R = \times_{i\in\Omega} R_i$ of a family of semirings $\{R_i \mid i \in \Omega\}$ has the structure of a semiring with the operations of addition and multiplication defined componentwise. This semiring is additively- [resp. multiplicatively-] idempotent [resp. zerosumfree, simple] if each of the R_i is additively- [resp. multiplicatively-] idempotent [resp. zerosumfree, simple]. It is not entire if Ω has order greater than 1.

If $\{R_i \mid i \in \Omega\}$ is a set of information algebras (i.e. entire zerosumfree semirings) then the **pseudodirect product** $R' = \bowtie_{i\in\Omega} R_i$ has the underlying set

$$\{0\} \cup \times_{i\in\Omega}(R_i \setminus \{0_{R_i}\}).$$

Operations between nonzero elements defined componentwise, and these operations are extended to all of R' by setting $0 + r' = r' + 0 = r'$ and $0r' = r'0 = 0$ for all $r' \in R'$. This is again an information algebra. If each of the R_i is additively- [resp. multiplicatively-] idempotent then so is $\bowtie_{i\in\Omega} R_i$. Similarly, it is simple if each of the R_i is.

In particular, we note that if A is a nonempty set and R is a semiring then R^A is a semiring, sometimes called the semiring of R-**valued subsets** of A. This name derives from the fact that each subset B of A defines a **characteristic function** $c_B \in \mathbb{B}^A$ given by

$$c_B : a \mapsto \begin{cases} 1 & \text{if } a \in B \\ 0 & \text{otherwise} \end{cases}$$

Thus $\mathbb{B}^A$ can be canonically identified with the semiring $sub(A)$ of all subsets of A. If $f \in R^A$ then the **support** of f is $supp(f) = \{a \in A \mid f(a) \neq 0\}$. If $im(f) \subseteq \{0, 1\}$ then f is **exact** (or **crisp**). The notion of sets with values in a semiring was considered in detail in [Eilenberg, 1974].

If S_i is a subsemiring of R_i for each $i \in \Omega$ then $\times_{i\in\Omega} S_i$ is a subsemiring of $\times_{i\in\Omega} R_i$. In particular, if S is a subsemiring of a ring R and if A is a nonempty set then S^A is a subsemiring of R^A. Thus, if R is an additively-idempotent semiring and A is a nonempty set then the set of all exact functions in R^A is a subsemiring of R^A.

If A is an infinite set and R is a semiring then $\{f \in R^A \mid f \text{ has finite support}\}$ is a subhemiring of R^A which is not a subsemiring, since it does not contain the multiplicative identity. If R is zerosumfree and entire then

$$\{0\} \cup \{f \in R \mid A \setminus supp(f) \text{ is finite}\}$$

is a subsemiring of R^A.

(2.1) APPLICATION. It is sometimes very important to allow an element of a set to appear in that set "more than once". For example, this happens when we are counting the zeroes of a function or the eigenvalues of a linear transformation. This has led to the theory of **multisets**, which were first formally studied by Knuth [1992] for use in computer science and have since been used extensively in many contexts. Thus, given a nonempty set A, a multisubset of A is defined by a **multiplicity function** in $\mathbb{N}^A$, where $\mathbb{N}$ is the set of all nonnegative integers. The theory of multisets has been formalized in [Blizard, 1989]. For a formalization of linear logic in terms of multisets, refer to [Troelstra, 1992]. Loeb [1992], concerned with various combinatorial problems, extended the notion of a multiset to that of a **hybrid set**, or "set with a negative number of elements" by considering multiplicity functions belonging to $\mathbb{Z}^A$.

Another extension of the notion of a multiset involves looking at multiplicity functions in R^A, where $R = \mathbb{N} \cup \{-\infty, \infty\}$. Elements of R^A are sometimes called **bags** on A. (On the other hand, the term "bag" is often used as a synonym for "multiset", so one has to be careful.) See [Andre, 1989] for an application of this construction to signal processing.

(2.2) APPLICATION. For any nonempty set A, we have the semiring $\mathbb{I}^A$ of all **fuzzy subsets** of A, which has been extensively studied by Zadeh, beginning with [Zadeh, 1965], and his disciples. Literally thousands of papers have been written on fuzzy set theory. See [Kaufmann, 1975] or [Dubois & Prade, 1980] for details. In [Gierz et al., 1980], which is based on a more geometric point of view, fuzzy subsets are called **cubes**. Thus, for example, fuzzy subsets of $\mathbb{R}^+$ can be considered as coinciding with nonnegative probability distribution functions. See [Klement, 1982] for details of this approach. The semiring $\mathbb{I}^A$ of all fuzzy subsets of a set A is in fact a frame [De Luca & Termini, 1972] in which meets and joins are defined componentwise: if $U \subseteq \mathbb{I}^A$ and $a \in A$ then $\vee U: a \mapsto sup\{f(a) \mid f \in U\}$ and $\wedge U: a \mapsto inf\{f(a) \mid f \in U\}$. These are not the only operations which can be defined on $\mathbb{I}^A$. Various other operations and their properties are discussed in detail in [Mizumoto & Tanaka, 1981]. Many of these are defined componentwise by various triangular norms and conorms on $\mathbb{I}$. If f and g are elements of $\mathbb{I}^A$ then $(g : f)$ is the function defined by

$$(g : f): a \mapsto \begin{cases} 1 & \text{if } f(a) \leq g(a) \\ g(a) & \text{otherwise} \end{cases}.$$

If $g = c_B$ is an exact subset of A then $(g : f) = c_D$, where $D = B \cup [A \setminus supp(f)]$ and this is an exact subset of A.

(2.3) EXAMPLE. For nonempty sets A and B, we have the semiring $sub(B)^A$ of all **multifunctions** from A to B. In [Manes & Arbib, 1986], multifunctions are the basis for the treatment of the semantics of computer programs. If R is the schedule algebra and $A = R^2$, then a semiring of interest in analysis is the subsemiring of R^A consisting of all functions $f: A \to R$ satisfying the condition that $supp(f)$ is an open connected subset (= domain) of A on which f is subharmonic. Another subsemiring of this semiring is obtained by substituting "upper semicontinuous" for "subharmonic".

(2.4) EXAMPLE. If X is a topological space, then the family $(\mathbb{R}^+)^X$ of all nonnegative real-valued functions on X is a commutative semiring under the usual pointwise operations. The family of all continuous nonnegative real-valued functions on X is a subsemiring of this semiring.

This notion can be generalized. Let A be a nonempty set and let $\mathcal{A}$ be a nonempty family of nonempty subsets of A satisfying the condition that if $B, B' \in \mathcal{A}$ then $B \cap B' \in \mathcal{A}$. If R is a semiring then $R^B \cap R^{B'} = \varnothing$ for all $B \neq B'$ in $\mathcal{A}$. Set $R^{\mathcal{A}} = \cup\{R^B \mid B \in \mathcal{A}\}$. Thus, for each $f \in R^{\mathcal{A}}$ there exists a unique $B \in \mathcal{A}$ such that $f \in R^B$. This subset B of A is called the **domain** of f and will be denoted by $dom(f)$. We now define operations of addition and multiplication on $R^{\mathcal{A}}$ as follows:

(1) If $f, g \in R^{\mathcal{A}}$ then $dom(f+g) = dom(f) \cap dom(g)$ and $(f+g)(a) = f(a)+g(a)$ for all $a \in dom(f+g)$.

(2) If $f, g \in R^{\mathcal{A}}$ then $dom(fg) = dom(f) \cap dom(g)$ and $(fg)(a) = f(a)g(a)$ for all $a \in dom(f+g)$.

(2.5) PROPOSITION. *Let A be a nonempty set and let $\mathcal{A}$ be a nonempty family of nonempty subsets of A satisfying the condition that if $B, B' \in \mathcal{A}$ then $B \cap B' \in \mathcal{A}$. If R is a semiring then $R^{\mathcal{A}}$ is also a semiring.*

PROOF. It is straightforward to check all of the conditions in the definition of a semiring. Notice that the additive identity of $R^{\mathcal{A}}$ is the function $a \mapsto 0$ having domain A and the multiplicative identity in $R^{\mathcal{A}}$ is the function $a \mapsto 1$ having domain A. □

Thus, in particular, we see that if A is a nonempty set then $R^A = R^{\{A\}}$.

If A and B are nonempty sets then $R^{A \times B}$ is called the semiring of all **R-valued relations between** A and B. The use of these concepts to define the notion of an R-**valued language** then follows the lines given in [Kim, Mizumoto, Toyoda & Tanaka, 1975] and [Wechler, 1975]. **Fuzzy relations**, namely relations with values in $\mathbb{I}$, are considered in [Dubois & Prade, 1980], [Fang, 1993], [Kawahara & Furusawa, 1999], [Murali, 1989], and [Ovchinnikov, 1981, 1993]. If A is a nonempty set then an **R-valued relation on** A is an element of $R^{A \times A}$.

(2.6) EXAMPLE. If V is a nonempty set then an element g of $\mathbb{B}^{V \times V}$ is called a **(directed) graph** on V. The elements of V are called the **vertices** (or **nodes**) of g and the elements (v, v') of $V \times V$ satisfying $g(v, v') \neq 0$ are called the **arcs** of the graph. If V is finite then the number of elements of V is the **order** of V. This notion can be generalized: if V is a nonempty set and R is an arbitrary semiring then an element g of $R^{V \times V}$ is an **R-valued graph** on V. The arcs of g are those elements (v, v') of $V \times V$ satisfying $g(v, v') \neq 0$. In this case, $g(v, v')$ is the **weight** of the arc in R. Graphs with values in I are called **fuzzy graphs**. A **path** of **length** n in an R-valued graph $g \in R^{V \times V}$ is a finite ordered subset $(v_1, v_2, \ldots, v_n)$ of vertices of g such that (v_i, v_{i+1}) is an arc of g for all $1 \leq i < n$. It is often convenient to extend the definition of g and consider it as a function from the set of all paths of g to R by setting

$$g(v_1, \ldots, v_n) = \prod_{i=1}^{n-1} g(v_i, v_{i+1}).$$

The problem of finding efficient algorithms for computing this value, or often the value $\sum g(p)$, where the sum ranges over all paths p of given length n from a fixed vertex v to a fixed vertex v', is often of great importance, as we shall see. Refer to [Fletcher, 1980] for an example of such an algorithm. Also consult [Gondran & Minoux, 1984a] for further details.

Another way of looking at things is to consider $\mathbb{B}^{V \times V}$ to be the set of **nondeterministic programs** on the set V of **states**. Here it is understood that if $f \in \mathbb{B}^{V \times V}$ then $f(v, v') = 1$ if the program f may transform v into v'. See [Main & Benson, 1985].

(2.7) EXAMPLE. If R is a semiring and if A and B are nonempty sets then an R-valued relation $h \in R^{A \times B}$ is sometimes called a **Chu space**. Such spaces have been studied intensively by Pratt [1986, 1993, 1994, 1995a, 1995b, 1996, 1997] and his students, with an eye on applications in computer science. In this approach, A is the set of **events** (or **values**, **locations**, **variables**, **points**) and B is the set of **states** ("possible worlds"). The value $f(a, b)$ represents the extent (or complexity) of the event a happening at state b. In particular, if $R = \mathbb{B}$ then $f(a, b) = 1$ if event a has happened at state b and $f(a, b) = 0$ if it has not. This interpretation has been used in [V. Gupta, 1994] and [Pratt, 1995b] to build models of concurrent systems.

The Chu space approach is basically categorical, and so it leads to the idea of a transform between R-valued relations. Let $f \in R^{A \times B}$ and $g \in R^{A' \times B'}$ be R-valued relations. A **transform** $(u, v): f \to g$ consists of a pair of functions $u: A \to A'$ and $v: B' \to B$ satisfying the condition that $f(a, v(b')) = g(u(a), b)$ for all $a \in A$ and $b' \in B'$. Note that if $(u, v): f \to g$ and $(u', v'): g \to h$ are transforms then $(u'u, vv'): f \to h$ is also a transform. If there exists a transform $(u, v): f \to g$ then we say that f is a **left adjoint** of g and g is a **right adjoint** of f. In the model of concurrent systems proposed in [V. Gupta, 1994], a transform $(u, v): f \to g$ determines a **simulation** of g by f.

An R-valued relation h on a nonempty set A is **transitive** if and only if, whenever $a, a', a'' \in A$ there exists an element r of R such that $h(a, a')h(a', a'') + r = h(a, a'')$. If in fact $h(a, a')h(a', a'') + h(a, a'') = h(a, a'')$ then h is **strongly transitive**. It is **reflexive** if and only if $h(a, a) = 1$ for each $a \in A$ and it is **symmetric** if and only if $h(a, a') = h(a', a)$ for all $a, a' \in A$. A [strongly] transitive, reflexive, and symmetric R-valued relation h on A is a [**strong**] **R-valued equivalence relation** on A. For example, if R is an arbitrary semiring and A is a nonempty set,then any

$g \in R^A$ defines a $\mathbb{B}$-equivalence relation $\sim_g$ on A by setting $a \sim_g a'$ if and only if $g(a) = g(a')$.

(2.8) EXAMPLE. If $R = (\mathbb{I}, \vee, \wedge)$ then the function $h \in R^{\mathbb{N}\times\mathbb{N}}$ defined by

$$h\colon (m,n) \mapsto \begin{cases} 1 & \text{if } m = n \\ \frac{1}{2} & \text{if } m+n \text{ is even} \\ 0 & \text{otherwise} \end{cases}$$

is an R-valued equivalence relation on $\mathbb{N}$.

(2.9) EXAMPLE. If $R = (\mathbb{I}, max, \cdot)$ and if A is a nonempty set then any function $f \in R^A$ defines an R-valued equivalence relation $h \in R^{A\times A}$ by

$$h\colon (a,b) \mapsto \begin{cases} \frac{min\{f(a),f(b)\}}{max\{f(a),f(b)\}} & \text{if } f(a) \neq f(b) \\ 1 & \text{otherwise} \end{cases}.$$

Indeed, if $R = (\mathbb{I}, max, \sqcap)$, where $\sqcap$ is a continuous triangular norm, then the family of all R-valued equivalence relations on a nonempty set A has been characterized by Valverde [1985]. For further results, refer to [Jacas and Valverde, 1996].

Let R be a semiring and let A be a nonempty set. If $f, g \in R^{A\times A}$ and if $fg \in R^{A\times A}$ is the function defined by $fg\colon (a,a') \mapsto f(a,a')g(a,a')$ then fg is surely symmetric and reflexive whenever both f and g are. Moreover, if both f and g are transitive and the image of at least one of them is contained in the center of R then it is straightforward to verify that fg is transitive as well. Thus we see that if f and g are R-valued equivalence relations on A, the image of one of which is in the center of R, then fg is also an R-valued equivalence relation on A. In particular, if R is commutative then the set of all R-valued equivalence relations on any nonempty set A is closed under taking products.

(2.10) PROPOSITION. *Let R be a simple semiring and let $h \in R^{A\times A}$ be a strong R-valued equivalence relation on a nonempty set A. Then the relation $\sim$ on A defined by $a \sim a'$ if and only if $h(a,a') = 1$ is an equivalence relation.*

PROOF. It is easy to see that $\sim$ is symmetric and reflexive. If $a \sim a'$ and $a' \sim a''$ then $1 = h(a,a')h(a,a'')$ so $h(a,a'') = h(a,a'') + h(a,a')h(a',a'') = h(a,a'') + 1 = 1$. Thus $a \sim a''$. □

(2.11) EXAMPLE. Let $R = (\mathbb{R}^+ \cup \{\infty\}, min, +)$ and let A be a nonempty set. Then an **(extended) pseudometric** on A is just an R-valued equivalence relation on A. Such a function is an **(extended) metric** if and only if the relation $\sim$ on A which it defines is trivial. Thus, for example, if A is the set of all continuous functions from $\mathbb{I}$ to $\mathbb{R}$ then we have an R-valued equivalence relation h on A defined by

$$h\colon (\varphi, \psi) \mapsto \int_0^1 |\varphi(t) - \psi(t)|dt.$$

Extended pseudometrics with values in $(\mathbb{N}^+ \cup \{\infty\}, \wedge, +)$ (namely $(\mathbb{N}^+ \cup \{\infty\})$-valued equivalence relations), also play an important role in theoretical computer science and formal language theory. For example, if A is a nonempty set we have the following pseudometrics on A^* with values in $\mathbb{N}^+ \cup \{\infty\}$ are among those given in [Choffrut & Pighizzini, 1997]: If $x = x_1 \cdot \ldots \cdot x_k$ and $y = y_1 \cdot \ldots \cdot y_n$ then

(1) The **Hamming distance** between x and y is defined by

$$d_H(x,y) = \begin{cases} \#\{i \mid x_i \neq y_i\} & \text{if } k = n \\ \infty & \text{otherwise} \end{cases}.$$

(2) The **subword distance** between x and y is defined by

$$d_S(x,y) = |x| + |y| - 2max\{|z| \mid z \text{ is a subword of both } x \text{ and } y\}.$$

(2.12) APPLICATION. Let A be a nonempty set of "states" and let L be a nonempty set which is the "language" in which we make statements about the elements of A. We assume that there is a distinguished subset $\models$ of $A \times L$ and we say that a state $a \in A$ **satisfies** a statement $\lambda \in L$ if and only if $(a, \lambda) \in \models$. In this case we write $a \models \lambda$. If $L' \subseteq L$ then the set of **models** for L') is the set

$$Mod(M') = \{a \in A \mid a \models \lambda \text{ for all } \lambda \in L'\}$$

and if $A' \subseteq A$ then the **theory** of A' is the set

$$Th(A') = \{\lambda \in L \mid a \models \lambda \text{ for all } a \in A'\}.$$

Now assume that L has a special element $\perp$ satisfying $Mod(\{\perp\}) = \varnothing$ and that there is an operation $\vee$ defined on L satisfying $Mod(\{\lambda \vee \lambda'\}) = Mod(\{\lambda\}) \cup Mod(\{\lambda'\})$ for all $\lambda, \lambda' \in L$. In case $L = \{\lambda_1, \lambda_2, \ldots\}$ is countable then we have an $(\mathbb{R}^+ \cup \{\infty\}, \wedge, +)$-valued equivalence relation h defined on A as follows:

(1) $h(a,a) = 0$ for all $a \in A$;
(2) If $a \neq a'$ in A then $h(a,a') = \frac{1}{n}$, where $n = min\{k \mid a \models \lambda_k \text{ and } a' \not\models \lambda_k)\}$.

These examples suggest that, for a general semiring R, we can treat R-valued equivalence relations in the same way we treat duals of pseudometrics. Thus, for example, if R is a semiring and A is a nonempty set, we say that R-valued equivalence relations $h, k \in R^{A \times A}$ are **Lipschitz equivalent** if and only if there exists $s_1, s_2, r_1, r_2 \in R$ satisfying $s_1 h(a,a') = r_1 + k(a,a')$ and $s_2 k(a,a') = r_2 + h(a,a,')$ for all $a, a' \in A$. It is easily checked that this is in fact an equivalence relation.

(2.13) PROPOSITION. *Let R be a semiring and let A be a nonempty set. If $h \in R^{A \times A}$ is a strong R-valued equivalence relation on A then the following conditions are equivalent for $a, b \in A$:*

(1) $h(a,c) = h(b,c)$ for all $c \in A$;
(2) $h(a,b) = 1$.

PROOF. Assume (1). Then, in particular, $h(a,b) = h(b,b) = 1$ and so we have (2). Conversely, assume (2). If $c \in A$ then $h(a,c) = h(a,c) + h(a,b)h(b,c) = h(a,c) + h(b,c)$ and similarly $h(b,c) = h(b,c) + h(a,c)$, proving (1). □

(2.14) PROPOSITION. *Let R be a semiring and let A be a nonempty set. If $h \in R^{A \times A}$ is a strong R-valued equivalence relation on A then:*

(1) *$h(a,c)h(b,c) = 0$ for all $c \in A$ if and only if $h(a,b) = 0$;*
(2) *$h(a,c) = h(b,c)$ for all $c \in A$ if and only if $h(a,b) = 1$.*

PROOF. (1) Assume that $h(a,c)h(b,c) = 0$ for all $c \in A$. Then

$$h(a,b) = 1 \cdot h(a,b) = h(a,a)h(a,b) = h(a,a)h(b,a) = 0.$$

Conversely, assume $h(a,b) = 0$. Then for all $c \in A$ we have

$$h(a,c)h(b,c) = h(a,c)h(c,b) + h(a,b) = h(a,b) = 0.$$

(2) Assume $h(a,c) = h(b,c)$ for all $c \in A$. Then, in particular, $1 = h(a,a) = h(b,a) = h(a,b)$. Conversely, assume that $h(a,b) = 1$. Then for any $c \in A$ we have $h(a,c) = h(a,c) + h(a,b)h(b,c) = h(a,c) + h(b,c)$ and similarly $h(b,c) = h(b,c) + h(a,c)$, proving that $h(a,b) = h(b,c)$. □

In particular, if $h \in R^{A \times A}$ is an R-valued equivalence relation on a nonempty set A then, for each $a \in A$, then **equivalence class** of a with respect to that relation is the R-valued subset h_a of A defined by $h_a : a' \mapsto h(a,a')$. Propositon 2.14 then says that, in the given situation,

(1) $h_a h_b$ is the 0-map if and only if $h(a,b) = 0$; and
(2) $h_a = h_b$ if and only if $h(a,b) = 1$.

The set of all equivalence classes of A with respect to an R-valued equivalence relation h is the R-**valued partition** P_h of A defined by h. Note that if we have a canonical surjection from A to P_h given by $a \mapsto h_a$.

3. BUILDING NEW SEMIRINGS FROM OLD

We now consider a material from the previous chapter from a different angle. Let R be a semiring and let A be a nonempty set which is either finite or countably-infinite. Then the set $R^{A\times A}$ of functions from $A \times A$ to R is denoted by $\mathcal{M}_A(R)$, and such functions are called $(A \times A)$-**matrices** on R. If A is a finite set of order n we write $\mathcal{M}_n(R)$ instead of $\mathcal{M}_A(R)$; if A is countably-infinite we sometimes write $\mathcal{M}_\omega(R)$ instead of $\mathcal{M}_A(R)$. If A is a finite or countably-infinite set we will often denote matrices in the usual matrix notation rather than in functional notation. In particular, we will sometimes employ "block notation" for such matrices. We have already noted that addition of such matrices, defined componentwise, turns $\mathcal{M}_A(R)$ into a commutative additive monoid, the identity element of which is the function which takes every element of $A \times A$ to 0.

A matrix $f \in \mathcal{M}_A(R)$ is **row finite** [resp. **column finite**] if and only if for each $i \in A$ [resp. $j \in A$] all but finitely-many values of $f(i,j)$ are equal to 0. If $f, g \in \mathcal{M}_A(R)$ such that either f is row-finite or g is column finite then we can define the product fg by setting $fg\colon (i,j) \mapsto \sum_{k\in A} f(i,k)g(k,j)$ for all $i,j \in A$. It is easy to verify that the set $\mathcal{M}_{A,r}(R)$ of all row-finite matrices in $\mathcal{M}_A(R)$, the set $\mathcal{M}_{A,c}(R)$ of all column-finite matrices in $\mathcal{M}_A(R)$, and the set $\mathcal{M}_{A,rc}(R)$ of all row-finite and column-finite matrices in $\mathcal{M}_A(R)$ are all semirings under the given operations of addition and multiplication. (The multiplicative identity is the function f defined by $f(i,i) = 1$ for all $i \in A$ and $f(i,j) = 0$ for $i \neq j$ in A.) If A has order greater than 1, these semirings are not entire. If A is finite then, needless to say, $\mathcal{M}_{A,r}(R) = \mathcal{M}_{A,c}(R) = \mathcal{M}_{A,rc}(R) = \mathcal{M}_A(R)$. If S is a subsemiring of a semiring R and A is a nonempty set then $\mathcal{M}_{A,r}(S)$, $\mathcal{M}_{A,c}(S)$, and $\mathcal{M}_{A,rc}(S)$ are subsemirings of $\mathcal{M}_A, r(R)$, $\mathcal{M}_{A,c}(R)$, and $\mathcal{M}_{A,rc}(R)$ respectively.

If R is a semiring and A is a nonempty set then the following are subsemirings of $\mathcal{M}_{A,rc}(R)$:

(1) $\{f \in \mathcal{M}_{A,rc}(R) \mid f(i,j) = 0 \text{ for } i \neq j\}$;
(2) $\{f \in \mathcal{M}_{A,rc}(R) \mid f(i,j) = 0 \text{ for } i > j\}$;
(3) $\{f \in \mathcal{M}_{A,rc}(R) \mid f(i,j) = 0 \text{ unless } i = j \text{ or } i = 1\}$.

Thus we see that, if A is a finite set, the elements of $R^{A\times A}$ can be considered in two different ways: as element of the semiring of R-valued graphs on A or as elements of the semiring of $(A \times A)$-matrices on R. These two semirings have the same addition but different multiplications. In the literature the distinction between the two is often marked by speaking of finite graphs and of their corresponding **transition matrices**. Matrix powers have a very natural graph-theoretic interpretation. Indeed, if $g \in R^{A\times A}$ and if $i, j \in A$ then, for each $k > 0$, the power g^k in $\mathcal{M}_A(R)$ is just $\sum\{g(p) \mid p$ a path of length k from i to $j\}$. In addition to the above, there are other multiplication operations on $R^{A\times A}$ which, together with the componentwise addition, turn it into a hemiring which is not necessarily a semiring. Thus, for example, for each $f \in R^{A\times A}$, let the **trace** of f be given by $tr(f) = \sum_{a\in A} f(a,a)$ and define the operation $*$ on $R^{A\times A}$ by $f * g\colon (i,j) \mapsto tr(f)g(i,j)$. Then $(R^{A\times A}, +, *)$ is a hemiring. For a general mechanism to construct such hemirings, refer to [Birkenmeier & Heatherly, 1987].

Note that if $n > 1$ is an integer, then $\mathcal{M}_n(R)$ is not simple for any semiring R. For the properties of semirings of matrices of the form $\mathcal{M}_n(R)$, where R is a bounded distributive lattice, see [Give'on, 1964]; for semirings of matrices of the form $\mathcal{M}_n(R)$ where R ranges over various other types of ordered algebraic structures, see [Blyth, 1964]. Semirings of matrices of the form $\mathcal{M}_n(\mathbb{B})$ and their many applications are discussed in detail in [Kim, 1982]; the structure of $I^{\times}(\mathcal{M}_n(\mathbb{B}))$ is completely described in [Chaudhuri & Mukherjea, 1980]. Semirings of matrices of the form $\mathcal{M}_n(\mathbb{I})$ and their applications are discussed in [Kim & Roush, 1980]; in particular, for a multiplicatively-regular element A of $\mathcal{M}_n(\mathbb{I})$, one finds there algorithms to find a generalized inverse and a Thierrin-Vagner inverse of A. Semirings of matrices over the semiring $(\mathbb{R}^+, max, \cdot)$ are considered in [Vorobjev, 1963]. Matrices over the semiring $(\mathbb{R}^+, +, \cdot)$ have played an important part in linear algebra since the work of Frobenius, and have important applications in such areas as the study of Markov chains. For an introduction to the research in this area, refer to [Gantmacher, 1959] or [Minc, 1988].

Semigroups of matrices over semirings are interesting in their own right; see [Straubing, 1983b] for example. In addition, they can be used with advantage as a basis for algorithms to compute general finite semigroups. Refer to [Froidure & Pin, 1998]. Semirings of matrices over the the semiring $(\mathbb{R} \cup \{\infty\}, min, +)$ and their applications in operations research are discussed in [Pandit, 1961] and [Gaubert, 1996b].

(3.1) EXAMPLE. In [1996b], Gaubert solves the Burnside problem in this context by showing that a finitely-generated torsion semigroup in $\mathcal{M}_n(R)$ is finite, where R is the schedule algebra $(\mathbb{R} \cup \{-\infty\}, max, +)$. Moreover, it is decidable whether a finitely-generated semigroup of $\mathcal{M}_n(R)$ is torsion. If one wants to extend this result to additively-idempotent semirings which are not semifields, the matter becomes more difficult. However, it is shown there that if R is a commutative additively-idempotent semiring and if, for each $r \in R$, the set

$$A(r) = \{r' \in R \mid r = a' + r' \text{ for some } a' \in R\}$$

is finite, then every finitely-generated torsion semigroup in $\mathcal{M}_n(R)$ is finite.

As we will see later, matrices over a semiring have important applications in the theory of finite automata. These applications give rise to certain problems which can be formulated in $\mathcal{M}_n(R)$ for an arbitrary semiring R. Some such problems are considered in [Simon, 1988]; among them

(1) **The Finite Section Problem**: Let U be a nonempty finite subset of $\mathcal{M}_n(R)$ and let G be the subsemigroup of $(\mathcal{M}_n(R), \cdot)$ generated by U. Given $1 \leq h, k \leq n$, is the set $\{r \in R \mid r = a_{hk}\}$ for some matrix $[a_ij] \in G$ finite?
(2) **The Finite Closure Problem**: If U is a nonempty finite subset of $\mathcal{M}_n(R)$, is the subgroup of $(\mathcal{M}_n(R), \cdot)$ generated by U finite?

It is clear that if we can decide (1) then we can decide (2), but the converse is not true in general. The decidability of these problems for the case of $R = \mathbb{N}$ is shown in [Mandel & Simon, 1977]. For the case of $R = (\mathbb{N} \cup \{\infty\}, min, +)$, it is considered in [Hashiguchi, 1982] and [Simon, 1988].

Since matrices over semirings have important applications, as we will see, the speed of computation of matrix multiplication is often very important. For the case of multiplication of finite matrices the entries of which come from a finite semiring R, this problem has been studied in [Rosenkrantz & Hunt, 1988], where it is shown that such matrix multiplication is linear-time reducible to integer matrix multiplication. Thus, any fast algorithm for integer matrix multiplication can be converted into a fast algorithm for multiplication of matrices over a finite semiring R. This is true both for computation on one-processor machines and for parallel computation. Refer also to [Mehlhorn, 1984].

(3.2) APPLICATION. A **matrix iteration theory** is an algebraic theory the objects of which are natural numbers and the morphisms $k \longrightarrow n$ in which are $k \times n$ matrices over some fixed semiring R. Such theories are studied in detail in [Bloom & Ésik, 1993]. Such theories have important applications in the analysis of flowchart schemes and automata. Refer also to [Ying, 1991].

Let $(M, *)$ be a monoid with identity e and let R be a semiring. The family $R[M]$ of all functions $f \in R^M$ having finite support is a semiring under the operations of addition $+$ and multiplication $\langle * \rangle$ defined as follows:

(1) $(f+g)(m) = f(m) + g(m)$ for all $m \in M$;
(2) $(f\langle * \rangle g)(m) = \sum\{f(m')g(m'') \mid (m', m'') \in supp(f) \times supp(g)$ and $m' * m'' = m\}$.

The additive identity of $R[M]$ is the function which takes every element of M to 0_R. The multiplicative identity of $R[M]$ is the function which takes e to 1_R and all other elements of M M to 0_R. The operation $\langle * \rangle$ is called **$*$-convolution**. If R and M are commutative, then surely the semiring $R[M]$ is commutative. Note too that if R is additively idempotent, so is $R[M]$.

(3.3) EXAMPLE. Let $M = \{1, m\}$ be a group of order 2 and let R be a semiring. Then we can identify $R[M]$ with the semiring $R \times R$ on which addition and multiplication are defined by

$$(a, b) + (c, d) = (a + c, b + d)$$

and

$$(a, c)\langle * \rangle(c, d) = (ac + bd, ad + bc).$$

If $x = (a, b) \in R[M]$, set $\ominus x = (b, a)$. Then we note that for each $x, y \in R[M]$ we have

(1) $\ominus(\ominus x) = x$;
(2) $\ominus(x + y) = (\ominus x) + (\ominus y)$;
(3) $\ominus(x\langle * \rangle y) = (\ominus x)\langle * \rangle y$.

(3.4) APPLICATION. For the application of semirings of the form $\mathbb{B}[A^*]$, where A is any alphabet, to the design of arithmetic circuits, see, for example, [Allender et al., 1996].

We can extend the above notion of convolution as follows: if $(M, *)$ is a monoid then a family $\mathcal{C}$ of subsets of M is **favorable** if and only if the following conditions are satisfied:

(1) $\mathcal{C}$ is closed under taking subsets and finite unions;
(2) If $m \in M$ then $\{m\} \in \mathcal{C}$;
(3) If $A, B \in \mathcal{C}$ then $A * B = \{m * m' \mid (m, m') \in A \times B\} \in \mathcal{C}$.

Moreover, if R is a semiring then a favorable family $\mathcal{C}$ of subsets of M is R-**favorable** if and only if it satisfies the additional condition:

(5) If $m \in M$ and if $A, B \in \mathcal{C}$ then for all functions $f: A \to R$ and $g: B \to R$, the sum $\sum\{f(m')g(m'') \mid m' \in A, m'' \in B \text{ and } m' * m'' = m\}$ is defined in R.

Note that the family $\mathcal{C}_0$ of all finite subsets of M is R-favorable for any semiring R and, by (1) and (3), it is in fact the unique minimal favorable family of $sub(M)$. If the monoid M is finitary then the family of all subsets of M is R-favorable for any semiring R.

If R is a semiring and $\mathcal{C}$ is an R-favorable family of subsets of a monoid $(M, *)$ then we will denote by $R[\mathcal{C}]$ the family of all functions $f \in R^M$ satisfying the condition that $supp(f) \in \mathcal{C}$. Note that conditions (1) and (2) insure that $R[\mathcal{C}]$ is a submonoid of the commutative monoid $(R^M, +)$. Also, if $f \in R^M$ has finite support then $f \in R[\mathcal{C}]$ for each R-favorable family $\mathcal{C}$ of subsets of M. In particular, for each element r of R and each element m of M we have the function $e_{r,m}: M \to R$ defined by

$$e_{r,m}: m' \mapsto \begin{cases} r & \text{if } m' = m \\ 0 & \text{otherwise} \end{cases}.$$

and this function belongs to $R[\mathcal{C}]$ for all R-favorable families $\mathcal{C}$ of subsets of M. For any R-favorable family $\mathcal{C}$ of subsets of M we can define the $*$-convolution operation $\langle * \rangle$ by setting

$$f\langle * \rangle g: m \mapsto \sum\{f(m')g(m'') \mid (m', m'') \in supp(f) \times supp(g) \text{ and } m' * m'' = m\}.$$

Then $(R[\mathcal{C}], +, \langle * \rangle)$ is a semiring, called the **convolution algebra** on R defined by M and $\mathcal{C}$.

If $\mathcal{C} \subseteq \mathcal{D}$ are R-favorable families of subsets of M then it is clear that $R[\mathcal{C}]$ is a subsemiring of $R[\mathcal{D}]$. Thus, in particular, $R[M] = R[\mathcal{C}_0]$ is a subsemiring of $R[\mathcal{C}]$ for every R-favorable family $\mathcal{C}$ of subsets of M.

Let R be a semiring, let $\mathcal{C}$ be an R-favorable family of subsets of a monoid $(M, *)$, and let $\theta: M \to \mathcal{C}$ be a function satisfying $m \in \theta(m)$ for each element m of M. For each $f \in R^M$ and each $m \in M$, let $f|_{m,\theta}$ be the function from M to R defined by

$$f|_{m,\theta}: m' \mapsto \begin{cases} f(m') & \text{if } m' \in \theta(m) \\ 0 & \text{otherwise} \end{cases}.$$

Functions of this type certainly belong to $R[\mathcal{C}]$. Now define an operation, $\langle *|_\theta \rangle$ on R^M, called **local convolution**, by setting $(f\langle *|_\theta\rangle g)(m) = (f|_{m,\theta}\langle * \rangle g|_{m,\theta})(m)$. Then $(R^M, +, \langle *|_\theta \rangle)$ is a hemiring.

(3.5) EXAMPLE. In Example 1.10 we considered a monoid $(M, *)$ and a semiring structure on $sub(M)$ in which addition is union and multiplication is given by $AB = \{a * b \mid a \in A, b \in B\}$. As already noted, we can identify $sub(M)$ with $\mathbb{B}^M$ by assigning to each subset A of M its characteristic function c_A. Then $c_{A \cup B} = c_A + c_B$ and $c_{AB} = c_A \langle * \rangle c_B$. Thus semirings of the type given in Example 1.10 – and in particular the important semirings given in Example 1.11 – are convolution semirings whenever M is finitary.

(3.6) EXAMPLE. One of the most useful semirings in number theory is $R = \mathbb{N}^M$, where M is the finitary multiplicative monoid $\mathbb{P}$ of positive integers. The convolution operation $\langle \cdot \rangle$ on R is usually denoted by $*$ and is often called the **Dirichlet convolution**. That is to say, $(f * g)(n) = \sum\{f(n')g(n'') \mid n = n'n''\}$ for each positive integer n. One checks that this operation is commutative. A function $f \in R$ is **multiplicative** if and only if $f(nn') = f(n)f(n')$ for all $n, n' \in \mathbb{P}$. The set S of all multiplicative functions in R is closed under $*$ and componentwise products and it is easy to see componentwise multiplication distributes over Dirichlet convolution so that $(S, *, \cdot)$ is a commutative semiring.

Let M' be the submonoid of $\mathbb{P}$ consisting of 1 and all those positive integers which can be written as a product of an even number of primes. Then $R' = \mathbb{N}^{M'}$ is a subsemiring of $(R, +, *)$. Let $\mu: M' \to \mathbb{N}$ be the function defined by $\mu(n) = 1$ if $n = 1$ or n can be written as a product of distict primes and $\mu(n) = 0$ otherwise. The Möbius Inversion Formula states that $g = \mu * f$ in R' if and only if $f = \iota * f$, where $\iota \in R'$ is the function defined by $\iota(n) = 1$ for all $n \in M'$. Note that $\iota * \mu$ is the multiplicative identity of R'.

Similarly, if M is the finitary monoid $(\mathbb{N}, +)$ then the convolution $\langle + \rangle$ is the **Cauchy convolution** on $\mathbb{N}^M$. For a detailed study of these and other convolutions of importance in number theory, refer to [Sivaramakrishnan, 1989].

(3.7) EXAMPLE. If $(M, *)$ is a finitary monoid then we have a $*$-convolution operation $\langle * \rangle$ defined among the fuzzy subsets of M. A fuzzy subset f of M is a **fuzzy submonoid** if and only if $f\langle * \rangle f \leq f$. Liu [1982] has justified this name by showing that:

(1) If g_1, g_2, g_3 are fuzzy singletons satisfying the condition that $g_i(x_i) = f(x_i)$ whenever $\{x_i\} = supp(g_i)$ then $(g_1\langle * \rangle g_2)\langle * \rangle g_3 = g_1\langle * \rangle(g_2\langle * \rangle g_3)$;

(2) If e is the fuzzy singleton with $supp(e) = \{1_M\}$ and $e(1_M) = 1$ then for any fuzzy singleton g satisfying $g(x) = f(x)$ whenever $\{x\} = supp(g)$ we have $e\langle * \rangle g = g = g\langle * \rangle e$.

(3.8) EXAMPLE. Let $(M,+)$ be an abelian group and let R be the idempotent semiring $(\mathbb{R}\cup\{\infty\}, min, +)$. Let S be the set of all functions in R^G bounded from below. Define addition on S componentwise and multiplication by convolution:

$$f\langle+\rangle g\colon m \mapsto inf\{f(m')+g(m-m') \mid m' \in M\}.$$

The boundedness of f and g insures that this is well defined. This semiring has important use in convex analysis. See [Aubin, 1993].

(3.9) EXAMPLE. If A is a nonempty set and A^* is the free monoid defined by A, then we denote the semiring $R[A^*]$ by $R\langle A\rangle$. This semiring is called the **semiring of formal polynomials in** A **over** R. It is additively idempotent if R is.

(3.10) EXAMPLE. [Kolokol'tsov & Maslow, 1987] Let X be a normal locally compact topological space and let R be one the following semirings

(1) The semiring $(R\cup\{-\infty\}, max, +)$ on which we have a metric d defined by $d(a,b) = |e^a - e^b|$;
(2) The semiring $(R\cup\{-\infty,\infty\}, max, min)$ on which we have a metric d defined by $d(a,b) = |arctan(a) - arctan(b)|$.

The subset S of R^X consisting of all continuous functions having compact support is a topological semiring with the topology coming from the metric d defined by $d(f,g) = sup\{d(f(x), g(x)) \mid x \in X\}$.

If, in addition, X is abelian additive topological group (for example, if $X = \mathbb{R}^n$) then S also has the structure of a convolution semiring $(S, +, \langle+\rangle)$.

(3.11) APPLICATION. Let M be the monoid $(\mathbb{Z}\times\mathbb{Z}, +)$ and let R be a semiring. A **(two-dimensional) cellular automaton** on R is a function α from R^M to itself satisfying the condition that, for each $f \in R^M$, the value of $\alpha(f)(i,j)$ depends only on the values $f(h,k)$ for $i-1 \leq h \leq i+1$ and $j-1 \leq k \leq j+1$. In most applications, R is taken to be $\mathbb{N}$ or $\mathbb{Z}/(k)$ for some positive integer k. Cellular automata were first developed by von Neumann [1966], and later Ulam, in the study of self-reproducing machines. The most well-known cellular automaton is John Conway's game of "Life", popularized by the columnist Martin . Cellular automata are now used extensively in computerized picture processing, pattern recognition, models of the human nervous system, and in the design of multiple-processor computers. Many operations of a cellular automaton can be regarded as application of local convolutions $f \mapsto f\langle+|_\theta\rangle g$, where

$$\theta(i,j) = \{(h,k) \mid i-1 \leq h \leq i+1; j-1 \leq k \leq j+1\}.$$

See [Preston & Duff, 1984] or [Martin, Odlyzko & Wolfram, 1984] for further details. Cellular automata provide a beautiful graphic example (often illustratable in beautiful graphics) of how repeated application of a simply-defined operation in a relatively-simple semiring can lead to very complex behavior.

If $(M,*)$ is a semigroup rather than a monoid then we can still define the notion of a favorable family $\mathcal{C}$ of subsets of M and the notion of $*$-convolution in $R[\mathcal{C}]$. However, in this case $(R[\mathcal{C}\ ], +, \langle*\rangle)$ turns out to be a hemiring rather than a semiring.

(3.12) EXAMPLE. Let V be a nonempty set and define the associative operation $*$ on $V \times V$ by setting $(v, v') * (w, w') = (v, w')$ for all $v, v', w, w' \in V$. Let R be a semiring and let $\mathcal{C}$ be an R-favorable family of subsets of the semigroup $(V \times V, *)$. Then $R[\mathcal{C}]$ is a subset of the set of all R-valued graphs on V. The $*$-convolution operation $\langle * \rangle$ is just the operation of **composition** of graphs. See [Peeva, 1983] for details. In particular, if V is finite or countable, then this is just the operation of matrix multiplication as defined above.

(3.13) EXAMPLE. Let s and m be integers greater than 1 and let A be the set of all m-tuples $(a_1, \ldots, a_m)$, where the a_i are integers satisfying $0 \leq a_i \leq s-1$. We can define an operation $\wedge$ on A as follows: $(a_1, \ldots, a_m) \wedge (b_1, \ldots, b_m) = (c_1, \ldots, c_m)$, where c_i equals 0 if $a_i \neq b_i$ and equals their common value otherwise. Then $(A, \wedge)$ is a semigroup. Let $\mathcal{C} = sub(A)$. The hemiring $(\mathbb{N}\mathcal{C}], +, \langle \wedge \rangle)$ has important applications in design theory and combinatorial geometry. See [Deza & Rosenberg, 1986], for details.

(3.14) PROPOSITION. *If R is a semiring and if $\mathcal{C}$ is an R-favorable family of subsets of a semigroup $(M, *)$ then the hemiring $(R[\mathcal{C}], +, \langle * \rangle)$ is a semiring if and only if there exists an element e of $R[\mathcal{C}]$ such that, for all elements $m \neq n$ of M, the following conditions are satisfied:*

(1) $\sum_{m' * m = m} e(m') = 1 = \sum_{m * m'' = m}$;
(2) $\sum_{m' * m = n} e(m') = 0 = \sum_{m * m'' = n} e(m'')$.

PROOF. Assume that $R[\mathcal{C}]$ is a semiring with multiplicative identity e. Then, for each element m of M, we have

$$1 = e_{1,m}(m) = (e\langle * \rangle e_{1,m})(m) = \sum_{m' * m'' = m} e(m')e_{1,m}(m'') = \sum_{m' * m = m} e(m')$$

and similarly

$$1 = e_{1,m}(m) = (e_{1,m}\langle * \rangle e)(m) = \sum_{m' * m'' = m} e_{1,m}(m')e(m'') = \sum_{m * m'' = m} e(m'').$$

Thus we have (1). If $m \neq n$ are elements of M then

$$0 = e_{1,m}(n) = (e\langle * \rangle e_{1,m})(n) = \sum_{m' * m'' = n} e(m')e_{1,m}(m'') = \sum_{m' * m = n} e(m')$$

and similarly,

$$0 = e_{1,m}(n) = (e_{1,m}\langle * \rangle e)(n) = \sum_{m' * m'' = n} e_{1,m}(m')e(m'') = \sum_{m * m'' = m} e(m'').$$

Thus we have (2).

Now, conversely, assume that e is an element of $R[\mathcal{C}]$ satisfying conditions (1) and (2). Then for each $f \in R[\mathcal{C}]$ and any element m of M we have

$$(e\langle * \rangle f)(m) = \sum_{m' * m'' = m} e(m')f(m'') = \sum_{m' * m = m} e(m')f(m) = f(m)$$

and so $e\langle * \rangle f = f$. Similarly, $f\langle * \rangle e = f$. Thus $R[\mathcal{C}]$ is a semiring with multiplicative identity e. □

If $R[\mathcal{C}]$ is not a semiring then it can always be embedded in a semiring, namely its Dorroh extension.

Very close to the notion of a convolution is the following construction. Let A be a partially-ordered set satisfying the condition that for each $a \in A$ the set $(a] = \{b \in A \mid b \leq a\}$ is finite and let $*$ be an operation defined on $\{(a,b) \in A \times A \mid b \leq a\}$. If R is a semiring then we can define the **Wiegandt convolution** $\circledast$ on R^A by setting

$$f \circledast g : a \mapsto \sum \left\{ f(b)g(a * b) \;\middle|\; b \in (a] \right\}.$$

(3.15) PROPOSITION. *Let R be a semiring and let A be a partially-ordered set satisfying the condition that $(a]$ is finite for each $a \in A$ and endowed with an operation $*$ defined on $\{(a,b) \in A \times A \mid b \leq a\}$. If addition is defined on R^A componentwise then in order for $(R^A, +, \circledast)$ to be a semiring it suffices that the following conditions are satisfied:*

(1) *If $a \geq b$ then $a \geq a * b$ and $a * (a * b) = b$;*
(2) *If $a \geq b \geq c$ then $a * c \geq b * c$ and $(a * c) * (b * c) = a * b$;*
(3) *If $a > b \geq c$ then $a * c > b * c$.*

PROOF. We have already noted that $(R^A, +)$ is an additive monoid the identity element of which is the function z defined by $z(a) = 0$ for all $a \in A$. Moreover, if $f, g, h \in R^A$ and $a \in A$ then

$$\begin{aligned}
([f+g] \circledast h)(a) &= \sum_{b \leq a} [f(b) + g(b)]h(a * b) \\
&= \sum_{b \leq a} f(b)h(a * b) + g(b)h(a * b) \\
&= \sum_{b \leq a} f(b)h(a * b) + \sum_{b \leq a} g(b)h(a * b) \\
&= ([f \circledast h] + [g \circledast h])(a)
\end{aligned}$$

and so $[f+g] \circledast h = [f \circledast h] + [g \circledast h]$. Similarly, $f \circledast [g+h] = [f \circledast g] + [f \circledast h]$, showing that $\circledast$ distributes over addition from either side. If $f \in R^A$ and $a \in A$ then $[f \circledast z](a) = \sum\{f(a)z(a*b) \mid b \leq a\} = 0 = [z \circledast f](a)$ and so $f \circledast z = z = z \circledast f$.

We are left to show that $(R^A, \circledast)$ is a monoid with identity element not equal to z. First, we note some consequences of conditions (1) - (3) of the hypothesis. If b and b' are distinct elements of $(a]$ then $a*b$ and $a*b'$ must also be distinct. Indeed, if $a * b = a * b'$ then, by (1), we have $b = a * (a * b) = a * (a * b') = b'$. We now claim that

(4) If $a \geq b > c$, then $a * c > a * b$.

Indeed, from (2) and (1) we obtain $a * b = [a * (b * c)] * [b * (b * c)] = [a * (b * c)] * c$. Moreover, $a \geq a * c \geq b * c$. Therefore $a \geq a * (b * c)$ so $a * c \geq [a * (b * c)] * c = a * b$. Since $b \neq c$, we in fact have $a * c > a * b$, establishing (4). By the choice of A, we know that it has minimal elements and, indeed, for each $a \in A$ there is a minimal element b of A satisfying $b \leq a$. Let A' denote the set of all minimal elements of A. If b is an element of A then, by condition (1), we have $b * b \leq b$ and so, if $b \in A'$, we conclude that $b * b = b$. If $a \in A$ we claim that $(a] \cup A' = \{a * a\}$. Indeed, we surely have $a * a \in (a]$. If $b \in (a] \cup A'$ then $b = a * (a * b)$. By (4) we see that $a \geq a * b$ implies that $b = a * (a * b) \geq a * a$ and so, by minimality, $b = a * a$. As a consequence, we see that $(a] = \{c \in A \mid a \geq c \geq a * a\}$. We also note that, as a consequence of (1), $a * (a * a) = a$ for each $a \in A$.

If $b \leq a$ in A let $[b, a] = \{d \in A \mid b \leq d \leq a\}$. Then we have a function $\varphi\colon [b, a] \to [a * a, a * b]$ defined by $\varphi\colon d \mapsto a * d$. By what we have already noted above, this function is injective. Moreover, by (4), it is order-reversing. We claim that it is also surjective. Indeed, suppose that $a * a \leq d' \leq a * b$ and set $d = a * d'$. Then $a * d = a * (a * d') = d'$. Moreover, $a \geq a * d$ while $a * b \geq d'$ implies that $d = a * d' \geq a * (a * b) = b$. Thus $d' = \varphi(d)$, proving that φ is bijective. Another order-reversing function from $[b, a]$ to $[a * a, a * b]$ is given by $\psi\colon d \mapsto d * b$. Indeed, by (2) we have $a \geq a * b \geq d * b$ and so $d * b \geq a * a$ by the minimality of $a * a$. To show that ψ is bijective, we must show that every element d' of $[a * a, a * b]$ can be uniquely represented in the form $d * b$ for some $d \in [b, a]$. Indeed, if d' is such an element then there exists a unique element d'' of $[b, a]$ such that $d' = \varphi(d'') = a * d''$. Hence $d' = a * d'' = [a * (d'' * b)] * [d'' * (d'' * b)] = [a * (d'' * b)] * b = \psi(d)$, where $d = a * (d'' * b)$, and this is uniquely determined since d'' is. Clearly $a \geq d$ while $b = d'' * (d'' * b) \leq a * (d'' * b) = d$ so $d \in [b, a]$.

We now return to prove the associativity of $\circledast$. If $f, g, h \in R^A$ and $a \in A$ then, by the above,

$$[f \circledast (g \circledast h)](a) = \sum_{d \leq a} f(d) \left[\sum_{c \leq a * d} g(c) h((a * d) * c) \right].$$

But, by the above, we note that every such c is of the form $d' * d$, where d' is a unique element of $[d, a]$ which ranges over all of $[d, a]$ as c ranges over $[a * a, a * d] = (a * d]$. Thus we have

$$\begin{aligned}
[f \circledast (g \circledast h)](a) &= \sum_{d \leq a} f(d) \left[\sum_{d \leq d' \leq a} g(d' * d) h([a * d] * [d' * d]) \right] \\
&= \sum_{d \leq a} f(d) \left[\sum_{d \leq d' \leq a} g(d' * d) h(a * d') \right] \\
&= \sum_{d' \leq a} \left[\sum_{d \leq d'} f(d) g(d' * d) \right] h(a * d') \\
&= [(f \circledast g) \circledast h](a),
\end{aligned}$$

proving associativity. Finally, let $y \in R^A$ be the characteristic function on A'. If $f \in R^A$ and $a \in A$ then $[f \circledast y](a) = \sum\{f(b)y(a * b) \mid b \leq a\} = f(a)y(a * a) = f(a)$ so $f \circledast y = a$. Similarly, $[y \circledast f](a) = y(a * a)f(a * (a * a)) = f(a)$ so $y \circledast f = f$. Since $f \neq z$, this proves that $(R^A, +, \circledast)$ is a semiring. $\square$

(3.16) EXAMPLE. If $A = sub(B)$ for some nonempty set B then the operation $*$ on $\{(a, b) \in A \times A \mid b \subseteq a\}$ defined by $a * b = a \setminus b$ satisfies conditions (1) - (3) of Proposition 3.15. If $A = \mathbb{P}$ is partially-ordered by the relation $b \subseteq a$ if and only if b divides a, and if we define the operation $*$ on $\{(a, b) \in \mathbb{P} \times \mathbb{P} \mid b \subseteq a\}$ by $a * b = a/b$ then then Wiegandt convolution on $\mathbb{N}^A$ coincides with the Dirichlet convolution as defined in Example 3.6. For an example of a Wiegandt convolution defined on the lattice of subgroups of a finite abelian group, see [Delsarte, 1948]. This was applied in [Rédei, 1967] to construct group-theoretic ζ-functions. For examples of applications of such convolutions to combinatorics and computing, see [Birkhoff, 1971].

Now let t be an indeterminate and consider the finitary multiplicative monoid $M = \{t^i \mid i \in \mathbb{N}\}$. In this case, for a semiring R, we follow the usual convention and write $R[t]$ instead of $R[M]$. This semiring is the **semiring of polynomials in the indeterminate** t over R. We will often follow the usual convention of denoting a polynomial in $f \in R[t]$ by $\sum f(i)t^i$ rather than as a function. If $0 \neq f \in R[t]$ and if h is a maximal element of the support of f, then h is the **degree** of f and $f(h) \in R$ is called the **leading coefficient** of f. We denote the degree of f by $deg(f)$. If $f = 0$ we set $deg(f) = -\infty$. Semirings of polynomials over the schedule algebra and their applications to graph theory and discrete-event dynamical systems are discussed in detail in [Baccelli et al., 1992].

(3.17) EXAMPLE. Let R be a semiring and t an indeterminate. If R' is an entire zerosumfree subsemiring of R then the set of all polynomials in $R[t]$ having leading coefficient in R' and the set of all polynomials in $R[t]$ having the coefficient of the lowest nonzero term in R' are both subsemirings of $R[t]$.

Note that if S is an entire zerosumfree subsemiring of a semiring R and if t is an indeterminate then $\{p(t) \in R[t] \mid \text{the leading coefficient of } p \text{ belongs to } S\} \cup \{0_R\}$ is a subsemiring of $R[t]$. Similarly, the set of all polynomials $p(t) \in R[t]$ the coefficient of the lowest nonzero term in which belongs to S, together with 0_R, is a subsemiring of $R[t]$.

A **derivation** on a semiring R is a function $d: R \to R$ satisfying $d(r + r') = d(r) + d(r')$ and $d(rr') = d(r)r' + rd(r')$ for all $r, r' \in R$. If d and d' are derivations on a semiring R then for all $r, r' \in R$ we have

$$(d + d')(r + r') = d(r) + d(r') + d'(r) + d'(r') = (d + d')(r) + (d + d')(r')$$

and

$$\begin{aligned}(d + d')(rr') &= d(rr') + d'(rr') = d(r)r' + rd(r') + d'(r)r' \\ &= rd'(r') = [(d + d')(r)]r' + r[(d + d')(r')]\end{aligned}$$

and so $d + d'$ is also a derivation on R. Since the function $r \mapsto 0$ is surely a derivation on R, we see that the family of all derivations on R is a monoid under the operation of componentwise addition. If R is an additively-idempotent semiring, then the identity map $r \mapsto r$ from R to itself is also a derivation on R.

(3.18) EXAMPLE. Let R be a semiring and let S be the subsemiring of $\mathcal{M}_2(R)$ consisting of all matrices of the form $\begin{bmatrix} a & b \\ 0 & a \end{bmatrix}$. Then the function $d: S \to S$ defined by $\begin{bmatrix} a & b \\ 0 & a \end{bmatrix} \mapsto \begin{bmatrix} 0 & b \\ 0 & 0 \end{bmatrix}$ is a derivation on S. Since R can be identified with the subsemiring of S consisting of all matrices of the form $\begin{bmatrix} a & 0 \\ 0 & a \end{bmatrix}$, we see that any semiring can be embedded in a semiring with a nontrivial derivation. For the computational implications of this fact, see [Ball, 1986].

If $R[t]$ is the semiring of polynomials in the indeterminate t over the semiring R, and if d is a derivation on R, we can define a new multiplication on $R[t]$ with the aid of the rule $tr = rt + d(r)$ for all $r \in R$ and the distributivity of regular multiplication over addition from both sides. Thus we obtain a hemiring, which, as we have seen, can be embedded in a semiring. This semiring, denoted by $R[t; d]$, is called the **differential polynomial semiring** defined by d over R. The necessity for using a Dorroh extension can be avoided if we insist that our derivations satisfy the additional condition that $d(1) = 0$.

Let A be a nonempty totally-ordered set. We can induce a partial order on the free monoid A^* inductively as follows:

(*) If $w = au$ and $w' = bv$ are nonempty words with $u, v \in A^*$, then $w < w'$ if and only if $a < b$ in A or $a = b$ and $u < v$.

Note that if $w < w'$ and if u, v are arbitrary words then $uw < vw'$ and $wv < w'v$. Also note that this order induces a total order on the set of all words of a given length.

If $w = a_1a_2 \ldots a_n$ is a word in A^* then there is a permutation σ of $\{1, \ldots, n\}$ such that $\sigma(1) \leq \sigma(2) \leq \cdots \leq \sigma(n)$. Denote the word $a_{\sigma(1)}a_{\sigma(2)} \ldots a_{\sigma(n)}$ by w_σ. Then we can define a new operation $\odot$ on A^* by $w \odot w' = (w \cdot w')_\sigma$. Moreover, $(A^*, \odot)$ is a commutative monoid, called the **symmetric free monoid** on the totally-ordered set A. If R is a semiring then the semiring $R[(A, \odot)]$ is called the **semiring of symmetric formal polynomials in** A **over** R. It is not a subsemiring of $R\langle A \rangle$. For the use of such semirings in number theory see [Cashwell & Everett, 1959].

Let A be a nonempty set and let A^* be the free monoid of A. If R is a semiring then the **semiring of formal power series in** A **over** R, denoted $R\langle\langle A \rangle\rangle$, is defined to be the set R^{A^*} on which addition is defined componentwise and multiplication is defined by the **Cauchy product** $(fg)(w) = \sum\{f(w')g(w'') \mid w'w'' = w\}$. This sum is finite and so the product is well-defined. The additive identity of $R\langle\langle A \rangle\rangle$ is the function which takes every element of A^* to 0_R. The multiplicative identity is the function f defined by $f(w) = 1_R$ if $w = \square$ and $f(w) = 0_R$ otherwise. An element f of $R\langle\langle A \rangle\rangle$ is **quasiregular** if and only if $f(\square) = 0$. The semiring $R\langle A \rangle$ subsemiring of $R\langle\langle A \rangle\rangle$. If R is additively idempotent or zerosumfree, then so is $R\langle\langle A \rangle\rangle$ for all A. If $\varnothing \neq B \subseteq A$ then we can consider $R\langle\langle B \rangle\rangle$ as a subsemiring of $R\langle\langle A \rangle\rangle$ by identifying $R\langle\langle B \rangle\rangle$ with the family of those functions in $R\langle\langle A \rangle\rangle$ the support of which is contained in B^*.

If R is a semitopological semiring then $R\langle\langle A \rangle\rangle$ is also semitopological under the product topology induced by the topology on R. If $f \neq g$ are distinct elements of $R\langle\langle A \rangle\rangle$, set $m(f, g) = min\{|w| \mid w \in A^* \text{ and } f(w) \neq g(w)\}$. Pick a real number c satisfying $0 < c < 1$ and define a function $d: R\langle\langle A \rangle\rangle^2 \to \mathbb{R}^+$ by setting $d(f, f) = 0$

for all $f \in R\langle\langle A\rangle\rangle$ and $d(f,g) = c^{m(f,g)}$ for $f \neq g$. It is straightforward to verify that this is a complete ultrametric on $R\langle\langle A\rangle\rangle$ (i.e. $d(f,g) \leq max\{d(f,h), d(h,g)\}$ for all $f, g, h \in R\langle\langle A\rangle\rangle$) which turns $R\langle\langle A\rangle\rangle$ into a topological semiring. If $f \in R\langle\langle A\rangle\rangle$ is quasiregular, then the sequence $\langle f, f^2, f^3, \ldots\rangle$ converges to 0 in this topology.

Thus we have seen that there are in fact two ways of defining multiplication on R^{A^*} in order to turn it into a semiring: the Cauchy product as given above and the pointwise product (in various contexts also called the **Hadamard product** which we considered before.

Formal power series over various semirings are an important tool in several fields of applied mathematics and computer science. For an example of their application to formal language theory, see [Chomsky & Schützenberger, 1963], [Eilenberg, 1974], [Kuich & Salomaa, 1986], [Salomaa & Soittola, 1978], [Stanat, 1972], and [Wang, 1998].

(3.19) APPLICATION. One of the most important applications of semirings in theoretical computer science is to automata theory. The use of semirings to study automata goes back to [Conway, 1971] and was given its major impetus in [Eilenberg, 1974]. There are several ways of defining automata over semirings, and we will use the approach given in [Kuich & Salomaa, 1986]. Also refer to [Lallement, 1979].

Let R be a semiring and let A be a nonempty set. An $R\langle\langle A\rangle\rangle$**-automaton** $\mathcal{A} = (S, M, s_0, P)$ consists of:

(1) A countable set S of **states** of $\mathcal{A}$;
(2) A matrix $M \in \mathcal{M}_{S,r}(R\langle\langle A\rangle\rangle)$, called the **transition matrix** of $\mathcal{A}$;
(3) An element s_0 of S called the **initial state** of $\mathcal{A}$;
(4) A column vector $P \in (R\langle\square\rangle)^S$, called the **final state vector** of $\mathcal{A}$.

In variants of this definition, the automaton is allowed to have any of a finite number of initial states and the transition matrix is restricted to having entries in some predesignated subset of R containing 0 and 1. See, for example, [Kuich, 1987].

An automaton $\mathcal{A}$ is **finite** if the set S is finite. Initially, all automata studied were finite automata. However, countably-infinite automata turn out to be useful in certain situations, such as describing machines with pushdown stacks. As we noted previously, the matrix M can also be thought of as a directed graph on the set S. If $M[s,t] \neq 0$ then there is an arc of the graph from s to t having **label** $M[s,t] \in R\langle\langle A\rangle\rangle$. More generally, if s and t are elements of S then to any path $p = (s = s_1, \ldots s_n = t)$ we assign the label $||p|| = M[s_1, s_2] \cdot \ldots \cdot M[s_{n-1}, s_n]$. If p is a path from s to t and q is a path from t to u then pq is a path from s to u and $||pq|| = ||p|| \cdot ||q||$.

If $s \in S$ and $k \in \mathbb{N}$ let $b_{k,s}$ be $\sum\{||p|| \mid p$ a path from s_0 to s of length $k\}$. Set $b_s = \sum_{k=0}^{\infty} b_{k,s}$. The **behavior** of the automaton $\mathcal{A}$ is the formal power series $||\mathcal{A}|| \in R\langle\langle A\rangle\rangle$ defined by $||\mathcal{A}|| = \sum\{b_s P(s) \mid s \in S\}$. Note that this power series may not exist! We will come back to the existence of behaviors for automata later. If $f \in R\langle\langle A\rangle\rangle$ is the behavior of some $R\langle\langle A\rangle\rangle$-automaton $\mathcal{A}$, then the power series f is **accepted** by $\mathcal{A}$. A language $B \subseteq A^*$ is **recognized** by an $R\langle\langle A\rangle\rangle$-automaton $\mathcal{A}$ if and only if $||\mathcal{A}||$ exists and has support B.

An extension of the above construction to formal power series over trees rather than over words is considered in [Berstel & Reutenauer, 1982] and is utilized in

[Wechler, 1986a] for the study of fuzzy program schemata in connection with the mathematical semantics of nondeterministic programs. Automata over semirings other than semirings of formal power series are considered in [Mizumoto, Toyoda & Tanaka, 1975], where several examples are given. Task resource modules using automata over the schedule algebra are described in [Gaubert & Mairesse, 1998], while representation of safe timed Petri nets by automata over the schedule algebra is described in [Gaubert & Mairesse, 1999]. Such representations allow the authors to obtain automata-based performance evaluations for such nets. For general performance evaluation of automata over the schedule algebra, refer to [Gaubert, 1995].

If R is an arbitrary semiring and if $-\infty$ is an element not in R then we can define the structure of a semiring on the set $S = R \cup \{-\infty\}$ as follows:

(1) If $a, b \in R$ then $a + b$ and ab are the same as in R;
(2) $a + -\infty = -\infty + a = a$ for all $a \in S$;
(3) $a(-\infty) = (-\infty)a = -\infty$ for all $a \in S$.

Moreover, one immediately sees that this new semiring, in which $-\infty$ is now the zero element, is in fact entire and zerosumfree. Note that R is not a subsemiring of S since the two semirings do not have the same zero element. We will denote the semiring S by $R\{-\infty\}$.

This same construction can be used to construct entire zerosumfree semirings when we are lacking is an additive identity. That is to say, if R is a nonempty set on which we have operations of addition and multiplication defined so that $(R, +)$ is a commutative semigroup, $(R, \cdot)$ is a monoid, and multiplication distributes over addition from either side, and if $-\infty$ is an element not in R then, $(R\{-\infty\}, +, \cdot)$ defined as above is a zerosumfree semiring. Thus, for example, $(\mathbb{R}^+, max, +)$ is not a semiring since 0 acts as both "additive" and "multiplicative" identity. However, as we have seen, $(\mathbb{R}^+ \cup \{-\infty\}, max, +)$ is a semiring, which is a subsemiring of the schedule algebra.

Similarly, let $R = \{r \in \mathbb{R} \mid r > 0\}$ and define operations $\oplus$ and $\odot$ on R by setting $a \oplus b = ab/(a+b)$ and $a \odot b = ab$. Then $(R, \oplus, \odot)$ lacks an additive identity but $(R \cup \{-\infty\}, \oplus, \odot)$ is an entire zerosumfree semiring.

(3.20) Example. A **partial function** from a set A to a semiring R is a function f from a subset $dom(f)$ of A to R. If f is such a function then we can extend f to a function f^+ from A to $R\{-\infty\}$ by setting $f^+(a) = -\infty$ for all $a \in A \setminus dom(f)$. Thus the family of all partial functions from A to R can be identified with the semiring $R\{-\infty\}^A$.

(3.21) Example. In [Park, 1981] and [Izumi, Inagaki & Honda, 1984] the construction in Example 1.11 is extended to the set $(sub(A^\infty), +, \cdot)$ of all formal ∞-languages on A in order to deal with automata which allow for the possibility of concurrent interpretation of commands. If L and L' are subsets of A^∞ then we set $L + L'$ to be $L \cup L'$, while $LL' = \{ww' \mid w \in L \cap A^* \text{ and } w' \in L'\} \cup (L \setminus A^*)$. With respect to these operations, $sub(A^\infty)$ satisfies all of the conditions for being a semiring except that $L \cdot \varnothing$ is not necessarily L, as one would need, but rather $L \cdot \varnothing = L \cap A^*$. However, this can be embedded in a zerosumfree semiring by adding a new additive identity $-\infty$ as above.

If R is an entire zerosumfree semiring and ∞ is an element not in R, then we can extend the semiring structure on R to a semiring structure on the set $R \cup \{\infty\}$ by setting

(1) $a + \infty = \infty + a = \infty + \infty = \infty$ for all $a \in R$,
(2) $a\infty = \infty a = \infty\infty = \infty$ for all $0 \neq a \in R$, and
(3) $0\infty = \infty 0 = 0$.

We will denote this semiring by $R\{\infty\}$. Clearly ∞ is a strongly-infinite element of this semiring. Since $R\{\infty\}$ is again zerosumfree and entire, this process can be iterated. Let $W_h = \{\infty_i \mid i < h\}$ be a family of indeterminates, where h is some ordinal. Define $R\{W_h\}$ by transfinite induction as follows:

(1) If h is a limit ordinal, then $R\{W_h\} = \cup_{i<h} R\{W_i\}$;
(2) If h is not a limit ordinal then $R\{W_h\} = (R\{W_{h-1}\})\{\infty_{h-1}\}$.

Let R be the semiring $(\mathbb{N}\{\infty\}, max, min)$. If $(A, *)$ is a monoid then the convolution $\langle * \rangle$ on R^A is called the **multiproduct**. See [Lake, 1976]; also see [Wongseelashote, 1976, 1979] for further details concerning this semiring and for its uses in graph-theoretic problems, including specific computational algorithms. For uses of R in the theory of formal languages, see [Mascle, 1986].

A function δ from a semiring R to itself is a **reduction** if and only if the following conditions are satisfied:

(1) $\delta(0) = 0$;
(2) $\delta(1) = 1$;
(3) $\delta(a + b) = \delta(\delta(a) + b)$ for all $a, b \in R$;
(4) $\delta(ab) = \delta(\delta(a)b) = \delta(a\delta(b))$ for all $a, b \in R$.

Such a function is necessarily idempotent. Indeed, if $a \in R$ then $\delta^2(a) = \delta(\delta(a)) = \delta(\delta(a) \cdot 1) = \delta(a \cdot 1) = \delta(a)$. It is straightforward to see that if δ is a reduction of a semiring R then $im(\delta) = \{a \in R \mid \delta(a) = a\}$ is a semiring with respect to the operations $\oplus$ and $\odot$ defined by $a \oplus b = \delta(a + b)$ and $a \odot b = \delta(ab)$. Note that this is not necessarily a subsemiring of R, though it has the same additive and multiplicative identities. If A is an additively- [resp. multiplicatively-] idempotent element of R then $\delta(a)$ is an additively- [resp. multiplicatively-] idempotent element of $im(\delta)$.

A special case of this construction was considered in [Wongseelashote, 1979] in his analysis of various path problems on graphs and the construction of semirings suitable for solving these problems, beginning from the semiring of all subsets of the set of vertices of a given graph. For example, let R be a zerosumfree entire semiring and let A be a nonempty set. Define a function δ from R^A to itself by setting $\delta(f)$ to be the characteristic function on $supp(f)$ for each $f \in R^A$. Then δ is a reduction on R^A. This example was considered in [Wongseelashote, 1979] for the special case of $R = \mathbb{N}\{\infty\}$.

Finally, we observe that if R is a hemiring and if $0 \neq a \in R$ then we can define a new operation $*_a$ on R by $r *_a r' = rar'$. Then $(R, +, *_a)$ is a hemiring, which is not necessarily a semiring, called the **shift** of R by a. We can embed it in a semiring by taking its Dorroh extension.

Let A and B be nonempty sets and let $h \in R^{A \times B}$ be an R-valued relation between A and B. If $f \in R^A$ has finite support, we define $h[f] \in R^B$ by setting

$$h[f]: b \mapsto \sum_{a \in A} f(a)h(a,b)$$

for all $b \in B$ and if $g \in R^B$ has finite support, we define $h^{-1}[g] \in R^A$ by setting

$$h^{-1}[g]: a \mapsto \sum_{b \in B} h(a,b)g(b)$$

for all $a \in A$. We note immediately that if $f = f' + f''$ in R^A then $h[f] = h[f'] + h[f'']$ in R^B while if $g = g' + g''$ in R^B then $h^{-1}[g] = h^{-1}[g'] + h^{-1}[g'']$ in R^A.

Functions $h \mapsto h[f]$ can be considered as inference schemes in an uncertain environment and as such include the fuzzy implication operators used in designing fuzzy controllers and fuzzy microprocessors [Gupta & Yamakawa, 1988]. There are several ways of doing this. For example, we can consider the following construction, based [De Baets & Kerre, 1993a]: an **implication** on a semiring R is an operation $\triangleright$ on R satisfying the boundary conditions $0 \triangleright 0 = 0 \triangleright 1 = 1 \triangleright 1 = 1$ and $1 \triangleright 0 = 0$. If A and B are nonempty sets and if $f_0 \in R^A$ and $g_0 \in R^B$ are given R-valued subsets of A and B respectively, then each implication $\triangleright$ on R defines an R-valued relation $f_0 \triangleright g_0 \in R^{A \times B}$ by $(f_0 \triangleright g_0): (a,b) \mapsto f_0(a) \triangleright g_0(b)$. The R-valued modus ponens rule then becomes: if $f_0(a)$ then $g_0(b)$ and if $f(a)$ then $(f_0 \triangleright g_0)[f](b)$. The case of $R = \mathbb{I}$ was first considered, in several papers, by Lofti Zadeh. Also refer also to [Fuller & Zimmermann, 1992] and [Hellendoorn, 1990].

Let R be a semiring. Any function $u: A \to B$ between nonempty sets defines an R-valued function h_u between A and B by setting

$$h_u(a,b) = \begin{cases} 1 & \text{if } u(a) = b \\ 0 & \text{otherwise} \end{cases}.$$

If $u^{-1}(b)$ is finite for $b \in B$, then

$$h_u[f]: b \mapsto \sum_{a \in A} f(a)h_u(a,b) = \sum_{u(a)=b} f(a)$$

for each $f \in R^A$. On the other hand, if $g \in R^B$ then

$$h_u^{-1}[g]: a \mapsto \sum_{b \in B} h_u(a,b)g(b) = gu(a)$$

for each $a \in A$.

We note, in particular, that if the function $u: A \to B$ is bijective then

$$h_u[f]: b \mapsto (fu^{-1})(b)$$

for all $b \in B$ for which $u^{-1}(b)$ is finite.

Let R be a semiring and let $u: A \to B$ be a function between nonempty sets. Then $f \in R^A$ is **u-stable** if and only if $f(a_1) = f(a_2)$ whenever $u(a_1) = u(a_2)$.

(3.22) PROPOSITION. *Let R be an additively-idempotent semiring and let A and B be finite sets. If $u: A \to B$ is a surjective map then there exists a bijective correspondence between R^B and the set of all u-stable elements of R^A.*

PROOF. If $g \in R^B$ then $h_u^{-1}[g] \in R^A$ is easily seen to be u-stable and, moreover, we have $h_u[h_u^{-1}[g]] = g$ since $h_u[h_u^{-1}[g]](b) = \sum_{u(a)=b} gu(a) = g(b)$ for all $b \in B$, by the u-stability of g and the additive idempotence of R.

Now suppose that $f_1, f_2 \in R^A$ are u-stable functions satisfying $h_u(f_1) = h_u(f_2)$. If $a_0 \in A$ and $b_0 = u(a_0) \in B$ then, by the additive idempotence of R, we have

$$f_1(a_0) = \sum_{u(a)=b_0} f_1(a) = h_u(f_1)(b_0) = h_u(f_2)(b_0) = f_2(a_0)$$

and so $f_1 = f_2$. □

4. SOME CONDITIONS ON SEMIRINGS

We usually consider semirings on which some sort of additional conditions have been imposed. Many such conditions were given in Chapter 1 and examples given of semirings which do or do not satisfy them. We now want to consider consequences of imposing some of these conditions on a semiring. In particular we will first look at the condition of being an additively-idempotent semiring and at the stronger condition of being a simple semiring. Then we will consider some weaker versions of the condition that elements have additive or multiplicative inverses. Finally, we will take up a condition which guarantees the existence of "enough" multiplicative units.

First, however, we must state a number of standard notational conventions: if n is a positive integer and a is an element of a semiring R, then we denote the sum $a + \cdots + a$ of n copies of a by na and the product $a \cdot \ldots \cdot a$ of n copies of a by a^n. We set $a^0 = 1_R$ for each element a of R. The semiring R is **algebraically closed** if for each $b \in R$ and for each positive integer n there exists an $a \in R$ satisfying $a^n = b$. Thus, for example, the schedule algebra $(\mathbb{R} \cup \{-\infty\}, max, +)$ is algebraically closed.

An element a of R is **nilpotent** if and only if there exists a positive integer n satisfying $a^n = 0$. The smallest such positive integer n is called the **index of nilpotency** of a. We will denote the set of all nilpotent elements of R by $N_0(R)$. Then $N_0(R) \neq \varnothing$ for any semiring R, since 0 is always nilpotent. If the semiring R is commutative then $N_0(R)$ is a submonoid of $(R, +)$. Indeed, if $a, b \in N_0(R)$ are nilpotent elements of R satisfying $a^n = b^k = 0$ then

$$(a + b)^{n+k} = \sum_{j=0}^{n+m} \binom{n+k}{j} a^j b^{n+m-j}.$$

since $a^j = 0$ if $j \geq n$ and $b^{n+m-j} = 0$ if $j \leq n$, we see that each summand is 0 and so $(a + b)^{n+k} = 0$.

If a and b are elements of a semiring R and if n and m are nonnegative integers we define the symbol $a^{[n]}b^{[m]}$ inductively as follows:

(1) $a^{[0]}b^{[m]} = b^m$ for all $m \geq 0$;

(2) $a^{[n]}b^{[0]} = a^n$ for all $n \geq 0$;
(3) $a^{[n+1]}b^{[m+1]} = (a^{[n]}b^{[m+1]})a + (a^{[n+1]}b^{[m]})b$.

Intuitively, $a^{[n]}b^{[m]}$ is the sum of all possible products of n of the a's and m of the b's.

(4.1) PROPOSITION. *If a and b are elements of a semiring R while n and m are nonnegative integers then:*

(1) $(a+b)^n = \sum_{i=0}^{n} a^{[i]}b^{[n-i]}$;
(2) $a^{[n]}b^{[m]} = \sum_{i=0}^{n}(a^{[i]}b^{[m-1]})ba^{n-i}$.

PROOF. This follows by a straightforward induction argument. □

If A and B are nonempty subsets of a semiring R, we define the subsets $A+B$ and AB of R as follows:

$$A + B = \{a + b \mid a \in A, b \in B\}$$

and

$$AB = \{a_1b_1 + \cdots + a_nb_n \mid n < \infty;\ a_i \in A, b_i \in B\}.$$

We now begin by looking at simple and additively-idempotent semirings. Simple semirings, as already observed in Chapter 1, are additively idempotent but the converse is not true. Thus, for example, the semiring $(\mathbb{R} \cup \{\infty\}, min, +)$ mentioned in Example 1.22 is additively idempotent but not simple.

(4.2) PROPOSITION. *If a, b, c, and d are elements of an additively-idempotent semiring R satisfying $a + c = b$ and $b + d = a$ then $a = b$.*

PROOF. By additive idempotence we have $a = a+a = a+b+d = a+a+c+d = a+c+d = b+d+c+d = b+d+c = a+c = b$. □

We now turn to simple semirings.

(4.3) PROPOSITION. *The following conditions on a semiring R are equivalent:*

(1) *R is simple;*
(2) *$a = ab + a$ for all $a, b \in R$;*
(3) *$a = ba + a$ for all $a, b \in R$;*
(4) *$ab = ab + acb$ for all $a, b, c \in R$.*

PROOF. Assume (1). If $a, b \in R$ then $a = a1 = a(1 + b) = a + ab$, proving (2). Conversely, if (2) holds then $1 + b = 1 + 1b = 1$ for all $b \in R$, proving (1). Similarly, (1) $\Leftrightarrow$ (3) and (1) $\Leftrightarrow$ (4). □

The identities (2) and (3) of Proposition 4.3 are noncommutative versions of the "absorption laws" familiar from the axiomatic algebraic definitions of lattices. They were studied separately in [Jordan,1949] in connection with the study of quantum logic. Because of conditions (2) and (3) of Proposition 4.3, simple semirings are sometimes refered to as **distributive pseudolattices**.

(4.4) Corollary. *For a semiring R the following conditions are equivalent:*

(1) *R is simple and multiplicatively idempotent;*
(2) *$(a+b)(a+c) = a+bc$ for all $a, b, c \in R$;*
(3) *If $a, b \in R$ then $a+b = a \Leftrightarrow ab = b = ba$.*

Proof. (1) $\Leftrightarrow$ (2): Assume (1). By Proposition 4.3 we have

$$(a+b)(a+c) = a^2 + ba + ac + bc = a + ba + ac + bc = a + bc.$$

Thus we have (2). Conversely, assume (2). If $a \in R$ then, by (2),

$$a^2 = (a+0)(a+0) = a + 0 \cdot 0 = a$$

so $I^{\times}(R) = R$. If $a, b \in R$ then $ab + a = (a+0)(b+1) = a + 0 \cdot 1 = a$ and so, by Proposition 4.3, R is simple.

(1) $\Leftrightarrow$ (3): Assume (1) and let a and b be elements of R. If $a+b=a$ then, by (2), we have $ab = (b+a)(b+0) = b+a0 = b$. Similarly $ba = (b+0)(b+a) = b+0a = b$. Conversely, if $ab = b$ then, by Proposition 4.3, $a+b = a+ab = a$. Now assume (3). If $b \in R$ then $1b = b$ so $1+b = 1$. Therefore R is simple. In particular, it is additively idempotent. Hence for each $a \in R$ we have $a+a = a$ and so $a^2 = a$. Thus R is multiplicatively idempotent as well. □

(4.5) Corollary. *A commutative semiring is a bounded distributive lattice if and only if it is a simple multiplicatively idempotent semiring.*

Proof. This is a direct consequence of Proposition 4.3 and the remarks in Example 1.5. □

(4.6) Corollary. *If R is a simple semiring then $(I^{\times}(R), +)$ is a submonoid of $(R, +)$ and $I^{\times}(R) \cap C(R)$ is a bounded distributive lattice.*

Proof. If a and b belong to $I^{\times}(R)$ then $(a+b)^2 = a^2+ab+ba+b^2 = a+ab+ba+b = a+b$ by Proposition 4.3. Therefore $I^{\times}(R)$ is closed under addition and hence, since it contains 0, is a submonoid of $(R, +)$. Furthermore, if $a, b \in C(R)$ then $a+b \in C(R)$.

We also note that $I^{\times}(R) \cap C(R)$ is nonempty since it contains both 0 and 1. If a and b belong to $I^{\times}(R) \cap C(R)$ then surely so does ab. By the above, $a+b \in I^{\times}(R) \cap C(R)$, proving that $I^{\times}(R) \cap C(R)$ is a subsemiring of R, which is simple since 1 is infinite in it. The result now follows from Corollary 4.5. □

(4.7) Proposition. *For each element a of a simple semiring R, let $S(a) = \{0\} \cup \{r \in R \mid r + a = 1\}$. Then:*

(1) *$S(a)$ is a subsemiring of R for each $a \in R$;*
(2) *$S(a) \cap S(b) = S(ab)$ for all $a, b \in R$.*

Proof. (1) Since R is simple, we clearly have $1 \in S(a)$. Therefore we must show that if $r, r' \in S(a)$ then $r + r'$ and rr' belong to $S(a)$. This is immediate if one of r, r' is 0, and so we can assume that both are nonzero. In that case, $r+a = 1 = r'+a$ and so $(r+r')+a = (r+r')+a+a = 1+1 = 1$, establishing that $r+r' \in S(a)$. Moreover, $1 = 1+a = (r+a)(r'+a)+a = rr'+ra+ar'+a^2+a$.

By Proposition 4.3, this equals $rr' + a$, proving that $rr' \in S(a)$. Thus $S(a)$ is a subsemiring of R.

(2) If $0 \neq r \in S(ab)$ then $r + ab = 1$ so, by Proposition 4.3, $1 = 1 + a = r + ab + a = r + a$, proving that $r \in S(a)$. Similarly $r \in S(b)$ and so $r \in S(a) \cap S(b)$. Conversely, assume that $0 \neq r \in S(a) \cap S(b)$. Then

$$1 = 1 + r = (r + a)(r + b) + r = r^2 + ar + rb + ab + r = ab + r$$

and so $r \in S(ab)$. Thus $S(ab) = S(a) \cap S(b)$. □

Note that for any simple semiring R we have $S(0) = \mathbb{B}$ and $S(1) = R$.

(4.8) PROPOSITION. *If a and b are elements of a simple semiring R and $m, n \in \mathbb{N}$ then there exist elements c and d of R satisfying $a^n = a^{[n]}b^{[m]} + c$ and $b^m = a^{[n]}b^{[m]} + d$.*

PROOF. We will first claim that for any $h, k \in \mathbb{N}$ there exists an element c of R such that $a^n = b^h a^n b^k + c$. Indeed, this is trivial if $h = k = 0$. By Proposition 4.3 we have $a^n = ba^n + a^n$ and so the result is true if $h = 1$ and $k = 0$. Now assume that there exists an element c' of R such that $a^n = b^h a^n + c'$. Then $a^n = ba^n + a^n = b(b^h a^n + c') + a^n = b^h + 1a^n + c''$, where $c'' = bc' + a^n$. Thus the result is true for all values of h when $k = 0$. Similarly, the result is true for all values of k when $h = 0$. Finally, assume that both h and k are nonzero. Let d and d' be elements of R satisfying $b^h a^n + d = a^n$ and $a^n b^k + d' = a^n$. Then $a^n = b^h a^n b^k + d''$, where $d'' = b^h d' + d$. This establishes the claim.

We next note that an arbitrary summand r in $a^{[n]}b^{[m]}$ is of the form

$$b^{m(1)}a^{n(1)} \cdot \ldots \cdot a^{n(t)}b^{m(t+1)},$$

where $m(1) + \cdots + m(t+1) = m$ and $n(1) + \cdots + n(t) = n$. By repeated applications of the claim, we see that there exists an element d_r of R such that $r + d_r = a^n$. Finally, we note that $a^{[n]}b^{[m]} + \sum d_r = a^n + \cdots + a^n = a^n$, since simple semirings are additively idempotent.

The second equality is proven similarly. □

(4.9) PROPOSITION. *Let R be a simple semiring for which there exists an integer n satisfying $r^n = r^n + 1$ for all $r \in R$. Then:*

(1) *$r^n + s^n = (r + s)^n$ for all $r, s \in R$;*

(2) *If $\odot$ is the operation on $I^{\times}(R)$ defined by $a \odot b = (ab)^n$, then $(I^{\times}(R), +, \odot)$ is a commutative simple semiring.*

PROOF. (1) Let $r, s \in R$. By expansion, we see that there exists an element d of R satisfying $(r + s)^n = r^n + s^n + d$. On the other hand, $(r + s)^n = (r + s)^{2n-1}$ and this can be expanded in the form $\sum r^{[h]}s^{[k]}$ where, in each summand, either $h \geq n$ or $k \geq n$. By Proposition 4.8, there exists an element e of R satisfying $r^n + s^n + (r + s)^n + e$. By Proposition 4.2, this implies that $r^n + s^n = (r + s)^n$, proving (1).

(2) Clearly 0 and 1 both belong to $I^{\times}(R)$. By Corollary 4.6, we see that $(I^{\times}(R), +)$ is a commutative monoid. Moreover, $a + 1 = 1$ for all $a \in I^{\times}(R)$ since this is true in R.

If $a, b, c \in I^{\times}(R)$ then, by (1), we have

$$a \odot [b+c] = (a[b+c])^n = (ab+ac)^n = (ab)^n + (ac)^n = a \odot b + a \odot c,$$

and similarly $[b+c] \odot a = b \odot a + c \odot a$. Thus $\odot$ distributes over $+$ from either side. Moreover, by the proof of Proposition 4.8, we see that there exists an element d of R satisfying $(abc)^n = (ab)^n + d$ and by Proposition 4.3 we have $(ab)^n = (ab)^n c + (ab)^n$. Therefore there exists an element d' of R with $(abc)^n = (ab)^n c + d'$ and so there exists an element d'' of R satisfying $(abc)^n = [(ab)^n c]^n + d''$. An analogous argument shows that there exists an element e of R satisfying $(ab)^n c^n = [(abc)^n (abc)]^n + e = (abc)^n + e$ and so $(abc)^n = [(ab)^n] c^n$ by Proposition 4.2. A similar argument shows that $[a(bc)^n]^n = (abc)^n$ and so $a \odot (b \odot c) = [a(bc)^n]^n = [(ab)^n c^n] = (a \odot b) \odot c$. Thus the operation $\odot$ is associative. Finally, we note that $(ab)^2 = abab = ba + d$ for some $d \in R$ and so there exists an element d' of R satisfying $(ba)^n = (ab)^{2n} + d' = (ab)^n + d'$. Similarly, there exists an element d'' of R satisfying $(ab)^n = (ba)^n + d''$ and so, by Proposition 4.2, $a \odot b = (ab)^n = (ba)^n = b \odot a$. This shows that the operation $\odot$ is commutative and so $(I^{\times}(R), +, \odot)$ is a commutative simple semiring. □

While every simple semiring is additively idempotent there are, as we have seen, additively-idempotent semirings which are not simple. We do, however, have the following result.

(4.10) Proposition. *Every additively-idempotent semiring has a simple subsemiring.*

Proof. Let R be an additively-idempotent semiring and let $S = \{a \in R \mid a + 1 = 1\}$. Clearly 0 and 1 belong to S. If $a, b \in S$ then $a + b + 1 = a + 1 = 1$ and $ab + 1 = ab + a + b + 1 = (a+1)(b+1) = 1$. Therefore S is a subsemiring of R, which is clearly simple. □

(4.11) Corollary. *Every additively-idempotent semiring has a subsemiring which is a bounded distributive lattice.*

Proof. This is a direct consequence of Proposition 4.10 and Corollary 4.6. □

In Proposition 4.10 we saw that if R is an additively-idempotent semiring then $\{a \in R \mid a + 1 = 1\}$ is a subsemiring of R. The following proposition complements this result.

(4.12) Proposition. *If R is an additively-idempotent semiring then $S = \{0\} \cup \{a \in R \mid a + 1 = a\}$ is a subsemiring of R.*

Proof. Clearly $0 \in S$, while $1 \in S$ since R is additively idempotent. If $0 \neq a, b \in S$ then $(a+b)+1 = a + (b+1) = a + b$ so $a + b \in S$. Moreover, $ab + 1 = a(b+1) + 1 = ab + a + 1 = (a+1)b + a + 1 = ab + b + a + 1 = (a+1)(b+1) = ab$ and so $ab \in S$. This proves that S is a subsemiring of R. □

Finally, we want to establish something about the structure of the additive monoid of an additively-idempotent semiring by proving an analog of a well-known result for semigroups. First we need some notation: for an element a of a semiring R, set $H(a) = \{b \in R \mid$ there exists an element c of R such that $a + b + c = a\}$. Note that $H(a) \neq \varnothing$ for each $a \in R$ since $a + 0 + 0 = a$ implies that $0 \in H(a)$. Moroever, if $a \in I^{+}(R)$ then $a \in H(a)$.

(4.13) PROPOSITION. *If R is a semiring and $a, a' \in I^+(R)$ then:*

(1) *$(H(a), +)$ is a commutative semigroup;*
(2) *$G(a) = \{a + b \mid b \in H(a)\}$ is a subgroup of $H(a)$;*
(3) *$G(a)G(a') \subseteq G(aa')$.*

PROOF. (1) If $b, b' \in H(a)$ then there exist elements c and c' of R satisfying $a = a + b + c = a + b' + c'$ and so $a = a + a = a + (b + b') + (c + c')$. Thus $b + b' \in H(a)$. Since addition in R is associative and commutative, this implies that $H(a)$ is a commutative semigroup.

(2) We note that $a = a + 0 \in G(a)$ and that, by (1) and the additive idempotence of a, the sum of elements of $G(a)$ is again in $G(a)$. If $a + b \in G(a)$ then $a + (a + b) = (a + a) + b = a + b$ so a is the additive identity of $G(a)$. Finally, if $a + b \in G(a)$ then there exists an element c of R (and hence of $H(a)$) satisfying $a + b + c = a$ and so $a = (a + b) + (a + c)$. Thus $a + b$ has an inverse in $G(a)$.

(3) If $a + b \in G(a)$ and $a' + b' \in G(a')$ then there exist elements c and c' of R such that $a = a + b + c$ and $a' = a' + b' + c'$. Therefore

$$\begin{aligned} aa' &= (a + b + c)(a' + b' + c') = aa' + ab' + ac' + ba' + bb' + bc' + ca' + cb' + cc' \\ &= aa' + (a + b)(a' + b') + (ac' + bc' + ca' + cb' + cc'), \end{aligned}$$

proving that $(a + b)(a' + b')$ belongs to $H(aa')$. But $aa' + (a + b)(a' + b') = (a + b)(a' + b')$ so it in fact belongs to $G(aa')$. Since $G(aa')$ is closed under taking finite sums, this proves (3). □

As a consequence of Proposition 4.13, we see that an additively-idempotent semiring R is the union of additive groups which are not, however, subgroups of $(R, +)$.

We now turn to the matter of additive inverses. Let a be an element of a semiring R. An element b of R is an **additive inverse** of a if and only if $a + b = 0$. If a has an additive inverse, then such an inverse is unique for if $a + b = 0 = a + b'$ then $b = b + 0 = b + a + b' = 0 + b' = b'$. We will denote the additive inverse of an element a, if it exists, by $-a$. Denote the set of all elements of R having additive inverses by $V(R)$; this set is nonempty since $0 \in V(R)$, with $-0 = 0$ and, indeed, it is a submonoid of $(R, +)$ since it is closed under taking sums. Moreover, if $a + b \in V(R)$ then both a and b belong to $V(R)$. Clearly R is a ring if and only if $V(R) = R$ and R is zerosumfree if and only if $V(R) = \{0\}$. An infinite element of R cannot belong to $V(R)$.

(4.14) EXAMPLE. [Gardner, 1993] In an elementary calculus course one studies partial functions on $\mathbb{R}$, i.e. functions $f: A \to \mathbb{R}$, where A is a nonempty subset of $\mathbb{R}$ called the **domain** of f. We will denote this domain by *dom*(f). Let S be the set of all such functions. If $f, g \in S$ then $f + g$ is the function having domain *dom*$(f) \cap$ *dom*(g) on which it is defined by the rule $x \mapsto f(x) + g(x)$. Similarly, fg is the function having the same domain on which is defined by the rule $x \mapsto f(x)g(x)$. It is easy to check that $(S, +, \cdot)$ is a semiring the additive identity in which is the function $x \mapsto 0$ with domain $\mathbb{R}$ and the multiplicative identity in which is the function $x \mapsto 1$ with domain $\mathbb{R}$. Howeover, note that if $f \in S$ then and if f^- is the function from $dom(f)$ to $\mathbb{R}$ defined by $x \mapsto -f(x)$, then f^- is an additive inverse of f only if $dom(f) = \mathbb{R}$. Thus $V(R) = \{f \in S \mid dom(f) = \mathbb{R}\}$.

For a more general approach to analysis using this approach, albeit without explicit mention of semirings, refer to [Prezeworska-Rolewicz, 1988, 1998].

Since not every element of a semiring has an additive inverse, we look for a weaker condition. An element a of a semiring R is **cancellable** if and only if $a + b = a + c \Rightarrow b = c$ in R. We will denote the set of all cancellable elements of R by $K^+(R)$. This set is nonempty since $V(R) \subseteq K^+(R)$. An infinite element of a semiring is never cancellable. Moreover, $K^+(R)$ is easily seen to be closed under addition. Thus $K^+(R)$ is a submonoid of the additive monoid $(R, +)$. If $K^+(R) = R$ then the semiring R is **cancellative**. Note that $I^+(R) \cap K^+(R) = \{0\}$ so that additively-idempotent semirings have no nontrivial cancellable elements and are thus as far away from being cancellative as possible.

(4.15) EXAMPLE. The semiring $\mathbb{N}$, which is not a ring, is cancellative. Thus we may have $R = K^+(R) \supset V(R) = \{0\}$.

(4.16) EXAMPLE. If X is a set having more than one element then the semiring $(sub(X), \cup, \cap)$ is not cancellative.

(4.17) EXAMPLE. Let ∞ be an element not in $\mathbb{N}$ and let $R = \mathbb{N}\{\infty\}$. Then $K^+(R) = \mathbb{N}$. This example is noted in [Smith, 1966].

(4.18) EXAMPLE. A subsemiring of a cancellative semiring is again cancellative. If $\{R_i \mid i \in \Omega\}$ is a family of cancellative semirings then $\times_{i\in\Omega} R_i$ is also cancellative. Similarly, if R is a cancellative semiring and A is a nonempty set then $R\langle\langle A\rangle\rangle$ and $R\langle A\rangle$ are cancellative.

(4.19) EXAMPLE. [H. E. Stone, 1977] If R is a cancellative semiring then $\mathcal{M}_n(R)$ is cancellative for every positive integer n. This is an immediate consequence of the fact that addition in $\mathcal{M}_n(R)$ is defined componentwise. Similarly, if A is a countably-infinite subset then $\mathcal{M}_{A,r}(R)$, $\mathcal{M}_{A,c}(R)$, and $\mathcal{M}_{A,rc}(R)$ are cancellative semirings.

We now present another weak version of the condition of having an additive inverse - one which is also satisfied by infinite elements. If R is a semiring, set $W(R) = \{a \in R \mid$ if $b \in R$ then there exists an element r of R such that $a + r = b$ or $b + r = a\}$. Clearly $W(R)$ is nonempty since $V(R) \subseteq W(R)$. Moreover, if $a \in R$ is infinite then $a \in W(R)$ since $a = b + a$ for all $b \in R$. If $R = W(R)$ then the semiring R is a **yoked** semiring.

(4.20) EXAMPLE. $\mathbb{N}$ and $\mathbb{Q}^+$ are surely yoked semirings. Similarly, if R is a totally-ordered set with unique minimal element 0 and unique maximal element 1 then (R, max, min) is a yoked semiring. Thus $\mathbb{I}$ and $\mathbb{N} \cup \{\infty\}$ are yoked semirings.

(4.21) PROPOSITION. *If I and H are subhemirings of a yoked semiring R satisfying the condition that $IH \subseteq V(R)$ then either $I^2 \subseteq V(R)$ or $H^2 \subseteq V(R)$.*

PROOF. Assume that $I^2 \nsubseteq V(R)$. Then there exist elements a and a' of I such that $aa' \notin V(R)$. Let $b, b' \in H$. If there exists an element r of R such that $a + r = b$

then $a'a + a'r = a'b \in IH \subseteq V(R)$ and so $a'a \in V(R)$, which is a contradiction. Hence, since R is a yoked semiring, there must exist an element r of R such that $a = b + r$. But then $bb' + rb' = ab' \in IH \subseteq V(R)$ and so $bb' \in V(R)$. This proves that $H^2 \subseteq V(R)$. □

The **zeroid** of a semiring R is $Z(R) = \{r \in R \mid r + a = a \text{ for some } a \in R\}$. Thus if $a \in R$ then $b \in H(a)$ implies that $b + c \in Z(R)$ for some $c \in R$. Clearly $I^+(R) \subseteq Z(R)$ so $Z(R) \neq \varnothing$. If $Z(R) = R$ then the semiring R is **zeroic**. Otherwise it is **nonzeroic**. If R has an infinite element then it is surely zeroic. A semiring R is **plain** if and only if $Z(R) = \{0\}$. If R is a ring then it is surely plain. If S is a subsemiring of a semiring R then $Z(S) \subseteq Z(R)$. Thus, in particular, subsemirings of plain semirings are plain. If R is cancellative then it is surely plain. The following result provides a partial converse of this fact.

(4.22) PROPOSITION. *A yoked semiring is cancellative if and only if it is plain.*

PROOF. This is an immediate consequence of the definition. □

As a consequence, we see that a semiring R is plain precisely when $(R, +)$ is a valuation monoid.

The size of the zeroid is a measure of how far, in some sense, a semiring is from being a ring, and it will play a very important part in our considerations later on.

We now turn from additive inverses to multiplicative inverses. An element r of a semiring R is a **unit** if and only if there exists an element r' of R satisfying $rr' = 1 = r'r$. The element r' is called the **inverse** of r in R. If such an inverse r' exists for a unit r, it must be unique. Indeed, if $rr' = rr'' = 1 = r''r = r'r$ then $r' = r'1 = r'(rr'') = (r'r)r'' = 1r'' = r$"; we will normally denote the inverse of r by r^{-1}. It is straightforward to see that if r and r' are units of R then $(rr')^{-1} = r'^{-1}r^{-1}$. Thus, in particular, $(r^{-1})^{-1} = r$. This implies that if $r^{-1} = r'^{-1}$ then $r = r'$. We will denote the set of all units of R by $U(R)$. This set is nonempty since it contains 1 and is not all of R, since it does not contain 0. Moreover, $N_0(R) \subseteq U(R)$. Indeed, if $r \in N_0(R)$ satisfies $r^n = 0$ for some positive integer n, then r^{n+1} also equals 0. Therfore, replacing n by $n+1$ if necessary, we can assume that n is odd. In this case, we have

$$(1 + r)(1 - r + r^2 - \cdots - r^{n+2} + r^{n-1}) = 1 + r^n = 1$$

and so $r \in U(R)$.

(4.23) EXAMPLE. Let R be a semiring and let t be an indeterminate over R. If the leading coefficient of $0 \neq f \in R[t]$ is not a zero divisor in R, then $f \in U(R[t])$ if and only if $deg(f) = 0$ and $f(0) \in U(R)$. The proof is essentially the same as that for rings.

The following result is found in [LaGrassa, 1995].

(4.24) PROPOSITION. *Let t be an indeterminate over a commutative semiring R. Then $U(R[t]) = \{p \in R[t] \mid p(0) \in U(R) \text{ and } p(i) \in N_0(R) \cap V(R) \text{ for all } i > 0\}$.*

PROOF. Assume $p \in R[t]$ satisfies the conditions that $p(0) \in U(R)$ and $p(i)$ is nilpotent for all $i > 0$. Set $b = p(0)^{-1}$ and let $deg(p) = n \geq 0$. For each $0 \leq i \leq n$, we know that $p(i) \in N_0(R) \cap V(R)$ and so $q = bp - p_1 \in V(R[t])$, where $p_1 \in R[t]$ is defined by

$$p_1(i) = \begin{cases} 1 & \text{if } i = 0 \\ 0 & \text{otherwise} \end{cases}.$$

Also, $bp(i)t^i \in N_0(R[t])$ for all $i > 0$ and so, $q \in N_0(R[t])$. Since p_1 is the multiplicative identity of $R[t]$, this implies that $p_1 + q = bp \in U(R[t])$.

Conversely, assume that $p \in U(R[t])$. Then $deg(p) = n \geq 0$ and there is a $g \in R[t]$ satisfying $pg = p_1$. Say $deg(g) = m$. If Then $(pg)(k) = \sum_{i+j=k} p(i)g(j)$ for all $k \leq m+n$, and $(pg)(k) = 0$ if $k > n+m$. In particular, $1 = p(0)g(0)$ and so $p(0) \in U(R)$. If $n = 0$ we are done, and so assume that this is not the case. Then we must have $m \neq 0$ as well. If $1 \leq k \leq n$ then $0 = p_1(k) = \sum_{i+j=k} p(i)g(j) = d + p(k)g(0)$ for some element d of R and so

$$0 = p(0)0 = p(0)d + p(0)p(k)g(0) = p(0)d + p(k),$$

proving that $p(k) \in V(R)$.

We also know that $0 = p_1(n+m) = p(n)g(m)$ and

$$0 = p_1(n+m-1) = p(n)g(m-1) + p(n-1)g(m).$$

Then

$$p(n)^2g(m-1) + p(n)g(m)p(n-1) = p(n)[p(n)g(m-1) + p(n-1)g(m)] = 0.$$

Since $p(n)g(m-1) = 0$ we have $p(n)^2g(m-1) = 0$. Now assume inductively that we have shown that $p(n)^h g(m-(h-1)) = 0$ for all $1 \leq h \leq k$. Then $p_1(n-m-k) = p(n)g(m-k) + p(n-1)g(m-k+1) + \cdots + p(n-k)g(m) = 0$ and so $p(n)^k[p(n)g(m-k) + p(n-1)g(m-k+1) + \cdots + p(n-k)g(m)] = 0$, proving, by induction, that $p(n)^{k+1}g(m-k) = 0$. Thus we see that $p(n)^h g(m-(h-1)) = 0$ for all $1 \leq h \leq m+1$ and so, in particular, $p(n)^{m+1}g(0) = 0$. But $g(0) \in U(R)$ and so we must have $p(n)^{m+1} = 0$. Thus $p(n) \in N_0(R)$. Similarly, $g(m) \in N_0(R)$.

Now suppose that we have already establisdhed that $p(n-h)$ and $g(m-h)$ belong to $N_0(R)$ for all $0 \leq h < k$. We must show that $p(n-k)$ is nilpotent. We know that $g(m)$ is nilpotent and hence so is $p(n-k)g(m)$. Therefore

$$p_1(n-m-(k+1)) = p(n-k-1)g(m) + p(n-k)g(m-1) + \cdots + p(n)g(m-(k+1)) = 0$$

and so $p(n-k)g(m-1) = -p(n-(k+1))g(m) + [-p(n-(k-1))g(m-2)] + \cdots + [-p(n-1)g(m-k)] + [-p(n)g(m-(k+1))]$. By the induction hypothesis, each of the summands on the right-hand side is nilotent and so $p(n-k)g(m-1)$ is nilpotent. Now suppose inductively that we have already shown that $p(n-k)^h g(m-h)$ is nilpotent for $1 \leq h \leq m-1$. Then $0 = p_1(n-k) = p(n-k)g(0) + p(n-(k+1))g(1) + \cdots + p(n-(k+m))g(m)$ so $p(n-k)^{m-1}p_1(n-k) = 0$ Thus $p(n-k)^m g(0) + \cdots + p(n-k)^{m-1}p(n-(k+m))g(m) = 0$. So $p(n-k)^m g(0) = -p(n-k)^{m-1}p(n-(k+1))g(1) + \cdots + [-p(n-k)^{m-1}p(n-(k-m))g(m)]$. Since each of the summands on the right-hand side of this equation is nilpotent, we conclude that $p(n-k)^m g(0)$ is nilpotent. But $g(0)$ is a unit in R and so $p(n-k)$ is nilpotent.

Thus $p(i)$ is nilpotent for all $1 \leq i \leq n$, as desired. □

(4.25) EXAMPLE. If S is the semiring in Example 4.14 then $U(S)$ consists of all functions $f \in S$ satisfying the condition that $dom(f) = \mathbb{R}$ and $f(r) \neq 0$ for all $r \in \mathbb{R}$.

The above comments immediately show that $U(R)$ is a submonoid of $(R, \cdot)$ which is in fact a group. If $U(R) = R \setminus \{0\}$ then R is a **division semiring**. Division semirings are surely entire. A commutative division semiring is a **semifield**. If $\{R_i \mid i \in \Omega\}$ is a collection of division semirings, where $|\Omega| > 1$, then $\times_{i \in \Omega} R_i$ is not a division semiring but $\bowtie_{i \in \Omega} R_i$ is.

Note that if R is a simple semiring then $U(R) = \{1\}$. Indeed, if $a \in U(R)$ then there exists an element b of R such that $ab = 1$. Hence, by Proposition 4.3, we have $a = a + ab = a + 1 = 1$.

(4.26) EXAMPLE. The semirings $(\mathbb{Q}^+, +, \cdot)$, $(\mathbb{Q}^+, max, \cdot)$, $(\mathbb{R}^+, +, \cdot)$, and $(\mathbb{R}^+, max, \cdot)$ are clearly semifields. A subring S of $(\mathbb{Q}^+, +, \cdot)$ is a semifield if and only if for each prime $p \in \mathbb{N}$ there exists an integer $n(p) \in \mathbb{N}$ such that $n(p)/p \in S \setminus \mathbb{N}$. See [H. E. Stone, 1977] for details. Yoked subsemifields of $(\mathbb{R}^+, +, \cdot)$ are considered in [Eilhauer, 1968]. There it is shown that no two distinct yoked subfields of $(\mathbb{R}^+, +, \cdot)$ are isomorphic. By the Krull-Kaplansky-Jaffard-Ohm Theorem [Gilmer, 1972], every additively-idempotent semifield is naturally isomorphic to the semifield of finitely-generated fractional ideals of a Bezout domain.

(4.27) EXAMPLE. Let G be a totally-ordered multiplicative group and let $R = G \cup \{0\}$. Extend the order of G to R by setting $0 \leq g$ for all $g \in G$. Moreover, define $0g = g0 = 0$ for all $g \in R$. Then $(R, max, \cdot)$ is a division semiring.

(4.28) EXAMPLE. The semiring $\mathbb{B}$ is an additively-idempotent division semiring. In fact, it is the only finite additively-idempotent division semiring. To see this, assume that R is a finite additively-idempotent division semiring and let d be the sum of all elements of R. By construction, $r + d = d$ for all $r \in R$ and so, in particular, $d^2 + d = d$. On the other hand, $1 + d = d$ and so $d + d^2 = d^2$. Thus $d = d^2$. Since R is a division semiring, this implies that $d = 1$. If $0 \neq r \in R$ then there exists an element $r' \in R$ satisfying $rr' = 1$. Since $r' + 1 = 1$ we have $1 = 1 + r = r(r' + 1) = r$. Therefore $R = \{0, 1\} = \mathbb{B}$.

(4.29) EXAMPLE. [Cuninghame-Green, 1984] The schedule algebra $R = (\mathbb{R} \cup \{-\infty\}, \oplus, \otimes)$ is a semifield. Indeed, if $a \neq -\infty$ then a has a multiplicative inverse $a^{(-1)} = -a$. If $n \in \mathbb{N}$ then the nth power of an element r of R is $r^{(n)} = nr$. If a and b are elements of R then $min\{a, b\} = a + b - max\{a, b\} = [a \otimes b][a \oplus b]^{(-1)}$.

If t is an indeterminate over R, then the elements of $R[t]$ are of the form $p(t) = \oplus_{i=0}^{n} b_i \otimes t^{(i)} = max\{b_i + it \mid 0 \leq i \leq n\}$. The algebra of such polynomials is considered in detail in [Cuninghame-Green & Meijer, 1980]. In particular, they note that each such polynomial $p(t)$ has a factorization in the form

$$p(t) = a \otimes p_1(t) \otimes \cdots \otimes p_n(t),$$

where $a \in R$ and each $p_i(t)$ either equals t or is of the form $t \oplus b_i$ for some $b_i \in R$. Apart from order, this factorization is unique. An algorithm for the construction of such a factorization is given, using the techniques of nonlinear programming.

We also note that $(\mathbb{N} \cup \{-\infty\}, \oplus, \otimes)$ is a subsemifield of R. If $S = \mathcal{M}_n(R)$ for some natural number n then $A = [a_{ij}]$ belongs to $U(S)$ if and only if every row and every column of A have precisely one element not equal to $-\infty$.

(4.30) EXAMPLE. If S is a bounded distributive lattice and $R = \mathcal{M}_n(S)$ for some positive integer n then the elements of $U(R)$ are characterized in several ways in [Skornjakov, 1986]. Thus, for example, a matrix $A = [a_{ij}]$ belongs to $U(R)$ if and only if $\sum_{j=1}^{n} a_i j = 1$ for all $1 \leq i \leq n$ and $a_{ij}a_{hj} = 0$ for all $i \neq h$. Moreover, this holds if and only if $A^k = 1_R$ for some positive integer k. Indeed, we always have $k \leq n!$. If S is entire then $U(R)$ consists precisely of those matrices $[a_{ij}]$ satisfying the condition that there exists a permutation $\sigma \in \mathcal{S}_n$ such that $a_{ij} = 1$ if $j = \sigma(i)$ and $a_{ij} = 0$ otherwise. Refer also to [Reutenauer & Straubing, 1984].

(4.31) EXAMPLE. [Kaashoek & West, 1974] Let B be a complex Banach space. A subhemiring A of B is a **semialgebra** if and only if $ra \in A$ for all $a \in A$ and all $0 \leq r \in \mathbb{R}$. A subalgebra $A \neq \{0\}$ of B is **locally compact** if and only if $A \cap \{b \in B \mid \|b\| \leq 1\}$ is a compact subset of B; it is **closed** if and only if it is a closed subset of B. If A is a locally compact semialgebra containing a right minimal idempotent element e then eAe is in fact a division semiring. The zerosumfree closed semialgebras of B which are division semirings are all of the form $\mathbb{R}^+e$ for some $e = e^2 \in B$.

(4.32) EXAMPLE. Let R be the set of pairs $(a, b) \in \mathbb{R} \times \mathbb{R}$ satisfying the conditions that either $a > 0$ and $b > 0$ or $a = b = 0$. Define operations $\oplus$ and $\otimes$ on R as follows:

$$(a, b) \oplus (a', b') = \begin{cases} (a, b) & \text{if } b > b' \\ (a', b') & \text{if } b < b' \\ (a + a', b) & \text{if } b = b' \end{cases}$$

and

$$(a, b) \otimes (a', b') = (aa', bb').$$

Then $(R, \oplus, \otimes)$ is a semifield with applications in the study of biopolymers. Refer to [Finkelstein & Roytberg, 1993] and [Akian, Bapat & Gaubert, 1998].

(4.33) PROPOSITION. *A division semiring R is cancellative if and only if $K^+(R) \neq \{0\}$.*

PROOF. If R is cancellative then $K^+(R) = R \neq \{0\}$. Conversely, assume that $0 \neq r \in K^+(R)$ and let $a, b, c \in R$ be elements of R satisfying $a + b = a + c$. If $a = 0$ then surely $b = c$. Otherwise, we multiply both sides of the equation on the left by ra^{-1} to obtain $r + ra^{-1}b = r + ra^{-1}c$ from which, by cancellability, we obtain $ra^{-1}b = ra^{-1}c$. Multiplying both sides of this equation on the left by ar^{-1}, we obtain the desired $b = c$. □

(4.34) PROPOSITION. *A division semiring R is either zerosumfree or is a division ring.*

PROOF. Assume R is not zerosumfree. Then there exists a nonzero element a of R having an additive inverse $-a$. If $0 \neq c \in R$ then $c + ca^{-1}(-a) = ca^{-1}(a + -a) =$

$ca^{-1}0 = 0$ and so c too has an additive inverse. Thus $(R, +)$ is a group and so R is a ring, which must be a division ring. □

As in the additive case, we can talk about cancellability as a weak version of having an inverse. However, since multiplication in an arbitrary semiring is not commutative, we must be careful to keep track of sides. Thus an element a of a semiring R is **right multiplicatively cancellable** if and only if $ba = ca$ only when $b = c$. **Left multiplicatively cancellable** elements are similarly defined. An element of R is **multiplicatively cancellable** if and only if it is both left and right multiplicatively cancellable. Clearly any unit of R is multiplicatively cancellable and no multiplicatively cancellable element of R is a zero divisor. We will denote the set of all multiplicatively cancellable elements of R by $K^{\times}(R)$. This set is nonempty since $1 \in U(R) \subseteq K^{\times}(R)$ and is not all of R since $0 \notin K^{\times}(R)$. Moreover, $K^{\times}(R)$ is a submonoid of $(R, \cdot)$. If every nonzero element of R is [left, right] cancellable then we say that the semiring R is **[left, right] multiplicatively cancellative**. Division semirings are surely multiplicatively cancellative.

(4.35) EXAMPLE. The semiring $\mathbb{N}$ is a multiplicatively cancellative semiring which is not a division semiring. Indeed, $U(\mathbb{N}) = \{1\}$.

(4.36) EXAMPLE. If R is a noetherian commutative integral domain then the additively-idempotent semiring *ideal*(R) is multiplicatively cancellative if and only if R is a Prüfer domain. More generally, a commutative integral domain R is a Prüfer domain if and only if every finitely-generated nonzero ideal of R is multiplicatively cancellable [Larsen & McCarthy, 1971].

(4.37) EXAMPLE. [Barbut, 1967] An element of a semiring R which is right multiplicatively cancellable is surely not a right zero divisor. The converse is true for rings but not necessarily true for semirings. Indeed, let R be the semiring $\mathcal{M}_2(\mathbb{Q}^+)$ and let $A = \begin{bmatrix} 1 & 1 \\ 1 & 1 \end{bmatrix}$. Then A is not a right zero divisor since $\begin{bmatrix} a & b \\ c & d \end{bmatrix} A = \begin{bmatrix} 0 & 0 \\ 0 & 0 \end{bmatrix}$ implies that $a + b = 0 = c + d$ and so $a = b = c = d = 0$. On the other hand, A is not right multiplicatively cancellable since $\begin{bmatrix} 1 & 2 \\ 1 & 2 \end{bmatrix} A = \begin{bmatrix} 2 & 1 \\ 2 & 1 \end{bmatrix} A$.

(4.38) EXAMPLE. [Duchamp & Thibon, 1988] Let R be a semiring and let A be a nonempty set on which we have defined a reflexive and symmetric relation $\sim$. Let M be the quotient monoid of the free monoid A^* with respect to the congruence generated by all pairs of the form (ab, ba) for $a \sim b$. Then the semiring $R[M]$ is cancellative and left multiplicatively cancellative if and only if R is. Moreover, it is entire if and only if R is entire.

(4.39) EXAMPLE. [H. E. Stone, 1977] Let R be the semiring of polynomials over $\mathbb{N}$ in noncommuting indeterminates x and y satisfying the condition that $yx = y$. Then R is a cancellative semiring which is right multiplicatively cancellative but not left multiplicatively cancellative.

(4.40) PROPOSITION. *If R is a right multiplicatively-cancellative semiring in which there exists an element other than 1_R having finite multiplicative multiplicative order, then R is a ring.*

PROOF. Assume that $1_R \neq a \in R$ satisfies $r^n = 1_R$ for some $n > 1$ and let $b = 1 + a + \cdots + a^{n-1}$. Then $ab = a + a^2 + \cdots + a^n = 1 + a + \ldots_a^{n-1} = b = 1_R b$. Since $a \neq 1_R$, this implies that $b = 0$ and so $1_R \in V(R)$, proving that $V(R) = R$ and hence R is a ring. □

(4.41) PROPOSITION. *If R is a cancellative yoked semiring then any element of R which is not a zero divisor is multiplicatively cancellable.*

PROOF. Let a be an element of R which is not a zero divisor and let b and c be elements of R satisfying $ba = ca$. Since R is a yoked semiring, there exists an element d of R such that $b = c + d$ or $c = b + d$. Say $b = c + d$. Therefore $ca + 0 = ca = ba = (c + d)a = ca + da$. Since R is cancellative, this implies that $da = 0$ and, since a is not a zero divisor, we must therefore have $d = 0$. Thus $b = c$, proving that a is right multiplicatively cancellative. A similar proof shows that a is also left multiplicatively cancellative. □

(4.42) PROPOSITION. *Each of the following conditions on an element a of a semiring R implies the next:*

(1) $a + 1 = 1$;
(2) $a^n + 1 = 1$ *for all* $n \in \mathbb{N}$;
(3) $a^n + 1 = 1$ *for some* $n \in \mathbb{N}$.

The conditions are are equivalent if R is additively idempotent and multiplicatively cancellative.

PROOF. (1) ⇒ (2): We will prove (2) by induction on n. If $n = 1$ the result follows from (1). Assume now that $n > 1$ and that $a^k + 1 = 1$ for all $k < n$. Then $1 = (a + 1)(a^{n-1} + 1) = a^n + a^{n-1} + a + 1 = a^n + a^{n-1} + 1 = a^n + 1$.

(2) ⇒ (3): This is immediate.

(3) ⇒ (1): Assume that R is additively idempotent and multiplicatively cancellative. If $a^n + 1 = 1$ then $(a+1)^n = a^n + a^{n-1} + \cdots + 1 = a^{n-1} + \cdots + 1 = (a+1)^{n-1}$. By multiplicative cancellation, we then obtain $a + 1 = 1$. □

(4.43) PROPOSITION. *If R is a multiplicatively-cancellative additively-idempotent commutative semiring then $(a + b)^n = a^n + b^n$ for all $a, b \in R$ and all positive integers n.*

PROOF. If $a + b = 0$ then $a = b = 0$ since additively-idempotent semirings are zerosumfree, and in this case the result is immediate. Hence we can assume that $a + b \neq 0$. The result is clearly true for $n = 1$. Moreover, since R is additively-idempotent, we have $(a + b)^3 = a^3 + a^2b + ab^2 + b^3 = (a^2 + b^2)(a + b)$ and so, by multiplicative cancellativity, $(a + b)^2 = a^2 + b^2$. Now assume that $n > 2$ and that the result has already been established for $n - 1$. Then

$$\begin{aligned}(a+b)^{n+1} &= (a+b)^2(a+b)^{n-1} = (a+b)^2(a^{n-1}+b^{n-1}) \\ &= (a^2+ab+b^2)(a^{n-1}+b^{n-1}) = a^{n+1}+a^nb+ab^n+b^{n+1} \\ &= (a+b)(a^n+b^n)\end{aligned}$$

and so, by multiplicative cancellativity, $(a+b)^n = a^n + b^n$. □

As a consequence of this result, we see that if R is a multiplicatively-cancellative additively-idempotent commutative semiring and if $a_1, \ldots, a_k \in R$ then

$$(a_1 + \cdots + a_k)^n = a_1^n + \cdots + a_k^n$$

for each positive integer n.

(4.44) PROPOSITION. *Let R be a multiplicatively-cancellative additively-idempotent commutative semiring. If $a \neq b$ are elements of R then $a^n \neq b^n$ for all positive integers n.*

PROOF. Assume that $a^n = b^n$, where $n > 1$. Then

$$\begin{aligned} a^n &= a^n + b^n = (a+b)^n \\ &= (a+b)(a+b)^{n-1} = (a+b)(a^{n-1} + b^{n-1}) \\ &= a^n + ab^{n-1} + a^{n-1}b \end{aligned}$$

so, adding $a^{n-1}b$ to both sides, we have $a^n = a^n + a^{n-1}b = a^{n-1}(a+b)$. Since R is multiplicatively cancellative, this implies that $a = a+b$. A similar argument shows that $b = a+b$ and so $a = b$, which is a contradiction. □

Thus, if R is a multiplicatively-cancellative additively-idempotent commutative algebraically-closed semiring then any equation of the form $X^n = b$ has a unique solution on R.

Now let us consider a situation slightly more general than the one considered in Proposition 4.42. In studying the Jacobson radical of a ring, it is important to consider the quasiregular elements of the ring, namely those elements a for which $1 + a$ is a unit. For a semiring R, let $G(R) = \{r \in R \mid 1 + r \in U(R)\}$. This set is nonempty since it contains 0 and, if R is a ring, contains the Jacobson radical of R. (The term "quasiregular", however, is used in the context of semirings in a different sense, as we shall see later.)

(4.45) EXAMPLE. Let R be a semiring and let $S = \mathcal{M}_n(R)$, where n is an integer greater than 1. If $1 \leq i \neq j \leq n$ and if $r \in V(R)$, let $e_{ij;r}$ be the element of S defined by

$$e_{ij;r}(h,k) = \begin{cases} r & \text{if } (h,k) = (i,j) \\ 0 & \text{otherwise} \end{cases}.$$

Then $[1_S + e_{ij;r}][1_S + e_{ij;-r}] = 1_S = [1_S + e_{ij;-r}][1_S + e_{ij;r}]$ and so $e_{ij;r} \in G(S)$ for all $1 \leq i \neq j \leq n$ and $r \in V(R)$.

(4.46) PROPOSITION. *The set $U(R) \cap G(R)$ is closed under taking inverses.*

PROOF. Assume $a \in U(R) \cap G(R)$. Then $a^{-1} \in U(R)$. Since $a \in G(R)$ we see that $1 + a \in U(R)$. Then $a(1 + a^{-1}) = a + aa^{-1} = 1 + a \in U(R)$ so $1 + a^{-1} = (a^{-1})(a+1) \in U(R)$. Thus $a^{-1} \in G(R)$. □

The semiring R is a **Gel'fand semiring** if and only if $R = G(R)$. By an easy induction argument, we see that if R is a Gel'fand semiring then $n1_R \in U(R)$ for each nonnegative integer n. Gel'fand [1941] first considered this condition for Banach algebras; the generalization to semirings first appeared in [Slowikowski & Zawadowski, 1955].

(4.47) EXAMPLE. Simple semirings are surely Gel'fand rings. Thus the semirings in Example 1.5 are Gel'fand semirings.

(4.48) EXAMPLE. The semiring $\mathbb{N}$ is a zerosumfree semiring which is not a Gel'fand semiring. Indeed, $G(\mathbb{N}) = \{0\}$ and $U(\mathbb{N}) = \{1\}$ so this example also shows that the set $G(R) \cap U(R)$ may be empty for some semirings R.

(4.49) EXAMPLE. Let A be a nonempty set having more that one element and let $R = (\mathbb{R}^+)^A$. This is a semiring which is not simple. On the other hand, $U(R) = \{f \in R \mid f(a) > 0 \text{ for all } a \in A\}$ and so R is a Gel'fand semiring.

The semiring $\mathbb{R}^+$ is, up to isomorphism, the only cancellative locally-compact connected topological Gel'fand semiring R having the property that translations of R are open mappings [Bourne, 1962].

As an immediate consequence of the definitions one sees that the family of Gel'fand semirings is closed under taking direct products.

(4.50) PROPOSITION. *A semiring R is a Gel'fand semiring if and only if $r + c \in U(R)$ for all $c \in U(R)$ and all $r \in R$.*

PROOF. If the stated condition holds, then R is a certainly a Gel'fand semiring. Conversely, let R be a Gel'fand semiring, let $r \in R$, and let $c \in U(R)$. Then $d = c^{-1}r + 1$ is a unit of R and so $r + c = cd \in U(R)$, which is what we wanted to prove. □

(4.51) PROPOSITION. *If R is a Gel'fand semiring for which there exist positive integers $n > m$ satisfying $n1_R = m1_R$, then R is additively idempotent.*

PROOF. Set $h = n - m$. Without loss of generality, we can assume that $h > 1$ since if $n1_R = m1_R$ then $(n+h)1_R = m1_R$. If $k \geq m$ then $k1_R = k1_R + th1_R$ for each nonnegative integer t. Choose a positive integer w such that $h^w > m$. Then $h^w 1_R = h^w 1_R + h^w - 1h1_R = 2h^w 1_R$ since $h^w > m$. Since R is a Gel'fand semiring, $h^w 1_R \in U(R)$ and so $1_R = 21_R$. Therefore $a = 2a$ for all $a \in R$, proving that R is additively idempotent. □

Since $(h+k)1_R = h1_R + k1_R$ and $hk1_R = (h1_R)(k1_R)$ for all nonnegative integers h and k, we see that, for any semiring R, the set $B(R) = \{h1_R \mid h \in \mathbb{N}\}$ is a subsemiring of $C(R)$, called the **basic subsemiring** of R. A semiring R is **basic** if and only if $R = B(R)$. Clearly $\mathbb{N}$ is basic. It is straightforward to check that the semirings $(B(n,i), \oplus, \odot)$ defined in Example 1.8 are basic.

By Proposition 4.51, we see that if R is a Gel'fand semiring then either $B(R) = \{0, 1\}$ or $B(R)$ is a copy of $\mathbb{N}$. Moreover, if R is a Gel'fand semiring then, as we have already noted, $B(R) \subseteq U(R)$.

(4.52) PROPOSITION. *Let R be a Gel'fand semiring and let h, k, m, n be nonnegative integers. Then $h1_R(k1_R)^{-1} + m1_R(n1_R)^{-1} = (hn1_R + km1_R)(kn1_R)^{-1}$.*

PROOF. Since R is a Gel'fand semiring we know that $n1_R \in U(R)$ for each natural number n. By distributivity and the commutativity of the elements of the form $n1_R$, we have $[h1_R(k1_R)^{-1} + m1_R(n1_R)^{-1}](kn1_R) = hn1_R + km1_R$ and so the result follows by multiplying both sides by $(kn1_R)^{-1}$. □

5. COMPLEMENTED ELEMENTS IN SEMIRINGS

Complemented elements play an important part in the study of lattices, and in particular in the study of frames. Since frames are examples of semirings, it is worth looking at this notion in the more general context of semirings. As it turns out, such elements play an important part in the semiring representation of the semantics of computer programs, as emphasized in the work of Manes and his collaborators.

If a and b are elements of a semiring R then a is **well inside** b, written $a \lhd b$, if and only if there exists an element c of R satisfying $ac = ca = 0$ and $c + b = 1$. This is a generalization of a notion discussed for frames in [Johnstone, 1982]; it formerly appeared in [Dowker & Strauss, 1974] in connection with the study of the T_3-separation axiom for frames. Also refer to [Golan & Simmons, 1988]. In any semiring R we have $0 \lhd 0$ and $a \lhd 1$ for each $a \in R$. If R is a simple semiring then we note immediately that $0 \lhd b$ for any element b of R. If $a \in C(R)$ then $a \lhd b$ implies that $ra \lhd b$ for all $r \in R$.

(5.1) PROPOSITION. *If a and b are elements of a semiring R satisfying $a \lhd b$ then $ab = a = ba$. Moreover, if R is simple then this also implies that $a + b = b$.*

PROOF. Since $a \lhd b$ there exists an element c of R satisfying $ac = ca = 0$ and $c + b = 1$. Hence $a = a(c + b) = ac + ab = ab$. Similarly $a = ba$. Now assume that R is simple. Then, by Proposition 4.3, we have $a + b = a(c + b) + b = ac + ab + b = ab + b = b$. □

An element a of R is **complemented** if and only if $a \lhd a$. That is to say, a is complemented if and only if there exists an element c of R satisfying $ac = ca = 0$ and $a + c = 1$. This element c of R is the **complement** of a in R. If a has a complement, it is unique. Indeed, if both b and c are complements of a then $b = (a + c)b = ab + cb = cb = cb + ca = c(b + a) = c$. We denote the complement of a complemented element a of R by $a^{\perp}$. Clearly, if a is complemented so is $a^{\perp}$ and $a^{\perp\perp} = a$.

Denote the set of all complemented elements of R is denoted by $comp(R)$. This set is nonempty since $0 \in comp(R)$ with $0^{\perp} = 1$. If $comp(R) = \{0, 1\}$ then R is

integral. If $a \in comp(R) \setminus \{0,1\}$ then $a^\perp \in comp(R) \setminus \{0,1\}$ and so we see that if R is entire then it is integral. Thus, for example, the semiring $(\mathbb{I}, max, min)$ is integral. Note that $comp(R) \subseteq I^\times(R)$. Indeed, if $a \in comp(R)$ then $a = a1 = a(a+a^\perp) = a^2 + aa^\perp = a^2$. If $a \in comp(R)$, set $a \sqcup b = a + a^\perp b$. Note that $a \sqcup a^\perp = a + a^\perp a^\perp = a + a^\perp = 1$ for all $a \in comp(R)$. Also, if $a + b = 1$ then $a^\perp = a^\perp(a+b) = a^\perp a + a^\perp b = a^\perp b$ so $a \sqcup b = a + a^\perp = 1$.

In passing, we note that if $a \in comp(R)$ and $b \in R$ then $a^\perp \sqcup b = a^\perp + ab$ corresponds to the **Saseki hook** implication operation in quantum logic. See [Román & Rumbos, 1991b] for details.

(5.2) EXAMPLE. If $R = \times_{i\in\Omega} R_i$ is a direct product of semirings and if Λ is a subset of Ω, then the element e_Λ of R defined by

$$e_\Lambda(i) = \begin{cases} 1 & \text{if } i \in \Lambda \\ 0 & \text{otherwise} \end{cases}.$$

is complemented. Indeed, $(e_\Lambda)^\perp = e_{\Omega\setminus\Lambda}$.

(5.3) EXAMPLE. If R is a semiring and if A is a nonempty set which is either countable or finite, then for each $B \subseteq A$ the element e_B of $\mathcal{M}_{A,rc}(R)$ defined by

$$e_B(i,j) = \begin{cases} 1 & \text{if } i = j \in B \\ 0 & \text{otherwise} \end{cases}.$$

is complemented, with $(e_B)^\perp = e_{A\setminus B}$.

More generally, if $0 \neq e \in I^\times(R)$ then e is **integral** if and only if the semiring eRe is integral. That is to say, e is integral if and only if there do not exist elements b and c of R such that ebe and ece are nonzero and satisfy $ebece = 0$ and $e = ebe + ece$. Thus, if R is entire then every nonzero element of $I^\times(R)$ is integral.

(5.4) EXAMPLE. If $\mathcal{T}$ is a topology on a nonempty set X then $(\mathcal{T}, \cup, \cap)$ is a multiplicatively-idempotent semiring with additive identity $\varnothing$ and multiplicative identity X. An element A of $\mathcal{T}$ is integral if and only if it is connected.

(5.5) EXAMPLE. We have noted that $comp(R) \subseteq I^\times(R)$ for any semiring R. If R is a plain simple yoked semiring then the converse is also true. Indeed, in such a situation let $e \in I^\times(R)$. Then there exists an element b of R satisfying $e = 1 + b$ or $e + b = 1$. In the first case, the simplicity of R yields $e = 1$ and so $e \in comp(R)$. In the second case, $e = e^2 = e(1+b) = e + eb$. By Proposition 4.22, this implies that $eb = 0$. Similarly, $be = 0$ and so $e \in comp(R)$ with $e^\perp = b$.

(5.6) PROPOSITION. *If R is a zerosumfree semiring and if $a, b \in comp(R)$ then:*

(1) $aba^\perp = 0$;

(2) *ab and $a \sqcup b$ belong to $comp(R)$;*

(3) $ab = ba$.

PROOF. (1) If $a, b \in comp(R)$ then $aba^\perp + ab^\perp a^\perp = a(b+b^\perp)a^\perp = aa^\perp = 0$ and so, since R is zerosumfree, we have $aba^\perp = 0$.

(2) We claim that $[a \sqcup b]^\perp = a^\perp b^\perp$. Indeed, $[a \sqcup b] + a^\perp b^\perp = a + a^\perp b + a^\perp b^\perp = a + a^\perp(b + b^\perp) = a + a^\perp = 1$. Also, by (1), $[a \sqcup b]a^\perp b^\perp = [\tilde{a} + a^\perp b]a^\perp b^\perp = aa^\perp b^\perp + a^\perp bab^\perp = 0$. Similarly $a^\perp b^\perp[a \sqcup b] = 0$, establishing the claim. Thus $a \sqcup b \in comp(R)$.

Finally, we claim that $[ab]^\perp = a^\perp \sqcup b^\perp = a^\perp + ab^\perp$. Indeed, $ab + (a^\perp \sqcup b^\perp) = ab + a^\perp + ab^\perp = a(b + b^\perp) + a^\perp = a + a^\perp = 1$ while $ab(a^\perp \sqcup b^\perp) = ab(a^\perp + ab^\perp) = aba^\perp + abab^\perp = 0$ and similarly $(a^\perp \sqcup b^\perp)ab = 0$.

(3) By (1), $aba^\perp = 0 = a^\perp ba^{\perp\perp} = a^\perp ba$ and so $ab = ab1 = ab(a + a^\perp) = aba + aba^\perp = aba = aba + a^\perp ba = (a + a^\perp)ba = 1(ba) = ba$. □

(5.7) Proposition. *The following conditions on a zerosumfree semiring R are equivalent:*

(1) *If $a, b \in comp(R)$ then $a + b \in comp(R)$;*
(2) *$1 + 1 \in comp(R)$;*
(3) *$comp(R) \subseteq I^+(R)$;*
(4) *If $a, b \in comp(R)$ then $a + b = a \sqcup b$;*
(5) *$(comp(R), +, \cdot)$ is a subsemiring of R.*

Proof. (1) $\Rightarrow$ (2): This is immediate.

(2) $\Rightarrow$ (3): If $a \in comp(R)$ then, by (2) and Proposition 5.6, we have $a + a \in comp(R)$. Set $b = (a+a)^\perp$. Then $ab + ab = (a+a)b = 0$ and so, by zerosumfreeness, $ab = 0$. Therefore $a = a1 = a(a + a + b) = a^2 + a^2 = a + a$, proving that $a \in I^+(R)$.

(3) $\Rightarrow$ (4): If $a, b \in comp(R)$ then, by Proposition 5.6, we have

$$\begin{aligned} a + b &= (a + b)(a + a^\perp)(b + b^\perp) \\ &= (a + b)(ab + a^\perp b + ab^\perp + a^\perp b^\perp) \\ &= ab + ab^\perp + ab + a^\perp b. \end{aligned}$$

By (3), $ab \in I^+(R)$ and so $a + b = ab + ab^\perp + a^\perp b = a + a^\perp b = a \sqcup b$.

(4) $\Rightarrow$ (1) $\Leftrightarrow$ (5): This is a direct consequence of Proposition 5.6(2). □

(5.8) Proposition. *If R is a zerosumfree semiring then $(comp(R), \sqcup, \cdot)$ is an idempotent commutative simple semiring.*

Proof. If $a, b, c \in comp(R)$ then

$$\begin{aligned} a \sqcup (b \sqcup c) &= a \sqcup (b + b^\perp c) = a + a^\perp(b + b^\perp c) \\ &= a + a^\perp b + a^\perp b^\perp c = a + a^\perp b + [a + a^\perp b]^\perp c \\ &= (a + a^\perp b) \sqcup c = (a \sqcup b) \sqcup c. \end{aligned}$$

Thus $\sqcup$ is associative. If $a \in comp(R)$ then $a \sqcup 0 = a + a^\perp 0 = a = 0 + 1a =$

$0 + 0^\perp a = 0 \sqcup a$. Finally, if $a, b \in comp(R)$ then

$$\begin{aligned} a \sqcup b &= a + a^\perp b = a + a^\perp ba + a^\perp ba^\perp \\ &= a + a^\perp ba^\perp = ba + b^\perp a + a^\perp ba^\perp \\ &= bab + bab^\perp + b^\perp ab + b^\perp ab^\perp + a^\perp ba = bab + b^\perp ab^\perp + a^\perp ba^\perp \\ &= bab + b^\perp ab^\perp + aba^\perp + a^\perp ba^\perp = bab + b^\perp ab^\perp + ba^\perp \\ &= bab + bab^\perp + ba^\perp + b^\perp ab^\perp = ba + ba^\perp + b^\perp ab^\perp \\ &= b + b^\perp ab^\perp = b + b^\perp ab + b^\perp ab^\perp \\ &= b + b^\perp a = b \sqcup a. \end{aligned}$$

Thus $\sqcup$ is commutative.

We already know that $comp(R)$ is closed under products and contains 1, so it is a monoid. Finally, if $a, b, c \in comp(R)$ then

$$\begin{aligned} a(b \sqcup c) &= a(b + b^\perp c) = ab + ab^\perp c \\ &= ab + ab \sqcup ac + ab^\perp a^\perp c = ab + ab^\perp ac \\ &= ab + a^\perp ac + ab^\perp ac = ab + (a^\perp + ab^\perp)ac \\ &= ab + (ab)^\perp ac = ab \sqcup ac \end{aligned}$$

and similarly $(b \sqcup c)a = ba \sqcup ca$. Thus $(comp(R), \sqcup, \cdot)$ is a semiring. We have already noted that every $a \in comp(R)$ is multiplicatively idempotent. Moreover, $a \in comp(R)$ implies that $a \sqcup a = a + a^\perp a = a$ and so $comp(R)$ is additively idempotent as well. It is commutative by Proposition 5.6(3). If $a \in comp(R)$ then $a \sqcup 1 = a + a^\perp 1 = a + a^\perp = 1$ and so $comp(R)$ is simple. □

(5.9) COROLLARY. *If R is a zerosumfree semiring then $(comp(R), \sqcup, \cdot)$ is a boolean algebra.*

PROOF. As was noted in Example 1.5, a commutative idempotent simple semiring is a bounded distributive lattice having unique minimal element 0 and unique maximal element 1. Hence, by Proposition 5.8, we note that $comp(R)$ is such a lattice which is complemented as well and so is a boolean algebra. □

(5.10) PROPOSITION. *If R is a zerosumfree semiring then the relation $\leq$ on R defined by the condition $r \leq s$ if and only if there exists an element $e \in comp(R)$ satisfying $r = es$ is a partial order relation on R.*

PROOF. Clearly $r \leq r$ for all $r \in R$ since $r = 1r$. If $r \leq s$ and $s \leq t$ then there exist $e, f \in comp(R)$ with $r = es$ and $s = ft$. Hence $r = eft$ with $ef \in comp(R)$ by Proposition 5.6(2), proving that $r \leq t$. Now assume that $r \leq s$ and $s \leq r$. Then there exist $e, f \in comp(R)$ such that $r = es$ and $s = fr$. This implies that $er = e^2 s = es = r$ and so, by Proposition 5.6(3), $r = es = efr = fer = fr = s$. □

For a semiring R, we define the **symmetric difference** of elements of $comp(R)$ by $a \triangle b = ab^\perp + a^\perp b$. In particular, if R is a zerosumfree semiring satisfying the condition that $comp(R)$ is a subsemiring of R (refer to Proposition 5.7) then by Corollary 5.9 we see that $comp(R)$ is a boolean algebra and this is just the symmetric difference in the usual sense. Note that, under these circumstances, the function $\delta\colon comp(R) \times comp(R) \to comp(R)$ defined by $\delta\colon (a,b) \mapsto a \bigtriangleup b$ defines a metric on $comp(R)$ with values in R.

If R is an arbitrary zerosumfree semiring then it is still clear that $\delta(a,b) = \delta(b,a) \geq 0$ for all $a,b \in R$ and $\delta(a,a) = 0$ for all $a \in R$. Conversely, assume that $\delta(a,b) = 0$. Since R is zerosumfree, this means that $ab^{\perp} = 0 = a^{\perp}b$ and so $a = a1 = a(b^{\perp} + b) = ab^{\perp} + ab = ab$, proving that $a \leq b$. Similarly, $b \leq a$ and so $a = b$. If $a,b,c \in R$ then $(a \bigtriangleup c)(a \bigtriangleup b + b \bigtriangleup c) = (ac^{\perp} + a^{\perp}c)(ab^{\perp} + a^{\perp}b + bc^{\perp} + b^{\perp}c)$. By Proposition 5.6, this equals $ab^{\perp}c^{\perp} + abc^{\perp} + a^{\perp}bc + a^{\perp}b^{\perp}c = a(b^{\perp} + b)c^{\perp} + a^{\perp}(b + b^{\perp})c = ac^{\perp} + a^{\perp}c = a \bigtriangleup c$ and so $\delta(a,c) = a \bigtriangleup c \leq a \bigtriangleup b + b \bigtriangleup c = \delta(a,b) + \delta(b,c)$. Thus we see that δ is a metric on $comp(R)$ with values in R.

Note that if $a \in comp(R)$ the $\delta(a,0) = a1 + a^{\perp}0 = a$. Also, $\delta(a,a^{\perp}) = aa + a^{\perp}a^{\perp} = a + a^{\perp} = 1$ for all $a \in comp(R)$.

6. IDEALS IN SEMIRINGS

Ideals play a fundamental role in ring theory and it is therefore natural to consider them also in the context of semiring theory. Here their role is no less important, though we will often have to restrict our consideration to special types of ideals. In particular, we will show that, as in the case of rings, the family of all ideals of a semiring is, in a natural way, a semiring. Formally, the definitions in the two situations are the same.

A **left ideal** I of a hemiring R is a nonempty subset of R satisfying the following conditions:

(1) If $a, b \in I$ then $a + b \in I$;
(2) If $a \in I$ and $r \in R$ then $ra \in I$;
(3) $I \neq R$.

Note that if R is a semiring then condition (3) is equivalent to the condition that $1 \notin I$. A **right ideal** of R is defined in the analogous manner and an **ideal** of R is a subset which is both a left ideal and a right ideal of R. Note that ideals are proper, namely R is not an ideal of itself. Also, 0 belongs to every [left, right] ideal of R and hence $\{0\}$ is an ideal of R contained in every [left, right] ideal of R. Moreover, $U(R) \cap I = \varnothing$ for every [left, right] ideal of R. Any ideal of a semiring R is a subhemiring of R which is not a subsemiring. We will denote the set consisting of R and all left ideals of R by $lideal(R)$, the set consisting of R and all right ideals of R by $rideal(R)$, and the set consisting of R and all ideals of R by $ideal(R)$.

(6.1) EXAMPLE. If R is a commutative semiring and if $I = R \setminus U(R)$ then for $r \in R$ and $a \in I$ we surely have $ra \in I$. Therefore I is an ideal of R if and only if it is closed under addition. A sufficient condition for this to happen is that if $a, b \in I$ then $a + b$ is either of the form ra or rb for some $r \in R$.

If I is an ideal of R then it surely contains every other ideal of R and so is the unique maximal ideal of R. In this case, the commutative semiring R is **quasi-local**. The semirings of the form $B(n, i)$ mentioned in Example 1.8 are quasi-local if $i = 0$ and $n = p^h$ for some prime integer p and natural number h, or if $i = 1$ and $n - 1 = p^h$ for some prime integer p and natural number h. Refer to [Alarcón & Anderson, 1994a].

CONVENTION: In general, when we prove that a certain result is true for left ideals of a hemiring, the corresponding result for right ideals and for ideals will also be assumed without specific mention.

A nonempty subset A of a hemiring R is **semisubtractive** if and only if $a \in A \cap V(R)$ implies that $-a \in A \cap V(R)$; it is **subtractive** if and only if $a \in A$ and $a + b \in A$ imply $b \in A$; it is **strong** if and only if $a + b \in A$ implies that $a \in A$ and $b \in A$. Every subtractive subset of R surely contains 0. Also, it is clear that every strong subset of R is subtractive and every subtractive subset of R is semisubtractive. The subtractive ideals of a semiring will be characterized in Chapter 9. If R is a hemiring then the ideal $\{0\}$ is always subtractive and, as we have noted, is contained in every other subtractive ideal of R. It is strong if and only if R is zerosumfree.

(6.2) EXAMPLE. If A is an infinite set then the family $fsub(A)$ of all finite subsets of A is a strong ideal of the simple idempotent zerosumfree semiring $(sub(A), \cup, \cap)$.

(6.3) EXAMPLE. In Chapter 2 we noted that multifunctions have important applications in describing the semantics of computer programs. If $f\colon A \to sub(B)$ is a multifunction from A to B then, in a natural manner, we can consider f as a multifunction from $A \cup B$ to itself by setting $f(b) = \emptyset$ for all $b \in B$. Therefore we can usually restrict ourselves to working with multifunctions from a set to the semiring of its subsets. Let A be a nonempty set and let $R = sub(A)^A$ be the set of all multifunctions on A with values in the semiring of subsets of A. Define operations $+$ and $\circ$ on R as follows: if $a \in A$ and $f, g \in R$ then $(f+g)(a) = f(a) \cup g(a)$ and $(f \circ g)(a) = \cup\{f(b) \mid b \in g(a)\}$. It is straightforward to check that R is a semiring with additive identity z defined by $z(a) = \emptyset$ for all $a \in A$ and multiplicative identity j defined by $j(a) = \{a\}$ for all $a \in A$. If B is a proper subset of A then $I_B = \{f \in R \mid f(a) \subseteq B \text{ for all } a \in A\}$ is a strong right ideal of R which is not a left ideal.

(6.4) EXAMPLE. [LaGrassa, 1995] Even in very small, "nice" semirings, not every ideal need be subtractive. For example, let $R = \{0, 1, u\}$ be the idempotent semiring in which $1 + u = u + 1 = u$. Then $\{0, u\}$ is an ideal of R which is not subtractive.

(6.5) EXAMPLE. If R is a ring then no ideal of R is strong. Indeed, if I is an ideal of R then $-1 + 1 \in I$ but $1 \notin I$. If R is a semiring which is not a ring then $V(R)$ is a strong ideal of R. If $\{0\}$ is the only ideal of R, R, this implies that either $V(R) = R$, in which case R is a ring, or $V(R) = \{0\}$, in which case R is zerosumfree. Thus we have another proof of Proposition 4.34.

(6.6) EXAMPLE. The set $2\mathbb{N}$ of all nonnegative even integers is a subtractive ideal of the semiring of all nonnegative integers. It is not strong since $3 + 5 \in 2\mathbb{N}$ while neither 3 nor 5 belong to $2\mathbb{N}$. A complete study of the subtractive ideals of $\mathbb{N}$ is given in [Noronha-Galvão, 1978a], where it is shown that these are precisely the sets of the form $k\mathbb{N}$ for some $k \in \mathbb{N}$. Refer also to [Noronha-Galvão, 1978b].

This result was generalized in [Alarcón & Anderson, 1994a]: let R be an integral domain with total order compatible with addition and multiplication and let R^+

be the semiring of nonnegative elements of R. Then R^+a is a subtractive ideal of R^+ for all $a \in R^+$.

(6.7) EXAMPLE. [Alarcón & Anderson, 1994a] Every ideal of the basic semiring $B(n, i)$ is subtractive if and only if $i \leq 1$.

(6.8) EXAMPLE. [Iséki & Miyanaga, 1956b] If X is a Hausdorff topological space then the set R of all continuous bounded functions from X to $\mathbb{R}^+$ is a commutative semiring, which is in fact a Gel'fand semiring. Moreover, there exists a bijective function Φ from $ideal(R)$ to the family of all filters of closed subsets of X defined as follows: if $I \in ideal(R)$ and if Y is a closed subset of X then $Y \in \Phi(I)$ if and only if for every closed subset W of X not meeting Y there exists a function $f \in I$ such that $inf\{f(w) \mid w \in W\} > 0$.

(6.9) EXAMPLE. If R is a semiring which is not additively idempotent then $I^+(R)$ is an ideal of R, since $1 \notin I^+(R)$. This ideal is not necessarily strong. Indeed, if $R = \mathbb{Z}$ then $I^+(R) = \{0\}$ so $-1 + 1 \in I^+(R)$ while $-1, 1 \notin I^+(R)$. If $I^+(R)$ is a strong ideal then the semiring R is **archimedian**. It is immediate to see that the family of all archimedian semirings is closed under taking products.

(6.10) EXAMPLE. If A is a nonempty subset of a semiring R set $(0 : A) = \{r \in R \mid ra = 0 \text{ for all } a \in A\}$. If $A \neq \{0\}$ then this is a left ideal of R, called the **left annihilator ideal** of A. Right annihilator ideals are defined similarly. If I is the left annihilator ideal of a nonempty subset A of R other than $\{0\}$ then I is a subtractive left ideal. Indeed, if r and r' are elements of R satisfying the condition that r and $r + r'$ are both elements of I then for each element a of A we have $0 = (r + r')a = ra + r'a = r'a$ and so $r' \in I$. Similarly, right annihilator ideals are subtractive right ideals. We note that if H is a left ideal of R then $(0 : H)$ is an ideal of R. If $a \in R$, we write $(0 : a)$ instead of $(0 : \{a\})$. Similarly, we note that if $a \neq b$ are elements of R then $\{r \in R \mid ra = rb\}$ is a left ideal of R.

(6.11) PROPOSITION. [LaGrassa, 1995] *The following conditions on an ideal I of a commutative semiring R are equivalent:*

(1) $H + (0 : I) = (HI : I)$ *for all ideals H of R;*
(2) $HI = KI$ *implies that* $(0 : I) + H = (0 : I) + K$ *for all ideals H and K of R.*

PROOF. Assume I satisfies (1). If $(0 : I) + H = (0 : I) + K$ for ideals H and K of R then $(0 : I) + H = (HI : I) = (KI : I) = (0 : I) + K$. Conversely, assume that I satisfies (2). If $a \in (HI : I)$ then $(a)I \subseteq HI$ and so $[(a) + H]I = (a)I + HI = HI$. By (2), this implies that $(a) + H + (0 : I) = H + (0 : I)$ and so $(a) \subseteq H + (0 : I)$. Therefore $a \in H + (0 : I)$ and so $(HI : I) \subseteq H + (0 : I) \subseteq (HI : I)$, establishing equality. □

In greater generality, if I is a left ideal of a semiring R and A is a nonempty subset of R, then $(I : A) = \{r \in R \mid ra \in I \text{ for each } a \in A\}$ is a left ideal of R provided that A is not a subset of I. The right-handed version of this is defined analogously. If $A = \{a\}$, we write $(I : a)$ instead of $(I : \{a\})$. It is easily seen that if I is a subtractive [resp. strong] left ideal of R then so is $(I : A)$ for any nonempty subset A of R not a subset of I. (If $A \subseteq I$ then, of course, $(I : A) = R$.)

(6.12) EXAMPLE. Let n be a positive integer. A nonempty subset K of $\mathbb{R}^n$ is a **proper cone** if and only if:

(1) $K + K \subseteq K$;
(2) $aK \subseteq K$ for all $a \in \mathbb{R}^+$;
(3) $K \cap (-K) = \{0\}$;
(4) $K + (-K) = \mathbb{R}^n$; and
(5) K is closed in the usual topology on $\mathbb{R}^n$.

A linear transformation φ from $\mathbb{R}^n$ to itself is a positive operator on K if and only if $K\varphi \subseteq K$. The set of all positive operators on K is clearly a semiring under the usual operations of addition and composition of linear transformations. The ideals of this semiring are studied in detail in [Tam, 1981].

(6.13) EXAMPLE. If R is a nonzeroic semiring then $Z(R)$ is a subtractive ideal of R. In general, it is not necessarily strong. The zeroid of the semiring $\mathcal{D}$ defined in Example 1.9 is strong. See [Pierce, 1972].

(6.14) EXAMPLE. An element a of a semiring R is **left absorbing** if and only if $ra = a$ for all $0 \neq r \in R$. Right absorbing elements are defined analogously. Clearly, if a is left absorbing then $\{0, a\}$ is a left ideal of R. The converse holds when R is entire.

The element 1 of a semiring R is left absorbing if and only if $R = \mathbb{B}$ or $R = \mathbb{Z}/2\mathbb{Z}$. Every semiring has at least one left absorbing element, namely 0. Moreover, if a is a strongly infinite element of R then a is also left absorbing. If a is a left absorbing element of R then either $a \in I^+(R)$ or $a + a = 0$. Thus, when R is zerosumfree, we conclude that every left absorbing element of R belongs to $I^+(R)$. The set of left absorbing [resp. right absorbing] elements of a semiring R is easily seen to be an ideal of R.

If A is a nonempty subset of a semiring R then the set RA consisting of all finite sums $\sum r_i a_i$ with $r_i \in R$ and $a_i \in A$ is either equal to R or is the smallest left ideal of R containing A. In the latter case, it is called the left ideal of R **generated** by A. If $A \subseteq B$ then surely $RA \subseteq RB$. Furthermore, as an immediate consequence of this observation and the defintions, we see that if A and B are nonempty subsets of R then $R(A \cup B) = R(RA \cup RB)$.

Similarly, AR is either equal to R or is the smallest right ideal of R containing A. The set (A) consisting of all finite sums of the form $\sum r_i a_i s_i$ with $r_i, s_i \in R$ and $a_i \in A$ is either equal to R or is the smallest ideal of R containing A. If $A = \{a\}$ we write Ra [resp. aR, (a)] instead of RA [resp. AR, (A)]. A left ideal [resp. right ideal, ideal] I of R is **finitely generated** if and only if there exists a finite subset A of R such that $I = RA$ [resp. $I = AR$, $I = (A)$]. It is **principal** if and only if there exists an element a of R such that $I = Ra$ [resp. $I = aR$, $I = (a)$].

(6.15) EXAMPLE. In general, if A is a nonempty subset of a semiring R then $\cup_{a \in A} Ra \subseteq RA$ but we do not necessarily have equality. A sufficient condition for equality is that the set $\{Ra \mid a \in R\}$ be linearly ordered. This condition is not sufficient since the semiring $R = (\mathbb{N} \cup \{-\infty\}, +, \cdot)$ does not satisfy it but does satisfy the condition that $\cup_{a \in A} Ra = RA$ for each nonempty subset A of R. The condition that every left ideal of R be principal is also insufficient, as the example of $(\mathbb{Z}, +, \cdot)$ shows. A necessary and sufficient condition for $\cup_{a \in A} Ra = RA$ to hold for every nonempty subset A of R is that for any $a, b \in R$ we have $a + b \in Ra \cup Rb$. See [LaGrassa, 1995] for details.

A semiring R is **left noetherian** if and only if it satisfies the ascending chain conditions on left ideals.

(6.16) PROPOSITION. *The following conditions on a semiring R are equivalent:*

(1) *R is left noetherian;*
(2) *Any nonempty collection of ideals of R has a maximal element;*
(3) *Every ideal of R is finitely generated.*

PROOF. (1) $\Rightarrow$ (2): Let $\mathcal{C}$ be a nonempty collection of left ideals of R and pick $I_1 \in \mathcal{C}$. If I_1 is not properly contained in any element of $\mathcal{C}$, we are done. If not, there exists an element I_2 of $\mathcal{C}$ properly containing I_1. If I_2 is not properly contained in any element of $\mathcal{C}$, we are done. If not, there exists an element I_3 of $\mathcal{C}$ properly containing I_2. Continue in this manner. By (1), the process must end after a finite number of steps, and so $\mathcal{C}$ has a maximal element.

(2) $\Rightarrow$ (3): Let I be a left ideal of R and let $\mathcal{C}$ be the collection of all finitely-generated ideals of R contained in I. This collection is nonempty and so, by (2), contains a maximal element $H = R\{a_1, \ldots, a_m\}$. For each $b \in I$, let $H_b = R\{a_1, \ldots, a_m, b\}$. But maximality, $H = H_b$ for all $b \in I$ and so, in particular, $b \in H$ for all $b \in I$. Thus $H = I$ and so I is finitely generated.

(3) $\Rightarrow$ (1): Let $I_1 \subseteq I_2 \subseteq \ldots$ be an ascending chain of left ideals of R and let $I = \cup_{j=1}^{\infty} I_j$. Then I is a left ideal of R and so, by (3), is finitely generated, say $I = R\{a_1, \ldots, a_k\}$. This means that there exists an index n such that $I \subseteq I_n \subseteq I$ and so $I_j = I_n$ for all $j \geq n$, proving (1). □

Thus, we note that if $a \notin U(R)$ then Ra and aR are a left ideal and a right ideal of R respectively. If $a \in C(R) \setminus U(R)$ then Ra is an ideal of R. If a and b are distinct elements of $I^{\times}(R) \cap C(R)$ then $Ra \neq Rb$. Indeed, if $Ra = Rb$ then there exist elements c and d of R satisfying $a = bc$ and $b = da$. But then $a = cb = cb^2 = ab = a^2b = ada = da^2 = da = b$.

(6.17) EXAMPLE. [Dale, 1976a] Let I be the ideal of $\mathbb{N}[t]$ generated by $t + 1$. Then $(t+1)^3 \in I$. But $(t+1)^3 = (t^3+1) + 3t(t+1)$, where $3t(t+1) \in I$ and $t^3 + 1 \notin I$. Therefore I is not subtractive.

Now let I be the principal ideal of $\mathbb{Z}[t]$ generated by $t + 1$. That is to say, $I = \{f(t)(t+1) \mid f(t) \in \mathbb{Z}[t]\}$. Then I contains no nonzero strong ideals. Indeed, assume that $H \neq \{0\}$ is a strong ideal of $\mathbb{Z}[t]$ contained in I. Then there exists a nonzero element a of $\mathbb{Z}$ and a positive integer k such that $at^k \in H$. Indeed, without loss of generality we can assume that k is even. Since $at^k \in I$, there must exists a polynomial $g(t)$ in $\mathbb{Z}[t]$ satisfying $at^k = g(t)(t+1)$ and so, evaluating at -1, we obtain $a = a(-1)^k = g(-1)(-1+1) = 0$, which is a contradiction.

The structure of ideals in semirings of the form $R\langle A \rangle$, and in particular in the semiring $\mathbb{N}[t]$, is discussed in [Dale & Allen, 1976]. In particular, they note that if

I is a subtractive left ideal of a semiring R and if A is a nonempty set then $I\langle A\rangle$ is a subtractive left ideal of the semiring $R\langle A\rangle$. The structure of ideals in polynomial semirings of the form $R[t]$, where R is a semiring, is discussed in [Dale, 1982]. Ideals in polynomial semirings in several variables are discussed in [Dale, 1976b].

(6.18) EXAMPLE. Let R be a semiring and A a nonempty set. The set of all quasiregular elements of $R\langle\langle A\rangle\rangle$ is clearly an ideal of $R\langle\langle A\rangle\rangle$, which is subtractive. It is strong if the semiring R is zerosumfree.

(6.19) EXAMPLE. The ideal $I = \mathbb{N} \setminus \{1\}$ of $\mathbb{N}$ is semisubtractive but is not subtractive. Indeed, $2 \in I$ and $3 = 3 + 1 \in I$ but $1 \notin I$.

(6.20) EXAMPLE. [Hilton, 1967] Let H be a boolean algebra and let e be an element not in H. Extend the addition on H to an operation on $R = H \cup \{e\}$ by setting $e + e = e + 0 = 0 + e = e$ and $a + e = e + a = 1$ for $a \notin \{0, e\}$. Similarly, extend the multiplication on H by setting $ae = ea = a$ for all $a \in R$. Then R is a commutative semiring with additive identity 0 and multiplicative identity e and H is an ideal of R. Note that H is not subtractive since $1 + e$ and 1 both belong to H but $e \notin H$.

(6.21) EXAMPLE. If R is a simple semiring and $1 \neq a \in R$ then $I_a = \{b \in R \mid b + a = a\}$ is an ideal of R. Indeed, this set is clearly closed under addition. If $b \in I_a$ and $r \in R$ then $rb + a = rb + b + a = (r + 1)b + a = 1b + a = b + a = a$ so $rb \in I_a$. Similarly $br \in I_a$.

(6.22) PROPOSITION. *If R is a division semiring and n is a positive integer then $S = \mathcal{M}_n(R)$ has no nonzero ideals.*

PROOF. For each $1 \leq i, j \leq n$ let e_{ij} be the element of S defined by

$$e_{ij}(m, n) = \begin{cases} 1 & \text{if } (i, j) = (m, n) \\ 0 & \text{otherwise} \end{cases}.$$

Then for each $f \in S$ we have $f = \sum\{f(i, j)e_ij \mid 1 \leq i, j \leq n\}$ in S.

Assume that I is a nonzero ideal of S and that g is a nonzero element of I. Then there exist $1 \leq r, s \leq n$ such that $g(r, s) \neq 0$. If f is a nonzero element of S then

$$f = \sum_{i,j} e_{ij} f(i, j) = \sum_{i,j} [e_{ir} g e_{sj}] g(r, s)^{-1} f(i, j) \in I.$$

In particular, the multiplicative identity of S belongs to I, which is a contradiction. Thus the semiring S can have no nonzero ideals. □

(6.23) PROPOSITION. *If R is a multiplicatively-cancellative semiring then $\{0\} \cup [R \setminus K^+(R)] \in ideal(R)$.*

PROOF. Set $A = \{0\} \cup [R \setminus K^+(R)]$. If $0 \neq a, a' \in A$ and $a + a' \in K^+(R)$ then $a + b = a + c \Rightarrow a + a' + b = a + a' + c \Rightarrow b = c$ and so $a \in K^+(R)$, which is a contradiction. Thus, sums of elements of A are again in A. If $0 \neq a \in A$ and if $0 \neq r \in R$ satisfies $ra \in K^+(R)$, then

$$a + b = a + c \Rightarrow ra + rb = ra + rc \Rightarrow rb = rc \Rightarrow b = c$$

and so $a \in K^+(R)$, which is a contradiction. Hence $ra \in A$ for all $r \in R$. Similarly $ar \in A$ for all $r \in R$. If $1 \in A$ then $A = R$. Otherwise, A is an ideal of R. In either case, $A \in ideal(R)$. □

If I is a left ideal of a semiring R then $N(I) = \{b \in R \mid ab \in I \text{ for all } a \in I\}$ is clearly a subsemiring of R containing I as an ideal. Indeed, if S is a subsemiring of R containing I as an ideal then surely $S \subseteq N(I)$ so $N(I)$ is the largest such subsemiring of R. The left ideal I is an ideal of R precisely when $N(I) = R$. For a right ideal H of R, we define $N(H) = \{b \in R \mid ba \in H \text{ for all } a \in H\}$ to obtain similar properties.

A semiring R having no nonzero subtractive left ideals is **left austere**. Right austere rings are defined similarly.

(6.24) EXAMPLE. The semiring defined in Example 1.6 is clearly left austere. Indeed, let I be a left ideal of R satisfying $\{u\} \neq I$. If $r \in R \notin I$ and $u \neq a \in I$ then $0r = 0 \in I$ so $r + a = 0 \in I$, showing that I is not subtractive.

(6.25) PROPOSITION. *If R is a left austere semiring then:*

(1) *R is entire;*
(2) *R is either zerosumfree or a ring;*
(3) *If R is cancellative then it is left multiplicatively cancellative as well.*

PROOF. (1) Assume that R has no nonzero subtractive left ideals and let a and b be nonzero elements of R satisfying $ab = 0$. Then $0 \neq a \in (0 : b)$ and so $(0 : b) = R$ since otherwise $(0 : b)$ would be a nonzero subtractive left ideal of R. But this is impossible since $1 \notin (0 : b)$. Thus R must be entire.

(3) If R is not zerosumfree then $V(R) \neq \{0\}$. Since $V(R)$ is clearly a subtractive left ideal of R, this means that $V(R) = R$ and so R is a ring.

(4) Assume that R is cancellative. Then for $a, b \in R$ we see that $I = \{r \in R \mid ra = rb\}$ is a subtractive left ideal of R or equals R itself. Indeed, if $I \neq \{0\}$ we must have $I = R$ and so $1 \in I$. This means that $a = b$, proving that R is left multiplicatively cancellative. □

(6.26) COROLLARY. *An austere commuative semiring which is not zerosumfree is a field.*

PROOF. If R is an austere commuative semiring which is not zerosumfree then, by Proposition 6.25, R is an integral domain. If $0 \neq a \in R$ then Ra is a subtractive ideal of R not equal to $\{0\}$ and so is all of R. Therefore there exists an element b of R satisfying $ba = 1$, proving that R is a field. □

(6.27) PROPOSITION. *Let R be a hemiring and let $S = R \times \mathbb{N}$ be the Dorroh extension of R by $\mathbb{N}$. Then a nonempty proper subset I of R is a [left, right] ideal of R if and only if $H = \{(a, 0) \mid a \in I\}$ is a [left, right] ideal of S. Moreover, I is subtractive if and only if H is too.*

PROOF. Assume that I is a left ideal of R. Then H is clearly closed under taking sums and $1_S = (0, 1) \notin H$. If $(a, n) \in S$ and $(b, 0) \in H$ then $(a, n) \cdot (b, 0) =$

$(nb+ab,0) \in H$ and so H is a left ideal of S. Conversely, if H is a left ideal of S and $a, b \in I$ then $(a+b,0) = (a,0)+(b,0) \in H$ and so $a+b \in I$. If $r \in R$ then $(ra,0) = (r,0)\cdot(a,0) \in H$ and so $ra \in I$. Thus I is a left ideal of R. The proof for right ideals and ideals is similar.

Now assume that I is a subtractive left ideal of R. If $(a,0) \in H$ and $(b,n) \in S$ is an element satisfying the condition that $(a,0)+(b,n) \in H$ then we must have $n = 0$ and $a+b \in I$. Since I is a subtractive left ideal, this implies that $b \in I$. Thus H is a subtractive left ideal. The converse is immediate. □

(6.28) EXAMPLE. [Barbut, 1967] Let $S = [\mathbb{R}^+ \times \{0\}] \cup [\{0\} \times \mathbb{R}^+]$ and define operations $\oplus$ and $\odot$ on S as follows:

(1) $(a,0) \oplus (a',0) = (a+a',0)$;
(2) $(0,b) \oplus (0,b') = (0,b+b')$;
(3) $(a,0) \oplus (0,b) = (0,a+b) = (0,b) \oplus (a,0)$;
(4) $(a,0) \odot (a',0) = (aa',0)$;
(5) $(0,b) \odot (0,b') = (0,bb')$;
(6) $(a,0) \odot (0,b) = (0,ab)$;
(7) $(0,b) \odot (a,0) = (ba,0)$.

Then $(S,\oplus,\odot)$ is a hemiring having Dorroh extension $R = S \times \mathbb{N}$. Moreover, $I = \{0\} \times \mathbb{R}^+$ is a left ideal of S and so, by Proposition 6.27, $H = I \times \{0\}$ is a nonzero left ideal of R. If H' is a nonzero left ideal of R contained in H and if $(0,b,0) \in H'$ for some nonzero element b of $\mathbb{R}^+$ then $(b,0,0)\oplus(0,b,0) = (0,2b,0) = (2,0,0) \odot (0,b,0) \in H'$ and so $(b,0,0)$ does not belong to H'. Thus H contains no nonzero subtractive left ideals of R.

We now note a generalization of Example 1.4.

(6.29) PROPOSITION. *For a semiring R, the sets lideal(R) and rideal(R) are zerosumfree hemirings under the operations of addition and multiplication of nonempty subsets of R, having infinite element R. Moreover, ideal(R) is a zerosumfree simple semiring. If R is commutative then these are commutative semirings which coincide.*

PROOF. This is a direct consequence of the definitions; the only reason that *lideal*(R) and *rideal*(R) are not semirings is that R is not a two-sided multiplicative identity in them. □

In particular, we note that, since the semiring *ideal*(R) is simple, we have $U(ideal(R)) = \{R\}$.

(6.30) EXAMPLE. The structure of *ideal*$(\mathbb{N})$ has been extensively studied in [Allen & Dale, 1975]. If $1 < n \in \mathbb{N}$ then $\{k \in \mathbb{N} \mid k \geq n\} \cup \{0\}$ is an ideal of $\mathbb{N}$ and the family of all such ideals is closed under taking unions and intersections. All elements of *ideal*$(\mathbb{N})$ are not necessarily principal, but for each $I \in ideal(\mathbb{N})$ there exists a finite subset A of $\mathbb{N}$ such that $I \cup A$ is a principal ideal of $\mathbb{N}$ or equals all of $\mathbb{N}$.

(6.31) EXAMPLE. If $R = \mathcal{M}_n(\mathbb{R}^+)$ for some integer $n \geq 1$, then $\{0_R\}$ is the only ideal of R. If S is a locally-compact zerosumfree subsemiring of a complex Banach space with $\{0_S\}$ as its only ideal, then S must be isomorphic to $\mathcal{M}_n(\mathbb{R}^+)$ for some $n \geq 1$. See Theorem 3.1 of [Kaashoek & West, 1974].

(6.32) PROPOSITION. *If R is a commutative semiring then the set S of all elements I of ideal(R) satisfying the condition that $a \in I$ implies $a \lhd b$ for some $b \in I$ is a subsemiring of ideal(R).*

PROOF. In Chapter 4 we have already noted that $0 \lhd 0$ and $a \lhd 1$ for each $a \in R$. Therefore $\{0\}$ and R belong to S. Assume that I and H are elements of S and let $a \in I$ and $a' \in H$. Then there exist elements $b \in I$ and $b' \in H$ satisfying $a \lhd b$ and $a' \lhd b'$. This means that there are elements c and c' of R satisfying $ac = a'c' = 0$ while $b + c = b' + c' = 1$. Therefore, if $d = bb' + bc' + cb' \in I + H$ we have $d + cc' = 1$ while $(a + a')cc' = 0$, proving that $a + a' \lhd d$. Thus $I + H \in S$.

Similarly, if I and H are elements of S and $a \in IH$ then $a \in I \cap H$ and so there exist elements $b \in I$ and $b' \in H$ such that $a \lhd b$ and $a \lhd b'$. In particular, there exist elements c and c' of R satisfying $ac = ac' = 0$ and $b + c = b' + c' = 1$. Then $bb' \in I$ and if $d = bc' + cb' + cc'$ we have $bb' + d = 1$ while $ad = 0$, proving that $a \lhd bb'$. Therefore $IH \in S$, proving that S is a subsemiring of *ideal*(R). □

(6.33) PROPOSITION. *If I and H are [left, right] ideals of a semiring R then $I + H$ is the unique minimal member of the family of all [left, right] ideals of R containing both I and H and $I \cap H$ is the unique maximal member of the family of all [left, right] ideals of R contained in both I and H.*

PROOF. Clearly $I + H$ contains both I and H. Conversely, if K is an ideal of R containing both I and H then K contains all elements of R of the form $a + b$, where $a \in I$ and $b \in H$, and hence K contains $I + H$. The proof of the second part is similar. □

If I and H are ideals of a semiring R then surely $IH \subseteq I \cap H$ but, in general, we do not have equality. If R is a commutative semiring and I, H are ideals of R satisfying $I + H = R$ then $I \cap H = (I + H)(I \cap H) \subseteq IH \subseteq I \cap H$ and so $IH = I \cap H$. In general, $(ideal(R), +, \cap)$ is not a semiring, even if R is commutative. If it is a semiring, then it must be simple. Therefore, we have the following result.

(6.34) PROPOSITION. *The following conditions on a commutative semiring are equivalent:*

(1) *$(ideal(R), +, \cap)$ is a semiring;*
(2) *$(ideal(R), \cap, +)$ is a semiring.*

PROOF. This is a direct consequence of Corollary 4.4. □

If R is a multiplicatively-regular semiring and if I and H are ideals of R with $a \in I \cap H$ then there exists an element b of R satisfying $a = aba = a(ba) \in I$ and so for multiplicatively-regular semirings we have $IH = I \cap H$ for all ideals I and H. Indeed, more generally, we have the following.

(6.35) PROPOSITION. *The following conditions on a semiring R are equivalent:*

(1) *R is multiplicatively regular;*

(2) $HI = H \cap I$ *for all left ideals* I *and right ideals* H *of* R;
(3) $I \cap H = \{a \in H \mid ba \in I$ *for all* $b \in H\}$ *for all left ideals* I *and right ideals* H *of* R;
(4) *ideal*(R) *is multiplicatively idempotent;*
(5) $H \cap K \subseteq HK$ *for all ideals* H *and right ideals* K *of* R;
(6) *If* K *is a right ideal of* R *contained in an ideal* H *of* R *then* $K \subseteq HK$.

PROOF. (1) $\Leftrightarrow$ (2): Assume (1). Let H be a right ideal of R and let I be a left ideal of R. Then surely $HI \subseteq H \cap I$. Conversely, let $a \in H \cap I$. Then there exists an element b of R satisfying $aba = a$. Since $ab \in H$, we have $aba \in HI$, proving that $H \cap I \subseteq HI$. Thus we have equality. Conversely, assume (2) and let $a \in R$. Then $a \in aR \cap Ra = (aR)(Ra)$ and so there exists an element b of R such that $a = aba$. Thus R is multiplicatively regular.

(2) $\Leftrightarrow$ (3): Assume (2) and let I and H be ideals of R. Then $G = \{a \in H \mid ba \in I$ for all $b \in H\}$ is an ideal of R and so, by (2), $G \cap H = HG \subseteq H \cap I$. The reverse inclusion is trivial and so we have equality. Conversely, assume (3). If I and H are ideals of R then, by (3), $HI \subseteq H \cap I \subseteq \{a \in H \mid ba \in HI$ for all $b \in H\} = HI \cap I = HI$.

(2) $\Leftrightarrow$ (4): Clearly (2) implies (4). Conversely, if (4) holds then for all ideals H and I of R we have $HI \subseteq (H \cap I)^2 = H \cap I \subseteq HI$ and so we have (2).

(2) $\Rightarrow$ (5): If H is an ideal of R and K is a right ideal of R then, by (2), $H \cap K \subseteq \{a \in H \mid ab \in HK$ for all $b \in H\} = HK \cap H = HK$.

(5) $\Rightarrow$ (6) $\Rightarrow$ (4): This is immediate. □

Proposition 6.33 can be extended to infinite sums. If $\{I_k \mid k \in \Omega\}$ is a set of [left, right] ideals of a semiring R then we define $\sum_{k\in\Omega} I_k$ to be the union of all possible sums $\sum_{k\in\Lambda} I_k$, where Λ is a finite subset of Ω. This is again a [left, right] ideal of R, which is the unique minimal [left, right] ideal of R containing all of the I_k. Similarly, $\cap_{k\in\Omega} I_k$ is the unique maximal [left, right] ideal of R contained in each of the I_k. We thus see that *lideal*(R), *rideal*(R), and *ideal*(R) are complete lattices. These lattices need not be modular, as the following example shows.

(6.36) EXAMPLE. The lattice *ideal*$(\mathbb{N})$ has a sublattice consisting of the following ideals:

(1) $I_1 = 2\mathbb{N} \setminus \{2\}$;
(2) $I_2 = 2\mathbb{N}$;
(3) $I_3 = \mathbb{N} \setminus \{1, 2, 5\}$;
(4) $I_4 = \mathbb{N} \setminus \{1, 2\}$;
(5) $I_5 = \mathbb{N} \setminus \{1\}$.

Moreover, $I_1 \subseteq I_2 \subseteq I_5$ and $I_1 \subseteq I_3 \subseteq I_4 \subseteq I_5$ and so this sublattice, and hence the lattice *ideal*$(\mathbb{N})$ is not modular.

From Example 6.36, we see that the ideal lattice of a semiring need not be modular, even if the semiring is commutative. This is the major difference between the ideal lattice of a semiring and that of a ring. On the other, *ideal*$(\mathbb{B})$ is trivially modular, even though $\mathbb{B}$ is a semiring which is not a ring.

(6.37) EXAMPLE. [Alarcón & Anderson, 1994a] For $i \leq 5$, the lattice *ideal*$(B(N, i))$ is distributive, but for $i \geq 6$ it is not even modular.

(6.38) Proposition. *If R is a semiring then a sufficient condition for the lattice lideal(R) [resp. rideal(R), ideal(R)] to be modular is that each of its members be subtractive.*

Proof. Assume that every element of *lideal*(R) is subtractive and let H, I, K be left ideals of R satisfying $I \cap H = I \cap K$ while $I + H = I + K$ and $H \subseteq K$. We must show that $H = K$. Indeed, if $a \in K$ we can write $a = b + c$, where $b \in I$ and $c \in H$. Since $c \in K$ we see, by subtractiveness, that $b \in I \cap K = I \cap H$. Therefore $a \in H$, establishing the desired equality. □

(6.39) Example. [Padmanabhan & Subramanian, 1968] The condition given in Proposition 6.38 is not necessary. To see this, consider the idempotent semiring $R = \{0, 1, a\}$ in which $1+a = a+1 = a$. Then *ideal*(R) has only two elements other than R itself: $\{0\}$ and $\{0, a\}$ and thus is modular. However, one sees immediately that the ideal $\{0, a\}$ is not subtractive.

(6.40) Example. [Alarcón & Anderson, 1994b] If R is a semiring and t is an indeterminate over R, then the lattice *lideal*($R[t]$) is modular if and only if each of its members is subtractive, and that is true if and only if R is in fact a ring.

Note that if $\{I_k \mid k \in \Omega\}$ is a set of semisubtractive [left, right] ideals of R then $\cap_{k\in\Omega} I_k$ is also semisubtractive. Similarly, if each I_k is subtractive then $\cap_{k\in\Omega} I_k$ is subtractive and if each I_k is strong then $\cap_{k\in\Omega} I_k$ is strong. Thus any subset of a semiring is contained in a **semisubtractive** [resp. **subtractive**, **strong**] **closure**, namely the intersection of R and all semisubtractive [resp. subtractive, strong] ideals containing it. Hence R is left austere when it is the subtractive closure of each of its nonzero left ideals.

(6.41) Example. It is easily verified that the subtractive closures of the ideals $I = 2\mathbb{N} \setminus \{2\}$ and $H = 2\mathbb{N} \setminus \{2, 4\}$ of $\mathbb{N}$ are both equal to $2\mathbb{N}$.

(6.42) Example. Subtractive closures of ideals in semirings of the form $R[t]$, where R is a commutative semiring, are studied in detail in [Dale, 1977a]. Thus, for example, if k and n are integers greater than 1 and H is the ideal of $\mathbb{N}[t]$ generated by k and $t^n + k$ then H is not subtractive since $t^n \notin H$. Its subtractive closure is the ideal generated by k and t^n.

Now assume that $1 < k < n$ in $\mathbb{N}$ and that n is not a multiple of k. The ideal I in $\mathbb{N}[t]$ generated by k, n, and $t + n$ is not subtractive since $t \notin I$. Its subtractive closure is the ideal generated by t and the greatest common divisor of k and n in $\mathbb{N}$.

Sums of subtractive ideals need not be subtractive. Indeed, $2\mathbb{N}$ and $3\mathbb{N}$ are subtractive ideals of $\mathbb{N}$ but $2\mathbb{N} + 3\mathbb{N} = \mathbb{N} \setminus \{1\}$ is not subtractive, as noted in Example 6.19. On the other hand, if $\{I_k \mid k \in \Omega\}$ is a set of semisubtractive [left, right] ideals of R and if $a \in (\sum_{k\in\Omega} I_k) \cap V(R)$ then there exists a finite subset Λ of Ω and elements $b_k \in I_k$ for each $k \in \Lambda$ such that $a = \sum_{k\in\Lambda} b_k$. If $h \in \Lambda$ then $b_h + (-a + \sum_{k\neq h} b_k) = 0$ so $b_h \in I_h \cap V(R)$. Since each I_h is semisubtractive, this implies that $-a = \sum_{k\in\Lambda} -b_k \in \sum_{k\in\Lambda} I_k$. Thus $\sum_{k\in\Omega} I_k$ is semisubtractive.

(6.43) EXAMPLE. [Dulin & Mosher, 1972] Define operations $\oplus$ and $\odot$ on $\mathbb{N}$ as follows:

(1) $a \oplus b = max\{a, b\}$ if $a \leq 6$ or $b \leq 6$ and $a + b$ otherwise;
(2) $a \odot b = min\{a, b\}$ if $a \leq 6$ or $b \leq 6$ and ab otherwise.

Then $(\mathbb{N}, \oplus, \odot)$ is a commutative hemiring having subtractive ideals $I = \{0,1,2,3,4,5,6\} \cup \{2t+4 \mid t \in \mathbb{N}\}$ and $H = \{0,1,2,3,4,5,6\} \cup \{3t+9 \mid t \in \mathbb{N}\}$. The ideal IH is not subtractive since 96 and $120 = 96 + 24$ belong to IH but $24 \notin IH$. Therefore, by Proposition 6.27, we see that if R is the Dorroh extension of this hemiring, then $I' = I \times \{0\}$ and $H' = H \times \{0\}$ are subtractive ideals of R but $I'H' = IH \times \{0\}$ is an ideal of R which is not subtractive. Similarly, $I' + H'$ is not a subtractive ideal of R.

(6.44) PROPOSITION. *An ideal I of a semiring R is complemented in ideal(R) if and only if $I = (a)$ for some $1 \neq a \in comp(R) \cap C(R)$.*

PROOF. Assume that $I = (a)$ for some $1 \neq a \in comp(R) \cap C(R)$ and let $H = (a^{\perp})$. Since $1 = a + a^{\perp}$, we have $I + H = R$. If $b \in I \cap H$ then $b = \sum c_i a d_i$ for some elements c_i and d_i of R. But since $b \in H$, we have $b = ba^{\perp} = \sum c_i a d_i a^{\perp} = \sum c_i a a^{\perp} d_i = 0$. Therefore $I + H = \{0\}$ and so $I \in comp(ideal(R))$.

Conversely, assume that $I \in comp(ideal(R))$ and let H be an ideal of R satisfying $I + H = R$ and $IH = HI = \{0\}$. Then there exist elements a of I and b of H satisfying $a + b = 1$. Moreover, $ab = ba = 0$ since $ab \in IH$ and $ba \in HI$. Therefore $a \in comp(R)$ and $b = a^{\perp}$. If $r \in R$ then $ra \in I$ so $bra \in I \cap H = \{0\}$. Thus $ra = 1ra = bra + ara = ara$. Similarly, $ar = ara$ and so $a \in C(R)$. If $r \in I$ then $rb \in I \cap H = \{0\}$ and so $r = r1 = ra + rb = ra$. Thus $r \in (a)$ and so $I = (a)$. □

(6.45) PROPOSITION. *If R is a semiring and n is a positive integer then there exists an inclusion-preserving bijection between the set of all ideals of R and the set of all ideals of $\mathcal{M}_n(R)$. Moreover, an ideal of R is subtractive if and only if the corresponding ideal of $\mathcal{M}_n(R)$ is subtractive.*

PROOF. We will denote the multiplicative identity of $\mathcal{M}_n(R)$ by E. For each $1 \leq h, k \leq n$, we will denote by E_{hk} the matrix $[a_{ij}]$ in $\mathcal{M}_n(R)$ defined by $a_{ij} = 1$ when $(i,j) = (h,k)$ and $a_{ij} = 0$ otherwise. If I is an ideal of R, set $\Psi(I) = \{[a_{ij}] \in \mathcal{M}_n(R) \mid a_{ij} \in I \text{ for all } 1 \leq i,j \leq n\}$. It is straightforward to verify that $\Psi(I)$ is an ideal of $\mathcal{M}_n(R)$. Moreover, $I \subseteq I'$ implies that $\Psi(I) \subseteq \Psi(I')$ so Ψ is inclusion-preserving. If K is an ideal of $\mathcal{M}_n(R)$, set $\Phi(K) = \{a \in R \mid aE \in K\}$. Then $\Phi(I) = I$ for each ideal I of R. Moreover, if K is an ideal of $\mathcal{M}_n(R)$ and if $A = [a_{ij}] \in K$ then for each $1 \leq i,j \leq n$ we have $a_{ij}E = \sum_{k=1}^{n} E_{ki}AE_{jk} \in K$ and so $\Psi\Phi(K) = K$. Thus Ψ is a bijection.

Now assume that I is a subtractive ideal of R and let $[a_{ij}]$ and $[b_{ij}]$ be elements of $\mathcal{M}_n(R)$ such that $[a_{ij}]$ and $[a_{ij}] + [b_{ij}]$ are elements of $\Psi(I)$. Then a_{ij} and $a_{ij} + b_{ij}$ belong to I for all $1 \leq i,j \leq n$ and so $b_{ij} \in I$ for all such i and j. Hence $[b_{ij}] \in \Psi(I)$, proving that $\Psi(I)$ is subtractive. Conversely, assume that $\Psi(I)$ is subtractive and let a and b be elements of R satisfying the condition that a and $a + b$ belong to I. Then aE and $(a+b)E = aE + bE$ belong to $\Psi(I)$ and so bE belongs to $\Psi(I)$, proving that $b \in I$. Hence I is subtractive. □

(6.46) PROPOSITION. *If R is a Gel'fand semiring then $K^+(R)$ is a strong ideal of R.*

PROOF. We have already noted that $K^+(R)$ is always closed under addition. Let $k \in K^+(R)$ and let $r \in R$. If $kr + a = kr + b$ then $k(1 + r) + a = k(1 + r) + b$ so $k + a(1 + r)^{-1} = k + b(1 + r)^{-1}$ and so $a(1 + r)^{-1} = b(1 + r)^{-1}$, whence $a = b$. Thus $kr \in K^+(R)$. A similar argument shows that $rk \in K^+(R)$, proving that $K^+(R)$ is an ideal of R. Now assume that $r + r' \in K^+(R)$. If $r + a = r + b$ then $r + r' + a = r + r' + b$ and so $a = b$. Thus $r \in K^+(R)$. Similarly $r' \in K^+(R)$, proving that $K^+(R)$ is strong. □

(6.47) EXAMPLE. If R is not a Gel'fand semiring then $K^+(R)$ need not even be an ideal of R. For example, if $R = \mathbb{N}\{\infty\}$ then, as noted in Example 4.15, $K^+(R) = \mathbb{N}$ and this is not an ideal of R.

An element a of a semiring R is **small** in R if and only if $a + b \notin U(R)$ for all $b \in R \backslash U(R)$. Note that if R is simple then $U(R) = \{1\}$ and so this definition reduces to the usual definition in the case of bounded distributive lattices. If $a \in U(R)$ then a is never small since $a + 0 \in U(R)$. On the other hand, 0 is always small. More generally, if $d \in R$ then an element a of R is d-**small** in R if and only if $a+b \in U(R)$ implies that $d + b \in U(R)$. Thus a is small in R if and only if it is 0-small in R. Clearly d is d-small in R for each $d \in R$.

(6.48) PROPOSITION. *If R is a Gel'fand semiring then the set I of all small elements of R is an ideal of R. If R is a simple semiring and $1 \neq d \in R$ then the set I_d of all d-small elements of R is a strong ideal of R.*

PROOF. If R is a Gel'fand semiring then clearly $I \cap U(R) = \emptyset$ and so $I \neq R$. Suppose that $a, a' \in I$ and that b is an element of R satisfying $a + a' + b \in U(R)$. By the smallness of a, we have $a' + b \in U(R)$ and then, by the smallness of a', we have $b \in U(R)$. Thus $a + a' \in I$. Now assume that $a \in I$ and that r and b are elements of R satisfying the condition that $ra + b = c \in U(R)$. Since R is a Gel'fand semiring, we know by Proposition 4.50 that $a + 1$ and $r + c$ belong to $U(R)$. Therefore $(r+c)a+b = c(a+1) \in U(R)$ and so $a+(r+c)^{-1}c(a+1) \in U(R)$. Since a is small, this implies that $(r + c)^{-1}b$ is in $U(R)$ and so $b \in U(R)$. Therefore $ra \in I$. A similar argument shows that $ar \in I$ and so I is an ideal of R.

Now assume that R is simple. In this case, as we observed previously, $U(R) = \{1\}$. If $a, a' \in I_d$ and $a + a' + b = 1$ then $d + a' + b = 1$ and so $d + d + b = 1$. Since simple semirings are additively idempotent, this implies that $d + b = 1$. If $a \in I_d$ and $c \in R$ satisfy $ca + b = 1$ then $a + b = (c + 1)a + b = ca + a + b = a + 1 = 1$ and so $d + b = 1$. Thus $ca \in I_d$. Similarly, $ac \in I_d$. Since $1 \neq d$ we see that $1 \notin I_d$, so I_d is an ideal of R. Finally, we note that if $a + a' \in I_d$ and $a + b = 1$ then $a + a' + b = 1 + a' = 1$ so $d + b = 1$. Thus $a \in I_d$. Similarly, $a' \in I_d$, showing that I_d is strong. □

(6.49) COROLLARY. *If I is an ideal of a simple semiring R then $I' = \cup_{d \in I} I_d$ is an ideal of R which is I-small in the semiring ideal(R).*

PROOF. Note first that $I' \neq R$ since $1 \notin I_d$ for each $d \in I$. Suppose that $a, a' \in I'$ and that d and d' are elements of I satisfying $a \in I_d$ and $a' \in I'_d$. If $a + a' + b = 1$

then $d+a'+b=1$ and so $d+d'+b=1$, proving that $a+a' \in I_{d+d'} \subseteq I'$. Moreover, if $r \in R$ then ra and ar both belong to I_d and hence to I'. Thus I' is an ideal of R. Now suppose that H is an ideal of R satisfying $I' + H = R$. Then there exist elements a of I' and b of H satisfying $a + b = 1$. Since $a \in I_d$ for some $d \in I$ we have $d + b = 1$, proving that $I + H = R$. Thus I' is I-small in $ideal(R)$. □

A [left, right] ideal I of a semiring R defines an equivalence relation $\equiv_I$ on R, called the **Bourne relation**, given by $r \equiv_I r'$ if and only if there exist elements a and a' of I satisfying $r + a = r' + a'$. Note that if $r \equiv_I r'$ and $s \equiv_I s'$ in R then $r + s \equiv_I r' + s'$. If I is an ideal, then this also implies $rs \equiv_I r's'$. We denote the set of all equivalence classes of elements of R under this relation by R/I and will denote the equivalence class of an element r of R by r/I. Note that if $I \subseteq H$ are [left, right] ideals of R then $r \equiv_I r'$ surely implies that $r \equiv_H r'$ for all elements r and r' of R. If I is additively idempotent then $r \equiv_I r'$ if and only if there exists an element $b \in I$ such that $r + b = r' + b$. Indeed, if $r \equiv_I r'$ then there exist elements a and a' of I satisfying $r+a = r'+a'$ and so $r+(a+a') = (r'+a')+a' = r'+a' = r+a = r+a+a = r'+(a+a')$.

Similarly, I defines an equivalence relation $[\equiv]_I$ on R, called the **Iizuka relation**, given by $r\ [\equiv]_I\ r'$ if and only if there exist elements a and a' of I and an element s of R satisfying $r + a + s = r' + a' + s$. Note that if $r[\equiv]_I r'$ and $s[\equiv]_I s'$ in R then $r + s[\equiv]_I r' + s'$ and, if I is an ideal, $rs[\equiv]_I r's'$. Also note that if $r \equiv_I r'$ then surely $r[\equiv]_I r'$. We denote the set of all equivalence classes of elements of R under this relation by $R[/]I$ and will denote the equivalence class of an element r of R by $r[/]I$. Again, if $I \subseteq H$ are [left, right] ideals of R then $r[\equiv]_I r'$ surely implies that $r[\equiv]_H r'$ for all elements r and r' of R.

If r and r' are elements of a semiring R and if I is an ideal of R then, as noted, $r \equiv_I r'$ implies that $r[\equiv]_I r'$. The converse does not necessarily hold. If R is a yoked semiring, it is easy to see that the converse holds for those ideals I containing $Z(R)$.

(6.50) PROPOSITION. *If I is a left ideal of a semiring R then $0/I$ is the subtractive closure of I in R.*

PROOF. If $r, r' \in 0/I$ then there exist elements a, a', b, b' of I satisfying $r + a = 0+b$ and $r'+a' = 0+b'$. Therefore $(r+r')+(a+a') = 0+(b+b')$ so $r+r' \in 0/I$. If $r'' \in R$ then $r''r+ra = 0+r''b$ so $r''r \in 0/I$. Similarly, $rr'' \in 0/I$. Thus $0/I = R$ or $0/I$ is an ideal of R, which clearly contains I.

If r and $r + r'$ belong to $0/I$ then there exist elements a, a', b, b' of I such that $r+a = 0+b$ and $(r+r')+a' = 0+b'$ so $0+(b'+a) = (r+r')+a+a' = r'+b+a'$, which proves that $r' \in 0/I$. Therefore the left ideal $0/I$ is subtractive.

Finally, let H be a subtractive ideal of R containing I. If $r \in 0/I$ then there exist elements a and b of I (and hence of H) such that $r + a = 0 + b \in H$ and so $r \in H$. Thus $0/I \subseteq H$. □

Thus we see, in particular, that an ideal I of a semiring R is subtractive if and only if $I = 0/I$. We can define operations $\oplus$ on the set S of all subtractive ideals on R by setting $I \oplus H = 0/(I+H)$ and $I \odot H = 0/IH$, and it is easily verified that $(S, \oplus, \odot)$ is itself a semiring.

(6.51) COROLLARY. *A semiring R is left austere if and only if for each $0 \neq r \in R$ there exist $a, b \in R$ satisfying $ar + 1 = br$.*

PROOF. Assume that R is left austere and let $0 \neq r \in R$. Then $0/Rr = R$ and so, by definition, the given elements a and b exist. Conversely, assume the stated condition holds. If I is a nonzero subtractive left ideal of R and $0 \neq r \in I$ then there exist elements a and b of R satisfying $ar+1 = br \in I$ and so, by subtractiveness, we have $1 \in I$. Thus $I = R$, which is a contradiction, proving that R is left austere. □

(6.52) PROPOSITION. *If I and H are left ideals of a semiring R then we have $0/R(I \cup H) = 0/R(0/I \cup 0/H)$.*

PROOF. Since $I \subseteq 0/I$ and $H \subseteq 0/H$ we have $R(I \cup H) \subseteq R(0/I \cup 0/H)$ and so $0/R(I \cup H) \subseteq 0/R(0/I \cup 0/H)$. To show the reverse containment it suffices, by Proposition 6.50, to show that the subtractive left ideal $0/R(I \cup H)$ of R contains $R(0/I \cup 0/H)$. Indeed, if $a \in 0/I$ then a belongs to every subtractive left ideal of R containing I and hence, in particular, to $0/R(I \cup H)$ Thus $0/I \subseteq 0/R(I \cup H)$. Similarly $0/H \subseteq 0/R(I \cup H)$ and so, since $0/R(I \cup H)$ is a left ideal of R, we have $0/R(I \cup H) \subseteq 0/R(I \cup H)$, as desired. □

(6.53) PROPOSITION. *A cancellative austere yoked semiring R is a division semiring.*

PROOF. Let r be a nonzero element of R. By Corollary 6.51, there exist elements a and b of R satisfying $ar + 1 = br$. Since R is a yoked semiring, there exists an element c of R satisfying $a = b + c$ or $a + c = b$. If $a = b + c$ then $ar = br + cr = ar + 1 + cr$ so $0 = 1 + cr$ since R is cancellative. Thus $1 \in V(R)$, proving that R is in fact a ring. Since every left ideal of a ring is subtractive, we conclude that R has no nonzero left ideals and this suffices to show that R is a division ring. If $a + c = b$ then $ar + cr = br = ar + 1$ and so $cr = 1$. Then $c \neq 0$ and so a similar argument shows that either R is a ring (in which case we are done) or there exists an element c' of R satisfying $c'c = 1$. But then $c' = c'(cr) = (c'c)r = r$, proving that $r \in U(R)$ with $c = r^{-1}$. Thus every nonzero element of R is a unit, proving that R is a division semiring. □

(6.54) PROPOSITION. *If I is a left ideal of a semiring R then the relations $\equiv_I$ and $\equiv_{0/I}$ coincide. Similarly, the relations $[\equiv]_I$ and $[\equiv]_{0/I}$ coincide.*

PROOF. Let r and r' be elements of R. Since $I \subseteq 0/I$, we note that $r \equiv_I r'$ implies that $r \equiv_{0/I} r'$. Conversely, assume that $r \equiv_{0/I} r'$. Then there exist elements b and b' of $0/I$ satisfying $r + b = r' + b'$. Moreover, since b and b' belong to $0/I$, there exist elements a and a' of I satisfying $b + a$ and $b' + a'$ both belong to I. Hence $b + a + a'$ and $b' + a + a'$ belong to I and $r + (b + a + a') = r' + (b' + a + a')$, proving that $r \equiv_I r'$.

The second part is proven similarly. □

(6.55) PROPOSITION. *Let R be a plain yoked semiring with descending chain condition on subtractive left ideals and having no nonzero nilpotent elements. Then every subtractive left ideal of R is of the form Re for some $e \in I^{\times}(R)$.*

PROOF. By Proposition 4.22, we note that R is cancellative. Let I be a subtractive left ideal of R. If $I = \{0\}$ then I $= R0$ and we are done. Hence we can assume that $I \neq \{0\}$. By the descending chain condition, I contains a minimal nonzero subtractive left ideal H. If $0 \neq c \in H$ then $c^2 \neq 0$ and so Hc is a nonzero left ideal of R, the subtractive closure of which is H. Hence $c \in 0/Hc$ and so there exist elements $h, h' \in H$ satisfying $hc = c + h'c$. Since R is a yoked semiring, there exists an element r of R satisfying $r + h = h'$ or $r + h' = h$. Since H is subtractive, we in fact have $r \in H$. If $r + h' = h$ then $c + h'c = hc = rc + h'c$ and so $c = rc$. If $r + h = h'$ then $rc + c + h'c = rc + hc = h'c$ so $rc + c = 0$. Hence $c = c + r(c + rc) = c + rc + r^2c = r^2c$. In either case, there exists a nonzero element e of H satisfying $c = ec$.

Since R is a yoked semiring, there exists an element d of R satisfying $d + e^2 = e$ or $e^2 = d + e$ and, again, we must have $d \in H$. By Example 6.10, the left ideal $(0 : c)$ of R is subtractive and hence so is $H \cap (0 : c)$. Since $c^2 \neq 0$, we see that $H \cap (0 : c) \subset H$ and so, by the minimality of H, we have $H \cap (0 : c) = \{0\}$. If $d + e^2 = e$ then $dc + ec = dc + e^2c = ec$ so $dc = 0$ and hence $d \in H \cap (0 : c)$, implying that $d = 0$. Similarly, if $e^2 = d + e$ then $d = 0$. Thus $e = e^2 \in I^{\times}(R) \cap H$, and so $I^{\times}(R) \cap H \neq \{0\}$.

We now claim that there exists an element f of $I \cap I^{\times}(R)$ satisfying the condition $I \cap (0 : f) = \{0\}$. Indeed, for each $e \in I \cap I^{\times}(R)$ let $M_e = I \cap (0 : e)$. This is a subtractive left ideal of R and so, by the descending chain condition, we can pick an element f of $I \cap I^{\times}(R)$ such that M_f is minimal. Suppose that $M_f \neq \{0\}$. Then, by the above, M_f contains an idempotent element g. Moreover, $gf = 0$ since $g \in M_f$. Since R is a yoked semiring, there exists an element H of R satisfying $h + fg = g + f$ or $fg = h + g + f$. Again, since I is subtractive we must in fact have $h \in I$. In the first case, we have $hf + fgf = gf + f^2$, which implies that $hf = f^2 = f$. Moreover, $hg + fg^2 = g^2 + fg$ so $hg = g$. Similarly $gh = g$. Thus $h^2 + fg = h^2 + hfg = hg + hf = g + f = h + fg$ and so $h^2 = h$. Furthermore, $M_h \subseteq M_f$. But this inclusion is proper since $g \in M_f \setminus M_h$, contradicting the minimality of M_f. Hence we must have $fg = h + g + f$. Set $k = hfg + g + f$. Then $k^2 = hfghfg + hfg + hfgf + ghfg + g + gf + fhfg + fg + f$. We know that $gf = 0$ and so $0 = gfg = gh + g^2 + gf = gh + g$, while $fg = f^2g = fh + fg + f$ implies that $fh + f = 0$. This, in turn, implies that $0 = ghf + gf = ghf$. Thus $k^2 = k$ and so $k \in I^{\times}(R) \cap I$. If $r \in M_k$ we have $rk = 0$ and hence

$$rg + rf = rhfhg + rhfg + rg + rf = rhfhg$$

so $rf = rgf + rf^2 = rhfhgf = 0$. Thus $r \in M_f$ and hence $M_k \subseteq M_f$ where, again, this containment is proper. Thus we have a contradiction in this situation too, implying that $M_f = \{0\}$ and establishing the claim.

If $a \in I$ then there exists an element b of I satisfying $b + a = af$ or $a = b + af$. If $b + a = af$ then $af = af^2 = bf + af$ so $bf = 0$ and hence $b \in M_f$, yielding $b = 0$. The other case yields the same result. Thus $a = af$ for all $a \in I$. Since $Rf \subseteq I = If \subseteq Rf$, we then have $I = Rf$, as desired. □

A [left, right] ideal of a semiring R is **maximal** if and only if it is not properly contained in any other [left, right] ideal of R.

(6.56) EXAMPLE. [Słowikowski & Zawadowski, 1955] Let X be a bicompact Hausdorff topological space and let R be the commutative semiring of all continuous functions from X to the semiring $\mathbb{R}^+$. Then for each $x \in X$, the set $\{f \in R \mid f(x) = 0\}$ is a maximal ideal of R and all maximal ideals of R are of this form.

(6.57) EXAMPLE. [Sancho de Salas, 1987] The set R of complements of bounded open sets in $\mathbb{R}^n$ is a basis for the usual topology on $\mathbb{R}^n$ and so, as we saw in Example 1.5, $(R, \cap, \cup)$ is a semiring. This semiring has a unique maximal ideal $R \setminus U(R)$.

(6.58) EXAMPLE. [Alarcón & Anderson, 1994a] If t is an indeterminate then the commutative cancellative semiring $\mathbb{B}^+[t]$ has a unique maximal ideal consisting of all polynomials of degree not equal to 0.

(6.59) PROPOSITION. *Every [left, right] ideal of a semiring R is contained in a maximal [left, right] ideal of R.*

PROOF. Let I be a [left, right] ideal of R. If I is maximal we are done. If not, there is a nonempty set $\mathcal{C}$ of [left, right] ideals of R properly containing I. If $\mathcal{C}'$ is a linearly-ordered subset of $\mathcal{C}$ then $\cup\mathcal{C}'$ is again a [left, right] ideal of R and so belongs to $\mathcal{C}$. By Zorn's Lemma, we then see that $\mathcal{C}$ has a maximal element. □

(6.60) EXAMPLE. The set $\mathbb{N} \setminus \{1\}$ is a maximal ideal of the semiring $\mathbb{N}$ which contains all ideals of $\mathbb{N}$. Note that this ideal is not principal. Similarly, if $\mathcal{A}$ is the commutative semiring defined in Example 1.9 then $\mathcal{A} \setminus \{[\mathbb{Z}]\}$ is a maximal ideal of $\mathcal{A}$ which contains all ideals of $\mathcal{A}$. See [Feigelstock, 1980] for details.

As an immediate consequence of Proposition 6.59 we see that an element a of a semiring R belongs to every maximal ideal of R if and only if (a) is a small element of the semiring *ideal*(R). Indeed, if a belongs to every maximal ideal of R and I is an arbitrary ideal of R then, by Proposition 6.59, I is contained in a maximal ideal H of R and so $(a) + I \subseteq H \subset R$. Conversely, if (a) is a small element of *ideal*(R) and H is a maximal ideal of R then $(a) + H \neq R$ so $(a) + H = H$, whence $a \in H$.

(6.61) PROPOSITION. *For an element a of a semiring R the following conditions are equivalent:*

(1) *$a \in U(R)$;*
(2) *a belongs to no maximal one-sided ideal of R.*

PROOF. Assume that $a \in U(R)$ and that H is a maximal left ideal of R. If $a \in H$ then $1 = a^{-1}a \in H$, which is a contradiction. Thus $a \notin H$. Similarly, $a \notin H$ for any maximal right ideal H of R. Conversely, assume that a belongs to no maximal one-sided ideal of R. By Proposition 6.59, this implies that Ra is not a left ideal of R and so $Ra = R$. Similarly $aR = R$. Thus there exist elements b and c of R satisfying $ba = 1 = ac$. But then $b = b1 = b(ac) = (ba)c = 1c = c$ so $a \in U(R)$ and $b = a^{-1}$. □

(6.62) PROPOSITION. *The following conditions on a semiring R are equivalent:*

(1) *R is a Gel'fand semiring;*
(2) *Every maximal one-sided ideal of R is strong.*

PROOF. (1) $\Rightarrow$ (2): Suppose that I is a maximal left ideal of R and that r and r' are elements of R satisfying $r + r' \in I$ and $r \notin I$. Then $H = \{a + br \mid a \in I, b \in R\}$ is a subset of R closed under addition and under multiplication from the left by arbitrary elements of R. By the maximality of I, we see that H is not an ideal of R

and so we must have $H = R$. Therefore, in particular, there exist elements a of I and b of R satisfying $a + br = 1$. Therefore $1 + br' = a + br + br' = a + b(r + r') \in I$, contradicting the fact that, by (1), $1 + br' \in U(R)$. Thus we see that $r + r' \in I$ implies that r and r' both belong to I. The proof for maximal right ideals is similar.

(2) $\Rightarrow$ (1): By (2) we see that if $r \in R$ then $1 + r \notin I$ for any maximal left ideal or maximal right ideal of R. By Proposition 6.61, we conclude that $1 + r \in U(R)$ for each element r of R. □

(6.63) PROPOSITION. *Let R be a Gel'fand ring.*

1. *An element a of R is small if and only if it belongs to every maximal one-sided ideal of R;*
2. *If R is simple and $1 \neq d \in R$ then an element a of R is d-small if and only if it belongs to every maximal one-sided left ideal of R containing d.*

PROOF. (1) Assume that a is a small element of R and let H be a maximal left ideal of R to which a does not belong. Then $Ra + H = R$ and so there exist $r \in R$ and $h \in H$ satisfying $ra + h = 1$. By Proposition 6.48, ra is also small in R and so $h \in U(R)$. Therefore $1 = h^{-1}h \in H$, which is a contradiction. Thus a belongs to every maximal left ideal of R. Similarly, it belongs to every maximal right ideal of R.

Conversely, assume that a belongs to every maximal one-sided ideal of R. Let $b \in R$ satisfy the condition that $a + b = c \in U(R)$. If Rb is a left ideal of R then, by Proposition 6.59, it is contained in a maximal left ideal H of R. But then $a \in H$ and so $c = a + b \in H$, which is a contradiction since $c \in U(R)$. Thus we must have $Rb = R$. Similarly, $bR = R$ and so there exist elements d and d' of R satisfying $bd = d'b = 1$. But then $d = (d'b)d = d'(bd) = d'$ and so $b \in U(R)$, proving that a is small in R.

(2) Now assume that R is simple and that $1 \neq d \in R$. Let a be a d-small element of R and assume that $a \notin H$, where H is a maximal left ideal of R containing d. Then $Ra + H = R$ and so there exists an element r of R satisfying $ra + h = 1$. Since a is d-small, so is ra by Proposition 6.48. Thus $d + h = 1$ and so $1 \in H$, which is a contradiction. Therefore a must be an element of H.

Conversely, assume that a is an element of R which belongs to every maximal left ideal of R containing d. Let b be an element of R satisfying $a + b = 1$. If $R(d + b)$ is a left ideal of R then it is contained in a maximal left ideal H of R. Moreover, by Proposition 6.62, H is strong and so $d, b \in H$. By the choice of a, this implies that $a \in H$ and so $1 = a + b \in H$, which is a contradiction. Hence we must have $R(d + b) = R$. In particular, there exists an element r of R satisfying $r(d + b) = 1$. Then, by Proposition 4.3, we have $1 = 1 + d + b = (rd + d) + (rb + b) = d + b$ and so a is d-small in R. □

A [left, right] ideal $I \neq \{0\}$ of a semiring R is **minimal** if and only if it does not contain any [left, right] ideal of R other than itself and 0.

(6.64) PROPOSITION. *If H is a minimal left ideal of a semiring R and if $0 \neq e \in I^{\times}(R) \cap H$ then eH is a division semiring with multiplicative identity e.*

PROOF. Clearly $(eH, +)$ is a commutative monoid and $(eH, \cdot)$ is a semigroup, and clearly multiplication in eH distributes over addition. Since He is a nonzero left ideal of R contained in H, we must have $He = H$ and so for each element a of H there exists an element b of H satisfying $a = be$. Hence $(ea)e = (ebe)e = e(be) = ea = e(ea)$, showing that e is the identity of $(eH, \cdot)$. Thus eH is a semiring.

If $0 \neq ea \in eH$ then $ea = e^2a \in Hea$ and so Hea is a nonzero left ideal of R contained in H. Thus $H = Hea$ and so $eH = eHea$. In particular, there exists an element d of H satisfying $(ed)(ea) = e$. Similarly, there exists an element H of H satisfying $(eh)(ed) = e$ and so $eh = ehe = eh(edea) = (ehed)ea = ea$. Therefore eH is a division semiring. □

(6.65) PROPOSITION. *If I is a minimal left ideal of R and $a \in R$ then Ia is a left ideal of R which is either minimal or $\{0\}$.*

PROOF. Clearly Ia is a left ideal of R. Assume that it is not equal to $\{0\}$ and that it properly contains a left ideal H of R not equal to $\{0\}$. Then $H' = \{r \in I \mid ra \in H\}$ is a left ideal of R properly contained in I and not equal to $\{0\}$, contradicting the minimality of I. Thus Ia must be minimal. □

(6.66) PROPOSITION. *If H is an ideal of a semiring R containing a minimal left ideal then the sum of all minimal left ideals of R contained in H is an ideal of R.*

PROOF. Let H' be the sum of all minimal left ideals of R contained in H. Then H' is a left ideal of R. If $a \in R$ and if I is a minimal left ideal of R contained in H, then $Ia \subseteq H$ and so, by Proposition 6.65, $Ia \subseteq H'$. Thus $H'a \subseteq H'$ for each $a \in R$, proving that H' is an ideal of R. □

A nonempty subset D of a semiring R is a **coideal** if and only if it is closed under multiplication and satisfies the condition that $d + r \in D$ whenever $d \in D$ and $r \in R$.

(6.67) EXAMPLE. If A is a nonempty subset of a semiring R then the set $F(A)$ of all elements of R of the form $a_1 \cdot \ldots \cdot a_n + r$, where the a_i belong to A and $r \in R$, is a coideal of R containing R and, in fact, is the unique smallest coideal of R containing A.

A zerosumfree semiring R must contain a maximal proper coideal. Indeed, let $\mathcal{C}$ be the set of all coideals of R not containing 0. This set is nonempty since it contains $F(\{1\})$ by zerofreeness. The set $\mathcal{C}$ is closed under taking unions of chains, and so the result follows using Zorn's Lemma.

7. PRIME AND SEMIPRIME IDEALS IN SEMIRINGS

As in the case of rings, an ideal I of a semiring R is **prime** if and only if whenever $HK \subseteq I$, for ideals H and K of R, we must have either $H \subseteq I$ or $K \subseteq I$. The set of all prime ideals of a semiring R is called the **spectrum** of R and will be denoted by $spec(R)$.

(7.1) EXAMPLE. [Feigelstock, 1980] Let $\mathcal{A}$ be the commutative semiring defined in Example 1.9. Then $\{[G] \mid G \text{ a torsion abelian group}\}$ is a prime subtractive ideal of $\mathcal{A}$. Moreover, for each prime integer p, $\{[G] \mid \text{ the torsion subgroup of } G \text{ is } p\text{-divisible}\}$ is a prime subtractive ideal of $\mathcal{A}$.

(7.2) EXAMPLE. [Sancho de Salas & Sancho de Salas, 1989] Let B be the family of all subsets of $\mathbb{I}$ which are finite unions of singletons and closed subintervals of $\mathbb{I}$. Then B is a basis for the closed sets of the usual topology on $\mathbb{I}$ and so $(B, \cup, \cap)$ is a commutative simple semiring. Refer to Example 1.5 for details. The maximal prime ideals of B are those of the form $I_r = \{b \in B \mid r \in b\}$ for each $r \in \mathbb{I}$. The other prime ideals of B are of the form $H_r = \{b \in B \mid [r, r+e] \subseteq b \text{ for some } e > 0\}$ for each $1 \neq r \in \mathbb{I}$ or of the form $K_r = \{b \in B \mid [re, r] \subseteq b \text{ for some } e > 0\}$ for each $0 \neq r \in \mathbb{I}$.

(7.3) EXAMPLE. [Alarcón & Anderson, 1994a] For each $A \subseteq \mathbb{N} \setminus \{0\}$ let $I(A)$ be the ideal of $\mathbb{B}[X]$ generated by X and $\{1 + X^h \mid h \in A\}$. A necessary and sufficient condition for I_A to be a prime ideal of $\mathbb{B}[X]$ is that $\mathbb{N} \setminus A$ be an ideal of $\mathbb{N}$. In particular, if $A_n = \mathbb{N} \setminus (2^n)$ for each nonnegative integer n then

$$(X) = I(A_0) \subset I(A_1) \subset \ldots$$

is an infinite ascending chain of prime ideals of $\mathbb{B}[X]$.

The following result generalizes the case for rings.

(7.4) PROPOSITION. *The following conditions on an ideal I of a semiring R are equivalent:*

(1) *I is prime;*

(2) $\{arb \mid r \in R\} \subseteq I$ *if and only if* $a \in I$ *or* $b \in I$;
(3) *If* a *and* b *are elements of* R *satisfying* $(a)(b) \subseteq I$ *then either* $a \in I$ *or* $b \in I$.

PROOF. (1) $\Rightarrow$ (2): Let $a, b \in R$ and set $I' = \{arb \mid r \in R\}$. If $a \in I$ or $b \in I$ then $I' \subseteq I$ since I is an ideal. Conversely, let $H = (a)$ and $K = (b)$. These are ideals of R and $I' \subseteq HK$. Indeed, HK is clearly contained in any ideal which contains I'. Therefore $I' \subseteq I$ implies, by (1), that $H \subseteq I$ or $K \subseteq I$. Since $a \in H$ and $b \in K$, this implies that $a \in I$ or $b \in I$.

(2) $\Rightarrow$(1): Let H and K be ideals of R satisfying $HK \subseteq I$. Assume that $H \not\subseteq I$ and let $a \in H \setminus I$. Then for each $b \in K$ we have $\{arb \mid r \in R\} \subseteq HK \subseteq I$ and so, by (2), we must have $b \in I$. Thus $K \subseteq I$.

(2) $\Rightarrow$ (3): This is immediate. □

(7.5) COROLLARY. *If* a *and* b *are elements of a semiring* R *then the following conditions on a prime ideal* I *of* R *are equivalent:*

(1) *If* $ab \in I$ *then* $a \in I$ *or* $b \in I$;
(2) *If* $ab \in I$ *then* $ba \in I$.

PROOF. Clearly ((1) implies (2). Conversely, assume (2). If $ab \in I$ then $abr \in I$ for all $r \in R$. By (2), this implies that $bra \in I$ for all $r \in R$ and so, by Proposition 7.4, we observe that $a \in I$ or $b \in I$. □

(7.6) COROLLARY. *An ideal* I *of a commutative semiring* R *is prime if and only if* $ab \in I$ *implies that* $a \in I$ *or* $b \in I$ *for all elements* a *and* b *of* R.

PROOF. Note that, by commutativity, $ab \in I$ if and only if $arb \in I$ for all $r \in R$. The result then follows from Proposition 7.4. □

(7.7) EXAMPLE. [Alarcón & Anderson, 1994a] The ideal $I = \mathbb{N} \setminus \{1\}$ of $\mathbb{N}$ is prime. However, if t is an indeterminate then the set $I[t]$ of all elements of $\mathbb{N}[t]$ with coefficients in I is not prime, since $(2+t)(1+3t) = 2+7t+3t^2 \in I[t]$ while $2+t, 1+3t \notin I[t]$. Note that I is semisubtractive but not subtractive.

(7.8) COROLLARY. *Every prime ideal of a semiring* R *is semisubtractive.*

PROOF. Let I be a prime ideal of R and let $a \in I \cap V(R)$. If $r \in R$ then $(-a)r(-a) + ar(-a) = 0$ and so $(-a)r(-a) = -[ar(-a)]$. On the other hand, $ara + ar(-a) = 0$ and so $ara = -[ar(-a)]$. By the uniqueness of additive inverses, this implies that $(-a)r(-a) = ara \in I$ for all $r \in R$ and so, by Proposition 7.4, $-a \in I$. □

(7.9) EXAMPLE. If R is a bounded distributive lattice then both $(R, \vee, \wedge)$ and $(R, \wedge, \vee)$ are commutative semirings. Moreover, an easy application of Corollary 7.6 shows that I is a prime ideal of $(R, \vee, \wedge)$ if and only if $R \setminus I$ is a prime ideal of $(R, \wedge, \vee)$. Thus there exists a bijective order-reversing correspondence between the spectra of these two semirings, given by complementation.

A nonempty subset A of a semiring R is an **m-system** if and only if $a, b \in A$ implies that there exists an element r of R such that $arb \in A$.

(7.10) Example. Since we assume that any semiring R has a multiplicative identity, any submonoid of $(R, \cdot)$ is an m-system. In particular, if R is a semiring then $U(R)$, $C(R)$ and $I^{\times}(R) \cap C(R)$ are m-systems. So, if R is a commutative semiring then $I^{\times}(R)$ is an m-system.

We now note the following immediate consequence of Proposition 7.4.

(7.11) Corollary. *An ideal I of a semiring R is prime if and only if $R \setminus I$ is an m-system.*

(7.12) Proposition. *If A is an m-system of elements of a semiring R and if I is an ideal of R maximal among all those ideals of R disjoint from A then I is prime.*

Proof. Let H, K be ideals of R not contained in I but satisfying $HK \subseteq I$. Then $H + I$ and $K + I$ properly contain I and so have nonempty intersection with A. In particular, there exist finite subsets $\{a_1, \ldots, a_n, b_1, \ldots, b_t\}$ of I, $\{h_1, \ldots, h_n\}$ of H, and $\{k_1, \ldots, k_t\}$ of K such that $a = \sum_{i=1}^{n} h_i + a_i \in A \cap (H + I)$ and $b = \sum_{j=1}^{t} k_j + b_j \in A \cap (K + I)$. Since A is an m-system, there exists an element r of R such that $arb \in A$. But

$$arb = \sum_{j=1}^{t} \left[\sum_{i=1}^{n} (a_i r b_j + h_i r b_j) + \sum_{i=1}^{n} (a_i r k_j + h_i r k_j) \right] \in I + HK \subseteq I,$$

contradicting the hypothesis that $I \cap A = \varnothing$. Thus I is prime. □

(7.13) Corollary. *Any maximal ideal of a semiring R is prime.*

Proof. This is a consequence of Proposition 7.12, Example 7.10, and the fact that an ideal of R is maximal if and only if it is maximal among all those ideals of R disjoint from $U(R)$. □

(7.14) Proposition. *Every prime ideal I of a semiring R contains a minimal prime ideal.*

Proof. Let $\{H_i \mid i \in \Omega\}$ be a descending chain of prime ideals of R (in other words, $i \geq j$ in Ω if and only if $H_i \subseteq H_j$) and set $H = \cap_{i \in \Omega} H_i$. Then H is an ideal of R. Let a and b be elements of R satisfying $\{arb \mid r \in R\} \subseteq H$ and suppose that $a \notin H$. Then there exists an element k of Ω such that $a \notin H_k$. By Proposition 7.4, this implies that $b \in H_k$ and so $b \in H_i$ for all $i \leq k$. Moreover, if $i > k$ then $H_i \subseteq H_k$ and so $a \notin H_i$. Again, by Proposition 7.4 this implies that $b \in H_i$. Thus $b \in H_i$ for all i in Ω, proving that $b \in H$. Thus, by Proposition 7.4, H is prime. The result now follows from applying Zorn's Lemma to the dual of the partially-ordered set of all prime ideals of R contained in I. □

(7.15) Proposition. *If I is an ideal of a semiring R and if H is an ideal of R minimal among those ideals of R properly containing I then $K = \{r \in R \mid rH \subseteq I\}$ is a prime ideal of R.*

Proof. It is straightforward to verify that K is an ideal of R. Let K' and K'' be ideals of R satisfying $K'K'' \subseteq K$ and assume that $K'' \not\subseteq K$. We must show that $K' \subseteq K$. Indeed, since $K'K'' \subseteq K$ and $K'' \not\subseteq K$ we have $K'K''H \subseteq I$ and

$K''H \not\subseteq I$. Therefore, $I \subset I + K''H \subseteq H$ and so, by the minimality of H, we have $I + K''H = H$. Therefore $K'I + K'K''H = K'H \subseteq H$ and so $K' \subseteq K$, as desired. □

(7.16) PROPOSITION. *If I is a subtractive ideal of a commutative semiring R which is maximal in the set of all ideals of R which are not finitely generated, then I is prime.*

PROOF. Assume that $a, b \in R \setminus I$ satisfy $ab \in I$. Then $I + (a)$ and $I + (b)$ are ideals of R properly containing I and so both are finitely generated, say $I + (a) = (\{d_1 + r_1a, \ldots, d_n + r_na\})$ and $I + (b) = (\{d'_1 + r'_1b, \ldots, d'_k + r'_kb\})$. The set $H = \{r \in R \mid ra \in I\}$ is an ideal of R. Moreover, if $1 \leq j \leq k$ we note that $(d'_j + r'_jb)a = d'_ja + r'_jab \in H$ and so $I \subset I + (b) \subseteq H$. By the maximality of I, this implies that H is finitely-generated, say $H = (\{e_1, \ldots, e_m\})$. If $c \in I$ then there exist elements $s_1, \ldots, s_n$ of R such that

$$c = \sum_{i=1}^{n}(d_i + r_ia) = \sum_{i=1}^{n} s_id_i + \sum_{i=1}^{n} s_ir_ia.$$

Since I is subtractive, $\sum_{i=1}^{n} s_ir_ia \in I$ and so $\sum_{i=1}^{n} s_ir_i \in H$. Thus there exist $t_1, \ldots t_n$ in R such that $\sum_{i=1}^{n} s_ir_i = \sum_{i=1}^{m} t_ie_i$ and $c = \sum_{i=1}^{n} s_id_i + \sum_{i=1}^{m} t_ie_ia$. Therefore I is generated by $\{d_1, \ldots, d_n, e_1a, \ldots, e_ma\}$, contradicting the assumption that I is not finitely-generated. Hence $ab \in I$ implies that $a \in I$ or $b \in I$, and so I is prime. □

(7.17) PROPOSITION. *If R is a commutative semiring every ideal of which is subtractive, then R is noetherian if and only if every prime ideal of R is finitely generated.*

PROOF. If R is noetherian then every prime ideal of R is finitely generated by Proposition 6.16. Conversely, assume that this condition holds and let $\mathcal{C}$ be the set of all ideals of R which are not finitely generated. By Proposition 6.16, we must show that $\mathcal{C}$ is empty. Assume that this is not the case and let $\{I_j \mid i \in \Omega\}$ be a chain of elements of $\mathcal{C}$. Then $I = \cup_{i \in \Omega} I_j$ is an ideal of R which cannot be finitely generated for, if it were, it would equal one of the I_j, contrary to the assumption that none of the I_j is finitely generated. Therefore, by Zorn's Lemma, $\mathcal{C}$ has a maximal element I_0. By Proposition 7.16, I_0 is prime, contradicting our hypothesis that all prime ideals of R are finitely-generated. Therefore R is noetherian. □

(7.18) PROPOSITION. *Let I be an ideal of a commutative semiring R and let t be an indeterminate over R. Then $I[t]$ is a prime ideal of $R[t]$ if and only if I is a subtractive prime ideal of R.*

PROOF. Assume that $I[t]$ is a prime ideal of $R[t]$ and let a and b be elements of R satisfying $ab \in I$. Then $a \in I[t] \cap R$ or $b \in I[t] \cap R$ But $I[t] \cap R = I$ and so we have shown that I is prime. Now suppose that $a, b \in R$ are elements satisfying $a + b \in I$ and $a \in I$. Then $b(a + b), a^2 + b(a + b), ab \in I$ and so

$$(bt + a)[(a + b)t + b] = (ab + b^2)t^2 + (b^2 + a^2 + ab)t + ab \in I[t].$$

Since $I[t]$ is prime, this implies that either $bt + a \in I[t]$ or $(a + b)t + b \in I[t]$, and either of these implies that $b \in I$. Thus I is subtractive as well.

Conversely, assume that I is a subtractive prime ideal of R and let $f, g \in R[t]$ with $deg(f) = n$ and $deg(g) = k$. Suppose that $fg \in I[t]$ and $f \notin I[t]$. Then there is some index h such that $f(h) \notin I$. If $f(0) \notin I$ then $(fg)(0) = f(0)g(0) \in I$ implies that $g(0) \in I$. Similarly, $(fg)(1) = f(0)g(1) + f(1)g(0) \in I$ and $f(1)g(0) \in I$ so, by subtractiveness, $f(0)g(1) \in I$. Since $f(0) \notin I$, this implies that $g(1) \in I$. Now assume inductively that we have shown that $g(0), \ldots, g(u) \in I$ for some $u < k$. Then

$$(fg)(u+1) = \sum_{i=0}^{u+1} f(i)g(u+1-i) = \left(\sum_{i=1}^{u+1} f(i)g(u+1-i)\right) + f(0)g(u+1) \in I$$

with $\sum_{i=1}^{u+1} f(i)g(u+1-i) \in I$ and so $f(0)g(u+1) \in I$, proving that $g(u+1) \in I$. Thus $g \in I[t]$.

Now suppose that $f(0), \ldots, f(m-1) \in I$ but $f(m) \notin I$. Then

$$(fg)(m) = \sum_{i=0}^{m-1} f(i)g(m-i) + f(m)g(0) \in I$$

and so $f(m)g(0) \in I$. Also,

$$(fg)(m+1) = \sum_{i=0}^{s-1} f(i)g(m+1-i) + f(m)g(1) + f(m+1)g(0)$$

which implies, as before, $g(1) \in I$. An induction argument similar to the one in the previous paragraph now shows that $g \in I[t]$. □

(7.19) Example. The structure of the prime ideals of $\mathbb{B}[t]$, where t is an indeterminate, is studied in detail by La Grassa, [1995]. In particular, she notes that every nonzero prime ideal of $\mathbb{B}[t]$ either contains t or $1+t$ but that the ideal $I = (1+t)$, itself, is not prime since $1 + t + t^3$ and $1 + t^2 + t^3$ do not belong to I whereas $(1 + t + t^3)(1 + t^2 + t^3) = (1+t)^6 \in I$.

For each ideal I of a semiring R let $\mathbb{V}(I) = \{H \in spec(R) \mid I \subseteq H\}$ and $\mathbb{D}(I) = spec(R) \setminus \mathbb{V}(I)$. Also set $\mathbb{V}(R) = \varnothing$ and $\mathbb{D}(R) = spec(R)$. It is easy to see that $\mathbb{V}(I) \cup \mathbb{V}(I') = \mathbb{V}(II')$ for all ideals I and I' of R and $\cap_{k\in\Omega}\mathbb{V}(I_k) = \mathbb{V}(\sum_{k\in\Omega} I_k)$ for every set $\{I_k \mid k \in \Omega\}$ of ideals of R. Therefore, $Zar(R) = \{\mathbb{V}(I) \mid I \in ideal(R)\}$ is the family of closed sets for a topology on $spec(R)$, called the **Zariski topology**. As a consequence of Corollary 7.13, we note that the set $mspec(R)$ of all maximal ideals of a semiring R is contained in $spec(R)$ and so the Zariski topology on $spec(R)$ induces a topology on $mspec(R)$. This topology is studied, for the case of commutative semirings, in [Iséki & Miyanaga, 1956a].

If $a \in R$ we will write $\mathbb{V}(a)$ and $\mathbb{D}(a)$ instead of $\mathbb{V}((a))$ and $\mathbb{D}((a))$ respectively. Note that $\{\mathbb{D}(a) \mid a \in R\}$ is a base of open sets for the Zariski topology. Indeed, if I is an ideal of R then $\mathbb{V}(I) = \cap\{\mathbb{V}(a) \mid a \in I\}$.

(7.20) Proposition. *If R is a semiring then spec(R), topologized with the Zariski topology, is a quasicompact T_0-space.*

Proof. We first note that $spec(R)$ is a T_0-space. Indeed, if $I \nsubseteq I'$ are elements of $spec(R)$ then $\mathbb{D}(I)$ is an open neighborhood of I' not containing I.

It is also quasicompact. Indeed, let $\{I_k \mid k \in \Omega\}$ be a family of ideals of R satisfying $\varnothing = \cap_{k\in\Omega}\mathbb{V}(I_k) = \mathbb{V}(\sum_{k\in\Omega} I_k)$. If $I = \sum_{k\in\Omega} I_k \neq R$ then, by Zorn's Lemma, I is contained in a maximal ideal of R which, by Corollary 7.13, is prime and so belongs to $\mathbb{V}(I)$. Since this is impossible, we must have $I = R$ and so $1 \in I$. Hence there exists a finite subset Λ of Ω such that $1 \in \sum_{k\in\Lambda} I_k$ and so $\varnothing = \cap_{i\in\Lambda}\mathbb{V}(I_k)$. □

An ideal I of a semiring R is **semiprime** if and only if, for any ideal H of R, we have $H^2 \subseteq I$ only when $H \subseteq I$. Prime ideals are surely semiprime.

(7.21) Proposition. *The following conditions on an ideal I of a semiring R are equivalent:*

(1) *I is semiprime;*

(2) *$\{ara \mid r \in R\} \subseteq I$ if and only if $a \in I$.*

Proof. (1) $\Rightarrow$ (2): Let $a \in R$ and set $I' = \{ara \mid r \in R\}$. If $a \in I$ then $I' \subseteq I$ since I is an ideal. Conversely, assume that $I' \subseteq I$ and let H be the set of all finite sums of elements of R of the form rar', where $r, r' \in R$. Then H is an ideal of R and H^2 consists of all finite sums of elements of the form $rar''ar'$, where $r, r', r'' \in R$. In particular, $I' \subseteq H$ and H^2 is contained in any ideal of R which contains I' and thus $H^2 \subseteq I$. By (1), this implies that $H \subseteq I$ and so $I' \subseteq I$.

(2) $\Rightarrow$ (1): Let H be an ideal of R satisfying $H^2 \subseteq I$ and let $a \in H$. Then $\{ara \mid r \in R \subseteq H^2\} \subseteq I$ and so, by (2), we must have $a \in I$. Thus $H \subseteq I$. □

(7.22) Corollary. *Every semiprime ideal of a semiring R is semisubtractive.*

Proof. The proof is the same as that of Corollary 7.8. □

Another way of stating Proposition 7.21 is the following: a nonempty subset A of a semiring R is a **p-system** if and only if $a \in A$ implies that there exists an element r of R such that $ara \in A$. Then we have the following immediate consequence of Proposition 7.21.

(7.23) Corollary. *An ideal I of a semiring R is semiprime if and only if $R \setminus I$ is a p-system.*

Any m-system of elements of a semiring R is a p-system. Also, it is clear that the union of p-systems is again a p-system. Conversely, we have the following result.

(7.24) Proposition. *A nonempty subset A of a semiring R is a p-system if and only if it is the union of m-systems.*

Proof. From the preceding remarks, we note that the union of m-systems is certainly a p-system. Conversely, let A be a p-system of elements of R and let $a_0 \in A$. Then there exists an element $r_0 \in R$ such that $a_1 = a_0r_0a_0 \in A$. Similarly, there exists an element $r_1 \in R$ such that $a_2 = a_1r_1a_1 \in A$. Continue in this manner to define the subset $B = \{a_0, a_1, a_2, \ldots\}$ of A. It is easily seen that B is in fact an m-system, containing a_0. Thus A is the union of m-systems. □

(7.25) PROPOSITION. *An ideal I of a semiring R is semiprime if and only if $I = \cap\mathbb{V}(I)$.*

PROOF. Let I be a semiprime ideal of a semiring R and let $A = R \setminus I$. Then, by Corollary 7.23, A is a p-system and so, by Proposition 7.24, $A = \cap_{i\in\Omega} B_i$, where each B_i is an m-system contained in A. Since $I \cap B_i = \varnothing$ for each $i \in \Omega$, we note by Zorn's Lemma that I is contained in an ideal K_i of R maximal with respect to being disjoint from B_i. By Proposition 7.12, each such K_i is prime. Therefore $I \subseteq \cap_{i\in\Omega} K_i \subseteq \cap_{i\in\Omega}(R \setminus B_i) = I$, and so I surely equals the intersection $\cap\mathbb{V}(I)$ of all prime ideals containing it.

Conversely, assume that $I = \cap\mathbb{V}(I)$. Then $R \setminus I = \cap\{R \setminus H \mid H \in \mathbb{V}(I)\}$. By Corollary 7.11, each $R \setminus H$ is an m-system and so, by Proposition 7.24, $R \setminus I$ is a p-system. Therefore, by Corollary 7.23, I is a semiprime ideal of R. □

As a consequence of Proposition 7.25 we see that every ideal I of a semiring R is contained in a unique minimal semiprime ideal of R, namely $\cap\mathbb{V}(I)$. If I is an ideal of a semiring R then the semiprime ideal $\cap\mathbb{V}(I)$ of R is denoted by $\sqrt{I}$. The ideal $\sqrt{(0)}$ is the **lower nil radical** of R.

(7.26) EXAMPLE. Let R be a semiring. If $r \in N_0(R)$ has index of nilpotency n then $r^n = 0 \in I$ for every prime ideal I of R and so, by primeness, $r \in I$. Thus $N_0(R) \subseteq \sqrt{(0)}$. Conversely, assume that $r \notin N_0(R)$. Then $A = \{r^i \mid i \in \mathbb{N}\}$ is an m-system not containing 0 so $(0) \cap A = \varnothing$. Then there exists an ideal I of R maximal among all ideals disjoint from A and, by Proposition 7.12, I is prime and $r \notin I$. Therefore $N_0(R) = \sqrt{(0)}$.

For an ideal I of R we see that $\sqrt{I}$ is precisely the set of all elements $r \in R$ such that every m-system in R which contains r has a nonempty intersection with I.

(7.27) PROPOSITION. *If I and H are ideals of a semiring R then:*

(1) *$I \subseteq H$ implies that $\sqrt{I} \subseteq \sqrt{H}$;*

(2) $\sqrt{\sqrt{I}} = \sqrt{I}$;

(3) $\sqrt{I+H} = \sqrt{\sqrt{I}+\sqrt{H}}$.

PROOF. (1) and (2) are immediate consequences of the definition. Moreover, by (1) we have $I + H \subseteq \sqrt{I} + \sqrt{H}$ and so $\sqrt{I+H} \subseteq \sqrt{\sqrt{I}+\sqrt{H}}$. Also by (1), we have $\sqrt{I}+\sqrt{H} \subseteq \sqrt{I+H}$ and so, using (2), $\sqrt{\sqrt{I}+\sqrt{H}} \subseteq \sqrt{\sqrt{I+H}} = \sqrt{I+H}$. Thus we have shown (3). □

(7.28) PROPOSITION. *(**Krull's Theorem**) If I is an ideal of a commutative semiring R then $\sqrt{I} = \{a \in R \mid a^n \in I$ for some positive integer $n\}$.*

PROOF. Set $K = \{a \in R \mid a^n \in I$ for some positive integer $n\}$. If $a, b \in K$ there exist $m, n \in \mathbb{P}$ such that a^m and b^n belong to I. Moreover, $(a+b)^{n+m-1} = \sum a^i b^j$, where either $i \geq m$ or $j \geq n$ in each summand. Therefore $(a+b)^{n+m-1} \in I$ and so $a + b \in K$. Similarly, if $r \in R$ then $(ra)^m = r^m a^m \in I$ and so $ra \in K$. Since $1 \notin K$, we conclude that K is an ideal of R.

Let $c \in R \setminus K$. If $c^2 \in K$ then there exists a positive integer n such that $c^n = (c^2)^n \in I$ and so $c \in K$, which is a contradiction. Thus $c \in R \setminus K$ and so $R \setminus K$ is a p-system, proving that the ideal K is semiprime.

Finally, let H be a prime ideal containing I. If $a \in K$ then there exists a positive integer n such that $a^n \in I \subseteq H$ and so $a \in H$ by Proposition 7.4. Hence $K \subseteq H$. This shows that $\mathbb{V}(I) \subseteq \mathbb{V}(K)$. The reverse containment is surely true and so we have equality. Since K is semiprime, this implies that $K = \sqrt{K} = \sqrt{I}$. □

As with rings, we say that an ideal I of a commutative semiring R is **primary** if and only if for each $a \in R \setminus I$ and $b \in R$ we have $ab \in I$ only when $b^k \in I$ for some positive integer k.

(7.29) COROLLARY. *If I is a primary ideal of a commutative semiring R then $\sqrt{I}$ is a prime ideal of R.*

PROOF. Let $a, b \in R$ satisfy $a \notin \sqrt{I}$ and $ab \in \sqrt{I}$. Then, by Proposition 7.28, there exists a positive integer n such that $a^n b^n = (ab)^n \in I$. Since I is primary, thee exists a positive integer k such that $b^{nk} = (b^n)^k \in I$ and so $b \in \sqrt{I}$. Therefore $\sqrt{I}$ is prime. □

(7.30) PROPOSITION. *If I and H are ideals of a commutative semiring R then $\sqrt{IH} = \sqrt{I \cap H} = \sqrt{I} \cap \sqrt{H}$.*

PROOF. Since $IH \subseteq I \cap H \subseteq I, H$ we have $\sqrt{IH} \subseteq \sqrt{I \cap H} \subseteq \sqrt{I} \cap \sqrt{H}$. Conversely, let $a \in \sqrt{I} \cap \sqrt{H}$. Then there exist positive integers n and m satisfying $a^n \in I$ and $a^m \in H$. Thus $a^{n+m} \in IH$ and so $a \in \sqrt{IH}$. This proves the desired equality. □

(7.31) PROPOSITION. *Let I be an ideal of a commutative semiring R satisfying the condition that $\sqrt{I}$ is finitely-generated. Then there exits a positive integer n satisfying $(\sqrt{I})^n \subseteq I$.*

PROOF. Suppose that $\sqrt{I} = (\{a_1, \ldots, a_k\})$. For each $1 \leq i \leq k$ there exists a positive integer n_i for which $a_i^{n_i} \in I$. Let $n = \sum_{i=1}^{k} n_i$. If $b = r_1a_1 + \cdots + r_ka_k \in \sqrt{I}$ then

$$b^n = \sum \frac{n!}{h_1!h_2! \cdot \ldots \cdot h_k!}(r_1a_1)^{h_1} \cdot \ldots \cdot (r_ka_k)^{h_k},$$

where the sum is taken over all k-tuples $(h_1, h_2, \ldots, h_k)$ satisfying $\sum_{i=1}^{k} h_i = n$. In each summand, we must have $h_j \geq n_j$ for at least one index j, and so each summand belongs to I. Therefore $b^n \in I$ for each $b \in \sqrt{I}$. □

(7.32) EXAMPLE. If R is a semiring satisfying $V(R) = \sqrt{V(R)}$, then LaGrassa [1995] has shown that an element $f \in R[t]$ is nilpotent if and only if $f(i) \in R$ is nilpotent for each $i \in \mathbb{N}$.

An ideal I of a semiring R is **irreducible** if and only if, for ideals H and K of R, we have $I = H \cap K$ only when $I = H$ or $I = K$. The ideal I is **strongly irreducible** if and only if, for ideals H and K of R, we have $H \cap K \subseteq I$ only when $H \subseteq I$ or $K \subseteq I$. A strongly irreducible ideal is surely irreducible.

A nonempty subset A of a semiring R is an **i-system** if and only if $a, b \in A$ implies that $(a) \cap (b) \cap A \neq \varnothing$.

(7.33) PROPOSITION. *The following conditions on an ideal I of a semiring R are equivalent:*

(1) *I is strongly irreducible;*
(2) *If $a, b \in R$ satisfy $(a) \cap (b) \subseteq I$ then $a \in I$ or $b \in I$;*
(3) *$R \setminus I$ is an i-system.*

PROOF. (1) $\Rightarrow$ (2): This is an immediate consequence of the definition.

(2) $\Rightarrow$ (3): If $a, b \in R \setminus I$ and $(a) \cap (b) \cap [R \setminus I] = \varnothing$ then $(a) \cap (b) \subseteq I$ and so, by (2), $(a) \subseteq I$ or $b \subseteq I$, which is a contradiction.

(3) $\Rightarrow$ (1): Let H and K be ideals of R not contained in I. Then there exist elements $a \in H \setminus I$ and $b \in K \setminus I$ and so, by (3), there exists an element $c \in [(a) \cap (b)] \setminus I$. In particular, $c \in H \cap K$ and so $H \cap K \nsubseteq I$. Thus we have (1). □

(7.34) PROPOSITION. *Let a be a nonzero element of a semiring R and let I be an ideal of R not containing a. Then there exists an irreducible ideal H of R containing I and not containing a.*

PROOF. If $\{H_i \mid i \in \Omega\}$ is a chain of ideals in R containing I and not containing a then $\cup_{i \in \Omega} H_i$ is an ideal of R not containing a. Therefore, by Zorn's Lemma, the set of all ideals of R not containing a has a maximal element H. Suppose that $H = H' \cap H''$, where H' and H'' are both ideals of R properly containing H. Then, by the choice of H, we have $a \in H'$ and $a \in H''$. Thus $a \in H' \cap H'' = H$, which is a contradiction. Hence H must be irreducible. □

(7.35) PROPOSITION. *Any ideal I of a semiring is the intersection of all irreducible ideals containing it.*

PROOF. Since $1 \notin I$, we know by Proposition 7.34 that there exists an irreducible ideal of R containing I. Let I' be the intersection of all irreducible ideals of R containing I. Then $I \subseteq I'$. If this inclusion is proper then there exists an element a of $I' \setminus I$. But, by Proposition 7.34, there exists an irreducible ideal H of R containing I but not a, which is a contradiction. Hence we must have $I = I'$. □

(7.36) PROPOSITION. *An ideal I of a semiring R is prime if and only if it is semiprime and strongly irreducible.*

PROOF. If I is prime then surely it is semiprime. Moreover, if H and K are ideals of R satisfying $H \cap K \subseteq I$ then $HK \subseteq H \cap K \subseteq I$ so $H \subseteq I$ or $K \subseteq I$. Therefore, I is strongly irreducible.

Conversely, assume that I is an ideal of R which is both semiprime and strongly irreducible. If H and K are ideals of R satisfying $HK \subseteq I$ then $(H \cap K)^2 \subseteq HK \subseteq I$ and so, by semiprimeness, $H \cap K \subseteq I$. Therefore, by strong irreducibility, $H \subseteq I$ or $K \subseteq I$, proving that I is prime. □

(7.37) PROPOSITION. *The following conditions are equivalent for an ideal I of a multiplicatively-regular semiring R:*

(1) *I is prime;*
(2) *I is irreducible.*

PROOF. By Proposition 7.36 we see that (1) implies (2). Conversely, assume (2) and let H and K be ideals of R satisfying $HK \subseteq I$. By Proposition 6.35 we see that

$$(H+I)\cap(K+I)=\{(h+a)(k+b)\mid h\in H;k\in K;a,b\in I\}\subseteq I.$$

Therefore, by (2), $H+I=I$ or $K+I=I$, namely $H\subseteq I$ or $K\subseteq I$. □

(7.38) PROPOSITION. *A commutative semiring R is multiplicatively regular if and only if every irreducible ideal of R is prime.*

PROOF. If R is multiplicatively regular then every irreducible ideal of R is prime, by Proposition 7.37. Conversely, assume that every irreducible ideal of R is prime. By Proposition 7.35, this implies that any ideal I of R satisfies $I=\sqrt{I}$. In particular, if H and K are ideals of R then, by Proposition 7.30 we have $HK=\sqrt{HK}=\sqrt{H\cap K}=H\cap K$ and so, by Proposition 6.35, R is multiplicatively regular. □

(7.39) PROPOSITION. *A semiring R is multiplicatively regular if and only if every ideal of R is semiprime.*

PROOF. If R is multiplicatively regular then every ideal of R is semiprime by Propositions 7.35 and 7.37. Conversely, assume that R satisfies the conditiont that every ideal of R is semiprime. Let I be an ideal of R. If $I^2=R$ then surely I is idempotent. If $I^2\subset R$ then $I^2=\cap\mathbb{V}(I^2)$. But this implies that $I\subseteq H$ for each $H\in\mathbb{V}(I^2)$ and so $I\subseteq I^2$, proving that $I=I^2$. By Proposition 6.35, this implies that R is multiplicatively regular. □

8. FACTOR SEMIRINGS

In the category of rings, factor objects are determined by ideals. In the category of semirings, as in the category of lattices, this is not so and we must look instead at congruence relations. An equivalence relation ρ defined on a semiring R which satisfies the additional condition that if $r\ \rho\ r'$ and $s\ \rho\ s'$ in R then $r+s\ \rho\ r'+s'$ and $rs\ \rho\ r's'$ is called a **congruence relation**. The congruence relation ρ defined by $r\ \rho\ r'$ if and only if $r=r'$ is the **trivial congruence relation** on R. All other congruence relations on R are **nontrivial**. The congruence relation ρ defined by $r\ \rho\ r'$ for all $r, r' \in R$ is the **improper congruence relation** on R. All other congruence relations are **proper**. Note that ρ is improper if and only if $1\ \rho\ 0$. Indeed, if ρ is improper this is clearly true. Conversely, if $1\ \rho\ 0$ then for each $r \in$R we have $r = r1\ \rho\ r0 = 0$ and so ρ is improper.

The family $Cong(R)$ of all congruence relations on R is a complete lattice with meets and joins defined as follows:

(1) If Y is a nonempty family of congruence relations on R then $\wedge Y$ is the congruence relation on R defined by $r(\wedge Y)r'$ if and only if $r\ \rho\ r'$ for all relations ρ in Y.

(2) If Y is a nonempty family of congruence relations on R then $\vee Y$ is the congruence relation on R defined by $r(\vee Y)r'$ if and only if there exist elements $r = s_0, s_1, \ldots, s_n = r'$ of R and elements $\rho_1, \ldots, \rho_n$ of Y such that $s_{i-1}\rho_i s_i$ for all $1 \leq i \leq n$.

Indeed, by an easy modification of a result of Funayama and Nakayama, $Cong(R)$ is in fact a frame, and hence a semiring. See [Birkhoff, 1973] for details.

(8.1) EXAMPLE. The Bourne relation $\equiv_I$ and the Iizuka relation $[\equiv]_I$ defined by an ideal I of a semiring R were shown in Chapter 5 to be congruence relations on R. If the semiring R is simple then $[\equiv]_I$ is improper for each ideal I of R.

(8.2) EXAMPLE. [Poyatos, 1977, 1980] An ideal I of a semiring R is **additively absorbing** if and only if $a+r \in I$ for all $0 \neq a \in I$ and $r \in R$. Thus, for example, if c is a strongly-infinite element of R then $\{0, c\}$ is an additively-absorbing ideal of R. An additively-absorbing ideal I of a semiring R defines a relation $\sim_{(I)}$ on R by setting $r \sim_{(I)} r'$ if and only if $r = r'$ or both r and r' belong to I. This is easily seen to be a congruence relation. Note that the family of additively-absorbing ideals of R

is closed under taking arbitrary intersections and unions. If I and H are additively-absorbing ideals of an entire semiring R then $\{0\}\cup\{a+b \mid 0 \neq a \in I \text{ and } 0 \neq b \in H\}$ is also an additively-absorbing ideal of R.

(8.3) EXAMPLE. Let R be an austere commutative semiring and define a relation ζ on R by the condition that $a\ \rho\ b$ if and only if $a = b = 0$ or $ab \neq 0$. By Proposition 6.25 it is easily seen that this is indeed a congruence relation on R whenever R is zerosumfree. Moreover, if R has more than two elements and if ρ is a congruence relation on R, then R must be zerosumfree. Indeed, in this situation, it is the unique maximal proper element of $Cong(R)$. See [Adhikari, Golan & Sen, 1994].

(8.4) EXAMPLE. If R is a simple semiring recall that, by Proposition 4.7, each element a of R defines a subsemiring $S(a)$ of R. Define a relation ρ on R by setting $a\ \rho\ b$ if and only if $S(a) = S(b)$. Clearly this is an equivalence relation; we claim that it is a congruence relation as well. Indeed, let a, b, c, and d be elements of R satisfying $a\ \rho\ c$ and $b\ \rho\ d$. If $0 \neq r \in R$ then

$$\begin{aligned} r \in S(a+b) &\Leftrightarrow r+a+b=1 \Leftrightarrow r+a \in S(b) \\ &\Leftrightarrow r+a \in S(d) \Leftrightarrow r+a+d=1 \\ &\Leftrightarrow r+d \in S(a) \Leftrightarrow r+d \in S(c) \\ &\Leftrightarrow r+c+d=1 \Leftrightarrow r \in S(c+d) \end{aligned}$$

and so $S(a+b) = S(c+d)$. Therefore $a+b\ \rho\ c+d$. Moreover, by Proposition 4.7(2), $S(ab) = S(a) \cap S(b) = S(c) \cap S(d) = S(cd)$ and so $ab\ \rho\ cd$, establishing our claim.

Let ρ be a congruence relation on R and, for each element r of R, let r/ρ be the equivalence class of r with respect to this relation. Set R/ρ equal to $\{r/\rho \mid r \in R\}$. If ρ is proper we can define a semiring structure on R/ρ by setting $(r/\rho)+(r'/\rho) = (r+r')/\rho$ and $(r/\rho)(r'/\rho) = rr'/\rho$. Note that, for any congruence relation ρ, at most one of the classes r/ρ can contain an ideal. Indeed, assume that r/ρ and r'/ρ contain ideals I and H respectively. Without loss of generality we can assume that, in fact, $r \in I$ and $r' \in H$. Then $rr' \in IH \subseteq I \cap H \subseteq (r/\rho) \cap (r'/\rho)$, which implies that this intersection is nonempty and so $r/\rho = r'/\rho$.

(8.5) APPLICATION. J. M. Anderson [1993] has formulated Mikusiński's operational calculus in a semiring context. Let S be the set of all continuous functions from $\mathbb{R}^+$ to $\mathbb{C}$ on which we have the operations of addition and convolution:

$$f * g \colon t \mapsto \int_0^t f(t-u)g(u)du$$

for all $t \in \mathbb{R}^+$. Then $(S,+,*)$ is a commutative and associative algebra over $\mathbb{C}$. If $h \in S$ is the constant function $t \mapsto 1$ then $h * f$ is the integral of f, for we see that $h * f \colon t \mapsto \int_0^t f(u)du$. For $n \geq 1$, let h^{*n} denote $h * \cdots * h$ (n times). Then it is easy to see, by induction, that $h^{*n} \colon t \mapsto \frac{1}{(n-1)!}t^{n-1}$. Let $H = \{h^{*n} \mid n \in \mathbb{N}\}$. The Little

Titchmarsh Theorem asserts that if $0 \neq f \in S$ and $k \in H$ then $k * f \neq 0$. Now consider the set $S \times H$ on which we have operations $\oplus$ and $\circledast$ defined as follows:

$$(f,k) \oplus (f',k') = (f * k' + f' * k, k * k')$$
$$(f,k) \circledast (f',k') = (f * f', k * k')$$

Then $(S \times H, \oplus, \circledast)$ is a semiring on which we can define a congruence relation ρ by setting $(f,k)\ \rho\ (f',k')$ if and only if $f * k' = f' * k$. Denote the factor semiring $(S \times H)/\rho$ by S_H and write the equivalence class of (f,k) as f/k. Then the identity element of S_H is k/k for any $k \in H$. Moreover, we have a monic function $\varphi: S \rightarrow S_H$ given by $\varphi: f \mapsto (f * k)/k$. This function is not surjective since $k/k \notin im(\varphi)$. The elements of $S_H \setminus im(\varphi)$ are called **hyperfunctions**. Multiplication by $s = h/h^{*2}$ in S_H behaves like differentiation, and so $s * f$ is the **generalized derivative** of $f \in S$.

(8.6) EXAMPLE. Let R be an additively-idempotent semiring and let M be a group of order 2. Let $R' = R[M]$ be the semiring discussed in Example 3.3. Define a relation ρ on R' by setting

$$(a,b)\ \rho\ (c,d) \Leftrightarrow \begin{cases} a+d = b+c & \text{if } a \neq b \text{ and } c \neq d \\ (a,b) = (c,d) & \text{otherwise} \end{cases}.$$

Then ρ is a congruence relation on R'. Baccelli et al. [1992] consider this relation for the special case of the schedule algebra $R = (\mathbb{R} \cup \{-\infty\}, max, +)$ and call $S = R'/\rho$ the **symmetrized algebra** over R. In particular, they distinguish three sorts of elements of S:

(1) classes of the form $(a,-\infty)/\rho = \{(a,b) \mid b < a\}$, called **positive elements** of $S/$;
(2) classes of the form $(-\infty,b)/\rho = \{(a,b) \mid a < b\}$, called **negative elements** of S;
(3) classes of the form $(a,a)/\rho = \{(a,a)\}$, called **balanced elements** of S.

They then associate each element a with the class $(a,-\infty)/\rho$. Note that $U(S)$ consists precisely of all non-balanced (i.e. positive or negative) classes in S.

We have already noted that an ideal I of a semiring R defines a congruence relation $\equiv_I$ on R. We denote the set of all such equivalence classes of elements of R by R/I and the equivalence class of an element r of R by r/I. Note that r/I is not necessarily equal to the set $r + I = \{r + a \mid a \in I\}$ but surely contains it! Then R/I is a semiring if $\equiv_I$ is proper, i.e. if $0/I \neq R$. A semiring of the form R/I is called the **Bourne factor semiring** of R by I. If $\varnothing \neq A \subseteq R$, then we set $A/I = \{r/I \mid r \in A\}$. By Proposition 6.54, we note that $R/I = R/(0/I)$ for each ideal I of R. Thus, in taking Bourne factor semirings we can always assume that we are doing so modulo a subtractive ideal. In a similar manner, if I is an ideal of R satisfying $0[/]I \neq R$ then $R[/]I$ is a semiring, called the **Iizuka factor semiring** of R by I. For any semiring R, we note that $Z(R) = 0[/]\{0\}$ and so the congruence relation $[\equiv]_{\{0\}}$ is proper if and only If R is nonzeroic.

If I is an additively-absorbing ideal of a semiring R then $R/\sim_{(I)}$ is just $(R \setminus I) \cup \{0,c\}$, where c is infinite.

It is in fact often convenient to represent a given semiring in the form R/ρ, where R is a semiring which is in some sense "simpler" than the one we are interested in studying. This is done, for example, in [Pierce, 1972] where the semiring $\mathcal{D}$, as defined in Example 1.9, is represented in the form $\mathbb{N}[M]/\rho$, where M is a suitable monoid.

(8.7) EXAMPLE. If Ω is a nonempty set then a **filter of subsets** $\mathcal{F}$ of Ω is a nonempty family of subsets of Ω satisfying the following conditions:

(1) $\varnothing \notin \mathcal{F}$;
(2) If $\Lambda \in \mathcal{F}$ and $\Lambda' \supseteq \Lambda$ then $\Lambda' \in \mathcal{F}$;
(3) If $\Lambda, \Lambda' \in \mathcal{F}$ then $\Lambda \cap \Lambda' \in \mathcal{F}$.

If $R = \times_{i\in\Omega} R_i$ is the product of a family of semirings indexed by a nonempty set Ω and if $\mathcal{F}$ is a filter of subsets of Ω, then we can define a congruence relation ρ on R by setting $\langle r_i \rangle \ \rho \ \langle s_i \rangle$ if and only if $\{i \in \Omega \mid r_i = s_i\} \in \mathcal{F}$. The semiring R/ρ is called the $\mathcal{F}$**-reduced product** of $\{R_i \mid i \in \Omega\}$ and is usually denoted by $R/\mathcal{F}$. If $\mathcal{F} = \{\Omega\}$ then $R/\mathcal{F} = R$.

Maximal filters of subsets of Ω are called **ultrafilters** and it is a well-known result in set theory that any filter of subsets of Ω is contained in an ultrafilter of subsets of Ω. If $\mathcal{F}$ is an ultrafilter of subsets of Ω then the semiring $R/\mathcal{F}$ is called an **ultraproduct** of the semirings R_i.

(8.8) EXAMPLE. If ρ is the relation on a semiring R defined by the condition that $a \ \rho \ b$ if and only if there exist elements r and s of R satisfying $a + r = b$ and $b + s = a$ then it is easy to verify that ρ is indeed a congruence relation. Note that $a \ \rho \ 0$ if and only if $a \in V(R)$ and so ρ is improper if and only if R is a ring. If ρ is trivial then the semiring R is **reduced**. If ρ is proper then the semiring R/ρ is the **reduced factor semiring** of R. From Proposition 4.22 it is clear that a semiring is clear if and only if its reduced factor semiring is clear.

(8.9) EXAMPLE. Let R be a commutative semiring and let Y be a nonempty family of strong prime ideals of R. Then we can define the relation ρ on R setting $a \ \rho \ b$ if and only if, for each $H \in Y$, both a and b either belong to H or do not belong to H. This is clearly an equivalence relation. Moreover, if $a \ \rho \ a'$ and $b \ \rho \ b'$ then for each H in Y we have $a + b \in H \Leftrightarrow a, b \in H \Leftrightarrow a', b' \in H \Leftrightarrow a' + b' \in H$ and similarly $ab \in H \Leftrightarrow a \in H$ or $b \in H \Leftrightarrow a' \in H$ or $b' \in H \Leftrightarrow a'b' \in H$ and so ρ is in fact a congruence relation on R.

A special case of this is considered in [Słowikowski & Zawadowski, 1955]. Let R be the semiring of all continuous functions from a bicompact topological space X to $\mathbb{R}^+$. This is a commutative Gel'fand semiring and so every maximal ideal of R is strong and prime. Indeed, the maximal ideals of R are all of the form $\{\varphi \in R \mid \varphi(x_0) = 0\}$ for some element x_0 of X. There exists a bijective correspondence θ between R/ρ and the lattice of all open subsets of X given by $\varphi/\rho \ \mapsto \{x \in X \mid \varphi(x) > 0\}$.

(8.10) EXAMPLE. [Vandiver, 1939] Let $1 < h < k$ be natural numbers and define a relation ρ on $\mathbb{N}$ as follows:

(1) If $i < h$ and $j \in \mathbb{N}$ then $i \ \rho \ j$ if and only if $i = j$;
(2) If $i \geq h$ and $j \in \mathbb{N}$ then $i \ \rho \ j$ if and only if $i \equiv j \pmod{k - h + 1}$.

Then ρ is a congruence relation and the semiring $\mathbb{N}/\rho$ is not cancellative.

The set of all congruences ρ on a semiring R such that R/ρ is a semilattice is studied in [Rodriquez, 1980].

(8.11) PROPOSITION. *If R is a commutative semiring having no nontrivial proper congruence relations then either $R = \mathbb{B}$ or R is a field.*

PROOF. If R has only two elements then either $R = \mathbb{B}$ or R is the field $\mathbb{Z}/(2)$ and so, in this case, the result is surely true. Hence we need only consider the case of R having more than two elements.

We first note that R is multiplicatively cancellative. Indeed, every element a of R defines a congruence relation ρ_a on R by $r \ \rho_a \ r'$ if and only if $ar = ar'$. This congruence relation is trivial when the element a is multiplicatively cancellable and is proper if $a \neq 0$. Since R has no nontrivial proper congruence relations, we see that it must be multiplicatively cancellative. This implies that $R \setminus \{0\}$ is a submonoid of $(R, \cdot)$.

Now assume that R is zerosumfree. Then $R \setminus \{0\}$ is closed under both addition and multiplication and so we have a nontrivial proper congruence relation ρ on R defined by the condition that $a \ \rho \ b$ if and only if $a = b$ or a and b are both nonzero. This is a contradiction and so R cannot be zerosumfree. Then $V(R)$ contains 0 and at least one nonzero element. Moreover, $V(R)$ is an ideal of R. The congruence relation $\equiv_{V(R)}$ defined on R is not trivial and hence, by assumption, it must be improper. In particular, $1 \equiv_{V(R)} 0$ and so there exists an element b of V(R) satisfying $1 + b = 0$. For any $r \in R$, this means that $r + br = (1 + b)r = 0$ and so every element of R has an additive inverse, proving that R is in fact a ring.

If $0 \neq a \in R$ and if $I = (a)$ is the principal ideal of R generated by a then the congruence relation $\equiv_I$ is nontrivial and so must be improper. In particular, $1 \equiv_I 0$ and so $1 \in I$, proving that a is a unit. Hence R is a field. □

By Proposition 8.11, we see that a division semiring or even a semifield may have proper nontrivial congruence relations: just consider $\mathbb{Q}^+$. If ρ is a proper congruence relation on a division semiring R, then R/ρ is surely again a division semiring.

We now turn to considering Bourne factor semirings.

(8.12) PROPOSITION. *If I is a subtractive maximal ideal of a commutative semiring R then R/I is a semifield.*

PROOF. Assume that $0/I \neq a/I \in R/I$. If $a^2 \in I$ then, by commutativity, $(a)^2 \subseteq I$ and so, by Corollary 7.13, we have $a \in I$, which contradicts the choice of a. Since $a^2 \in (a)$, this implies that $I \subset I + (a)$ and so, by the maximality of I, we have $R = I + (a)$. Hence there exist an element b of I and an element r of R such that $1 = b + ra$ and so $1/I = ra/I = (r/I)(a/I)$. Thus $a/I \in U(R/I)$, proving that R/I is a semifield. □

(8.13) PROPOSITION. *If I is an ideal of a semiring R satisfying the condition that $R \neq H = 0[/]I$ then the semirings $R[/]H$ and R/H are plain.*

PROOF. Let $r[/]H \in Z(R[/]H)$. Then there exists an element a of R such that $(r + a)[/]H = a[/]H$ and so there exist elements h and h' of H and an element s

of R with $r+h+(a+s)=r+a+h+s=a+h'+s=h'+(a+s)$, proving that $r[/]H=0[/]H$.

Let $r/H \in Z(R/H)$. Then there exists an element a of R with $(r+a)/H=a/H$ and so there exist elements h and h' of H satisfying $r+a+h=a+h'$. Since h belongs to H, there exist elements c and c' of I and s of R satisfying $h+c+s=c'+s$. Similarly, there exist elements d and d' of I and s' of R satisfying $h'+d+s'=d'+s'$. Therefore

$$\begin{aligned} r+(c'+d)+(a+s+s') &= r+a+c'+d+s+s' \\ &= r+a+h+c+s+d+s' \\ &= a+h'+c+s+d+s' \\ &= a+c+s+d'+s' \\ &= (c+d')+(a+s+s'), \end{aligned}$$

proving that $r[/]I=0[/]I$. Therefore $r/H=0/H$. □

(8.14) COROLLARY. *If R is a nonzeroic semiring then $R/Z(R)$ is plain.*

PROOF. This is a direct consequence of Proposition 8.13. □

(8.15) PROPOSITION. *If R is a yoked semiring then a subtractive ideal I of R contains $Z(R)$ if and only if R/I is cancellative.*

PROOF. Assume that $Z(R) \subseteq I$ and that $a/I+b/I=a/I+c/I$ in R/I. Then there exist elements d and d' of I satisfying $a+b+d=a+c+d'$ in R. Since R is a yoked semiring, there exists an element r of R satisfying $b+d+r=c+d'$ or $c+d'+r=b+d$. In the first case, we have $a+c+d'=a+b+d+r=a+c+d'+r$ and so $r \in Z(R) \subseteq I$. Thus $b+(d+r)=c+d'$ implies that $b \equiv_I c$ and so $b/I=c/I$. The second case yields the same result by a similar argument.

Conversely, assume that R/I is cancellative. If $r \in Z(R)$ then there exists an element a of R satisfying $r+a=a$ and so $r/I+a/I=a/I$. Therefore $r/I=0/I$ and so $r \in I$. □

(8.16) PROPOSITION. *If R is a cancellative semiring then R/I is cancellative for every ideal I of R.*

PROOF. If R is a cancellative semiring then R is plain and so this is an immediate consequence of Proposition 8.15. □

The following construction, found in [Bourne, 1962] and [Bleicher & Bourne, 1965], shows how to construct a ring R^{Δ} from any given nonzeroic semiring R. In a later chapter, we will show how to further construct a canonical morphism of semirings from R to R^{Δ}.

Let R be a semiring and let $S=R\times R$. Define operations of addition and multiplication on S by $(a,b)+(c,d)=(a+c,b+d)$ and $(a,b)(c,d)=(ac+bd,ad+bc)$ for all $a,b,c,d \in R$. These operations turn S into a semiring with additive identity $(0,0)$ and multiplicative identity $(1,0)$. If the semiring R is commutative, so is S. (Indeed, this is just the semiring $R[M]$, where M is a group of order 2; refer to Example 3.3.)

Now set $D = \{(a,a) \mid a \in R\}$. Clearly D is an ideal of S. We claim that $0/D \neq S$ if and only if R is nonzeroic. Indeed, if R is zeroic then there exists an element $r \in R$ such that $1 + r = r$ and so $(1,0) + (r,r) \in D$. Therefore, $(1,0) \in 0/D$. Since $(1,0)$ is the multiplicative identity of S, this implies that $S = 0/D$. Conversely, suppose that $S = 0/D$. Then there exists an element r of R such that $(1,0) + (r,r) = (r,r)$ and so $1 + r = r$, implying that $1 \in Z(R)$ and hence $Z(R) = R$. We also note that is a subtractive ideal of S if and only if the semiring R is cancellative and that $(a,b)/D = (c,d)/D$ if and only if there exist elements r and r' of R such that $(a,b) + (r,r) = (c,d) + (r',r')$. That is to say, $(a,b)/D = (c,d)/D$ if and only if there exist elements r and r' of R satisfying $a + r = c + r'$ and $b + r = d + r'$. In particular, $(0,0)/D = \{(a,b) \mid$ there exists an element r such that $a+r = b+r\} = \{(a,b) \mid a[\equiv]_{\{0\}}b\}$.

In particular, we see that If R is nonzeroic then S/D is a semiring. We claim that in this case S/D is in fact a ring. Indeed, if $(a,b) \in S$ then $(a,b)/D + (b,a)/D = (a+b, a+b)/D = (0,0)/D$ so $V(S/D) = S/D$ and hence S/D is a ring, which we will call the **ring of differences** of the semiring R. We will denote this ring by R^{Δ}. For the analogous construction for topological semirings, see [Botero & Weinert, 1971]. Note that $(a,b)/D = (c,d)/D$ if and only if there exists an element r'' of R satisfying $a + d + r'' = b + c + r''$. Indeed, if $(a,b)/D = (c,d)/D$ and r, r' are as above, take $r'' = r + r'$. Conversely, if such an element r'' exists, take $r = d + r''$ and $r' = b + r$". Thus the construction given here is the same as the one given in [Poyatos, 1971].

If H is a [left, right] ideal of R then $H^{\Delta} = \{(a,b)/D \mid a,b \in H\}$ is a [left, right] ideal of R^{Δ}. Conversely, if I is a [left, right] ideal of R^{Δ} then $\{a \in R \mid (a,0)/D \in I\}$ is a [left, right] ideal of R.

The above construction can be generalized. If I is an ideal of R and if $S' = R \times I$, then S' is a subsemiring of the semiring S defined above. Moreover, $D' = S' \cap d$ is an ideal of S'. If $0/D' \neq S'$ then we can construct the semiring $r^{\Delta I} = S'/D'$. In general, this is not a ring.

(8.17) EXAMPLE. [H. E. Stone, 1972] Let S be the ring of all functions from $\mathbb{N}$ to $\mathbb{Q}$ with the operations of elementwise addition and multiplication, and let R be the subsemiring of S consisting of the zero function and all functions f satisfying the condition that $f(i) > 0$ for all $i \in \mathbb{N}$. Then $S = R^{\Delta}$.

(8.18) EXAMPLE. [H. E. Stone, 1977] If R is a cancellative semiring and n is a positive integer then, as remarked in Example 4.19, $S = \mathcal{M}_n(R)$ is also cancellative. Moreover, $S^{\Delta} = \mathcal{M}_n(R^{\Delta})$.

Unlike the situation with rings, it is usally not very easily to visualize the structure of the Bourne factor ring. Under certain circumstances, however, it is easier to do so. We will now describe one such circumstance which generalizes the situation of rings. An ideal I of a semiring R is **partitioning** if and only if there exists a nonempty subset $Q(I)$ of R such that $R_{Q(I)} = \{q + I \mid q \in Q(I)\}$ is a partition of R into pairwise-disjoint subsets.

(8.19) EXAMPLE. If R is a ring then every ideal of R is partitioning.

(8.20) EXAMPLE. [Allen, 1969] If m is a positive integer then the ideal $m\mathbb{N}$ of the semiring $\mathbb{N}$ is partitioning. The ideal $\mathbb{N} \setminus \{1\}$ of $\mathbb{N}$ is not partitioning.

If I is a partitioning ideal of a semiring R and if $r \in R$ then it is easy to verify that there exists a unique element q of $Q(I)$ such that $r + I \subseteq q + I$. Thus we see that if I is a partitioning ideal of a semiring R there exists a surjective function $\varphi_I : R \to R_{Q(I)}$ which assigns to each element r of R the unique element $q + I$ of $R_{Q(I)}$ such that $r + I \subseteq q + I$.

(8.21) PROPOSITION. *If I is a partitioning ideal of a semiring R then there exists a unique element $q_0 \in Q(I) \cap V(R)$ satisfying $I = q_0 + I$.*

PROOF. Since I is partitioning, there exists a unique element q_0 of $Q(I)$ such that $0 \in q_0 + I$. Thus there exists an element a_0 of I satisfying $a_0 + q_0 = 0$, which shows that $q_0 \in V(R)$ as well.

If $b \in I$ and $b \in q + I$ for some $q \in Q(I)$ then there exists an element a of I satisfying $q + a = b = b + 0 = q_0 + (b + a_0)$ and so $b \in (q + I) \cap (q_0 + I)$. Since I is partitioning, this implies that $q = q_0$ and so $b \in q_0 + I$. Hence $I \subseteq q_0 + I$. Since I is partitioning, there exist an element q of $Q(I)$ and an element c of I such that $q_0 + q_0 = q + c$. Then

$$q_0 = q_0 + (q_0 + a_0) = q_0 + q + c + a_0 = q + c \in q + I.$$

Thus $(q_0 + I) \cap (a + I) \neq \varnothing$. This implies that $q = q_0$ and so $q_0 + q_0 = q_0 + c$. Therefore

$$q_0 + I = q_0 + 0 + I = q_0 + q_0 + a_0 + I = q_0 + c + a_0 + I = c + I \subseteq I$$

and so $q_0 + I = I$, as desired. □

(8.22) PROPOSITION. *Let I be a partitioning ideal of a semiring R. Then $r \equiv_I r'$ if and only if $\varphi_I(r) = \varphi_I(r')$.*

PROOF. If $r \equiv_I r'$ then there exist elements a and a' of I such that $r + a = r' + a'$. Hence $(r + I) \cap (r' + I) \neq \varnothing$. This implies that $\varphi_I(r) \cap \varphi_I(r') \neq \varnothing$. Since $R_{Q(I)}$ is a partition of R, this means that $\varphi_I(r) = \varphi_I(r')$. Conversely, assume that $\varphi_I(r) = \varphi_I(r') = q + I$. Then there exist elements a and a' of I such that $r = q + a$ and $r' = q + a'$. Thus $r + a' = r' + a$ and so $r \equiv_I r'$. □

(8.23) COROLLARY. *Any partitioning ideal I of a semiring R is subtractive.*

PROOF. By Proposition 8.21, we know that there exists an element q_0 of $Q(I)$ satisfying $I = q_0 + I$. If a and b are elements of R satisfying $a + b, b \in I$ then $a + b \equiv_I a$ and so, by Proposition 8.22, $q_0 = \varphi_I(a + b) = \varphi_I(a)$ so $a \in q_0 + I = I$, proving that I is subtractive. □

By Proposition 8.22, we see that, if I is a partitioning ideal of a semiring R, the function φ_I induces a bijective correspondence between R/I and structure on $R_{Q(I)}$ under the operations $\oplus$ and $\odot$ defined as follows:

(1) $(q + I) \oplus (q' + I) = q'' + I$, where q'' is the unique element of $Q(I)$ such that $(q + q') + I \subseteq q'' + I$;

(2) $(q + I) \odot (q' + I) = q'' + I$, where q'' is the unique element of $Q(I)$ such that $qq' + I \subseteq q'' + I$.

Note that the set $Q(I)$ is not uniquely determined by the partitioning ideal I. However, the above result shows that if $Q(I)$ and $Q'(I)$ are two possible such sets then the semirings $R_{Q(I)}$ and $R_{Q'(I)}$ are isomorphic, and so it is immaterial which of them we choose to work with.

9. MORPHISMS OF SEMIRINGS

If R and S are semirings then a function $\gamma: R \rightarrow S$ is a **morphism of semirings** if and only if:

(1) $\gamma(0_R) = 0_S$;
(2) $\gamma(1_R) = 1_S$; and
(3) $\gamma(r + r') = \gamma(r) + \gamma(r')$ and $\gamma(rr') = \gamma(r) \cdot \gamma(r')$ for all $r, r' \in R$.

A function γ satisfying conditions (1) and (3) is a **morphism of hemirings**. A morphism of semirings [hemirings] which is both injective and surjective is called an **isomorphism**. If there exists an isomorphism between semirings [hemirings] R and S we write $R \cong S$. If $\gamma: R \rightarrow S$ is a morphism of semirings [resp. hemirings] then $im(\gamma) = \{\gamma(r) \mid r \in R\}$ is a subsemiring [resp. subhemiring] of S.

(9.1) EXAMPLE. [Heatherly, 1974] Let R be a semiring and let $End_0(R)$ be the set of all endomorphisms α of the commutative monoid $(R, +)$ satisfying $\alpha(0) = 0$ which, as we have already noted in Example 1.14, is also a semiring. For each $r \in R$, let $\beta_r: R \rightarrow R$ be the function defined by $\beta_r: r' \mapsto rr'$. Then $\beta_r \in End_0(R)$ for each element r of R and the map $r \mapsto b_r$ is a morphism of semirings. Indeed, this morphism is injective since $\beta_r = \beta_{r'}$ implies that $r = \beta_r(1) = \beta_{r'}(1) = r'$.

(9.2) EXAMPLE. The semiring $(\mathbb{R}^+, max, \cdot)$ is isomorphic to the schedule algebra $(\mathbb{R} \cup \{-\infty\}, max, +)$ via the map $a \mapsto ln(a)$. Similarly, the semiring $(\mathbb{R} \cup \{\infty\}, min, +)$ is isomorphic to the schedule algebra via the map $a \mapsto -a$.

(9.3) APPLICATION. Let $R = \mathbb{R} \cup \{-\infty\}$ and let a be a positive real number. Define operations of $\oplus_a$ and $\odot_a$ on R by setting $r \oplus_a r' = a \cdot ln(e^{r/a} + e^{r'/a})$, where we take $e^{-\infty} = 0$, and $r \odot_a r' = r + r'$. Then $(R, \oplus_a, \odot_a)$ is a semiring and we have a morphism of semirings $\gamma: \mathbb{R}^+ \rightarrow (R, \oplus_a, \odot_a)$ given by $c \mapsto a \cdot ln(c)$. Note that

$$lim_{a \rightarrow 0} r \oplus_a r' = max\{r, r'\}$$

for all $r, r' \in R$. This construction is used to reduce problems in probability calculus to problems in optimal control. See [Akian, Quadrat & Viot, 1994] for further details.

(9.4) EXAMPLE. For any semiring R we have a canonical morphism from the semiring $\mathbb{N}$ to R given by $n \mapsto n1_R$. Note that the image of this morphism is a subsemiring of R which is clearly contained in every subsemiring of R. Thus it is just the basic subsemiring $B(R)$ of R, and that it is contained in $C(R)$.

(9.5) EXAMPLE. Let k be a positive integer and let M be the monoid $(\mathbb{N}^k, +)$. The semiring $sub(M)$, as defined in Example 1.10, is additively-idempotent. Moreover, Shubin [1992] has shown that it is a free additively-idempotent semiring with k generators, in the sense that if R is any additively-idempotent semiring and if $r_1, \ldots, r_k \in R$ then there exists a unique morphism of semirings $\gamma: sub(M) \rightarrow R$ satisying $\gamma(\{a_i\}) = r_i$ for all i, where $a_i = [0, \ldots, 0, 1, 0, \ldots, 0]$ (the 1 being in the ith position). Moreover, any other additively-idempotent semiring with k generators having this property is isomorphic to $sub(M)$.

(9.6) EXAMPLE. If R is a multiplicatively-cancellative additively-idempotent commutative semiring and if n is a positive integer then, by Proposition 4.43, the function $\gamma_n: a \mapsto a^n$ from R to itself is a morphism of semirings which, by Proposition 4.44, is in fact monic. This happens, for example if R is an additively-idempotent semifield, such as the schedule algebra.

(9.7) PROPOSITION. *If R is a semiring then $B(R)$ is isomorphic to $\mathbb{N}$ or to a semiring of the form $B(n, i)$ for some $n > 1$ and $n > i \geq 0$.*

PROOF. Let $\gamma: \mathbb{N} \rightarrow R$ be the morphism of semirings given by $\gamma: n \mapsto n1_R$. As we have already noted, $im(\gamma) = B(R)$. Three posibilities exist:

(1) The map γ is injective. In this case, $B(R)$ is isomorphic to $\mathbb{N}$.

(2) The map γ is not injective and there exists a positive integer k such that $\gamma(k) = 0_R$. Let n be the least such positive integer. Then one checks that $B(R)$ is isomorphic to $B(n, 0) = \mathbb{Z}/(n)$.

(3) The map γ is not injective and $\gamma(k) \neq 0_R$ for all $k > 0$, but there exist $m \neq m' \in \mathbb{N}$ such that $\gamma(m) = \gamma(m')$. Let n be the least positive integer for which there exists an integer $n > i > 0$ such that $\gamma(n) = \gamma(i)$. Then it is straightforward to check that $B(R)$ is isomorphic to $B(n, i)$. □

We will say that the **characteristic** of a semiring R equals 0 if $B(R)$ is isomorphic to $\mathbb{N}$ and equals (n, i) if $B(R)$ is isomorphic to $B(n, i)$. Note that if R has characteristic $B(n, 0)$ for some $n > 1$ then $1_R \in V(R)$ and so R is in fact a ring. Thus, as was observed in Chapter 3, a Gel'fand semiring must have characteristic 0 or characteristic $(2, 1)$.

We also can extend Example 9.4 by noting that if R is a subhemiring of a semiring S then we have a morphism of semirings γ from the Dorroh extension $R \times \mathbb{N}$ of R by $\mathbb{N}$ to S defined by $\gamma: (r, n) \mapsto r + n1_S$, the image of which is clearly the smallest subsemiring of S containing R as a subhemiring.

(9.8) EXAMPLE. If R is any semiring and M is a monoid with identity element e then we have an injective morphism of semirings $\gamma: R \rightarrow R[M]$ given by

$$\gamma(r): m \mapsto \begin{cases} r & \text{if } m = e \\ 0_R & \text{otherwise} \end{cases}.$$

(9.9) EXAMPLE. If R is a semiring then we have already seen that $sub(R)$ has the structure of a semiring, with addition and multiplication defined by $A + B = \{a + b \mid a \in A; b \in B\}$ and $AB = \{ab \mid a \in A; b \in B\}$. The function from R to $sub(R)$ defined by $a \mapsto \{a\}$ is then surely a morphism of semirings.

(9.10) EXAMPLE. Let R be a semiring. The function $\gamma: R \to \mathbb{B}$ defined by

$$\gamma: r \mapsto \begin{cases} 0 & \text{if } r = 0 \\ 1 & \text{otherwise.} \end{cases}$$

is a surjective morphism of semirings if and only if R is both zerosumfree and entire. In particular, such a morphism exists from $\mathbb{N}$, $\mathbb{Q}^+$, or $\mathbb{R}^+$ to $\mathbb{B}$. Conversely, the function $\delta: \mathbb{B} \to R$ from $\mathbb{B}$ to a semiring R defined by $\delta(0) = 0$ and $\delta(1) = 1$ is a morphism of semirings precisely when $1 + 1 = 1$ in R, i.e. precisely when R is additively idempotent. Note that in this case δ is the only possible morphism from $\mathbb{B}$ to R and that it is injective. Moreover, it is straightforward to verify that if S is a subhemiring of an additively-idempotent semiring R which is not a subsemiring then δ induces an injective morphism of semirings from the Dorroh extension of S by $\mathbb{B}$ to R defined by $(s, i) \mapsto s + \delta(i)$. See [Haftendorn, 1979] for details.

This construction can be generalized for commutative semirings. Indeed, in such a situation the condition that R be zerosumfree and entire is equivalent to the condition that the ideal $\{0\}$ be strong and prime. Thus, more generally, if H is any ideal of a commutative semiring R which is both strong and prime, then H defines a surjective morphism of semirings $\gamma: R \to \mathbb{B}$ by $\gamma(r) = 0$ if $r \in H$ and $\gamma(r) = 1$ if $r \notin H$.

(9.11) EXAMPLE. Let $M = \mathbb{R}^k$, partially-ordered with the Pareto partial order and if $A \in sub(M)$, let $min(A)$ be the set of all minimal elements of the closure of A in M. Let $R = \{A \in sub(M) \mid A = min(A)\}$. Then the operations $\oplus$ and $\odot$ on R defined by $A \oplus B = min(A \cup B)$ and $A \odot B = min(A + B)$ define the structure of a semiring on R.

Now let $L = \{[b_1, \ldots, b_k] \in \mathbb{R}^k \mid \sum_{j=1}^{k} b_j = 0\}$, which is a submonoid of $(M, +)$. Let S be the semiring of all functions from L to the semiring $(\mathbb{R} \cup \{\infty\}, min, +)$ under the operations of pointwise addition and convolution $\langle + \rangle$. Let $f \in S$ be the function defined by

$$f: [b_1, \ldots, b_k] \mapsto max\{-b_1, \ldots, -b_k\}.$$

Then $f\langle + \rangle f = f$. The set $S' = \{g \in S \mid f\langle + \rangle g = g\langle + \rangle f = g\}$ is a subsemiring of S which is isomorphic to R.

For applications of these semirings to multicriteria optimization, refer to [Kolokol'tsov & Maslov, 1998].

A morphism from a semiring R to $\mathbb{B}$ is called a **character** of R. The set of all characters on a semiring R will be denoted by $char(R)$.

(9.12) EXAMPLE. Let R be a semiring and let S be the semiring $(sub(char(R)), \cup, \cap)$. For each $a \in R$, set $\chi(a) = \{\gamma \in char(R) \mid \gamma(a) = 1\}$. Then for $a, b \in R$ we have $\chi(a+b) = \chi(a) \cup \chi(b)$ and $\chi(ab) = \chi(a) \cap \chi(b)$. Moreover, $\chi(0) = \varnothing$ and $\chi(1) = char(R)$. Thus χ is a morphism of semirings. It is injective if and only if R is a bounded distributive lattice [Priestly, 1970].

(9.13) EXAMPLE. If A and B are nonempty sets and if $\theta: A \to B$ is a function then any morphism of semirings $\gamma: R \to S$ defines a morphism of semirings $\gamma^{\theta}: R^B \to S^A$ by $(\gamma^{\theta} f)(a) = \gamma(f(\theta(a)))$. In particular, if $A \subseteq B$ are nonempty sets and if R is a semiring then we have a canonical morphism of semirings $R^B \to R^A$ given by restriction of functions. Similarly, for each nonempty set A and each morphism of semirings $\gamma: R \to S$, the identity map on A induces a morphism of semirings $\gamma^A: R^A \to S^A$ given by $f \mapsto \gamma f$. Also, we have a morphism of semirings $\gamma\langle\langle A\rangle\rangle: R\langle\langle A\rangle\rangle \to S\langle\langle A\rangle\rangle$ for every nonempty set A. If $f \in R\langle\langle A\rangle\rangle$ has finite support then so does γf and so this morphism restricts to a morphism of semirings $\gamma\langle R\rangle: R\langle A\rangle \to S\langle R\rangle$.

If A is a set which is either finite or countably-infinite and if $\mathcal{M}(\_)$ is $\mathcal{M}_{A,r}(\_)$, $\mathcal{M}_{A,c}(\_)$, or $\mathcal{M}_{A,rc}(\_)$, then a morphism of semirings $\gamma: R \to S$ defines a morphism semirings $\mathcal{M}(\gamma): \mathcal{M}(R) \to \mathcal{M}(S)$ by $f \mapsto gf$. If M is a monoid with identity e then γ defines a morphism of semirings $\gamma[M]: R[M] \to S[M]$ by $f \mapsto gf$. We note that if $f \in R^M$ has finite support then so does $\gamma f \in S^M$.

In particular, if R is a zerosumfree entire semiring then there exists a character δ of R defined by $\delta(a) = 1$ if $a \neq 0$ and $\delta(0) = 0$. If A is a set which is either finite or countably-infinite and if $\mathcal{M}(\_)$ is $\mathcal{M}_{A,r}(\_)$, $\mathcal{M}_{A,c}(\_)$, or $\mathcal{M}_{A,rc}(\_)$, then the image of a matrix C in $\mathcal{M}(R)$ under $\mathcal{M}(\delta)$ is called the **pattern** of C.

(9.14) PROPOSITION. *If R is a commutative zerosumfree semiring then $char(R) \neq \varnothing$.*

PROOF. We have already noted in Chapter 5 that if R is zerosumfree then it has a maximal proper coideal D and that $1 \in D$. Set $I = R \backslash D$. We claim that I is an ideal of R. As before, we denote the smallest coideal of R containing a set A by $F(A)$. If $a, b \in I$ then, by the maximality of D, we have $F(D \cup \{a\}) = R = F(D \cup \{b\})$. Thus there must exist elements $d, e \in D$, elements $r, s \in R$, and positive integers h, k such that $a^h d + r = 0 = b^k e + s$. Since R is zerosumfree, this implies that $a^h d = b^k e = 0$ and so $(a+b)^{h+k} de = 0$. Therefore $F(D \cup \{a+b\}) = R$, whence $a + b \in I$. Similarly, if $a \in I$ and $r \in R$ then $a^h d = 0$ for some positive integer h and $d \in D$. Therefore $(ra)^h d = 0$ and so $F(D \cup \{ra\}) = R$, proving that $ra \in I$.

We now define the function $\gamma: R \to \mathbb{B}$ by $\gamma(r) = 1$ if and only if $r \in D$. The proof that this is indeed a morphism of semirings is immediate. □

(9.15) EXAMPLE. [Golan & Wang, 1996] The commutativity condition in Proposition 9.14 is necessary. Ineed, let R be the noncommutative semiring $\mathcal{M}_2(\mathbb{B})$ and let us assume that $\gamma \in char(R)$. We claim that $\gamma\left(\begin{bmatrix} 1 & 0 \\ 0 & 0 \end{bmatrix}\right) = 0$. Indeed, if $\gamma\left(\begin{bmatrix} 1 & 0 \\ 0 & 0 \end{bmatrix}\right) = 1$ then $\gamma\left(\begin{bmatrix} 0 & 0 \\ 1 & 0 \end{bmatrix}\right) = 0$ since $\begin{bmatrix} 1 & 0 \\ 0 & 0 \end{bmatrix}\begin{bmatrix} 0 & 1 \\ 0 & 0 \end{bmatrix} = \begin{bmatrix} 0 & 0 \\ 0 & 0 \end{bmatrix}$. Thus

$$\gamma\left(\begin{bmatrix} 1 & 0 \\ 1 & 0 \end{bmatrix}\right) = \gamma\left(\begin{bmatrix} 1 & 0 \\ 0 & 0 \end{bmatrix} + \begin{bmatrix} 0 & 0 \\ 1 & 0 \end{bmatrix}\right) = 1 + 0 = 1.$$

But $\begin{bmatrix} 1 & 0 \\ 1 & 0 \end{bmatrix} = \begin{bmatrix} 1 & 1 \\ 0 & 1 \end{bmatrix} \begin{bmatrix} 0 & 0 \\ 1 & 0 \end{bmatrix}$ and so

$$\gamma\left(\begin{bmatrix} 1 & 0 \\ 1 & 0 \end{bmatrix}\right) = \gamma\left(\begin{bmatrix} 1 & 1 \\ 0 & 1 \end{bmatrix}\right) \cdot \gamma\left(\begin{bmatrix} 0 & 0 \\ 1 & 0 \end{bmatrix}\right) = 0,$$

which is a contradixtion that establishes the claim. Similarly, we must have $\gamma\left(\begin{bmatrix} 0 & 0 \\ 0 & 1 \end{bmatrix}\right) = 0$. Therefore

$$\gamma\left(\begin{bmatrix} 1 & 0 \\ 0 & 1 \end{bmatrix}\right) = \gamma\left(\begin{bmatrix} 1 & 0 \\ 0 & 0 \end{bmatrix}\right) + \gamma\left(\begin{bmatrix} 0 & 0 \\ 0 & 1 \end{bmatrix}\right) = 0,$$

which is also a contradiction.

We now return to other examples of morphisms of semirings.

(9.16) EXAMPLE. [Loh & Teh, 1966/7] More generally, if $\gamma: R \to S$ is a morphism of semirings and if $\theta: M \to M'$ is a morphism of monoids satisfying the condition that $\theta^{-1}(m')$ is finite for each $m' \in M'$ then we have a morphism of semirings $\gamma[\theta]: R[M] \to S[M']$ defined as follows: $\gamma[\theta](f): m' \mapsto \sum\{\gamma f(m) \mid m \in \theta^{-1}(m')\}$ for each $f \in R[M]$ and $m' \in M'$.

(9.17) EXAMPLE. If R is a semiring and if A and B are nonempty sets then there exists an isomorphism of semirings $\gamma: R\langle\langle A\rangle\rangle^B \to R^B\langle\langle A\rangle\rangle$ defined as follows: if $f \in R\langle\langle A\rangle\rangle^B$, $w \in A^*$, and $b \in B$ then $\gamma(f)(w)(b) = f(b)(w)$.

(9.18) EXAMPLE. [Thornton, 1972] Define a topology on $\mathbb{N}$ by taking the following as open sets: $\varnothing$, $\mathbb{N}$, and $\{0, 1, \ldots, k\}$ for each $k \in \mathbb{N}$. Let X be a finite T_0-space and let $C(X)$ be the family of all continuous functions from X to $\mathbb{N}$. This is a subsemiring of $\mathbb{N}^X$ which, moreover, uniquely characterizes the topology on X. If Y is another finite T_0-space, then any morphism of semirings $\gamma: C(Y) \to C(X)$ is induced by a unique continuous function $\varphi_\gamma: X \to Y$. If γ is an isomorphism then $\varphi_g\gamma$ is a homeomorphism.

(9.19) EXAMPLE. Let R be a semiring, let A be a nonempty set, and let φ be a function from A to the center $C(R)$ of R. Then φ defines a function $\epsilon_\varphi: R\langle A\rangle \to R$ given by

$$\epsilon_\varphi: f \mapsto \sum \left\{f(a_1a_2 \cdot \ldots \cdot a_n)\varphi(a_1) \cdot \ldots \cdot \varphi(a_n) \;\middle|\; a_1a_2 \cdot \ldots \cdot a_n \in A^*\right\}$$

(which is well-defined since f has finite support). Indeed, ϵ_φ is a morphism of semirings, called the **φ-evaluation morphism**. In particular, if R is a semiring and if $r \in C(R)$ then there exists a morphism of semirings $\epsilon_r: R[t] \to R$ given by $\sum a_it^i \mapsto \sum a_ir^i$. The complexity of computing the evaluation morphism for commutative semirings has been considered in detail in [Jerrum & Snir, 1982].

If we consider the special case of the function $\varphi: A \to C(R)$ defined by $\varphi(a) = 0$ for all $a \in A$ then ϵ_φ, defined by $\epsilon_\varphi: f \mapsto f(\square)$, is the **augmentation morphism** on $R\langle A\rangle$. In fact, this function can be extended to a map from $R\langle\langle A\rangle\rangle$ to R given by $f \mapsto f(\square)$.

(9.20) EXAMPLE. Let R be a subsemiring of a semiring S, let I be an ideal of R, and let H be an ideal of S satisfying $I \subseteq R \cap H$. Then we have a canonical morphism of semirings $\gamma: R/I \to S/H$ defined by $\gamma: r/I \mapsto r/H$. This map is well-defined since if $r \equiv_I r'$ in R then $r \equiv_H r'$ in S. As a consequence of this, it is also easy to see that if $I \subseteq H$ are ideals of a semiring R then we have a canonical morphism of semirings $R^{\Delta I} \to R^{\Delta H}$. If $S = R$, this morphism is surjective. As a special case of this, we note that for any ideal H of a semiring R we have a surjective morphism of semirings from R to R/H given by $r \mapsto r/H$.

(9.21) EXAMPLE. Let R be the schedule algebra $(\mathbb{R} \cup \{-\infty\}, max, +)$. For each $k \in \mathbb{N}$, let $\gamma_k: R \to R$ be the map given by $\gamma_k(a) = ka$ for each $a \in R$. Clearly γ_k is a morphism from R to itself which is both injective and surjective and so is an isomorphism. If $S = (\mathbb{R} \cup \{\infty\}, min, +)$ then the function $\gamma: R \to S$ defined by $\gamma(a) = -a$ is an isomorphism of semirings. Now let $S = \mathbb{N} \cup \{-\infty\}$, which is a subsemiring of R. For each positive integer n, let X_n be the semiring given in Example 1.8 and let $\delta_n: S \to X_n$ be the function given by

$$\delta_n: i \mapsto \begin{cases} i & \text{if } i \leq n \\ n & \text{otherwise} \end{cases}.$$

Then δ_n is a surjective morphism of semirings for each n.

(9.22) EXAMPLE. Let R be the semiring $(\mathbb{R}^+ \cup \{\infty\}, min, +)$ and let S be the semiring $(\mathbb{I}, \max, \cdot)$. Then we have a morphism of semirings $\gamma: R \to S$ defined by $\gamma: r \mapsto 2^{-r}$ (where, by definition, $2^{-\infty} = 0$).

(9.23) EXAMPLE. Let $f: X \to Y$ be a continuous function between topological spaces. Let $(R, \cap, \cup)$ be the semiring of all closed subsets of X and let $(S, \cap, \cup)$ be the semiring of all closed subsets of Y (see Example 1.5). Then the function $\gamma_f: S \to R$ defined by $\gamma_f: a \mapsto f^{-1}(a)$ is a morphism of semirings. If R' is a basis for the semiring $(R, \cap, \cup)$ and if X' is a subspace of X then $\{X' \cap b \mid b \in R'\}$ is a basis for the semiring $(R'', \cap, \cup)$ of all closed subsets of X' and the function $\gamma: b \mapsto X' \cap b$ is a morphism of semirings from R' to R''.

(9.24) EXAMPLE. In Proposition 6.29 we saw that if R is a semiring then the set $ideal(R)$ is a semiring under the operations of addition and multiplication of ideals. Similarly, in Chapter 6, we defined the set $Zar(R)$ of subsets of $spec(R)$. This was the family of closed subsets for the Zariski topology on $spec(R)$ and so it too is a semiring if we take addition to be intersection and multiplication to be union. Moreover, the map $ideal(R) \to Zar(R)$ given by $I \mapsto V(I)$ is clearly a surjective morphism of semirings.

(9.25) EXAMPLE. If I is a nonzero ideal of a Dedekind domain R, then to each prime ideal $H \in spec(R)$ we can assign a natural number $n(I, H)$ such that $I = \prod\{H^{n(I,H)} \mid H \in spec(R)\}$, where, by convention, $H^0 = R$ for all $H \in spec(R)$. We furthermore define $n(R, H) = 0$ and $n((0), H) = \infty$ for all $H \in spec(R)$. For a fixed element H of $spec(R)$ and for ideals I and I' of R, we then have:

(1) $n(I + I', H) = min\{n(I, h), n(I', H)\}$;

(2) $n(I \cap I', H) = max\{n(I, H), n(I', H)\}$;
(3) $n(II', H) = n(I, H) + n(I', H)$.

Therefore, for each fixed prime ideal H of R we have a function $n(\_, H)$ which is both a surjective morphism from the semiring $(ideal(R), +, \cap)$ to the semiring $(\mathbb{N} \cup \{\infty\}, min, max)$ and a surjective morphism from the semiring $(ideal(R), +, \cdot)$ to the semiring $(\mathbb{N} \cup \{\infty\}, min, +)$. By allowing negative exponents as well, we can extend this map to corresponding morphisms from $fract(R)$ to $\mathbb{Z} \cup \{\infty\}$. See [Gilmer, 1972] for details.

(9.26) EXAMPLE. If A is a nonempty set then the function $supp: \mathbb{B}\langle\langle A\rangle\rangle \rightarrow sub(A^*)$ is an isomorphism of semirings.

(9.27) EXAMPLE. If $h, k \in BN$, let $h \sqcup k$ be the greatest common divisor of h and k. By Example 1.17, we see that $(\mathbb{N}, \sqcup, \cdot)$ is a semiring. Moreover, we have an isomorphism of semirings $\gamma: \mathbb{N} \rightarrow ideal(\mathbb{Z})$ defined by $\gamma: n \mapsto \mathbb{Z}n$.

(9.28) EXAMPLE. Let R be the commutative semiring $(\mathbb{N} \cup \{\infty\}, max, min)$. Then $ideal(R) = \{R, \mathbb{N}\} \cup \{K_r \mid r \in R\}$, where $K_r = \{a \in R \mid a \leq r\}$. The function $\gamma: R \rightarrow ideal(R)$ given by $\gamma: R \mapsto K_r$ is a morphism of semirings which is injective but not surjective since $\mathbb{N}$ is not in the image of γ.

(9.29) EXAMPLE. [Cao, Kim & Roush, 1984] If R is a simple semiring for which there exists a positive integer n satisfying the condition that $r^n = r^{n+1}$ for all $r \in R$ then, by Proposition 4.9, we saw that $(I^{\times}(R), +, \odot)$ is a commutative simple semiring, where $\odot$ is the operation defined by $a \odot b = (ab)^n$. Moreover, it is easy to see that the function $\gamma: r \mapsto r^n$ is a morphism of semirings from R to $I^{\times}(R)$.

(9.30) EXAMPLE. Let X and Y be topological spaces and let R and S be the semirings of all closed subsets of X and Y respectively. If $R \cong S$ it does not necessarily follows that X and Y are homeomorphic. A sufficient condition for this to happen is that X and Y satisfy the T_1 separation axiom. The problem of when an isomorphism between R and S implies the existence of a homeomorphism between X and Y is studied in detail in [Thron, 1962].

We now begin considering some properties of morphisms of semirings.

(9.31) PROPOSITION. *If $\gamma: R \rightarrow S$ is a morphism of semirings then $\gamma(comp(R)) \subseteq comp(S)$.*

PROOF. If $a \in comp(R)$ then $\gamma(a) + \gamma(a^{\perp}) = \gamma(a + a^{\perp}) = \gamma(1_R) = 1_S$ while $\gamma(a)\gamma(a^{\perp}) = \gamma(aa^{\perp}) = \gamma(0_R) = 0_S$ and, similarly, $\gamma(a^{\perp})\gamma(a) = 0_S$. Thus $\gamma(a)$ is complemented, with $\gamma(a)^{\perp} = \gamma(a^{\perp})$. □

If $\{R_i \mid i \in \Omega\}$ is a family of semirings having direct product $R = \times_{i\in\Omega} R_i$ then for each $h \in \Omega$ we have a surjective morphism of semirings $\nu_h: R \rightarrow R_h$ which assigns to each element of R its hth component and an injective morphism of hemirings (but not of semirings!) $\lambda_h: R_h \rightarrow R$ which assigns to each element a of R_h the element of R the value of whose hth component is a and the value of all of whose other components is 0. Note that the image of the multiplicative identity of R_h is not 1_R but does belong to $C(R) \cap I^{\times}(R)$. A subring S of R is a **subdirect product** of the R_i if and only if the restriction of ν_h to S for each $h \in \Omega$ is still surjective.

A set $\{e_1, \ldots, e_n\}$ of nonzero elements of $C(R) \cap I^{\times}(R)$ is a **complete set of orthogonal central idempotents** of R if and only if $e_1 + \cdots + e_n = 1$ and $e_i e_j = 0$ for all $i \neq j$. Let $\{e_1, \ldots, e_n\}$ be a complete set of orthogonal central idempotents of R and set $e_U = \sum_{i \in U} e_i$ and $f_U = \sum_{i \notin U} e_i$ for any proper nonempty subset U of $\{1, \ldots, n\}$. Then $e_U + f_U = 1$ while $e_U f_U = 0 = f_U e_U$. Thus, for each such U, $e_U \in comp(R)$ and $e_U^{\perp} = f_U$. In particular, $e_i \in comp(R)$ for each $1 \leq i \leq n$.

(9.32) PROPOSITION. *The following conditions on a semiring R are equivalent:*

(1) *There exist semirings $R_1, \ldots, R_n$ and an isomorphism $\gamma: R \to \times_{i=1}^{n} R_i$;*
(2) *There exists a complete set $\{e_1, \ldots, e_n\}$ of orthogonal central idempotents of R.*

PROOF. (1) $\Rightarrow$ (2): Let $S = \times_{i=1}^{n} R_i$. For each $1 \leq i \leq n$, let 1_i be the multiplicative identity of R_i, let $\lambda_i: R_i \to S$ be the canonical injective morphism of hemirings defined above. Note that since γ is an isomorphism, it is both injective and surjective and so for each element s of S there is a unique element $r = \gamma^{-1}(s)$ satisfying $\gamma(r) = s$.

For each $1 \leq i \leq n$, let $e_i = \gamma^{-1}(\lambda_i(1_i))$. Then

$$\gamma(e_1 + \cdots + e_n) = \lambda_1(1_1) + \cdots + \lambda_n(1_n) = 1_S = \gamma(1_R)$$

and so $e_1 + \cdots + e_n = 1_R$. If $i \neq j$ then $\gamma(e_i e_j) = (\lambda_i(1_i))(\lambda_j(1_j)) = 0_S = \gamma(0_R)$ and so $e_i e_j = 0_R$. Also, $\gamma(e_i^2) = \gamma(e_i)\gamma(e_i) = \lambda_i(1_i)\lambda_i(1_i) = \lambda_i(1_i) = \gamma(e_i)$ and so $e_i^2 = e_i$, proving that $e_i \in I^{\times}(R)$. Finally, if $r \in R$ then $\gamma(re_i) = \gamma(r)\lambda_i(1_i) = \lambda_i(1_i)\gamma(r) = \gamma(e_i r)$ and so $e_i \in C(R)$.

(2) $\Rightarrow$ (1): If $r \in R$ then $r = r1_R = re_1 + \cdots + re_n$, where $re_i \in Re_i$ for all i. Thus every element of R can be written as a sum of the elements of the Re_i. This sum is unique in the sense that if $r = r_1 e_1 + \cdots + r_n e_n$ then $re_i = (r_i e_i)e_i = r_i e_i$ for each $1 \leq i \leq n$. Thus we have an injective and surjective function $\gamma: R \to \times R_i$ given by $\gamma(r)(i) = re_i$. As noted in Chapter 1, each Re_i is a semiring, and it is straightforward to show that γ is a morphism of semirings. □

(9.33) COROLLARY. *If R is an integral semiring then there do not exist semirings R' and R'' such that R is isomorphic to $R' \times R''$.*

PROOF. This is a direct consequence of Proposition 9.32 and the remarks before it. □

We now extend the notion of a derivation, introduced in Chapter 3. Let R be a semiring and let $\gamma: R \to R$ be a morphism from R to itself. A **γ-derivation** is a function $d: R \to R$ satisfying $d(r + r') = d(r) + d(r')$ and $d(rr') = \gamma(r)d(r') + d(r)r'$ for all $r, r' \in R$.

(9.34) EXAMPLE. [Brzozowski, 1964] Let A be a nonempty set, let $R = sub(A^*)$ be the semiring of all formal languages on A introduced in Example 1.11, and let $\gamma: R \to R$ be the morphism defined by $\gamma: L \mapsto L \cap \{\square\}$. Every word w in A^* defines a function $d_w: R \to R$ by $d_w: L \mapsto \{w' \in A^* \mid ww' \in L\}$ and it is easy to verify that this is a γ-derivation of R.

If $R[t]$ is the semiring of polynomials in the indeterminate t over R then we can define a new operation of multiplication on $R[t]$ with the aid of the rule $tr = \gamma(r)t + d(r)$ for all $r \in R$ and the distributivity of multiplication over addition from both sides. This semiring, denoted $R[t;\gamma,d]$, is called the **Øre extension** of R by γ and d. In the special case that d is taken to be the zero map, we obtain the semiring $R[t;\gamma]$, called the **skew polynomial semiring** over R. In the special case γ is taken to be the identity map we obtain the differential polynomial semiring $R[t;d]$ defined in Chapter 3.

We now return to the rings of differences introduced in Chapter 8. Let R be a nonzeroic semiring and let $R^\Delta = S/D$ be the ring of differences of R. Then we have a morphism of semirings $\nu: R \to R^\Delta$, called the **canonical morphism**, defined by $\nu: r \mapsto (r,0)/D$. This morphism need not be injective. Indeed, if $\nu(r) = \nu(r')$ then $(r,0)/D = (r',0)/D$ and so there exists an element a of R such that $(r+a,a) = (r'+a,a)$ and hence $r+a = r'+a$. Thus we see that a necessary and sufficient condition for ν to be injective is that R be cancellative. Furthermore, we note that an arbitrary element $(a,b)/D$ of R^Δ is $\nu(a)\nu(b)$ and so every element of R^Δ is the difference between two elements in $im(\nu)$. In particular, we conclude that if R is a cancellative semiring then R is isomorphic to a subsemiring of a ring R^Δ such that every element of R^Δ is the difference between two elements in the image of R. It is this property of cancellative semirings which leads some authors to call them **halfrings**. We will, in general, identify a cancellative semiring with its image in its ring of differences, and thus consider it as a subsemiring of that ring.

(9.35) PROPOSITION. *If I is a left ideal of a nonzeroic semiring R then $I^\Delta = \{\nu(a)\nu(b) \mid a,b \in I\}$ is a left ideal of R^Δ.*

PROOF. If $a,a',b,b' \in I$ then $[\nu(a)\nu(b)] + [\nu(a')\nu(b')] = \nu(a+a')\nu(b+b') \in I^\Delta$. If, furthermore, $r,r' \in R$ then

$$\begin{aligned}[\nu(r)\nu(r')][\nu(a)\nu(b)] &= \nu(r)\nu(a)\nu(r)\nu(b)\nu(r')\nu(a) + \nu(r')\nu(b) \\ &= \nu(ra+r'b)\nu(rb+r'a) \in I^\Delta\end{aligned}$$

Thus I^Δ is a left ideal of R^Δ. □

It is clearly true from the above construction that I^Δ is the smallest left ideal of R^Δ containing $\nu(I)$.

(9.36) PROPOSITION. *If R is a nonzeroic semiring then the function $\gamma: ideal(R) \to ideal(R^\Delta)$ defined by $I \mapsto I^\Delta$ is a morphism of semirings.*

PROOF. Clearly $\gamma(\{0\}) = \{0\}$ and $\gamma(R) = R^\Delta$. If I and H are ideals of R then $(I+H)^\Delta$ is the smallest ideal of R^Δ containing $I+H$ and hence surely $(I+H)^\Delta \subseteq I^\Delta + H^\Delta$. Conversely, suppose that $s = [\nu(a)\nu(a')] + [\nu(b)\nu(b')]$ belongs to $I^\Delta + H^\Delta$, where $a,a' \in I$ and $b,b' \in H$. Then $s = \nu(a+b)\nu(a'+b') \in (I+H)^\Delta$, proving equality. Similarly, $(IH)^\Delta \subseteq I^\Delta H^\Delta$. Conversely, if $s = [\nu(a)\nu(a')][\nu(b)\nu(b')]$ belongs to $I^\Delta H^\Delta$ then $s = \nu(ab+a'b')\nu(ab'+a'b) \in (IH)^\Delta$, again proving equality. □

(9.37) EXAMPLE. [Dale, 1981] If R is a nonzeroic semiring and I is an ideal of R, it does not follow that $I = \nu^{-1}(I^{\Delta})$, even if R is cancellative. For example, let $R = \mathbb{N}$ and let $I = \{0\} \cup \{2i+6 \mid i \in \mathbb{N}\}$. Then $I^{\Delta} = 2\mathbb{Z}$ and so $\nu^{-1}(I^{\Delta}) = 2\mathbb{N} \supset I$.

Example 9.37 shows that the morphism γ given in Proposition 9.36 need not be injective even if ν is injective.

If R is a cancellative semiring then its ring of differences is also a morphic image of the semiring of polynomials $R[t]$ over R. To see this, consider the element $u = (0,1)/D$ of R^{Δ}. This is an element of the center of R^{Δ} satisfying $1 + u = 0$. Therefore, by Example 9.19, there exists an evaluation morphism of semirings $\epsilon: R[t] \to R^{\Delta}$ determined by the function $t \mapsto u$. Note that if $a, b \in R$ then $ab = a + bu \in im(\epsilon)$. Since every element of R^{Δ} is the difference between two elements of R, we thus conclude that $im(\epsilon) = R^{\Delta}$.

Let R be a nonzeroic semiring having ring of differences $R^{\Delta} = S/D$ and let $\nu: R \to R^{\Delta}$ be the canonical morphism. If $\gamma: R \to R'$ is a morphism from R to a ring R' then γ defines a morphism of semirings γ' from S to R' given by $\gamma': (a,b) \mapsto \gamma(a)\gamma(b)$. Moreover, $\gamma'(a,a) = 0$ for all $a \in R$ and so γ' induces a ring homomorphism $\gamma'': R^{\Delta} \to R'$. If $a \in R$ then $\gamma''\nu(a) = \gamma''((a,0)/D) = \gamma'(a,0) = \gamma(a)$ and so $\gamma = \gamma''\nu$. The map γ'' is unique with this property. Indeed, if $\delta: R^{\Delta} \to R'$ is a ring homomorphism satisfying the condition that $\gamma = \delta\nu$ then for each element $(a,b)/D$ of R^{Δ} we have $\delta((a,b)/D) = \delta((a,0)/D)\delta((b,0)/D) = \delta\nu(a)\delta\nu(b) = \gamma(a)\gamma(b) = \gamma''((a,b)/D)$ so $\delta = \gamma''$.

In particular, if R_1 and R_2 are cancellative semirings contained in their respective rings of differences R_1^{Δ} and R_2^{Δ} then every morphism of semirings γ from R_1 to R_2 can be extended to a unique ring homomorphism γ^{Δ} from R_1^{Δ} to R_2^{Δ}. Moreover, γ^{Δ} is injective if and only if γ is injective and it is surjective if and only if γ is surjective.

(9.38) PROPOSITION. *If R is a cancellative semiring then there exists an injective morphism of semirings $\gamma: R \to S$ from R to an entire ring if and only if R satisfies the following condition:*

(*) *If $a, a', b, b' \in R$ satisfy $ab + a'b' = ab' + a'b$ then $a = a'$ or $b = b'$.*

PROOF. Assume that such an injective morphism γ exists and identify R with its image in S. If a, a', b, b' are elements of R satisfying $ab + a'b' = ab' + a'b$ then in S we have $(a-a')(b-b') = 0$. Since S is assumed to be entire, we deduce that $a = a'$ or $b = b'$.

Now, conversely, assume that R satisfies (*). It suffices to show that the ring R^{Δ} is entire. Indeed, suppose that $(a,a')/D$ and $(b,b')/D$ are two elements of R^{Δ} satisfying $0/D = [(a,a')/D][(b,b')/D] = (ab + a'b', ab' + a'b)/D$. Then we must have $ab + a'b' = ab' + a'b$ and so, by (*), $a = a'$ or $b = b'$, i.e. either $(a,a') \in D$ or $(b,b') \in D$. Thus R^{Δ} is entire. □

(9.39) EXAMPLE. [Mitchell & Sinutoke, 1982] Let $R = \{(a_1, a_2, a_3, a_4) \in \mathbb{N}^4 \mid a_i = 0 \text{ for all } i \text{ or } a_i \neq 0 \text{ for all } i\}$. Then R is a subsemiring of $\mathbb{N}^4$ which is cancellative and entire. On the other hand, R cannot be embedded in an entire ring. To see this, note that we have distinct elements $a = (2,1,1,1)$, $a' = (1,2,1,1)$,

$b = (1,1,2,1)$, and $b' = (1,1,1,2)$ satisfying $ab + a'b' = ab' + a'b$, so R does not satisfy condition (*) of Proposition 9.38.

(9.40) EXAMPLE. [H. E. Stone, 1972] If R is a cancellative semiring and if t is an indeterminate over R then there exists a canonical morphism of rings $\gamma: R[t]^{\Delta} \to R^{\Delta}[t]$ given by $\gamma: \sum a_i t^i - \sum b_i t^i \mapsto \sum (a_i b_i) t^i$ for all $\sum a_i t^i$ and $\sum b_i t^i$ in $R[t]$. It is straightforward to show that γ is in fact an isomorphism of rings.

(9.41) PROPOSITION. *If R is a cancellative semiring with ring of differences R^{Δ} then a proper subset I of R is a subtractive left ideal if and only if it is of the form $R \cap H$ for some left ideal H of R^{Δ}.*

PROOF. Assume that I is a subtractive left ideal of R and let $H = I^{\Delta} \subseteq R^{\Delta}$. Then H is an left ideal of R^{Δ} satisfying $I \subseteq R \cap H$. Conversely, if $r \in R \cap H$ then there exist elements a and b of I such that $r + b = a$. Since I is subtractive, this implies that $r \in I$. Hence $I = R \cap H$.

Conversely, assume that $I = R \cap H$ for some left ideal H of R^{Δ}. Then clearly I is an left ideal of R. If a and b are elements of R such that $a + b$ and b belong to I then $a = (a + b)b \in R \cap H = I$. Thus I is a subtractive left ideal of R. □

We have thus seen that cancellative semirings have very nice properties. The following result shows that there are "enough" such semirings around.

(9.42) PROPOSITION. *If R is a semiring then there exists a cancellative semiring S and a surjective morphism from S to R.*

PROOF. Let R be a semiring and let $A = \{a_r \mid r \in R\}$ be a set indexed by R. Let S be the free monoid on A, written additively. Define a new operation $\cdot$ on S by

$$\left(\sum_{r \in \Lambda} a_r\right) \cdot \left(\sum_{s \in \Omega} a_s\right) = \sum_{r \in \Lambda} \sum_{s \in \Omega} a_{rs}.$$

Then $(S, +, \cdot)$ is a cancellative semiring with additive identity □ and multiplicative identity a_1. Moreover, the function $\gamma: S \to R$ given by $\gamma(\square) = 0$ and $\gamma: \sum_{r \in \Lambda} a_r \mapsto \sum_{r \in \Lambda} r$ for $\Lambda \neq \varnothing$ is clearly a surjective morphism of semirings. □

(9.43) PROPOSITION. *A zerosumfree semiring R is either additively idempotent or contains a subsemiring isomorphic to $\mathbb{Q}^+$.*

PROOF. Let γ be the function $\mathbb{Q}^+ \to R$ defined by $\gamma: m/n \mapsto (m1_R)(n1_R)^{-1}$. Using Proposition 4.52, it is easy to verify that this is a morphism of semirings. Moreover, if $\gamma(h/k) = \gamma(m/n)$ then $h1_R(k1_R)^{-1} = m1_R(n1_R)^{-1}$ and so $hn1_R = mk1_R$. If R is not additively idempotent then, from Proposition 4.51, we conclude that $hn = mk$ and so $h/k = m/n$. Thus, in this case, γ is an isomorphism from $\mathbb{Q}^+$ to a subsemiring of R. □

Note that if R is a zerosumfree semiring which is not additively idempotent then the subsemiring of R constructed in Proposition 9.43 contains $B(R)$ and is contained in $U(R)$.

If $\gamma: R \to S$ is a morphism of semirings and if ρ is a congruence relation on S then, as an immediate consequence of the definitions, the relation ρ' on R defined by $r\ \rho'\ r'$ if and only if $\gamma(r)\ \rho\ g(r')$, is a congruence relation on R. In particular, each morphism of semirings $\gamma: R \to S$ defines a congruence relation $\equiv_\gamma$ on R by setting $r \equiv_\gamma r'$ if and only if $\gamma(r) = \gamma(r')$.

(9.44) EXAMPLE. If R is a nonzeroic semiring and $\nu: R \to R^\Delta$ is the canonical morphism then $\equiv_\nu$ and $[\equiv]_{\{0\}}$ are equal.

(9.45) PROPOSITION. *A morphism of semirings $\gamma: R \to S$ induces an injective morphism of semirings $\gamma': R/\equiv_\gamma \to S$ defined by $\gamma'(r/\equiv_\gamma) = \gamma(r)$. If γ is surjective then γ' is an isomorphism.*

PROOF. The function γ' is well-defined since $r/\equiv_\gamma = r'/\equiv_\gamma$ implies that $\gamma(r) = \gamma(r')$, and it is clearly a morphism of semirings. If $\gamma'(r/\equiv_\gamma) = \gamma'(r'/\equiv_\gamma)$ then $\gamma(r) = \gamma(r')$ and so $r/\equiv_\gamma = r'/\equiv_\gamma$. Thus γ' is injective. □

(9.46) PROPOSITION. *Let $\gamma: R \to S$ be a morphism of semirings.*

(1) *If H is a left ideal of S then $\gamma^{-1}(H)$ is a left ideal of R. Moreover, if H is subtractive then so is $\gamma^{-1}(H)$.*
(2) *If γ is a surjective morphism and if I is a left ideal of R then $\gamma(I)$ is a left ideal of S.*

PROOF. (1) Assume that H is a left ideal of S. If $a, b \in \gamma^{-1}(H)$ then $\gamma(a+b) = \gamma(a) + \gamma(b) \in H$ so $a + b \in \gamma^{-1}(H)$. If $r \in R$ and $a \in \gamma^{-1}(H)$ then $\gamma(ra) = \gamma(r)\gamma(a) \in H$ so $ra \in \gamma^{-1}(H)$. Finally, if $1_R \in \gamma^{-1}(H)$ then $1_S = \gamma(1_R) \in H$, which is impossible. Thus $1_R \notin \gamma^{-1}(H)$ and so $\gamma^{-1}(H)$ is a left ideal of R. Now assume that H is subtractive. If $a, a+b \in \gamma^{-1}(H)$ then $\gamma(a)$ and $\gamma(a)+\gamma(b) = \gamma(a+b) \in H$ and so $\gamma(b) \in H$. Hence $b \in \gamma^{-1}(H)$.

(2) Assume that I is a left ideal of R. If $a, b \in I$ then $\gamma(a)+\gamma(b) = \gamma(a+b) \in \gamma(I)$. If $a \in I$ and $s \in S$ then $s = \gamma(r)$ for some $r \in R$ and so $s\gamma(a) = \gamma(r)\gamma(a) = \gamma(ra) \in \gamma(I)$. Thus $\gamma(I)$ is a left ideal of S. The proof for right ideals and for ideals is similar. □

(9.47) PROPOSITION. *If $\gamma: R \to S$ is a surjective morphism of semirings and I is an ideal of R then:*

(1) *$\gamma(0/I) \subseteq 0/\gamma(I)$; and*
(2) *$\gamma(0[/]I) \subseteq 0[/]\gamma(I)$.*

PROOF. (1) If $r \in 0/I$ then there exist elements a and a' of I satisfying $r+a = a'$. Then $\gamma(r) + \gamma(a) = \gamma(a') = 0 + \gamma(a')$ and so $\gamma(r) \in 0/\gamma(I)$.

(2) If $r \in 0[/]I$ then there exist elements a and a' of I and r'' of R satisfying $r + a + r'' = a' + r''$. Therefore $\gamma(r) + \gamma(a) + \gamma(r'') = \gamma(a') + \gamma(r'')$ and so $\gamma(r) \in 0[/]\gamma(I)$. □

(9.48) PROPOSITION. *Let $\gamma: R \to S$ be a surjective morphism of semirings. If R is a yoked semiring then so is S.*

PROOF. Let s and s' be elements of S and let r and r' be elements of R satisfying $\gamma(r) = s$ and $\gamma(r') = s'$. Since R is a yoked semiring, there exists an element a of R satisfying $r + a = r'$ or $r' + a = r$. Hence $s + \gamma(a) = s'$ or $s' + \gamma(a) = s$, proving that S is also a yoked semiring. □

(9.49) PROPOSITION. *If R is a plain yoked semiring satisfying the descending chain condition on subtractive left ideals and having no nonzero nilpotent elements and if I is a nonzero subtractive ideal of R then I is itself a semiring and there exists a surjective morphism of semirings $\gamma: R \to I$.*

PROOF. By Proposition 4.22, the semiring R is cancellative. By Proposition 6.55, we know that there exists an element $e \in I^{\times}(R) \cap I$ satisfying $I = Re$. In fact, from the proof of that result we see that $a = ae$ for all $a \in I$. Let $H = \{a \in I \mid ea = 0\}$. Then H is a right ideal of R satisfying $He = H$ and so $H^2 = (He)H = H(eH) = \{0\}$. Since R has no nonzero nilpotent elements, this implies that $H = \{0\}$. If $a \in I$ then, since R is a yoked semiring, there exists an element b of R satisfying $b + a = ea$ or $a = b + ea$. Since I is subtractive, we in fact have $b \in I$. If $b + a = ea$ then $ea = e^2a = eb + ea$ and so, by cancellation, $eb = 0$. Thus $b \in H$ and so $b = 0$. Similarly, $b = 0$ if $a = b + ea$ as well. Therefore $a = ea$ for all $a \in I$, proving that $(I, +, \cdot)$ is a semiring with multiplicative identity e. If $a \in I$ then $a = eae$ and so the function $\gamma: R \to I$ defined by $\gamma: R \mapsto ere$ is a surjective morphism of semirings. □

The following is an adaptation of a well-known result for rings.

(9.50) PROPOSITION. *The following conditions on a semiring R are equivalent:*

(1) *There exist a positive integer n, a semiring S, and an isomorphism $\gamma: R \to \mathcal{M}_n(S)$.*
(2) *There exists a set $\{e_{ij} \mid 1 \leq i, j \leq n\}$ of elements of R satisfying the conditions that $\sum_{i=1}^{n} e_{ii} = 1$ and*

$$e_{ij}e_{kh} = \begin{cases} e_{ih} & \text{for } j = k \\ 0 & \text{otherwise} \end{cases}.$$

PROOF. (1) ⇒ (2): For each $1 \leq i, j \leq n$ let E_{ij} be the matrix $[a_{hk}] \in \mathcal{M}_n(S)$ defined by

$$a_{hk} = \begin{cases} 1 & \text{for } h = i \text{ and } k = j \\ 0 & \text{otherwise} \end{cases}.$$

Set $e_{ij} = \gamma^{-1}(E_{ij})$ for each $1 \leq i, j \leq n$. Then $\{e_{ij}\}$ clearly satisfies the conditions in (2).

(2) ⇒ (1): Let $S = \{\sum_{h=1}^{n} e_{h1}ae_{1h} \mid a \in R\}$. This set is clearly closed under addition. Moreover,

$$\left(\sum e_{h1}ae_{1h}\right)\left(\sum e_{h1}be_{1h}\right) = \sum e_{h1}(ae_{11}b)e_{1h}$$

and so S is closed under multiplication as well. Surely $0 \in S$, while we also have $1 = \sum e_{h1}1e_{1h} \in S$. Thus S is a subring of R. Define the function $\gamma: R \to \mathcal{M}_n(S)$ by $\gamma: r \mapsto [c_{ij}]$, where $c_{ij} = \sum e_{hi}re_{jh} = \sum e_{h1}(e_{1i}re_{j1})e_{1h}$. It is straightforward to check that $\gamma(r + r') = \gamma(r) + \gamma(r')$ for all $r \in R$. Moreover, if $r, r' \in R$ then $\gamma(rr') = [c_{ij}]$, where

$$c_{ij} = e_{ii}rr'e_{jj} = e_{ii}r\left(\sum_{h=1}^{n} e_{hh}\right)r'e_{j}j = \sum_{h=1}^{n}(e_{ii}re_{hh})(e_{hh}r'e_{jj}),$$

and this is the value of the (i,j)-entry of $\gamma(r)\gamma(r')$. Thus γ is a morphism of semirings. It is straightforward to see that γ is both surjective and injective, and so it is an isomorphism of semirings. □

We now generalize another well-known result for rings to the case of semirings. Ideals I and H of a semiring R are **comaximal** if and only if $I + H = R$. A family $\{I_j \mid j \in \Omega\}$ of ideals of R is **pairwise comaximal** if and only if every pair of distinct elements of the family is comaximal.

(9.51) PROPOSITION. (**Chinese Remainder Theorem**) *Let $\{I_1, \ldots, I_n\}$ be a finite set of pairwise comaximal ideals of a semiring R. Then the morphism of semirings $\gamma: R \to \times_{j=1}^{n} R/I_j$ given by $r \mapsto (r/I_1, \ldots, r/I_n)$ is surjective.*

PROOF. It suffices to show that for each $1 \leq k \leq n$, the element

$$(0, \ldots, 0, 1/I_k, 0, \ldots, 0)$$

belongs to $im(\gamma)$. We will show this for the case $k = 1$, the proof of the other cases being similar. Since the given ideals are pairwise comaximal, we know that for each $k > 1$ there exist elements $a_k \in I_1$ and $b_k \in I_k$ such that $1 = a_k + b_k$. Therefore $1 = 1^{k-1} = \prod_{h=2}^{k}(a_h + b_h)$. By distributivity, this product becomes $a' + r$, where $a' \in I_1$ and $r = b_2 b_3 \cdot \ldots \cdot b_n \in I_2 \cap \cdots \cap I_n$. Therefore $r/I_1 = 1/I_1$ and $r/I_k = 0/I_k$ for $1 < K \leq n$, proving that $\gamma(r) = (1/I_1, 0, \ldots, 0)$, as desired. □

A semiring R is **separative** if and only if $a + a = a + b = b + b$ in R implies that $a = b$. Cancellative semirings are certainly separative. This condition is defined for semigroups in [Clifford & Preston, 1961]. Moreover, it is shown there that a commutative semigroup is separative if and only if it is embeddable in a semigroup which is a union of groups. An analogous result can be proven for semirings.

(9.52) PROPOSITION. *A semiring R is separative if and only if there exists an injective morphism of semirings $\gamma: R \to S$, where S is a semiring satisfying the property that its additive monoid is the union of groups.*

PROOF. Assume that R is separative. Define a relation ρ on R by setting $a \;\rho\; b$ if and only if there exist positive integers m and n and elements r and s of R such that $a + r = mb$ and $b + s = na$. This can be easily checked to be a congruence relation. Let $R' = \{(a,b) \in R \times R \mid a \;\rho\; b\}$ and define operations $\oplus$ and $\odot$ on R' by setting $(a,b) \oplus (c,d) = (a+c, b+d)$ and $(a,b) \odot (c,d) = (ac+bd, ad+bc)$. Then $(R', \oplus, \odot)$ is a semiring with additive identity $(0,0)$ and multiplicative identity $(1,0)$. Define a relation ζ on R' by setting $(a,b) \;\zeta\; (c,d)$ if and only if $a \;\rho\; c$ and $a + d = b + c$. This is also a congruence relation and so $S = R'/\zeta$ is a semiring.

Define a function γ from R to S by $\gamma: a \mapsto (2a, a)/\zeta$. Clearly $\gamma(a+b) = \gamma(a) + \gamma(b)$ for all $a, b \in R$. Moreover, $ab \;\rho\; 2ab \;\rho\; 4ab \;\rho\; 5ab$ and so $2ab + 4ab = ab + 5ab$, proving that $(2ab, ab) \;\zeta\; (5ab, 4ab)$. Thus

$$\gamma(a)\gamma(b) = [(2a,a)/\zeta] \odot [(2b,b)/\zeta] = (5ab, 4ab)/\zeta = (2ab, ab)/\zeta = \gamma(ab).$$

Also, $\gamma(0) = 0_S$ and $\gamma(1) = (1+1, 1)/\zeta = (1,0)/\zeta = 1_S$. Hence γ is a morphism of semirings. It is injective since $\gamma(a) = \gamma(b)$ implies that $(2a, a) \;\zeta\; (2b, b)$. But then

$2a \;\rho\; a \;\rho\; b \;\rho\; 2b$ and $2a + b = a + 2b$, whence $a = b$, since R is separative. We are thus left to show that S has the desired property.

Indeed, if $(a,b)/\zeta \in S$ then $a \;\rho\; b \;\rho\; 2a \;\rho\; a+b$ so $(a,b)/\zeta \oplus (a,a)/\zeta = (a,b)/\zeta$. Similarly $(a,b)/\zeta \oplus (b,a)/\zeta = (a,a)/\zeta$. Therefore $(a,b)/\zeta$ generates an additive subgroup of $(S,\oplus)$ with identity element $(a,a)/\zeta$. Hence $(S,\oplus)$ is the union of groups.

Now, conversely, assume that there exists an injective morphism of semirings $\gamma: R \to S$, where S satisfies the property that its additive monoid is the union of groups. For each $a \in R$, let $H(a)$ be a maximal group contained in $(S,+)$ containing $\gamma(a)$. If a, b are elements of R satisfying $a + a = a + b = b + b$ then $\gamma(a) + \gamma(b) \in H(a) \cap H(b)$. Moreover, $\gamma(a) = [\gamma(a) + \gamma(b)]\gamma(b) \in H(b)$ and so $\gamma(a) \in H(b)$, which implies that $H(b) \subseteq H(a)$ by the maximality of $H(a)$. Similarly, $H(a) \subseteq H(b)$ and so we have equality. Therefore, by cancellation in the group $H(a)$ we have $\gamma(a) = \gamma(b)$ and hence $a = b$ since γ is injective. Thus R is separative. □

A **preordered set** is a nonempty set together with a reflexive and transitive relation, usually denoted by $\leq$, defined on it. If Ω is a preordered set then a **direct system** of semirings over Ω is a family $\{R_i \mid i \in \Omega\}$ of semirings together with morphisms of semirings $\gamma_{ij}: R_i \to R_j$ for all $i \leq j$ in Ω satisfying the following conditions:

(1) γ_{ii} is the identity map for all $i \in \Omega$;
(2) $\gamma_{jk}\gamma_{ij} = \gamma_{ik}$ for all $i \leq j \leq k$ in Ω.

If $\{R_i \mid i \in \Omega\}$ is a direct system of semirings the the **direct limit** $\varinjlim R_i$ of the system is a semiring R together with morphisms $\delta_i: R_i \to R$ for each $i \in \Omega$ such that:

(3) $\delta_j\gamma_{ij} = \delta_i$ for all $i \leq j$ in Ω and
(4) For any semiring S and any set of morphisms $\eta_i: R_i \to S$ $(i \in \Omega)$ satisfying the condition that $\eta_j\gamma_{ij} = \eta_i$ for all $i \leq j$ in Ω there exists a unique morphism of semirings $\eta: R \to S$ satisfying $\eta\delta_i = \eta_i$ for all $i \in \Omega$.

Directed limits of directed systems of semirings always exist. Indeed, if $\{R_i, \gamma_{ij}; \Omega\}$ is such a system let S be the disjoint union of the R_i and define a binary relation ζ on S by setting $a \;\zeta\; b$ if and only if there exists $i, j \leq k$ in Ω such that $a \in R_i$, $b \in R_j$, and $\gamma_{ik}(a) = \gamma_{jk}(b)$. Then $\gamma_{in}(a) = \gamma_{jn}(b)$ for all $n \geq k$, from which we can easily verify that ζ is an equivalence relation on S. Moreover, $R = S/\zeta$ can be checked to be a semiring and we have canonical morphisms of semirings $\delta_i: R_i \to R$ given by $\delta_i: a \mapsto a/\zeta$ which have the required properties.

(9.53) EXAMPLE. Let R be a semiring and, for each positive integer i, let $k(i) = 2^i$ and $S_i = \mathcal{M}_{k(i)}(R)$. Then there exists an injective morphism of semirings $\gamma_i: S_i \to S_{i+1}$ defined by $A \mapsto \begin{bmatrix} A & 0 \\ 0 & A \end{bmatrix}$ If $i \leq j$ are positive integers, define γ_{ii} to be the identity map and $\gamma_{ij} = \gamma_{j-1}\gamma_{j-2}\cdots\gamma_i$ if $i < j$. Then $\{S_i \mid i \in \mathbb{P}\}$, together with the morphisms $\{\gamma_{ij}\}$ is a directed system and so the semiring $S = \varinjlim S_i$ exists.

Dually, if Ω is a preordered set then an **inverse system** of semirings over Ω is a family $\{R_i \mid i \in \Omega\}$ of semirings together with morphisms of semirings $\gamma_{ij}: R_j \to R_i$ for all $i \leq j$ in Ω satisfying the following conditions:

(1) γ_{ii} is the identity map for all $i \in \Omega$;

(2) $\gamma_{ij}\gamma_{jk} = \gamma_{ik}$ for all $i \leq j \leq k$ in Ω.

If $\{R_i \mid i \in \Omega\}$ is an inverse system of semirings then the **inverse limit** $\varprojlim R_i$ of the system is a semiring R together with morphisms $\delta_i \colon R \to R_i$ for each $i \in \Omega$ such that:

(3) $\gamma_{ij}\delta_j = \delta_i$ for all $i \leq j$ in Ω; and

(4) For any semiring S and any set of morphisms $\eta_i \colon S \to R_i$ $(i \in \Omega)$ satisfying the condition $\gamma_{ij}\eta_j = \eta_i$ for all $i \leq j$ there exists a unique morphism of semirings $\eta \colon S \to R$ satisfying $\delta_i\eta = \eta_i$ for all $i \in \Omega$.

Inverse limits of semirings always exist. Indeed, if $\{R_i \mid i \in \Omega\}$ is an inverse system of semirings then we can take $\varprojlim R_i$ to be $\{\langle r_i \rangle \in \times_{i \in \Omega} R_i \mid r_i = \gamma_{ij}(r_j)\}$ for all $i \leq j\}$.

Let R be a semiring and let $I_1 \supseteq I_2 \supseteq \ldots$ be a descending chain of ideals of R. For each $j \geq 1$, set $R_j = R/I_j$. Then for all $i \leq j$ we have a canonical surjective morphism of semirings $\gamma_{ij} \colon R_j \to R_i$ and this turns $\{R_i \mid i \geq 1\}$ into an inverse system of semirings. Thus we can form the semiring $\varprojlim R_i$, which is called the **completion** of R with respect to the given chain of ideals. In particular, if there exists an ideal I of R such that $I_j = I^j$ for all $j \geq 1$ then this inverse limit is called the I**-adic completion** of the semiring R. The elements of the I-adic completion S of R are sequences of the form $\langle r + I^j \rangle$ for $r \in R$. Therefore we have a canonical morphism of semirings $\gamma \colon R \to S$ given by $r \mapsto \langle r/I^j \rangle$.

10. KERNELS OF MORPHISMS

By Proposition 9.8 we see that if $\gamma: R \to S$ is a morphism of semirings then $\gamma^{-1}(0)$ is an ideal of R, called the **kernel** of γ, and denoted by $ker(\gamma)$. By Proposition 9.46, $ker(\gamma)$ is an ideal of R. If R is a ring, we know that any ideal of R can be the kernel of a morphism from R to some ring S but, as we shall see, this is not the case for arbitrary semirings. Also, unlike the case of rings, we note that a morphism of semirings $\gamma: R \to S$ need not be monic when $ker(\gamma) = \{0\}$. To see an example of this, consider the totally-ordered set $R = \{0, a, 1\}$ on which we define addition to be *max* and multiplication to be *min*. This is a semiring by Example 1.5. Let $\gamma: R \to \mathbb{B}$ be the character of R defined by $\gamma(0) = 0$ and $\gamma(a) = \gamma(1) = 1$. This map has kernel $\{0\}$ but is not monic.

(10.1) EXAMPLE. [Shubin, 1992] Let S and S' be entire zerosumfree semirings and let $R = S \bowtie S'$. Then the function $\gamma: R \to S$ defined by $\gamma: 0 \mapsto 0_S$ and $\gamma: (s, s') \mapsto s$ is a morphism of semirings which is not monic, having kernel $\{0\}$.

We note that if R is a division semiring and $\gamma: R \to S$ is a morphism of semirings then $ker(\gamma) = \{0\}$. Indeed, if $0 \neq a \in ker(\gamma)$ then $1_S = \gamma(1_R) = \gamma(a)\gamma(a^{-1}) = 0_S$, which is a contradiction. In this case, the image of γ is again a division semiring. Nonetheless, γ may not be monic. Indeed, if the semirings S and S' in Example 10.1 are both division semirings then R is a division semiring and $im(\gamma) = S$, but nonetheless γ is not monic.

(10.2) EXAMPLE. If I is an ideal of a semiring R and if $\gamma: R \to R/I$ is the surjective morphism defined by $r \mapsto r/I$ then $ker(\gamma) = \{r \in R \mid r + a \in I$ for some $a \in I\} = 0/I$.

(10.3) EXAMPLE. Let R be a semiring and let $\{S_j \mid j \in \Omega\}$ be a family of semirings. For each $j \in \Omega$, let $\gamma_j: R \to S_j$ be a morphism of semirings. Then we have a morphism of semirings $\gamma: R \to \times_{j \in \Omega} S_j$ given by $r \mapsto \langle \gamma_j(r) \rangle$. The kernel of this morphism is $\cap_{j \in \Omega} ker(\gamma_j)$.

(10.4) EXAMPLE. If R is a nonzeroic semiring then the kernel of the canonical morphism $R \to R^{\Delta}$ is precisely $Z(R)$.

(10.5) EXAMPLE. Let $(S, \circ, \cdot)$ be the hemiring defined in Example 1.18. Then the function $\gamma: S \rightarrow S$ defined by $\gamma: a \mapsto a^3$ is a morphism of semirings the image of which is contained in $Z(S)$ since, for each $a \in S$, the self-distributivity condition implies that $a^3 \circ a^3 = a^3 + a^3a^6 = a^3$. The kernel of γ is precisely $\{a \in S \mid abc = 0 \text{ for all } b, c \in S\}$.

(10.6) EXAMPLE. We now generalize the construction given in Example 9.18. If R is a semiring and$M$ is a monoid then the function $\epsilon_M: R[M] \rightarrow R$ defined by $\epsilon_M: f \mapsto \sum\{f(m) \mid m \in M\}$ is a surjective morphism of semirings called the **augmentation morphism**. The kernel of ϵ_M is called the **augmentation ideal** of $R[M]$.

(10.7) PROPOSITION. *If R is a semiring and $\gamma \in char(R)$ then $ker(\gamma)$ is prime.*

PROOF. Let a and b be elements of R satisfying the condition that $arb \in ker(\gamma)$ for all $r \in R$. Then, in particular, $ab \in ker(\gamma)$. If $a \notin ker(\gamma)$ then $0 = \gamma(ab) = \gamma(a)\gamma(b) = \gamma(b)$ and so $b \in ker(\gamma)$. Thus, by Proposition 7.4, $ker(\gamma)$ is prime. □

(10.8) PROPOSITION. *If $\gamma: R \rightarrow S$ is a morphism of semirings then $ker(\gamma) \cap U(R) = \varnothing$.*

PROOF. If $a \in ker(\gamma) \cap U(R)$ then there exists an element b of R satisfying $ab = 1_R$ and so $0_S = 0_S\gamma(b) = \gamma(a)\gamma(b) = \gamma(ab) = \gamma(1_R) = 1_S$. This is a contradiction, and so $ker(\gamma) \cap U(R)$ must be empty. □

We have already seen that any morphism of semirings $\gamma: R \rightarrow S$ defines a congruence relation $\equiv_\gamma$ on R by setting $r \equiv_\gamma r'$ if and only if $\gamma(r) = \gamma(r')$. Another congruence relation defined on R by γ is the relation $\equiv_{ker(\gamma)}$. It is clearly true that $r \equiv_\gamma r'$ whenever $r \equiv_{ker(\gamma)} r'$ but the converse need not be true. If the relations $\equiv_\gamma$ and $\equiv_{ker(\gamma)}$ coincide, then the morphism γ is **steady**. A steady morphism $\gamma: R \rightarrow S$ is monic if and only if $ker(\gamma) = \{0\}$. Morever, by Proposition 9.45 we see that if $\gamma: R \rightarrow S$ is a steady surjective morphism of semirings then S is isomorphic to $R/ker(\gamma)$.

A surjective morphism of semirings $\gamma: R \rightarrow S$ is a **semiisomorphism** if and only if $ker(\gamma) = \{0\}$. Isomorphisms of semirings are clearly semiisomorphisms but the converse is not true, as we have seen. However, a steady semiisomorphism is an isomorphism. By combining Proposition 8.16 and Proposition 9.42, we see that for each semiring R there exists a cancellative semiring S and a semiisomorphism $S \rightarrow R$.

(10.9) EXAMPLE. Let R be the semiring $(\mathbb{N} \cup \{-\infty\}, max, +)$. Let t be an indeterminate over a zerosumfree semiring S and let $\gamma: S[t] \rightarrow R$ be the **degree function** given by $\gamma(p) = sup\{i \mid p(i) \neq 0\}$ if $p \neq 0$ and $\gamma(0) = -\infty$, which we defined previously. Then γ is a surjective morphism of semirings. Moreover, $ker(\gamma) = \{p \in S[t] \mid \gamma(p) = -\infty\} = \{0\}$ so γ is a semiisomorphism but is clearly not an isomorphism.

(10.10) PROPOSITION. *If $\gamma: R \rightarrow S$ is a semiisomorphism of semirings then:*

(1) R is entire if and only if S is entire; and

(2) R is a ring if and only if S is a ring.

PROOF. (1) Assume R is entire and let $s, s' \in S$ be elements satisfying $ss' = 0$. Then there exist elements r and r' of R satisfying $\gamma(r) = s$ and $\gamma(r') = s'$. Thus $\gamma(rr') = 0$ and so $rr' \in ker(\gamma) = \{0\}$ so $rr' = 0$. Since R is entire, this means that $r = 0$ or $r' = 0$, and hence $s = 0$ or $s' = 0$. Now assume that S is entire and that r and r' are elements of R satisfying $rr' = 0$. Then $\gamma(r)\gamma(r') = \gamma(rr') = 0$ and so $\gamma(r) = 0$ or $\gamma(r') = 0$. Thus either r or r' belongs to $ker(\gamma)$ and so $r = 0$ or $r' = 0$.

(2) Now assume that R is a ring. If $s \in S$ and if $r \in R$ is an element satisfying $\gamma(r) = s$, then $0 = \gamma(0) = \gamma(-r + r) = \gamma(-r) + \gamma(r) = \gamma(-r) + s$ and so $s \in V(S)$. Thus S is a ring. Conversely, assume S is a ring. If $r \in R$ then there exists an element $r' \in R$ satisfying $\gamma(r') = -\gamma(r)$. But then

$$\gamma(r + r') = \gamma(r) + \gamma(r') = \gamma(r) + -\gamma(r) = 0$$

so $r + r' \in ker(\gamma) = \{0\}$. This implies that $r \in V(R)$ and so R is a ring. □

We now characterize those ideals of a semiring which can be kernels of morphisms.

(10.11) PROPOSITION. *An ideal I of a semiring R is the kernel of a morphism of semirings if and only if it is subtractive.*

PROOF. Assume that I is the kernel of a morphism $\gamma: R \to S$. If a and b are elements of R satisfying $a, a + b \in I$ then $0 = \gamma(a + b) = \gamma(a) + \gamma(b) = \gamma(b)$ and so $b \in I$. Thus I is subtractive. Conversely, if I is a subtractive ideal and if $\gamma: R \to R/I$ is the surjective morphism of semirings defined by $r \mapsto r/I$ then surely $I \subseteq ker(\gamma)$. On the other hand, if $r \in ker(\gamma)$ then there exist elements a and a' of I such that $r + a = 0 + a' \in I$. Since I is subtractive, this means that $r \in I$ and so $I = ker(\gamma)$. □

In particular, we note that if R is an austere semiring then any morphism $\gamma: R \to S$ has kernel $\{0\}$.

(10.12) PROPOSITION. *Let R be a yoked semiring, let S be a plain semiring, and let $\gamma: R \to S$ be a surjective morphism of semirings. Then there exists a bijective correspondence between the set of all subtractive left ideals of R containing $ker(\gamma)$ and the set of all subtractive left ideals of S. This correspondence is given by $I \mapsto \gamma(I)$.*

PROOF. If I is a left ideal of R then, by Proposition 9.46, we know that $\gamma(I)$ is a left ideal of S. Now assume that I is subtractive and contains $ker(\gamma)$. If $s, s + t \in \gamma(I)$ then there exist elements $a, b \in I$ and $d \in R$ satisfying $\gamma(a) = s$, $\gamma(b) = s + t$, and $\gamma(d) = t$. Since R is a yoked semiring, there exists an element r of R such that $r + b = d$ or $r + d = b$.

Case I: Assume that $r + b = d$. Then $\gamma(r + a) + \gamma(d) = \gamma(r) + \gamma(a) + \gamma(d) = \gamma(r) + \gamma(a + d) = \gamma(r) + \gamma(b) = \gamma(r + b) = \gamma(d)$ and so $\gamma(r + a) \in Z(S) = \{0\}$, Hence $r + a \in ker(\gamma) \subseteq I$. Since I is subtractive, this implies that $r \in I$ and so $d = r + b \in I$. Hence $t \in \gamma(I)$.

Case II: Assume that $r + d = b$. Since R is a yoked semiring, there exists an element r' of R satisfying $r' + a = r$ or $a = r' + r$. If $r' + a = r$ then

$$\begin{aligned}\gamma(a + d) &= \gamma(b) = \gamma(r + d) = \gamma(r) + \gamma(d) \\ &= \gamma(r') + \gamma(a) + \gamma(d) = \gamma(r') + \gamma(a + d)\end{aligned}$$

so $\gamma(r') \in Z(S) = \{0\}$ and so $r' \in ker(\gamma) \subseteq I$. Thus $r = r' + a \in I$. Since $r + d = b \in I$ and I is subtractive, this implies that $d \in I$ and so $t \in \gamma(I)$. If $a = r' + r$ then $\gamma(b) = \gamma(a + d) = \gamma(r' + r + d) = \gamma(r') + \gamma(r + d) = \gamma(r') + \gamma(b)$ so, again, $r' \in ker(\gamma) \subseteq I$. Hence $r \in I$ and, as before, $t \in \gamma(I)$.

Thus we have shown that if I is a subtractive left ideal of R containing $ker(\gamma)$ then $\gamma(I)$ is a subtractive left ideal of S. Conversely, by Proposition 9.46(1), we see that every subtractive left ideal of S is of the form $\gamma(\gamma^{-1}(H))$, where $\gamma^{-1}(H)$ is a subtractive left ideal of R containing $ker(\gamma)$.

Finally, let I and I' be subtractive left ideals of R containing $ker(\gamma)$ and satisfying $\gamma(I) = \gamma(I')$. If $b \in I'$ then there exists an element a of I satisfying $\gamma(b) = \gamma(a)$. Since R is a yoked semiring, there exists an element r of R satisfying $r + a = b$ or $r + b = a$. In the first case, $\gamma(a) = \gamma(b) = \gamma(r + a) = \gamma(r) + \gamma(a)$ and so $\gamma(r) \in Z(S) = \{0\}$. Thus $r \in ker(\gamma) \subseteq I$ and so $b = r + a \in I$. In the second case, $\gamma(a) = \gamma(r) + \gamma(b) = \gamma(r) + \gamma(a)$ so, again, $\gamma(r) \in Z(S) = \{0\}$ and $r \in ker(\gamma) \subseteq I$. Since I is subtractive, this implies that $b \in I$. Thus, in both cases, we have shown that $I' \subseteq I$. A similar argument shows that $I \subseteq I'$ and so we have equality.

Thus the correspondence $I \mapsto \gamma(I)$ is bijective. □

(10.13) Example. [Dulin & Mosher, 1972] Let $R = (\mathbb{N} \cup \{\infty\}, max, min)$ and let S be the subsemiring of R given by $S = \{2i \mid i \in \mathbb{N}\} \cup \{1, \infty\}$. Define the function $\gamma: R \to S$ by

$$\gamma(i) = \begin{cases} i+1 & \text{if } 1 < i < \infty \text{ and } i \text{ is odd} \\ i & \text{otherwise} \end{cases}.$$

Then $\gamma: R \to S$ is a surjective morphism of semirings having kernel $\{0\}$. However, $I = \{0, 1, 2, 3\}$ and $H = \{0, 1, 2, 3, 4\}$ are subtractive ideals of R satisfying $\gamma(I) = \gamma(H) = \{0, 1, 2, 4\}$. We note that, since S is simple, we have $Z(S) = S \neq \{0\}$.

A morphism of semirings $\gamma: R \to S$ is **tame** if and only if the following conditions are satisfied:

(1) For each $s \in im(\gamma)$, the family $\{r + ker(\gamma) \mid r \in \gamma^{-1}(s)\}$, partially ordered by set inclusion, has a unique maximal member; and
(2) The unique maximal member of $\{r + ker(\gamma) \mid r \in \gamma^{-1}(1_S)\}$ is $1_R + ker(\gamma)$.

(10.14) Example. Any homomorphism from one ring to another is tame.

(10.15) Proposition. *If $\gamma: R \to S$ is a tame morphism of semirings then $ker(\gamma)$ is a partitioning ideal of R.*

Proof. Set $I = ker(\gamma)$. For each element $s \in im(\gamma)$, let q_s be the element of $\gamma^{-1}(s)$ satisfying the condition that $q_s + \gamma^{-1}(s)$ is maximal. Set $Q = \{q_s \mid s \in im(\gamma)\}$. Then surely $R = \cup\{q_s + I \mid s \in im(\gamma)\}$. Suppose that $s \neq t$ are distinct elements of $im(\gamma)$ satisfying the condition that $(q_s + I) \cap (q_t + I) \neq \varnothing$. Then there exist elements $a, a' \in I$ such that $q_s + a = q_t + a'$ and so $s = \gamma(q_s + a) = \gamma(q_t + a') = t$, which is a contradiction. Thus I is a partitioning ideal of R. □

(10.16) PROPOSITION. *If $\gamma: R \rightarrow S$ is a surjective morphism of semirings then there exists a semiisomorphism from $R/ker(\gamma)$ to S. If, in addition, γ is tame then there exists an isomorphism from $R/ker(\gamma)$ to S.*

PROOF. Set $I = ker(\gamma)$ and define the function $\delta: R/I \rightarrow S$ by $\delta: a/I \mapsto \gamma(a)$. This is well-defined since $a/b = b/I$ if and only if there exist elements c and d of I satisfying $a+c = b+d$ and in that case $\gamma(a) = \gamma(a)+\gamma(c) = \gamma(a+c) = \gamma(b+d) = \gamma(b)+\gamma(d) = \gamma(d)$. Moreover, δ is clearly a morphism of semirings which is surjective since γ is. If $\delta(a/I) = 0$ then $\gamma(a) = 0$ so $a \in I$. Thus $ker(\delta) = \{0/I\}$, proving that δ is a semiisomorphism.

Now assume that γ is tame. For each element s of S, let q_s be the unique element of $\gamma^{-1}(s)$ satisfying the condition that $q_s + \gamma^{-1}(s)$ is maximal. By the hypothesis of tameness, $q_1 = 1_R$ and it is easy to verify that $q_0 = 0_R$. Set $Q = \{q_s \mid s \in S\}$. By Proposition 10.15 and the discussion in Chapter 7, we see that I is a partitioning ideal of R and that the semiring R/I is isomorphic to R_Q.

Define a function $\delta: R_Q \rightarrow S$ by $\delta: q_s + I \mapsto s$. This function is well-defined and is clearly both monic and surjective. Therefore, all that remains for us to show is that it is an isomorphism. Indeed, let s and t be elements of S. Then $\delta((q_s+I) \oplus (q_t+I)) = \delta(q_u+I) = u$, where u is the unique element of S satisfying $(q_s+q_t)+I \subseteq q_u+I$. From this condition, we know that there exists an element a of I satisfying $q_s+q_t = q_u+a$. Thus $s+t = \gamma(q_s)+\gamma(q_t) = \gamma(q_s+q_t) = \gamma(q_u+a) = \gamma(q_u)+\gamma(a) = u$. Therefore $\delta(q_s+I)+\delta(q_t+I) = s+t = u = \delta((q_s+I)\oplus(q_t+I))$. A similar argument shows that $\delta(q_s+I)\delta(q_t+I) = \delta((q_s+I)\odot(q_t+I))$. Thus δ is an isomorphism of semirings. □

(10.17) EXAMPLE. [Cao, Kim & Roush, 1984] Let R be a commutative simple semiring and let n be a positive integer satisfying the condition that $a^n = a^{n+1}$ for all $a \in R$. Let $\gamma: R \rightarrow R$ be the function defined by $\gamma: a \mapsto a^n$. Then $\gamma(e) = e$ for each $e \in I^{\times}(R)$. In particular, this is so for $e = 0$ and $e = 1$. Since R is commutative, $\gamma(ab) = \gamma(a)\gamma(b)$ for all $a, b \in R$. Moreover, if $a, b \in R$ then

$$\gamma(a+b) = (a+b)^n = (a+b)^{2n} = \sum_{i+h=2n} a^i b^h.$$

Thus we see that if $i+h = 2n$ then either $i \geq n$ or $h \geq n$. In the first case, $a^{2n} + a^i b^h = a^n + a^n(a^{i-n}b^h) = a^n$ by Proposition 4.3; while in the second case, by similar reasoning, we have $b^{2n} + a^i b^h = b^n$. This implies that $\gamma(a+b) = a^n + b^n = \gamma(a)+\gamma(b)$ and so γ is a morphism of semirings, the image of which is $I^{\times}(R)$. The kernel of γ is the set N of all nilpotent elements of R which is thus an ideal of R. Therefore, by Proposition 10.16, there is a semiisomorphism from R/N to $I^{\times}(R)$.

(10.18) PROPOSITION. *A semiring S is subisomorphic to a subdirect product of a family $\{R_i \mid i \in \Omega\}$ of semirings if and only if for each $i \in \Omega$ there exists a surjective morphism of semirings $\gamma_i: S \rightarrow R_i$ such that $\cap_{i\in\Omega} ker(\gamma_i) = \{0\}$.*

PROOF. Set $R = \times_{i\in\Omega} R_i$. By Example 10.3, we have a morphism of semirings $\gamma: S \rightarrow R$ given by $\gamma: r \mapsto \langle\gamma_i(r)\rangle$, the kernel of which is $\{0\}$. Therefore S is subisomorphic to the subsemiring $R' = im(\gamma)$ of R. Since γ_i is surjective for each

i, we see that for each $h \in \Omega$ the restriction of the canonical projection $\nu_h: R \to R_h$ to R' is a surjection. Thus R' is a subdirect product of the R_i. $\square$

We now prove versions of the Second Isomorphism Theorem and Third Isomorphism Theorem for semirings.

(10.19) PROPOSITION. *If S is a subsemiring of a semiring R and I is an ideal of R then:*

(1) *$S + I$ is a subsemiring of R;*
(2) *$S \cap I$ is an ideal of S;*
(3) *There exists a surjective morphism of semirings $\gamma: S/(S \cap I) \to (S + I)/I$, which is a semiisomorphism if I is subtractive.*

PROOF. (1) and (2) are clear. Define the function γ by $\gamma: s/(S \cap I) \mapsto s/I$. This is clearly a surjective morphism of semirings. If I is subtractive and $s/(S \cap I) \in ker(\gamma)$ then there exist elements a and a' of I satisfying $s + a = 0 + a'$ and so, by subtractiveness, we have $s \in I$, which implies that $s/(S \cap I) = 0/(S \cap I)$. Therefore γ is a semiisomorphism. $\square$

(10.20) PROPOSITION. *If $I \subseteq H$ are ideals of a semiring R and if $H' = 0/H$, then R/H is isomorphic to $(R/I)/(H'/I)$.*

PROOF. Define a function $\gamma: R/I \to R/H$ by $\gamma: r/I \mapsto r/H$. This function is well-defined since $I \subseteq H$ and, indeed, it is straightforward to show that γ is a surjective morphism of semirings having kernel $\{r/I \in R/I \mid r/H = 0\} = \{I \in R/I \mid r \in H'\} = H'/I$. By Proposition 10.16, γ induces a semiisomorphism γ' from $(R/I)/(H'/I)$ to R/H. If $\gamma((r/I)/(H'/I)) = \gamma'((r'/I)/(H'/I))$ then $r/H = r'/H$ so $(r/I)/(H'/I) = (r'/I)/(H'/I)$. Therefore γ is monic and so is in fact an isomorphism. $\square$

If $\gamma: R \to S$ is a morphism of semirings then $\gamma^{-1}(1_S) = \{r \in R \mid \gamma(r) = 1_S\}$ is not, in general, closed under sums and so is not necessarily an ideal of R. If $r, r' \in g^{-1}(1_S)$ then $\gamma(rr') = \gamma(r)\gamma(r') = 1_S 1_S = 1_S$ and so $\gamma^{-1}(1_S)$ is closed under multiplication. Since 1_R also clearly belongs to this set, we see that it is a submonoid of $(R, \cdot)$. The following result shows that it is sometimes an ideal of R.

(10.21) PROPOSITION. *If $\gamma: R \to S$ is a morphism of semirings and if S is a strongly-infinite element of S then $\gamma^{-1}(s)$ is an ideal of R.*

PROOF. Note that $\gamma^{-1}(s) \neq R$ since $0_R \notin \gamma^{-1}(s)$. On the other hand, if $a, b \in \gamma^{-1}(s)$ and if $r \in R$ then $\gamma(a + b) = \gamma(a) + \gamma(b) = s + s = s$ while $\gamma(ra) = \gamma(r)s = s = s\gamma(r) = \gamma(ar)$. $\square$

(10.22) APPLICATION. Let A be a finite set, let M be the idempotent monoid $(sub(A), \cap)$, and let $R = \mathbb{R}^+[M]$. Let $\epsilon_M: R \to \mathbb{R}^+$ be the augmentation morphism of semirings. Then $\epsilon_M^{-1}(1)$ is the set of all probability distributions on $sub(A)$. These functions are called "unnormalized belief states" in [Hummel & Landy, 1988] and are used to form a space of "belief states" for a statistical theory of evidence used in the design of expert systems which is a modification of the Dempster/Shafer theory of evidence [Shafer, 1976].

If $\gamma: R \to S$ is a morphism of semirings we define the **multiplicative kernel** of γ to be $mker(\gamma) = \gamma^{-1}(1_S) \cap U(R) = \{a \in U(R) \mid \gamma(a) = 1_S\}$. This set is always nonempty since it containa 1_R. It is a proper subset of R since $0 \notin mker(\gamma)$ for any morphism γ.

(10.23) PROPOSITION. *If $\gamma: R \to S$ is a morphism of semirings then $mker(\gamma)$ is a normal subgroup of the group $(U(R), \cdot)$.*

PROOF. If $a, b \in mker(\gamma)$ then $\gamma(ab) = \gamma(a)\gamma(b) = 1_S \cdot 1_S = 1_S$ so $ab \in mker(\gamma)$. If $a \in mker(\gamma)$ then $1_S = \gamma(1_R) = \gamma(aa^{-1}) = \gamma(a)\gamma(a^{-1}) = \gamma(a^{-1})$ so $a^{-1} \in mker(\gamma)$. Thus $mker(\gamma)$ is a subgroup of $U(R)$. Finally, if $r \in U(R)$ and $a \in mker(\gamma)$ then $\gamma(rar^{-1}) = \gamma(r)\gamma(a)\gamma(r^{-1}) = \gamma(r)\gamma(r^{-1}) = \gamma(rr^{-1}) = \gamma(1_R) = 1_S$ so $rar^{-1} \in mker(\gamma)$. Hence $mker(\gamma)$ is normal in $U(R)$. □

If R is a division semiring then we see by Proposition 10.23 that $mker(\gamma)$ is a normal subgroup of the multiplicative group $R \setminus \{0\}$ for each morphism of semirings $\gamma: R \to S$. In such a situation a normal subgroup of $(R \setminus 0, \cdot)$ is called a **normal divisor** of R.

(10.24) PROPOSITION. *A normal divisor N of a division ring R is of the form $mker(\gamma)$ for some morphism of semirings $\gamma: R \to S$ if and only if for all elements $r, r' \in R$ satisfying $r + r' = 1$ and for all $a, b \in N$ we have $ar + br' \in N$.*

PROOF. If $N = mker(\gamma)$ for some morphism $\gamma: R \to S$ and if r, r', a, b are as stated then $\gamma(ar + br') = \gamma(a)\gamma(r) + \gamma(b)\gamma(r') = \gamma(r) + \gamma(r') = \gamma(r + r') = 1_S$ and so $ar + br' \in mker(\gamma) = N$.

Conversely, assume that N satisfies the desired condition. Define a relation $\equiv_N$ on R by setting $r \equiv_N r'$ if and only if $r = r'$ or $r'r^{-1} \in N$. This is clearly an equivalence relation, and we claim that it is a congruence relation as well. Indeed, if $a \equiv_N b$ and $c \equiv_N d$ in R then

$$r = (a + c)(b + d)^{-1} = (ab^{-1})[b(b + d)^{-1}] = (cd^{-1})[d(b + d)^{-1}].$$

But ab^{-1} and cd^{-1} belong to N while $b(b + d)^{-1} + d(b + d)^{-1} = 1$ so, by the assumed property of N, $r \in N$. Therefore $a + c \equiv_N b + d$. Finally, $ac(bd)^{-1} = acd^{-1}b^{-1} = (ab^{-1})b(cd^{-1})b^{-1}$ and this belongs to N since $ab^{-1} \in N$, $cd^{-1} \in N$, and $b(cd^{-1})b^{-1} \in N$ by normality. Thus $ac \equiv_N bd$.

The congruence relation $\equiv_N$ is proper since clearly 0 and 1 are not related under it. Therefore we can define the factor semiring $S = R/\equiv_N$ and the morphism of semirings $\gamma: R \to S$ given by $r \mapsto r/\equiv_N$. For this morphism, $mker(\gamma) = \{r \in R \mid r \equiv_N 1\} = N$. □

(10.25) PROPOSITION. *If R is a division semiring then a morphism of semirings $\gamma: R \to S$ is monic if and only if $mker(\gamma) = \{1\}$.*

PROOF. If γ is monic then surely $mker(\gamma) = \{1\}$. Conversely, assume that γ is not monic. Then there exist elements $a \neq b$ of R satisfying $\gamma(a) = \gamma(b)$. One of these, say a, must be nonzero. Therefore $1_S = \gamma(1) = \gamma(aa^{-1}) = \gamma(a)\gamma(a^{-1}) = \gamma(b)\gamma(a^{-1}) = \gamma(ba^{-1})$, and so $1 \neq ba^{-1} \in mker(\gamma)$. □

11. SEMIRINGS OF FRACTIONS

In this chapter we build the classical semiring of fractions of a semiring using a straightforward adaptation of the method used for rings. This is a special case of the more general method of constructing semirings and semimodules of quotients, to which we will return in Chapter 18.

Let R be a semiring. A **left Øre set** of elements of R is a submonoid A of $(R, \cdot)$ satisfying the following conditions:

(1) For each pair $(a, r) \in A \times R$ there exists a pair $(a', r') \in A \times R$ satisfying $a'r = r'a$;

(2) If $ra = r'a$ for some $r, r' \in R$ and $a \in A$ then there exists an element $a' \in A$ satisfying $a'r = a'r'$;

(3) $0 \notin A$.

Right Øre sets are defined analogously.

(11.1) EXAMPLE. If R is a semiring then any submonoid of $(C(R), \cdot)$ not containing 0 is a left and right Øre set. If $\gamma: R \to S$ is a morphism of semirings then this is true, for example, of $\gamma^{-1}(1_S) \cap C(R)$. In particular, if R is commutative then any submonoid of $(R \setminus \{0\}, \cdot)$ is a left and right Øre set. Thus, if I is a prime ideal of a commutative semiring R then $R \setminus I$ is a left and right Øre set.

A semiring R is a **left** [resp. **right**] **Øre semiring** if and only if $R \setminus \{0\}$ is a left [resp. right] Øre set. Note that Øre semirings are necessarily entire.

(11.2) EXAMPLE. Any commutative entire semiring is a left and right Øre semiring; thus, in particular, $\mathbb{N}$ is a left and right Øre semiring.

For a left Øre set A of elements of a semiring R define the relation $\sim$ on $A \times R$ by setting $(a, r) \sim (a', r')$ if and only if there exist elements u and u' of R such that $ur = u'r'$ and $ua = u'a' \in A$. If $(a, r) \in A \times R$ and if u is an element of R satisfying $ua \in A$ then by taking $u' = 1$ we see that $(a, r) \sim (ua, ur)$. We also note that if $r, r' \in R$ then $(1, r) \sim (1, r')$ if and only if there exists an element $a \in A$ such that $ar = ar'$.

(11.3) PROPOSITION. *Let A be a left Øre set of elements of a semiring R. If $(a_1, r_1) \sim (a_2, r_2)$ in $A \times R$ and if there exist elements u and u' of R satisfying $ua_1 = u'a_2 \in A$ then there exists an element v of R satisfying $vur_1 = vu'r_2$ and $vua_1 = vu'a_2 \in A$.*

PROOF. Since $(a_1, r_1) \sim (a_2, r_2)$, there exist elements R and r' in R satisfying $rr_1 = r'r_2$ and $ra_1 = r'a_2 \in A$. Since A is a left Øre set, there exist elements r'' of R and a'' of A such that $(a''r)a_1 = a''(ra_1) = r''(ua_1) = (r''u)a_1$ and so there exists an element b_1 of A satisfying $b_1(a''r) = b_1(r''u)$. This implies that $(b_1r''u')a_2 = b_1r''ua_1 = b_1a''ra_1 = (b_1a''r')a_2$ and so, again, there exists an element b_2 of A satisfying $b_2(b_1r''u') = b_2(b_1a''r')$ If we now set $v = b_2b_1r''$ then it is straightforward to verify that this element has the desired property. □

(11.4) PROPOSITION. *For each left Øre set A of elements of a semiring R, the relation $\sim$ on $A \times R$ defined above is an equivalence relation.*

PROOF. Clearly $(a, r) \sim (a, r)$ for all $(a, r) \in A \times R$ and $(a, r) \sim (a', r')$ implies that $(a', r') \sim (a, r)$. We are therefore left to show transitivity. Indeed, assume that $(a_1, r_1) \sim (a_2, r_2)$ and $(a_2, r_2) \sim (a_3, r_3)$ in $A \times R$. Then there exist elements $u, u', v, v' \in R$ satisfying $ur_1 = u'r_2$, $vr_2 = v'r_3$, $ua_1 = u'a_2 \in A$, and $va_2 = v'a_3 \in A$. Since A is a left Øre set of elements of R, there exist elements A of A and R of R such that $av = ru'$. Then $(ru)a_1 = r(u'a_2) = (av)a_2 = a(va_2) \in A$. By Proposition 11.3, there exists an element w of R satisfying $w(ru)r_1 = w(av)r_2$ and $w(ru)a_1 = w(av)a_2 \in A$. Then $(wru)r_1 = (wav)r_2 = (wav')r_3$ and similarly $(wru)a_1 = (wav')a_3 \in A$, and we are done. □

We will denote the set $(A \times R)/\sim$ by $A^{-1}R$ and the equivalence class of each pair (a, r) in $A \times R$ by $a^{-1}r$. Note that if u is an element of R satisfying $ua \in A$ then $a^{-1}r = (ua)^{-1}ur$. If B is a nonempty subset of R then we set $A^{-1}B$ to be equal to $\{a^{-1}b \mid a \in A, b \in B\}$.

Now define operations of addition and multiplication on $A^{-1}R$ as follows:

(1) $(a_1^{-1}r_1)+(a_2^{-1}r_2) = (aa_1)^{-1}[ar_1+rr_2]$, where $r \in R$ and $a \in A$ are elements chosen such that $aa_1 = ra_2$;

(2) $(a_1^{-1}r_1)(a_2^{-1}r_2) = (aa_1)^{-1}rr_2$, where $a \in A$ and $r \in R$ are chosen so that $aa_1 \in A$ and $ar_1 = ra_2$.

We must, of course, establish that these operations are indeed well-defined. This will be done in three stages:

Stage I: First, we show that sums and products are independent of the choice of the elements R and a. Indeed, since A is a left Øre domain there exist elements a_0 of A and r_0 of R satisfying $a_0a_1 = r_0a_2$. Now choose elements R of R and A of A such that $aa_1 = ra_2$. Then there exist elements r' of R and a' of A satisfying $r'a_0 = a'a$ and so $(a'r)a_2 = a'(ra_2) = a'(aa_1) = (a'a)a_1 = (r'a_0)a_1 = r'(a_0a_1) = r'(r_0a_2) = (r'r_0)a_2$ and so there exists an element b of A satisfying $b(a'r) = b(r'r_0)$. Moreover, $br'a_0a_1 = ba'aa_1 \in A$ and so $(a_0a_1)^{-1}[a_0r_1 + r_0r_2] = (br'a_0a_1)^{-1}[br'a_0r_1 + br'r_0r_2] = (ba'aa_1)^{-1}[ba'ar_1 + ba'rr_2] = (aa_1)^{-1}[ar_1 + rr_2]$. Thus, this last expression is independent of the choice of a and r.

Similarly, there exist elements c_0 of A and s_0 of R satisfying $c_0r_1 = s_0a_2$. Now choose elements r of R and a of A such that $ar_1 = ra_2$. Then there exist elements s' of R and c' of A satisfying $s'c_0 = c'a$ and so $(c'r)a_2 = c'(ra_2) = c'(ar_1) = (c'a)r_1 = (r'c_0)r_1 = (r's_0)a_2$ and so there exists an element c of A satisfying $cc'r = cr's_0$. Moreover, $cs'c_0a_1 = cc'aa_1 \in A$ and so $(c_0a_1)^{-1}s_0r_2 = (cs'c_0a_1)^{-1}cs'r_0r_2 = (cc'aa_1)^{-1}cc'rr_2 = (aa_1)^{-1}rr_2$ and so this last expression is independent of the choice of A and r.

Stage II: Next, we must show that these operations are independent of the choice of representative of the equivalence class $a_1^{-1}r_1$. Indeed, suppose that $a_1^{-1}r_1 = b_1^{-1}s_1$ in $A^{-1}R$. Then there exist elements u and u' of R such that $ur_1 = u's_1$ and $ua_1 = u'b_1 \in A$. Select elements $r \in R$ and $a \in A$ such that $a(ua_1) = ra_2$. Then $(au)a_1 = ra_2$ and so

$$\begin{aligned}(a_1^{-1}r_1) + (a_2^{-1}r_2) &= (aua_1)^{-1}[aur_1 + rr_2] = (ua_1)^{-1}ur_1 + a_2^{-1}r_2 \\ &= (u'b_1)^{-1}u's_1 + a_2^{-1}r_2 = (au'b_1)^{-1}[au's_1 + rr_2] \\ &= (ab_1)^{-1}[as_1 + rr_2] = (b_1^{-1}s_1) + (a_2^{-1}r_2).\end{aligned}$$

Similarly, select $s \in R$ and $b \in A$ such that $b(ur_1) = sa_2$. Then $b(ua_1) \in A$ so

$$\begin{aligned}(a_1^{-1}r_1)(a_2^{-1}r_2) &= (ua_1)^{-1}sr_2 = [(ua_1)^{-1}ur_1][a_2^{-1}r_2] \\ &= [(u'b_1)^{-1}u's_1][a_2^{-1}r_2] = (bu'b_1)^{-1}sr_2 \\ &= (b_1^{-1}s_1)(a_2^{-1}r_2).\end{aligned}$$

Stage III: Finally, we must show that these operations are independent of the choice of representative of the equivalence class $a_2^{-1}r_2$. Indeed, suppose that $a_2^{-1}r_2 = b_2^{-1}s_2$. Then there exist elements u and u' of R satisfying $ur_2 = u's_2$ and $ua_2 = u'b_2 \in A$. Select elements $r \in R$ and $a \in A$ such that $aa_1 = r(ua_2)$. Then $aa_1 = r(u'b_2)$ and we have $(a_1^{-1}r_1) + (a_2^{-1}r_2) = (aa_1)^{-1}[ar_1 + (ru)r_2] = (aa_1)^{-1}[ar_1 + r(u's_2)] = (a_1^{-1}r_1) + [(u'b_2)^{-1}u's_2] = (a_1^{-1}r_1) + (b_2^{-1}s_2)$. Similarly, select $s \in R$ and $b \in A$ such that $ba_1 \in A$ and $br_1 = s(ua_2)$. Then $br_1 = s(u'b_2)$ so $(a_1^{-1}r_1)(a_2^{-1}r_2) = (ba_1)^{-1}sur_2 = (ba_1)^{-1}su's_2 = (a_1^{-1}r_1)[(u'b_2)^{-1}u'r_2] = (a_1^{-1}r_1)(b_2^{-1}r_2)$.

Thus the operations of sum and product in $A^{-1}R$ are well-defined. We also note that if $a \in A$ and if $r_1, r_2 \in R$ then $(a^{-1}r_1) + (a^{-1}r_2) = a^{-1}(r_1 + r_2)$. Moreover, if $a_1^{-1}r_1$ and $a_2^{-1}r_2$ are elements of $A^{-1}R$ and if $r' \in R$ and $a' \in A$ are elements satisfying $a'a_2 = r'a_1$ then $a_1^{-1}r_1 = b^{-1}(r'r_1)$ and $a_2^{-1}r_2 = b^{-1}(a'r_2)$, where $b = a'a_2 \in A$. Repeating this process finitely-many times, we conclude that any finite set of elements of $A^{-1}R$ can be represented with a "common denominator".

(11.5) PROPOSITION. *If A is a left Øre set of elements of a semiring R then $(A^{-1}R, +, \cdot)$ is a semiring.*

PROOF. The verification of the semiring axioms is routine. The additive identity of $A^{-1}R$ is $1^{-1}0$ and the multiplicative identity of $A^{-1}R$ is $1^{-1}1$. □

The semiring $A^{-1}R$ defined in Proposition 11.5 is called the **classical left semiring of fractions** of R with respect to the left Øre set A. The **classical right semiring of fractions** RB^{-1} of R with respect to a right Øre set B of elements of R is defined in an analogous manner. This construction can also be accomplished for topological semirings. See [Botero & Weinert, 1971].

(11.6) EXAMPLE. Clearly $\mathbb{P}^{-1}\mathbb{N}$ is isomorphic to $\mathbb{Q}^+$.

(11.7) EXAMPLE. Let R be a commutative entire semiring and let $A = R \setminus \{0\}$ which, as we noted in Example 11.2, is an Øre set. If $a^{-1}r$ is a nonzero element of $A^{-1}R$ then we must have $r \neq 0$ and so $r^{-1}a \in A^{-1}R$ as well. Moreover, $a^{-1}r \cdot r^{-1}a = 1^{-1}1$ and so $a^{-1}r \in U(A^{-1}R)$. Thus every nonzero element of $A^{-1}R$ is a unit, proving that $A^{-1}R$ is in fact a semifield.

(11.8) EXAMPLE. Let R be a semiring, let $A = \{t\}$, and let $S = R\langle\langle A\rangle\rangle$ be the ring of formal power series in an indeterminate t. Each $t^i \in A^*$ corresponds to the element $f_i \in S$ defined by

$$f_i : t^h \mapsto \begin{cases} 1 & \text{if } i = h \\ 0 & \text{otherwise} \end{cases}.$$

Thus we can consider A^* as a subset of the center of S and, indeed, it is an Øre set. Moreover, $(A^*)^{-1}S$ is just the **semiring of Laurent series** over R in the indeterminate t.

(11.9) EXAMPLE. If $R = (\mathbb{R} \cup \{-\infty\}, \oplus, \otimes)$ is the schedule algebra (where $\oplus = max$ and $\otimes = +$) and if t is an indeterminate over R, then $A = R \setminus \{0\}$ is a left Øre set of elements of R and so we can consider the semiring $A^{-1}R$. This is done in [Cuninghame-Green & Meijer, 1980]. In particular, they give necessary and sufficient conditions for an element $q(t)^{-1}p(t)$ to have a resolution into partial fractions of the form $\oplus_{i=1}^{n}[t \oplus a_i]^{-1}b_i$. Such resolutions have important applications in optimization theory.

For every left Øre set A of elements of a semiring R we have a function $\gamma_A: R \to A^{-1}R$ defined by $\gamma_A: r \mapsto 1^{-1}r$. Clearly γ_A takes the additive and multiplicative identities of R to the additive and multiplicative identities of $A^{-1}R$ respectively. Moreover, $\gamma_A(r + r') = 1^{-1}(r + r') = 1^{-1}r + 1^{-1}r' = \gamma_A(r) + \gamma_A(r')$ and $\gamma_A(rr') = 1^{-1}(rr') = \gamma_A(r)\gamma_A(r')$ and so γ_A is a morphism of semirings. The kernel of this morphism is precisely $\{r \in R \mid ur = 0 \text{ for some } u \in A\}$. Thus the kernel of this morphism is $\{0\}$ whenever $ar = 0 \Rightarrow r = 0$ for all $a \in A$. This is so if every element of A is left multiplicatively cancellable and hence certainly so if R is entire. In fact, if every element of A is left multiplicatively cancellable then γ_A is easily seen to be injective since $\gamma_A(r) = \gamma_A(r')$ implies that $1^{-1}r = 1^{-1}r'$ and so there exists an element a of A satisfying $ar = ar'$, which in turn implies that $r = r'$. Moreover, if $a \in A$ then $(1^{-1}a)(a^{-1}1) = (a^{-1}1)(1^{-1}a) = 1^{-1}1$ and so $\gamma_A(A) \subseteq U(A^{-1}R)$. As a consequence, we see that if R is a left Øre semiring then $A^{-1}R$ is a division semiring.

(11.10) EXAMPLE. [Vandiver, 1940; Murata, 1950] Let R be a semiring and let c be a multiplicatively cancellable element of $C(R)$. Then $A = \{c^i \mid i \in \mathbb{N}\}$ is an Øre set satisfying the condition that γ_A is injective.

(11.11) PROPOSITION. *Let R be a semiring and let A be a left Øre set of elements of R contained in $I^{\times}(R)$. Then γ_A is surjective.*

PROOF. If $a \in A$ and $r \in R$ then $a^{-1}r = (a^2)^{-1}(ar) = a^{-1}(ar) = 1^{-1}r = \gamma_A(r)$. □

(11.12) EXAMPLE. [Sancho de Salas, 1987] If R is a bounded distributive lattice then $\gamma_A: R \rightarrow A^{-1}R$ is surjective for every Øre subset A of R. Thus, for example, if Y is a subspace of a topological space X, if R is the semiring of all closed subsets of X, and if A is the Øre subset of all closed subsets of X which do not intersect Y then $A^{-1}X$ is the lattice of "germs of closed sets along Y", namely the lattice obtained from R by identifying closed sets which agree in a neighborhood of Y.

(11.13) PROPOSITION. *Let A be a left Øre set of elements of a semiring R and let $\gamma: R \rightarrow S$ be a morphism of semirings satisfying $\gamma(A) \subseteq U(S)$. Then there exists a morphism of semirings $\delta: A^{-1}R \rightarrow S$ satisfying $\delta\gamma_A = \gamma$ and $ker(\delta) = A^{-1}ker(\gamma)$.*

PROOF. Define δ by $\delta(a^{-1}r) = [\gamma(a)]^{-1}\gamma(r)$. This is well-defined since if $a^{-1}r = b^{-1}s$ then there exist elements u and u' of R satisfying $ur = u's$ and $ua = u'b \in A$. Then

$$\begin{aligned}
[\gamma(a)]^{-1}\gamma(r) &= [\gamma(ua)]^{-1}\gamma(ua)\gamma(a)^{-1}\gamma(r) = [\gamma(ua)]^{-1}\gamma(u)\gamma(a)\gamma(a)^{-1}\gamma(r) \\
&= [\gamma(ua)]^{-1}\gamma(u)\gamma(r) = [\gamma(ua)]^{-1}\gamma(ur) \\
&= [\gamma(u'b)]^{-1}\gamma(u's) = [\gamma(u'b)]^{-1}\gamma(u'b)\gamma(b)^{-1}\gamma(s) \\
&= [\gamma(b)]^{-1}\gamma(s).
\end{aligned}$$

Straightforward verification shows that δ is a morphism of semirings satisfying $\delta\gamma_A = \gamma$. Clearly $A^{-1}ker(\gamma) \subseteq ker(\delta)$. Conversely, if $a^{-1}r \in ker(\delta)$ then $\gamma(a)^{-1}\gamma(r) = 0$ and so $\gamma(r) = 0$, proving that $a^{-1}r \in A^{-1}ker(\gamma)$. □

(11.14) APPLICATION. Let R be the semifield $(\mathbb{N} \cup \{-\infty\}, max, +)$ and let S be a zerosumfree semiring. In Example 10.9 we considered the semiisomorphism $\gamma: S[t] \rightarrow R$ which assigns to each polynomial its degree. If A is a left Øre set of elements of $S[t]$ then, by Proposition 11.13, this map can be extended to a morphism of semirings $\delta: A^{-1}S[t] \rightarrow R$ which is also a semiisomorphism.

This construction was considered in [Cuninghame-Green, 1984] for the special case of $S = \mathbb{N}$ and $A = S \setminus \{0\}$ and applied to various problems in optimization theory, linear programming, and quadratic programming. His method is to formulate a problem as a computation in R, to consider its preimage in $S[t]$, solve the problem there using regular polynomial computation, and then translate the solution back to R via γ.

(11.15) PROPOSITION. *If A is a left Øre set of elements of a semiring R and if I is a left ideal of R then:*

(1) *$A^{-1}I$ is a left ideal of $A^{-1}R$ if and only if $A \cap I = \emptyset$;*
(2) *$A^{-1}I = A^{-1}R$ otherwise.*

PROOF. If $a, b, c \in A$, $r \in R$, and $s, t \in I$ then it is easy to verify that $b^{-1}s + c^{-1}t$ and $(a^{-1}r)(b^{-1}s)$ both belong to $A^{-1}I$. Therefore, $A^{-1}I$ is either a left ideal of $A^{-1}R$ or equal to all of $A^{-1}R$. If $a \in A \cap I$ then $1^{-1}a \in A^{-1}I \cap U(A^{-1}R$. so for each element $b^{-1}r$ of $A^{-1}R$ we have $b^{-1}r = [(b^{-1}r)(a^{-1}1)](1^{-1}a) \in A^{-1}I$. Thus $A^{-1}I = A^{-1}R$. Conversely, assume that $A^{-1}I = A^{-1}R$. Then $1^{-1}1 = 1^{-1}b$ for some element b of I. Thus there exists an element u of A satisfying $ub = u$ and so $ub \in I \cap A$, proving that $I \cap A \neq \emptyset$. □

(11.16) PROPOSITION. *If A is a left Øre set of elements of a semiring R which satisfies the condition that γ_A is injective then:*

(1) *$A^{-1}R$ is cancellative whenever R is;*
(2) *$A^{-1}R$ is plain whenever R is.*

PROOF. (1) Assume that R is cancellative and let u_1, u_2, and u_3 be elements of $A^{-1}R$ satisfying $u_1 + u_3 = u_2 + u_3$. By the remark before Proposition 11.5, we see that there exist an element a of A and elements r_1, r_2, and r_3 of R such that $u_i = a^{-1}r_i$ for $i = 1, 2, 3$. Hence

$$\begin{aligned}\gamma_A(r_1 + r_2) &= 1^{-1}(r_1 + r_2) = (1^{-1}a)(a^{-1}[r_1 + r_2]) \\ &= (1^{-1}a)[a^{-1}r_1 + a^{-1}r_2] = (1^{-1}a)[a^{-1}r_1 + a^{-1}r_3] \\ &= (1^{-1}a)(a^{-1}[r_1 + r_3]) = 1^{-1}(r_1 + r_3) \\ &= \gamma_A(r_1 + r_3)\end{aligned}$$

and so $r_1 + r_2 = r_1 + r_3$. Since R is cancellative, this implies that $r_2 = r_3$ and so $u_2 = u_3$. Thus $A^{-1}R$ is also cancellative.

(2) Assume that R is plain and that u and v are elements of $A^{-1}R$ satisfying $u+v = v$. By the remark before Proposition 11.5, there exist an element a of A and elements R and r' of R such that $u = a^{-1}r$ and $v = a^{-1}r'$. Thus $a^{-1}[r+r'] = a^{-1}r'$ so $\gamma_A(r+r') = 1^{-1}(r+r') = (1^{-1}a)[a^{-1}(r+r')] = (1^{-1}a)(a^{-1}r') = 1^{-1}r' = \gamma_A(r')$ and hence $r + r' = r'$. This implies that $r \in Z(R)$ and so $r = 0$, whence $u = 0$. Thus $A^{-1}R$ is plain. □

If R is a commutative semiring and A is an Øre set of elements of R, then LaGrassa [1995] has shown that $A^{-1}I$ is a prime ideal of $A^{-1}R$ for every prime ideal I of R disjoint from A. Also, conversely, if H is a prime ideal of $A^{-1}R$ then $\gamma_A^{-1}(H)$ is a prime ideal of R. Thus there in fact exists an order-preserving bijection between $spec(A^{-1}R)$ and $\{I \in spec(R) \mid I \cap A = \varnothing\}$.

If R is a subsemiring of a semiring S then S is a **left semiring of fractions** of R if and only if for all $s \neq s' \in S$ and $s'' \in S$ there exists an element R of R such that $rs \neq rs'$ and $rs'' \in R$. This condition was studied for rings in [Lambek, 1966] and for bounded distributive lattices in [Schmid, 1983].

(11.17) PROPOSITION. *If A is a left Øre set of elements of a semiring R then $A^{-1}R$ is a left semiring of fractions of $\gamma_A(R)$.*

PROOF. If $a^{-1}r \neq b^{-1}s$ and $c^{-1}t$ are elements of $A^{-1}R$ then, from the definitions, one sees that $(1^{-1}c)(c^{-1}t) = 1^{-1}t \in \gamma_A(R)$ and $(1^{-1}c)(a^{-1}r) \neq (1^{-1}c)(b^{-1}s)$. □

The construction of other semirings of fractions of a semiring is tied in with general localization theory for semimodules over semirings, and we will therefore defer its consideration until Chapter 18.

12. EUCLIDEAN SEMIRINGS

If a is an element of a semiring R then we denote by $RD(a)$ the set of all right divisors of a in the monoid $(R, \cdot)$. That is to say, $RD(a) = \{b \in R \mid a \in Rb\} = \{b \in R \mid Ra \subseteq Rb\}$. Since $b \in RD(b)$ for all $b \in R$, it is clearly true that $b \in RD(a)$ if and only if $RD(b) \subseteq RD(a)$. Note that if R is a simple semiring and if $b \in RD(a)$ then there exists an element r of R such that $a = rb$ and so, by Proposition 4.3, we have $a + b = rb + b = b$. Thus we see that if a is an element of a simple semiring R then $RD(a) \neq \varnothing$ implies that $a \in Z(R)$.

If a is an element of a semiring R then $U(R) \subseteq RD(1_R) \subseteq RD(a)$. If $a \notin U(R)$ and $RD(a) = U(R) \cup \{a\}$ then a is said to be **irreducible from the right**. Irreducibility from the left is defined similarly.

(12.1) EXAMPLE. [Jacobson & Wisner, 1966] If $R = \mathcal{M}_2(\mathbb{N})$ then the only elements of R having determinant 1 which are irreducible from the right are $\begin{bmatrix} 1 & 1 \\ 0 & 1 \end{bmatrix}$ and $\begin{bmatrix} 1 & 0 \\ 1 & 1 \end{bmatrix}$.

If A is a nonempty subset of a semiring R then the set of **common right divisors** of A is $CRD(A) = \cap\{RD(a) \mid a \in A\} = \{b \in R \mid RA \subseteq Rb\}$. An element $b \in CRD(A)$ is a **greatest common right divisor** of A if and only if $CRD(A) = RD(b)$.

(12.2) PROPOSITION. *If A is a nonempty subset of a semiring R then an element b of R is a greatest common right divisor of A if and only if the following conditions are satisfied:*

(1) *$RA \subseteq Rb$;*
(2) *If $c \in R$ satisfies $RA \subseteq Rc$ then $Rb \subseteq Rc$.*

PROOF. Assume that b is a greatest common right divisor of A. Then $b \in CRD(A)$ and so $b \in RD(a)$ for each $a \in A$. Thus $Ra \subseteq Rb$ for each $a \in A$, implying that $RA \subseteq Rb$. Moreover, if $RA \subseteq Rc$ for some element c of R then $c \in CRD(A) = RD(b)$ and so $Rb \subseteq Rc$.

Conversely, assume conditions (1) and (2) are satisfied. By (1), $b \in CRD(A)$ and so $RD(b) \subseteq CRD(A)$. By (2), if $c \in CRD(A)$ then $RA \subseteq Rc$ and so $Rb \subseteq Rc$. Hence $c \in RD(b)$, proving that $CRD(A) \subseteq RD(b)$ and thus yielding equality. □

(12.3) COROLLARY. *If every left ideal of a semiring R is principal, then every nonempty subset of R has a greatest common right divisor.*

PROOF. Let A be a nonempty subset of R. Then $RA = R$ or RA is a left ideal of R. Hence, by hypothesis, there exists an element b of R satisfying $RA = Rb$. By Proposition 12.2, b is a greatest common right divisor of A. □

(12.4) PROPOSITION. *Let a, b, and c be elements of a semiring R. If d is a greatest common right divisor of $\{a,b\}$ and e is a greatest common right divisor of $\{c,d\}$ then e is a greatest common right divisor of $\{a,b,c\}$.*

PROOF. By definition, $RD(e) = RD(d) \cap RD(c) = RD(a) \cap RD(b) \cap RD(c) = CRD(\{a,b,c\})$. □

If a and b are elements of a semiring R then $CRD(\{a,b\})$ is clearly contained in $CRD(\{a+b,b\})$. We now investigate the conditions for having equality.

(12.5) PROPOSITION. *The following conditions on a semiring R are equivalent:*

(1) *$CRD(\{a,b\}) = CRD(\{a+b,b\})$ for all $a, b \in R$;*

(2) *Every principal left ideal of R is subtractive.*

PROOF. Assume (1) and let Rd be a principal left ideal of R. If a and $a+b$ belong to Rd then $d \in CRD(\{a+b,a\}) = CRD(\{a,b\})$ and so $b \in Rd$. Therefore Rd is subtractive. Conversely, assume (2) and let $a, b \in R$. If $d \in CRD(\{a+b,b\})$ then $a+b$ and b both belong to Rd and so, by (2), $a \in Rd$. Therefore $d \in CRD(\{a,b\})$. □

A semiring for which the equivalent conditions of Proposition 12.5 hold will be called a **PLIS-semiring**.

(12.6) EXAMPLE. Recall that in Example 6.28 we presented a semiring R having a nonzero left ideal H containing no nonzero subtractive left ideals. In particular, if $0 \neq h \in H$ then Rh is not subtractive. Hence R is not a PLIS-semiring.

Elements a and b of a semiring R are **right associates** if there exists an element $u \in U(R)$ satisfying $a = ub$. Note that in this case $b = u^{-1}a$ and $Ra = Rb$.

A **left euclidean norm** δ defined on a semiring R is a function $\delta\colon R \setminus \{0\} \to \mathbb{N}$ satisfying the following condition:

(*) If a and b are elements of R with $b \neq 0$ then there exist elements q and r of R satisfying $a = qb + r$ with $r = 0$ or $\delta(r) < \delta(b)$.

A **right euclidean norm** is defined similarly, except that in condition (*) we have $a = bq + r$. A semiring R is **left [resp. right] euclidean** if and only if there exists a left [resp. right] euclidean norm defined on R. For commutative semirings, needless to say, the notions of left and right euclidean norm coincide. If we want to emphasize the role of δ, we will speak of the euclidean semiring (R, δ).

If δ is a left euclidean norm on a semiring R we can extend δ to a function δ' from R to $\mathbb{N} \cup \{\infty\}$ by setting $\delta'(0) = \infty$. This function satisfies the condition that if a and b are elements of R satisfying $\delta'(a) \geq \delta'(b)$ then there exist elements q and r of R satisfying $a = qb + r$, where $r = 0$ or $\delta'(r) < \delta'(b)$. Conversely, if $\delta'\colon R \to \mathbb{N} \cup \{\infty\}$ is a function satisfying this condition then its restriction is a left euclidean norm on R.

(12.7) Example. The semiring $\mathbb{N}$ is euclidean if we define the euclidean norm δ by $\delta\colon n \mapsto n$ or $\delta\colon n \mapsto n^2$.

(12.8) Example. [Hebisch & Weinert, 1987] Let $S[t]$ be the semiring of polynomials in the indeterminate t over a division semiring S and let ρ be the congruence relation on $S[t]$ defined by $\sum a_i t^i \;\rho\; \sum b_i t^i$ if and only if $a_1 t + a_0 = b_1 t + b_0$. Let R be the factor semiring $S[t]/\rho$. Then the there exists a left euclidean norm $\delta\colon R \setminus \{0\} \to \mathbb{N}$ defined by setting

$$\delta\left(\sum a_i t^i/\rho\right) = \begin{cases} 1 & \text{if } a_0 \neq 0 \\ 0 & \text{otherwise} \end{cases}$$

(12.9) Example. [Hebisch & Weinert, 1987] Let R be the subsemiring of $\mathbb{Q}^+$ defined by $R = \{q \in \mathbb{Q}^+ \mid q = 0 \text{ or } q \geq 1\}$ and suppose that we have a left euclidean norm $\delta\colon R \setminus \{0\} \to \mathbb{N}$. Let $0 < a < b$ be elements of R. If $\delta(a) \geq \delta(b)$ then there would have to exist elements q and r of R satisfying $a = qb + r$, where $r = 0$ or $\delta(r) < \delta(b)$. But $a < b$ implies that $a < qb$ for all $0 \neq q \in R$ and $a = 0b + r$ leads to the contradiction $\delta(a) = \delta(r) < \delta(b)$. Thus $a < b$ implies that $\delta(a) < \delta(b)$ for all $0 \neq a, b \in R$. Therefore $R \setminus \{0\}$ is order-isomorphic to the subset $im(\delta)$ of $\mathbb{N}$, which is impossible. Thus no left euclidean norm can be defined on R, and so R is not a left euclidean semiring.

(12.10) Proposition. *If δ is a left euclidean norm defined on a semiring R then there exists another left euclidean norm δ^* defined on R and satisfying:*

(1) *$\delta^*(a) \leq \delta(a)$ for all $a \in R \setminus \{0\}$; and*
(2) *$\delta^*(b) \leq \delta(rb)$ for all $b, r \in R$ satisfying $rb \neq 0$.*

Proof. For each $0 \neq a \in R$, set $\delta^*(a) = min\{\delta(ra) \mid ra \neq 0\}$. The function δ^* clearly satisfies (1) and (2), so all we have to show is that it is indeed a left euclidean norm on R. Let a and b be nonzero elements of R satisfying $\delta^*(a) \geq \delta^*(b)$. Then there exists an element s of R such that $\delta^*(sb) = \delta(sb)$. Then $\delta(a) \geq \delta(sb)$ and so there exist elements q and r of R such that $a = qsb + r$, where $r = 0$ or $\delta(r) < \delta(sb)$. In the second case, we have $\delta^*(r) \leq \delta(r) < \delta(sb) = \delta^*(b)$. Thus δ^* is a left euclidean norm on R. □

Thus, if (R, δ) is a left euclidean semiring we can, without loss of generality, assume that δ satisfies the condition that $\delta(b) \leq \delta(rb)$ for all $0 \neq b \in R$ and all $r \in R$ such that $rb \neq 0$. A left euclidean norm satisfying this condition is said to be **submultiplicative**. A left euclidean norm δ defined on a semiring R is **multiplicative** if and only if $\delta(ab) = \delta(a)\delta(b)$ for all $a, b \in R$ satisfying $ab \neq 0$. That is to say, δ is multiplicative if and only if it is a semigroup homomorphism from $R \setminus \{0\}, \cdot)$ to $(\mathbb{N}, \cdot)$.

(12.11) Proposition. *If R is a semiring on which we have defined a submultiplicative euclidean norm $\delta\colon R \setminus \{0\} \to \mathbb{N}$ and if $M_\delta = \{r \in R \mid \delta(r)\}$ is a minimal element of $im(\delta)$ then:*

(1) *$1_R \in M_\delta$;*

(2) *If $a \in M_\delta$ then there exists an element q of R satisfying $1 = qa$;*
(3) $M_\delta \cap I^+(R) = \{1_R\}$;
(4) $U(R) \subseteq M_\delta$, *with equality holding if R is commutative.*

PROOF. (1) If $0 \neq a \in R$ then, by submultiplicity, $\delta(1_R) \leq \delta(a1_R) = \delta(a)$ and so $1_R \in M_\delta$.

(2) If $a \in M_\delta$ then there exist elements q and r of R satisfying $1_R = qa + r$ with $r = 0$ or $\delta(r) < \delta(a)$. The latter case is impossible by minimality, and so $1_R = qa$.

(3) If $c \in M_\delta \cap I^+(R)$ then, by (2), there exists an element q of R satisfying $1_R = qc$ and so $c = 1_R c = qc^2 = qc = 1_R$.

(4) If $a \in U(R)$ there exists an element b of R such that $1_R = ba$ and so $\delta(a) \leq \delta(ba) = \delta(1_R)$. Since $1_R \in M_\delta$, this means we have equality and $a \in M_\delta$. Hence $U(R) \subseteq M_\delta$. If R is commutative, then by (2) the reverse containment is true. □

(12.12) PROPOSITION. *Let R be a commutative cancellative semiring and let δ be a submultiplicative euclidean norm defined on R. Then $\delta(a) = \delta(-a)$ for all $a \in V(R)$.*

PROOF. Assume the result is false and let A be the nonempty set of all of nonzero elements a' of $V(R)$ satisfying $\delta(a') > \delta(-a')$. Choose a to be an element of A for which $\delta(-a)$ is minimal. Then there exist elements q and r of R satisfying $a = q(-a)+r$, where $r = 0$ or $\delta(r) < \delta(-a)$. Assume $r \neq 0$. Then $q(-a)+(-a)+r = 0$ and so $c = q(-a) + (-a) \in V(R)$ and $-[q(-a) + (-a)] = r$. Moreover, $c \notin A$, since otherwise we would contradict the choice of a. Thus $\delta(c) = \delta(r)$. But this is impossible since then $\delta(r) = \delta(c) = \delta([q+1](-a)) \geq \delta(-a)$ by submultiplicity. Thus we must have $r = 0$ and hence $a = q(-a)$. Then $qa+a = q(a+(-a)) = q0 = 0$ and so $-a = qa + a + (-a) = qa$. This implies that $\delta(-a) \geq \delta(a)$, contradicting the assumption that $a \in A$. Thus A must be empty, proving the proposition. □

(12.13) PROPOSITION. *If $\gamma: R \to S$ is a surjective morphism of semirings and if δ is a left euclidean norm on R then there exists a left euclidean norm δ' on S defined by $\delta'(c) = min\{\delta(a) \mid a \in \gamma^{-1}(c)\}$ for all $0 \neq a \in S$.*

PROOF. Let c and d be elements of S with $d \neq 0$. Then there exist elements a and $b \neq 0$ of R such that $\gamma(a) = c$ and $\gamma(b) = d$. Moreover, we can choose b so that $\delta'(d) = \delta(b)$. Since δ is a left euclidean norm on R, there exist elements q and r of R satisfying $a = qb+r$, where $r = 0$ or $\delta(r) < \delta(b)$. Hence $c = \gamma(a) = \gamma(q)d+\gamma(r)$, where $\gamma(r) = 0$ or $\delta'(\gamma(r)) \leq \delta(r) < \delta(b) = \delta'(d)$. This proves that δ' is a left euclidean norm on S. □

(12.14) PROPOSITION. *If R is a left Euclidean semiring then every subtractive left ideal of R is principal.*

PROOF. Let δ be a left euclidean norm on R and let I be a subtractive left ideal of R. Then $\{\delta(a) \mid a \in I\}$ has a minimal element, say $\delta(b)$. Assume that $a \in I \setminus Rb$. Then there exists an element $r \in R \setminus \{0\}$ such that $a = qb + r$ and $\delta(r) < \delta(b)$. But $r \in I$ since I is subtractive, contradicting the minimality of $\delta(b)$. Hence we must have $I = Rb$. □

(12.15) Proposition. *The following conditions on a left Euclidean semiring R are equivalent:*

(1) *R is a PLIS-semiring;*

(2) *There exists a left Euclidean norm δ defined on R satisfying the condition that if $a = qb + r$ for $r \in R \setminus \{0\}$ and $\delta(r) < \delta(b)$ then $a \notin Rb$.*

Proof. (1) $\Rightarrow$ (2): By Proposition 12.10, we know that there exists a left euclidean norm δ on R satisfying the condition that $\delta(s) \leq \delta(rs)$ for all $r, s \in R \setminus \{0\}$. Assume that $a = qb + r$ for $r \in R \setminus \{0\}$ and $\delta(r) < \delta(b)$. If $a \in Rb$ then by (1) we must have $r = cb$ for some $c \in R$ and so $\delta(r) \geq \delta(b)$, which is a contradiction. Thus $a \notin Rb$.

(2) $\Rightarrow$ (1): Assume that $a, b \in R$ and that $t \in CRD(\{a + b, b\})$. Then we can write $a + b = dt$ and $b = et$ for elements d and e of R. By the choice of δ, we know that $\delta(a) \geq \delta(t)$ and so either $a = qt$ or $a = qt + r$ for some $0 \neq r \in R$ satisfying $\delta(r) < \delta(t)$. But in the latter case we have $dt = (e + q)t + r$, which again contradicts the stated condition. Thus we must have $a = qt$ and so $t \in RD(a)$. Since $t \in RD(b)$ by the choice of t, we have $t \in CRD(\{a, b\})$. Thus R is a PLIS-semiring by Proposition 12.5. □

(12.16) Proposition. *If R is a left Euclidean PLIS-semiring then any nonempty finite subset A of R has a greatest common right divisor.*

Proof. By Proposition 12.4, it suffices to consider the case of $A = \{a, b\}$. If $a = b = 0$ then 0 is a greatest common right divisor of $\{a, b\}$ and we are done. Hence, without loss of generality, we can assume that $b \neq 0$. Since R is a PLIS-semiring, we know by Proposition 12.15 that there exists a left Euclidean norm δ defined on R satisfying the condition that if $a = qb + r$ for $r \in R \setminus \{0\}$ satisfying $\delta(r) < \delta(b)$ then $a \notin Rb$.

By repeated applications of δ, we can find elements $q_1, \ldots, q_{n+1}$ and $r_1, \ldots, r_n$ of $R \setminus \{0\}$ such that $a = q_1 b + r_1$, $b = q_2 r_1 + r_2$, $\ldots$, $r_{n-2} = q_n r_{n-1} + r_n$, $r_{n-1} = q_{n+1} r_n$ and $\delta(b) > \delta(r_1) > \cdots > \delta(r_n)$. (The process of selecting the q_i and r_i must indeed terminate after finitely-many steps, since there are no infinite decreasing sequences of elements of $\mathbb{N}$.) Working backwards, we then see that

$$r_{n-2} = [q_n q_{n+1} + 1] r_n$$
$$r_{n-3} = [q_{n-1} q_n q_{n+1} + q_n - 1 + q_{n+1}] r_n$$

etc. until we establish that $r_n \in CRD(\{a, b\})$. Conversely, assume that $d \in CRD(\{a, b\})$. By Proposition 12.15, we see that $d \in RD(r_1)$, $d \in RD(r_2)$, $\ldots$, $d \in RD(r_n)$ and so $RD(r_n) = CRD(\{a, b\})$. Thus r_n is a greatest common right divisor of $\{a, b\}$. □

Closely related to the notion of a Euclidean norm is that of a Dale norm. If R is a commutative antisimple semiring then a function $\delta: R \to \mathbb{N}$ is a **Dale norm** if and only if the following conditions are satisfied:

(1) $\delta(a) = 0$ if and only if $a = 0_R$;

(2) If $0_R \neq a + b \in R$ then $\delta(a + b) \geq \delta(a)$;

(3) $\delta(ab) = \delta(a)\delta(b)$ for all $a, b \in R$;

(4) If $a \in R$ and $0 \neq b \in R$ then there exist elements q and r of R such that $a = qb + r$, where $r = 0$ or $\delta(r) < \delta(b)$.

One sees immediately that a Dale norm is a left and right euclidean norm. Also, if R is a semiring on which a Dale norm is defined then R must be entire.

(12.17) EXAMPLE. If R is a division semiring then we can define a Dale norm δ on R by $\delta(0) = 0$ and $\delta(a) = 1$ for all $0 \neq a \in R$.

(12.18) EXAMPLE. Let R be the semiring $(\mathbb{N} \cup \{-\infty\}, max, +)$ and let $1 < c \in \mathbb{R}$. Then the function $\delta \colon R \to \mathbb{N}$ defined by $\delta(-\infty) = 0$ and $\delta(i) = c^i$ for $i \in \mathbb{N}$ is a Dale norm on R.

(12.19) EXAMPLE. The functions $n \mapsto n$ and $n \mapsto n^2$ are Dale norms defined on $\mathbb{N}$.

(12.20) EXAMPLE. A left euclidean norm need not be a Dale norm, even if R is a commutative ring (which is surely antisimple as a semiring). For example, consider $R = \mathbb{Z}/(4)$ and define the function $\delta \colon R \setminus \{0\} \to \mathbb{N}$ by $\delta(1) = \delta(3) = 2$ and $\delta(2) = 3$. Then δ is a left euclidean norm which cannot be converted into a Dale norm since the ring R is not entire.

(12.21) PROPOSITION. *If R is a commutative antisimple semiring on which we have defined a Dale norm δ then:*

(1) *$U(R) = \{a \in R \mid \delta(a) = 1\}$;*
(2) *R is a division semiring if and only if $\delta(R)$ is finite.*

PROOF. (1) Note first that $0 \neq \delta(1_R) = \delta(1_R \cdot 1_R) = \delta(1_R)^2$ and so $\delta(1_R)$ is a nonzero idempotent of $\mathbb{N}$, which implies that $\delta(1_R) = 1$. If $a \in U(R)$ then there exists an element b of R satisfying $ab = 1$ so $1 = \delta(ab) = \delta(a)\delta(b)$, which implies that $\delta(a) = \delta(b) = 1$ since $\mathbb{N}$ has only one unit. Conversely, assume that $a \in R$ satisfies the condition that $\delta(a) = 1$. Then there exist elements q and r of R such that $1_R = qa + r$ and either $r = 0$ or $\delta(r) < \delta(a)$. Since $\delta(a) = 1$ we must have $r = 0$ and so $1_R = qa$, proving that $a \in U(R)$.

(2) If R is a division semiring then $\delta(R) = \{0, 1\}$ and this is finite. Conversely, if R is not a division semiring then there exists a nonzero element r of R which is not a unit. By (1), this means that $\delta(r) > 1$. Moreover, since r is not a unit, neither is r^k for all $k \geq 1$ and, for each $k > 1$, we have $\delta(r^k) = \delta(r)\delta(r^{k-1}) > \delta(r^{k-1})$. Thus $\delta(R)$ is infinite. □

(12.22) PROPOSITION. *If R is a commutative antisimple semiring on which we have a Dale norm $\delta \colon R \to \mathbb{N}$ then R is isomorphic to one of the following:*

(1) *$\mathbb{N}$;*
(2) *$(\mathbb{N} \cup \{-\infty\}, max, +)$; or*
(3) *A division semiring.*

PROOF. Let $A = \{n1_R \mid n \in \mathbb{N}\}$ be the basic subsemiring of R. First left us consider the case that $R = A$. In this case, we have a surjective morphism of semirings $\gamma \colon \mathbb{N} \to R$ given by $n \mapsto n1_R$. If γ is injective, we have shown that R is isomorphic to $\mathbb{N}$ and we are done. If γ is not injective, then $im(\gamma) = R$ is finite and hence $\delta(R)$ is a finite subset of $\mathbb{N}$. By Proposition 12.21(2), R is a division semiring and we are done.

Now assume that $R \neq A$ and let $a \in R \setminus A$. Since R is antisimple, there exists an element a_1 of R satisfying $a = a_1 + 1_R$. Clearly $a_1 \notin A$. By an easy induction, we see that for each $n \geq 1$ there exists an element a_n of $R \setminus A$ satisfying $a = a_n + n1_R$. Hence $\delta(a) \geq \delta(n1_R)$ for all natural numbers n. Thus, for each $k, n \in \mathbb{N}$ we have $\delta(a) \geq \delta(n^k 1_R) = \delta((n1_R)^k) = \delta(n1_R)^k$, which forces $\delta(n1_R) \leq 1$ for all $n \in \mathbb{N}$. We now consider two cases:

Case I: $1_R + 1_R = 0$. If $0 \neq a \in R$ then $\delta(a) \neq 0$ and $1 = \delta(1_R) = \delta(1_R + a1_R + a1_R) \geq \delta(a1_R) = \delta(a)\delta(1_R) = \delta(a)$ so $\delta(a) = 1$. By Proposition 12.21(2), this implies that $a \in U(R)$ and so R is a division semiring and we are done.

Case II: $1_R + 1_R = b \neq 0$. Then $\delta(b) = 1$ by the above and so $b \in U(R)$. Let c be an element of R satisfying $bc = 1_R$. Then $c + c = cb = 1_R$. Moreover, since R is antisimple there exists an element y of R satisfying $c = y + 1_R$. If $z = c + y$ then $z + 1_R = c + y + 1_R = c + c = 1_R$ and $\delta(z) \leq \delta(1_R) = 1$. Thus $z = 0$ or $z \in U(R)$. If $z = 0$ then $c + y = 0$ so $0 = 1_R + by$. If $0 \neq a \in R$ then $1 = \delta(1_R) = \delta(1_R + a1_R + aby) \geq \delta(a1_R) = \delta(a)$ and so $\delta(a) = 1$. Thus, again, R is a division semiring and we are done. Hence assume that $z \in U(R)$. Then $z + z = c + c + y + y = 1_R + y + y = c + y = z$ and so $z \in I^+(R)$. If $z' \in R$ satisfies $z'z = 1_R$ then $1_R = z'z = z'(z + z) = z'z + z'z = 1_R + 1_R$. This implies that R is additively idempotent.

Thus we are in the situation in which $A = \{0, 1\}$ is a proper subsemiring of the semiring R. If $0 \neq a \in R$ then, by antisimplicity, there exists an element b of R satisfying $a = b + 1_R$ and so $a + 1_R = b + 1_R + 1_R = b + 1_R = a$. If $a \in U(R)$ and if a' is an element of R satisfying $aa' = 1_R$ then $1_R = aa' = (a + 1_R)a' = aa' + a' = 1_R + a' = a'$ so $a = 1_R$. Thus we see that $U(R) = \{1_R\}$. Thus, in particular, we note that if $r \in R \setminus A$ then $r \notin U(R)$ so $\delta(r) > 1$. Pick an element a of $R \setminus A$ having the property that $\delta(a)$ is minimal in $\mathbb{N}$. Then $\delta(a) < \delta(a^2) < \ldots$ in $\mathbb{N}$. Let $G = \{1_R, a, a^2, \ldots\}$. Then G is an infinite cyclic semigroup. Since $1_R + a^h = a^h$ for all $h \in \mathbb{N}$, we see that $a^i + a^{i+h} = a^{i+h}$ for all $i, h \in \mathbb{N}$. Hence $R' = G \cup \{0\}$ is a subsemiring of R and there is an isomorphism γ from $(\mathbb{N} \cup \{-\infty\}, max, +)$ to R' satisfying $\gamma(-\infty) = 0$ and $\gamma(i) = a^i$ for all $i \in \mathbb{N}$. We will therefore complete the proof if we can show that $R = R'$.

Assume that $R \neq R'$ and let $c \in R \setminus R'$. Then $\delta(c) > 1$ and we can pick c among those elements of $R \setminus R'$ such that $\delta(c)$ is minimal. By the choice of a and by the fact that $\delta(a^i) = \delta(a^{i+1})$ for all natural numbers i, there exists a natural number n such that $\delta(a^n) \leq \delta(c) < \delta(a^{n+1})$. Moreover, since R is a euclidean semiring there exist elements q and r of R such that $c = qa^n + r$, where $r = 0$ or $\delta(r) < \delta(a^n)$. If $r \neq 0$ then the choice of c implies that $r = a^k$ for some $1 \leq k \leq n - 1$ and hence $c = (qa^{n-k} + 1_R)a^k$. If $qa^{n-k} = 0$ we obtain $c = a^k \in G$, which is a contradiction. Hence $qa^{n-k} \neq 0$. By the above, we then have $qa^{n-k} + 1_R = qa^{n-k}$ and so $c = qa^n$, corresponding to the case $r = 0$. Moreover, we now have $\delta(a)^n = \delta(a^n) \leq \delta(c) = \delta(q)\delta(a)^n < \delta(a^{n+1}) = \delta(a)^{n+1}$ and so $1 \leq \delta(q) < \delta(a)$, implying that $\delta(q) = 1$ and so $q \in U(R)$. But this means that $q = 1_R$, which is a contradiction. Thus we must have $R = R'$, as desired. □

13. ADDITIVELY-REGULAR SEMIRINGS

An element a of a semiring R is **additively regular** if and only if there exists an element $a^{\#}$ of R satisfying $a + a + a^{\#} = a$ and $a^{\#} + a^{\#} + a = a^{\#}$. Actually, as in the case of multiplicatively-regular elements, it suffices to assume that there exists an element b of R satisfying $a + a + b = a$ for, if such an element exists, the element $a^{\#} = b + b + a$ satisfies both of the above conditions. If $a \in I^{+}(R)$ then a is additively regular with $a^{\#} = a$. If ρ is a congruence relation on R and a is an additively-regular element of R then surely a/ρ is an additively-regular element of R/ρ.

(13.1) PROPOSITION. *If a is an additively-regular element of a semiring R then the element $a^{\#}$ is unique.*

PROOF. Assume that b and c are elements of R satisfying $a+a+b = a = a+a+c$, $b+b+a = b$, and $c+c+a = c$. Then $b = b+b+a = b+b+a+a+c = b+a+c = c+b+a = c+c+b+a+a = c+c+a = c$. □

We will denote the set of all additively-regular elements of a semiring R by $reg(R)$. This set is nonempty since $0 \in reg(R)$ with $0^{\#} = 0$. Also, if $a \in reg(R)$ then $a^{\#} \in reg(R)$ with $a^{\#\#} = a$. Note that we clearly have $I^{+}(R) \subseteq reg(R) \cap Z(R)$.

(13.2) PROPOSITION. *If R is a semiring then $reg(R) \in ideal(R)$.*

PROOF. If $a, b \in reg(R)$ then $(a^{\#} + b^{\#}) + (a + b) + (a + b) = a + b$ and

$$(a^{\#} + b^{\#}) + (a^{\#} + b^{\#}) + (a + b) = a^{\#} + b^{\#}$$

so $a+b$ is additively regular with $(a+b)^{\#} = a^{\#} + b^{\#}$. If $a \in reg(R)$ and $r \in R$ then $ra^{\#} + ra + ra = ra$ and $ra^{\#} + ra^{\#} + ra = ra^{\#}$, so ra is additively regular, with $(ra)^{\#} = ra^{\#}$. Similarly ar is additively regular with $(ar)^{\#} = a^{\#}r$. Thus $reg(R)$ is either all of R or is an ideal of R. □

Note, in particular, that if $a, b \in reg(R)$ then $a^{\#}b^{\#} = (a^{\#}b)^{\#} = (ab)^{\#\#} = ab$. Thus, if $a \in I^{\times}(R)$ we have $a = aa = (a^{\#})^2$.

If a is an additively-regular element of a semiring R we set $a^{\circ} = a + a^{\#}$. This is an additively-idempotent element of R. Conversely, if a is an additive-idempotent element of R then $a = a + a = a + a^{\#} = a^{\circ}$. We note that if $a, b \in R$ then $a^{\circ}b = (a + a^{\#})b = ab + a^{\#}b = ab + (ab)^{\#} = (ab)^{\circ}$. Similarly, $ab^{\circ} = (ab)^{\circ}$.

The semiring R is **additively regular** if and only if $R = reg(R)$. If R is additively regular then clearly so is R^A for any nonempty set A and $\mathcal{M}_n(R)$ is also additively regular for every positive integer n.

(13.3) EXAMPLE. [Tirasupa, 1979] A sufficient condition for R to be additively regular is that $R = V(R) + I^+(R)$. Indeed, if this condition holds and if $a \in R$ then we can write $a = b + e$, where b has an additive inverse and e is additively idempotent. If $a^{\#} = -b+e$ we then have $a+a+a^{\#} = b+e+b+e+(-b)+e = b+e = a$ while $a^{\#} + a^{\#} + a = (-b) + e + (-b) + e + b + e = (-b) + e = a^{\#}$.

(13.4) EXAMPLE. A ring R is additively regular, with $a^{\#} = -a$ for all $a \in R$. A generalization of this observation is due to [Lee, 1971]. Let $(\Omega, \leq)$ be a join semilattice having a unique minimal element u and, for each $i \in \Omega$, let $(R_i, +_i, \cdot_i)$ be a ring, where we assume that $R_i \cap R_j = \mathbb{N}$ for all $i \neq j$ in Ω. Assume furthermore that for each $i \leq j$ we have a ring homomorphism $\gamma_{ij}: R_i \to R_j$ satisfying

(1) γ_{ii} is the identity map for each $i \in \Omega$; and

(2) $\gamma_{jk}\gamma_{ij} = \gamma_{ik}$ for all $i \leq j \leq k$ in Ω.

Set $R = \cup\{R_i \mid i \in \Omega\}$ and define on it operations of addition and multiplication as follows: if $a \in R_i$ and $b \in R_{ij}$ and if $k = i \vee j$ in Ω, then $a+b = \gamma_{ik}(a) +_k \gamma_{jk}(b)$ and $a \cdot b = \gamma_{ik}(a) \cdot_k \gamma_{jk}(b)$. Under these definitions, $(R, +, \cdot)$ is a semiring with additive identity 0_u and multiplicative identity 1_u. (In fact, $R = \varinjlim R_i$.) Moreover, R is additively regular where, for $a \in R_i$, we let $a^{\#}$ be the negation of a in R_i.

(13.5) EXAMPLE. If R is an additively-idempotent semiring then R is additively regular with $a^{\#} = a$ for all $a \in R$.

(13.6) EXAMPLE. [Sen & Adhikari, 1992] Let $S = (\mathbb{P} \cup \{\infty\}, \sqcup, \sqcap)$, where $\mathbb{P}$ is the set of positive integers and where the operations are given by

$$a \sqcup b = \begin{cases} lcm(a,b) & \text{if } a, b \in \mathbb{P} \\ \infty & \text{if } a = \infty \text{ or } b = \infty \end{cases}$$

and

$$a \sqcap b = \begin{cases} gcd(a,b) & \text{if } a, b \in \mathbb{P} \\ a & \text{if } b = \infty \\ b & \text{if } a = \infty \end{cases}.$$

Then $R = \mathbb{Z} \times S$ is an additively-regular commutative semiring in which $(k, s)^{\#} = (-k, s)$. Moreover, $H = \{(0, s) \mid s \in S\}$ is a subtractive ideal of R containing $I^+(R)$.

(13.7) EXAMPLE. Let R be a ring and let S be a subsemiring of $ideal(R)$. Set $R' = \{(a, I) \mid a \in I \in S\}$ and define operations $\oplus$ and $\odot$ on R' by setting $(a, I) \oplus (b, H) = (a + b, I + H)$ and $(a, I) \odot (b, H) = (ab, IH)$. Then R' is an additively-regular semiring where, for each $(a, I) \in R'$, we have $(a, I)^{\#} = (-a, I)$. Moreover, $I^+(R') = \{(0, I) \mid I \in S\}$.

(13.8) Proposition. *If R is an additively-regular semiring which is not a ring, then there does not exist a semiisomorphism from R to a cancellative semiring.*

Proof. Assume that there exist a cancellative semiring S and a semiisomorphism $\gamma: R \rightarrow S$. Since R is not a ring, there exists an element $r \in R \setminus V(R)$. Then $r^{\circ} \in I^{+}(R)$ and so $\gamma(r^{\circ}) \in I^{+}(S)$. But S is cancellative so $\gamma(r^{\circ}) = 0$ and hence $r^{\circ} \in ker(\gamma)$. Since γ is a semiisomorphism, this means that $r^{\circ} = 0$ and so $r \in V(R)$, contradicting our assumption. □

(13.9) Corollary. *If $\gamma: R \rightarrow S$ is a surjective morphism of semirings with R additively regular and S cancellative then the congruence relations $\equiv_{ker(\gamma)}$ and $[\equiv]_{ker(\gamma)}$ on R coincide.*

Proof. Set $I = ker(\gamma)$. By Proposition 10.16 we know that there exists a semiisomorphism $\gamma': R/I \rightarrow S$ induced by γ. Moreover, R/I is additively regular since R is and so, by Proposition 13.8, we conclude that R/I is a ring. If $r, r' \in R$ satisfy $r[\equiv]_I r'$ then there exist elements a and a' of I and an element s of R satisfying $r+a+s = r'+a'+s$. Therefore $(r+s)/I = (r'+s')/I$ and so $r/I = r'/I$ since R/I is a ring. Hence $r \equiv_I r'$. The converse is always true, as remarked in Chapter 5, and so the relations $\equiv_I$ and $[\equiv]_I$ coincide. □

If R is an additively-regular semiring then we have a congruence relation ρ on R defined by $a \, \rho \, b$ if and only if $a^{\circ} = b^{\circ}$. Since $a \, \rho \, 0$ if and only if $a + a^{\#} = 0$, we can deduce easily that this relation is improper if and only if R is a ring.

(13.10) Proposition. *If R is an additively-regular semiring then $(r/\rho, +)$ is a group for each $r \in R$.*

Proof. Let $r \in R$ and let $G = r/\rho$. If $a, b \in G$ then $(a+b)^{\circ} = a^{\circ} + b^{\circ} = r^{\circ} + r^{\circ} = r^{\circ}$ and so $a + b \in G$. In particular, $r^{\circ} = r + r^{\#} \in G$. If $a \in G$ then $a + r^{\circ} = a + a^{\circ} = a$. Furthermore, $a^{\#} \in G$ since $(a^{\#})^{\circ} = a^{\circ} = r^{\circ}$ and $a + a^{\#} = a^{\circ} = r^{\circ}$. Thus $(G, +)$ is an additive group with identity element r°. □

Thus, in particular, we see that if R is an additively-regular semiring then $(R, +)$ is the union of groups.

(13.11) Proposition. *Let R be an additively-regular semiring which is not a ring and let $S = R/\rho$. Then*

(1) *$S = I^{\times}(S)$ if and only if $a^{\circ} = aa^{\circ}$ for all $a \in R$;*
(2) *S is commutative if and only if $ab^{\circ} = b^{\circ}a$ for all $a, b \in R$;*
(3) *S is a lattice if and only if $S = I^{\times}(S)$, S is commutative, and $a + a^{\circ}b = a$ for all $a, b \in R$.*

Proof. (1) Assume that $S = I^{\times}(S)$. If $a \in R$ then $a \, \rho \, a^2$ and so $a^{\circ} = (a^2)^{\circ} = aa^{\circ}$. Conversely, suppose that $a^{\circ} = aa^{\circ}$. Then $(a^2)^{\circ} = aa^{\circ} = a^{\circ}$ and so $a^2 \, \rho \, a$ for all $a \in R$. This shows that $S = I^{\times}(S)$.

(2) Assume that S is commutative. Then for $a, b \in R$ we have $ab^{\circ} = (ab)^{\circ} = (ba)^{\circ} = b^{\circ}a$. Conversely, assume that $ab^{\circ} = b^{\circ}a$. Then $(ab)^{\circ} = (ba)^{\circ}$ and so $ab \, \rho \, ba$ for all $a, b \in R$, proving that S is commutative.

(3) Assume that S is commutative, that $S = I^{\times}(S)$, and that the given condition is satisfied. If $a \in R$ then $(a+a)^{\circ} = a^{\circ} + a^{\circ} = a^{\circ}$ and so $a \, \rho \, a+a$. Thus $S = I^{+}(S)$.

If $a, b \in R$ then $a^\circ = a^\# + a = a^\# + a^\circ b = a^\circ + a^\circ b = a^\circ + (ab)^\circ = [a + ab]^\circ$ so $a \ \rho \ a + ab \ \rho \ a^2 + ab \ \rho \ a(a+b)$ and hence $(S, +, \cdot)$ is a lattice by Example 1.5. Conversely, if S is a lattice then S is commutative, $S = I^\times(S)$, and $a^\circ = a^\circ + a^\circ b$ so $a = a + a^\circ = a + a^\circ b$ for all $a, b \in R$. □

If R is an additively-regular semiring we now define a relation ζ on R by setting $a \ \zeta \ b$ if and only if $a + b^\circ = a^\circ + b$.

(13.12) EXAMPLE. If $a, b \in I^+(R)$ then $a = a^\circ$ and $b = b^\circ$ so $a + b^\circ = a^\circ + b$ and hence $a \ \zeta \ b$.

(13.13) PROPOSITION. *If a and b are elements of an additively-regular semiring R then $a \ \zeta \ b$ if and only if $a + b^\# \in I^+(R)$.*

PROOF. If $a \ \zeta \ b$ then

$$\begin{aligned}(a + b^\#) + (a + b^\#) &= a + b^\# + b^\# + b + a + b^\# \\ &= a + b^\circ + (a + b^\#) + b^\# \\ &= a^\circ + b + a + b^\# + b^\# \\ &= a + a^\# + b + a + b^\# + b^\# \\ &= a + b^\#\end{aligned}$$

and so $a + b^\# \in I^+(R)$. Conversely, if $a + b^\# \in I^+(R)$ then $a + b^\# = (a + b^\#)^\# = a^\# + b$ so

$$\begin{aligned}a + b^\circ &= a + b^\# + b = (a + b^\#) + (a + b^\#) + b \\ &= a + b^\# + a^\# + b + b \\ &= a + a^\# + b = a^\circ + b,\end{aligned}$$

proving that $a \ \zeta \ b$. □

In particular, $a \ \zeta \ 0$ if and only if $a \in I^+(R)$.

(13.14) PROPOSITION. *The following conditions on an additively-regular semiring R are equivalent:*

(1) *ζ is a congruence relation on R;*
(2) *$Z(R) = I^+(R)$.*

PROOF. (1) ⇒ (2): Clearly $I^+(R) \subseteq Z(R)$. Conversely, assume that $b \in Z(R)$. Then there exists an element a of R satisfying $a + b = a$. Hence $b + a^\circ = a^\circ$. By (1), $a^\circ = (a+b)^\circ = a^\circ + b^\circ$ so $a^\circ \ \zeta \ b$ and $a^\circ \ \zeta \ b^\circ$, whence $b \ \zeta \ b^\circ$ by (1). This implies that $b = b + b^\circ = b^\circ + b^\circ = b^\circ$ so $b \in I^+(R)$.

(2) ⇒ (1): If a and b are elements of R then clearly $a \ \zeta \ a$, and $a \ \zeta \ b$ when and only when $b \ \zeta \ a$. Assume that a, b, and c are elements of R satisfying $a \ \zeta \ b$ and $b \ \zeta \ c$. Then, by Proposition 13.13,

$$\begin{aligned}&(a + c^\#) + (a + c^\# + b^\circ) \\ &\quad = (a + c^\# + b^\circ) + (a + c^\# + b^\circ) + (a + b^\#) + (b + c^\#) + (a + b^\#) + (b + c^\#) \\ &\quad = (a + b^\#) + (b + c^\#) = a + c^\# + b^\circ.\end{aligned}$$

By (2), this implies that $a + c^{\#} \in I^{+}(R)$. Therefore $a + c^{\#} = (a + c^{\#})^{\circ}$ and so $(a+c^{\#})+(a+c^{\#}) = (a+c^{\#})+(a+c^{\#})^{\#}+(a+c^{\#}) = a+c^{\#}$ which, by Proposition 13.13, implies that $a \zeta c$. Thus ζ is an equivalence relation.

If $a \zeta c$ and $b \zeta d$ in R then $a+b+(c+d)^{\circ} = a+b+c^{\circ}+d^{\circ} = a^{\circ}+b^{\circ}+c+d = (a+b)^{\circ}+c+d$ and so $a+b \zeta c+d$. Similarly, $ab+(cb)^{\circ} = ab+c^{\circ}b = (a+c^{\circ})b = (a^{\circ}+c)b = a^{\circ}b+cb = (ab)^{\circ}+cb$ and so $ab \zeta cb$. In a like manner, $cb \zeta cd$ and so $ab \zeta cd$. Thus ζ is a congruence relation on R. □

An additively-regular semiring satisfying the equivalent conditions of Proposition 13.14 will be called a **Bandelt semiring**.

(13.15) COROLLARY. *If R is a Bandelt semiring which is not additively idempotent then R/ζ is a ring.*

PROOF. By Proposition 13.14 we know that ζ is a congruence relation on R and so R/ζ is a semiring. If $a \in R$ then $a + a^{\#} \in I^{+}(R)$ and so, by Example 13.12, $a+a^{\#} \zeta 0$. Thus $a/\zeta + a^{\#}/\zeta = (a+a^{\#})/\zeta = 0/\zeta$, showing that a/ζ has an additive inverse in R/ζ. Thus R/ζ is a ring. □

(13.16) PROPOSITION. *A semiring R is isomorphic to a subdirect product of a ring and a lattice if and only if it is a Bandelt semiring satisfying the following conditions:*

(1) $aa^{\circ} = a^{\circ}$ *for all* $a \in R$;
(2) $ab^{\circ} = b^{\circ}a$ *for all* $a, b \in R$;
(3) $a + a^{\circ}b = a$ *for all* $a, b \in R$.

PROOF. If R is isomorphic to a subdirect product of a ring and a lattice then surely it is Bandelt semiring satisfying the given conditions. Conversely, assume that R is a Bandelt semiring satisfying the given conditions. By Proposition 13.11 we see that R/ρ is a lattice and by, Corollary 13.15, we see that R/ζ is a ring. We also a morphism of semirings $\gamma: R \rightarrow R/\rho \times R/\zeta$ given by $r \mapsto (r/\rho, r/\zeta)$, and all we need to show is that this map is injective. Indeed, assume that $\gamma(a) = \gamma(b)$. Then $a^{\circ} = b^{\circ}$ and $a + b^{\circ} = a^{\circ} + b$ so $a = a + a^{\circ} = a + b^{\circ} = a^{\circ} + b = b^{\circ} + b = b$. □

14. SEMIMODULES OVER SEMIRINGS

The modules over a ring are an important tool in characterizing properties of the ring and so it is only natural that we should look at the corresponding construction over semirings. And, indeed, many of the constructions from ring theory can be transfered, at least partially, to this more general setting. Moreover, many important constructions in pure and applied mathematics can, as we shall see, be understood as semimodules over appropriate semirings. In this chapter we lay the foundations for the study of semimodules.

Let R be a semiring. A **left R-semimodule** is a commutative monoid $(M, +)$ with additive identity 0_M for which we have a function $R \times M \to M$, denoted by $(r, m) \mapsto rm$ and called **scalar multiplication**, which satisfies the following conditions for all elements r and r' of R and all elements m and m' of M:

(1) $(rr')m = r(r'm)$;
(2) $r(m + m') = rm + rm'$;
(3) $(r + r')m = rm + r'm$;
(4) $1_R m = m$;
(5) $r0_M = 0_M = 0_R m$.

Right semimodules over R are defined in an analogous manner. In what follows, we will generally work with left semimodules, with the corresponding results for right semimodules being assumed without explicit mention.

If R and S are semirings then an (R, S)-**bisemimodule** $(M, +)$ is both a left R-semimodule and right S-semimodule satisfying the additional condition that $(rm)s = r(ms)$ for all $m \in M$, $r \in R$, and $s \in S$. If M is a left R-semimodule then it is in fact an $(R, C(R))$-bisemimodule, with scalar multiplication on the right being defined by $m \cdot r = rm$. In particular, if R is commutative then any left R-semimodule is an (R, R)-bisemimodule.

If R is a semitopological semiring then a left R-semimodule M is **semitopological** if and only if it has the additional structure of a topological space such that the function $M \times M \to M$ defined by $(m, m') \mapsto m + m'$ and the function $R \times M \to M$ defined by $(a, m) \mapsto am$ are continuous. If the underlying topological space is Hausdorff, then the semimodule is **topological**. Thus every semitopological semiring is a semitopological left semimodule over itself.

If m is an element of a R-module M then an element m' of M satisfying $m+m' = 0_M$ is an **additive inverse** of m. Clearly additive inverses, if they exist, are unique, and we will denote the additive inverse of m, if it exists, by $-m$. The set $V(M)$ of all elements of M having additive inverses is nonempty, since $0 \in V(M)$. An R-semimodule M is **zerosumfree** if and only if $V(M) = \{0\}$. At the other extreme, an R-semimodule M satisfying $V(M) = M$ is an R-**module**.

A nonempty subset N of a left R-semimodule M is a **subsemimodule** of M if and only if N is closed under addition and scalar multiplication. Note that this implies that $0_M \in N$. Subsemimodules of right semimodules and subbisemimodules are defined analogously. For example, if A is a nonempty subset of a left R-semimodule M and if $I \in lideal(R)$ then the set IA of all finite sums of the form $r_1m_1+\cdots+r_km_k$ ($r_i \in I$ and $m_i \in A$) is a subsemimodule of M. A subsemimodule which is an R-module is a **submodule**. Thus $V(M)$ is a submodule of any left or right R-semimodule M containing all other submodules of M. We will denote the poset of all subsemimodules of a left R-semimodule M by $ssm(M)$. An atom of $ssm(M)$ is a **minimal subsemimodule** of M.

If $N \in ssm(M)$ and $a \in C(R)$ then $aN = \{an \mid n \in N\}$ is also a subsemimodule of M. Moreover, if $a, b \in C(R)$ and $N, N' \in ssm(M)$ we have $a(N+N') = aN+aN'$ and $a(bN) = (ab)N$. Thus $ssm(M)$ is itself a left $C(R)$-semimodule.

We note that if N is a subsemimodule of a left R-semimodule M and if $m \in M$ then $(N : m) = \{a \in R \mid am \in N\}$ is a left ideal of R. More generally, if A is a nonempty subset of M we set $(N : A) = \cap\{(N : m) \mid m \in A\}$. Following the usual convention, we will write $(0 : A)$ instead of $(\{0\} : A)$. Since the intersection of an arbitrary family of left ideals is again a left ideal, this too is a left ideal of R.

(14.1) PROPOSITION. *If N and N' are subsemimodules of a left R-semimodule M and if A and B are nonempty subsets of M then:*

(1) *$A \subseteq B$ implies that $(N : B) \subseteq (N : A)$;*

(2) *$(N \cap N' : A) = (N : A) \cap (N' : A)$;*

(3) *$(N : A) \cap (N : B) \subseteq (N : A + B)$, with equality holding if $0_M \in A \cap B$.*

PROOF. (1) This is an immediate consequence of the definition.

(2) By definition, if $r \in R$ then $r \in (N \cap N' : A) \Leftrightarrow rm \in N \cap N'$ or all $m \in A \Leftrightarrow rm \in N$ and $rm \in N'$ for all $m \in A \Leftrightarrow r \in (N : A) \cap (N' : A)$.

(3) If $r \in (N : A) \cap (N : B)$ then $r(m + m') \in N$ for all $m \in A$ and $m' \in B$ and so $r \in (N : A + B)$. Conversely, if $0_M \in A \cap B$ then $A \cup B \subseteq A + B$ and so the reverse inclusion holds. □

If $\gamma: R \to S$ is a morphism of semirings and if M is a left S-semimodule then it is also canonically a left R-semimodule, with scalar multiplication defined by $r \cdot m = \gamma(r)m$ for all $r \in R$ and $m \in M$. In particular, if M is a left S-semimodule then M is a left R-semimodule for every subsemiring R of S.

(14.2) EXAMPLE. If R is a bounded distributive lattice then the left R-semimodules have been studied by Fofanova [1971, 1982] under the name of **polygons**. More generally, the structure of R-semimodules over commutative simple semirings R has been studied in [Kearnes, 1995].

(14.3) EXAMPLE. The $\mathbb{N}$-semimodules are precisely the commutative additive monoids. Thus, for example, $\mathbb{Z} \setminus \mathbb{P}$ is an $\mathbb{N}$-semimodule. Also, every semiring R is an $\mathbb{N}$-semimodule. If R is a semifield which is not a field then, by Proposition 4.34, R is zerosumfree and so R is also a $\mathbb{Q}^+$-semimodule, where, for $r \in R$ and $\frac{m}{n} \in \mathbb{Q}^+$, we set

$$\left(\frac{m}{n}\right) r = (n1_R)^{-1}(mr).$$

If $(M, +)$ is an idempotent commutative monoid then M is a left $\mathbb{N}$-semimodule with scalar multiplication defined by $0m = 0_M$ for all $m \in M$ and $im = m$ for all $m \in M$ and all $0 < i \in \mathbb{N}$.

(14.4) EXAMPLE. If M is a left R-semimodule and A is a nonempty set then M^A is a left R-semimodule with addition and scalar multiplication defined elementwise: if $f, g \in M^A$ and $r \in R$ then $(f+g)(a) = f(a) + g(a)$ and $(rf)(a) = r[f(a)]$ for all $a \in A$. Moreover, $M^{(A)} = \{f \in M^A \mid f \text{ has finite support}\}$ is a subsemimodule of M^A.

Similarly, if M is an (R, R)-bisemimodule then so are M^A and $M^{(A)}$. This is true certainly for $M = R$. Thus we see that the set of all R-valued relations between nonempty sets A and B is an (R, R)-bisemimodule, as is the set of all R-valued graphs on a set of vertices V.

If B is a boolean ring then a **measure** on B is a function m from B to the semiring $\mathbb{R}^+ \cup \{\infty\}$ satisfying the following conditions:

(1) $m(b \vee b') = m(b) + m(b')$ whenever $b \wedge b' = 0$ in B;
(2) $m(b) = 0$ if and only if $b = 0_B$;
(3) If $\{b_i\}$ is a sequence converging to b in B then $m(b) = sup\{m(b_i)\}$ in $\mathbb{R}^+$.

The family of all measures on B is a subbisemimodule of the $(\mathbb{R}^+, \mathbb{R}^+)$-bisemimodule $(\mathbb{R}^+)^B$.

(14.5) APPLICATION. Let $R = (\mathbb{R} \cup \{\infty\}, min, +)$ and consider $M = R^{\mathbb{R}}$ as a left R-semimodule. Elements of M are **signals**. Addition in M corresponds to **parallel composition** of signals, and scalar multiplication corresponds to **amplification** of signals. See [Baccelli et al., 1992] for an analysis of this situation and its applications to systems theory and signal processing.

(14.6) EXAMPLE. Let A be a nonempty set and let A^* be the free monoid of A. Let R be a semiring and let M be a left semimodule over R. Then, in a manner analogous to that used in Chapter 2, we can define $M\langle\langle A\rangle\rangle$ to be the set M^{A^*} together with an operation of addition defined componentwise. This is a commutative monoid. If $f \in R\langle\langle A\rangle\rangle$ and $q \in M\langle\langle A\rangle\rangle$ then we can define $fq \in M\langle\langle A\rangle\rangle$ by setting $fq \colon w \mapsto \sum_{w'w''=w} f(w')q(w'')$ for all $w \in A^*$. This operation of scalar multiplication turns $M\langle\langle A\rangle\rangle$ into a left $R\langle\langle A\rangle\rangle$-semimodule. Since $R\langle A\rangle$ is a subsemiring of $R\langle\langle A\rangle\rangle$, this means that $M\langle\langle A\rangle\rangle$ is also a left $R\langle A\rangle$-semimodule. Moreover, $M\langle A\rangle = \{q \in M\langle\langle A\rangle\rangle \mid q \text{ has finite support}\}$ is clearly an $R\langle A\rangle$-subsemimodule of $M\langle\langle A\rangle\rangle$.

(14.7) EXAMPLE. Let Ω be a nonempty set which is either finite or countably-infinite, let R be a semiring, and let M be a left semimodule. In a manner analogous to that used in Chapter 2, we can define the set $\mathcal{M}_\Omega(M)$ of all $(\Omega \times \Omega)$-matrices on M to be the set of all functions from $\Omega \times \Omega$ to M. Again, addition can be defined componentwise on this set to turn $(\mathcal{M}_\Omega(M), +)$ into an additive monoid which is a left semimodule over $\mathcal{M}_{\Omega,r}(R)$ or $\mathcal{M}_{\Omega,rc}(R)$.

(14.8) EXAMPLE. If R is an entire zerosumfree semiring, if M is a left R-semimodule, and if ∞ is an element not in M then we can define the left R-semimodule $M\{\infty\}$ to be the set $M \cup \{\infty\}$ on which the operations of addition and scalar multiplication from M have been extended by setting $m' + \infty = \infty + m' = \infty$ for all $m' \in M\infty$, $r\infty = \infty$ for all $0 \neq r \in R$, and $0\infty = 0_M$.

(14.9) EXAMPLE. If $\gamma: R \to S$ is a morphism of semirings then S is an (R, R)-bisemimodule in which we define $r \cdot s = \gamma(r)s$ and $s \cdot r = s\gamma(r)$ for all $r \in R$ and $s \in S$. Thus, in particular, if R is a semiring and A is a nonempty set then $R\langle\langle A\rangle\rangle$ is an (R, R)-bisemimodule in which we define rf and fr by $rf: w \mapsto rf(w)$ and $fr: w \mapsto f(w)r$ for all $r \in R$, $f \in R\langle\langle A\rangle\rangle$, and $w \in A^*$. Also, by Proposition 9.10, we see that every additively-idempotent semiring is a $(\mathbb{B}, \mathbb{B})$-bisemimodule.

(14.10) EXAMPLE. Let R be a semiring and let M be a left R-semimodule. Then $(0 : M) = \{r \in R \mid rm = 0_M \text{ for all } m \in M\}$ is an ideal of R. Moreover, if I is any ideal of R contained in $(0 : M)$ then M is a left (R/I)-semimodule, with scalar multiplication defined by $(r/I)m = rm$ for all $r \in R$ and $m \in M$.

(14.11) EXAMPLE. An $\mathbb{R}^+$-subsemimodule of $\mathbb{R}^n$ for some positive integer n is called a **convex cone** in $\mathbb{R}^n$. If C is any nonempty convex subset of $\mathbb{R}^n$ then $\{rv \mid r \in \mathbb{R}^+ \text{ and } v \in C\}$ is a convex cone in $\mathbb{R}^n$. For the relations between these semimodules and barycentric algebras, see [Romanowska & Smith, 1985].

(14.12) EXAMPLE. A **quemiring** is a structure of the form $R \times M$, where R is a semiring and M is a left R-semimodule, on which addition is defined componentwise, while multiplication is given by $(a, m) \cdot (a', m') = (aa', am' + m)$. This is not a semiring since $(a, m) \cdot (0, 0) = (0, 0)$ only when $m = 0$. Also note that while right distributivity of multiplication over addition always holds, left distributivity holds only sometimes. However, the quemiring $R \times M$ certainly contains $R' = \{(a, 0) \mid a \in R\}$ as a subsemiring, and can be profitably be studied as a left R'-semimodule.

An element m of a left R-semimodule M is **idempotent** if and only if $m+m = m$. The set of all idempotent elements of M is nonempty since it contains 0_M and, indeed, is clearly a subsemimodule of M, denoted by $I(M)$. If $I(M) = M$ then M is **additively idempotent**. If M is a left R-semimodule then the left $C(R)$-semimodule $ssm(M)$ is additively idempotent. As with semirings, every additively-idempotent semimodule is zerosumfree.

(14.13) EXAMPLE. If $(M, +)$ is an additively-idempotent commutative monoid then M is a left $\mathbb{B}$-semimodule with scalar multiplication defined by $0m = 0_M$ and $1m = m$ for all $m \in M$. By Proposition 9.10 we know that if R is a zerosumfree entire semiring then there exists a surjective morphism $\gamma: R \to \mathbb{B}$. Thus M is also a left R-semimodule with scalar multiplication defined by $r \cdot m = \gamma(r)m$ for each $r \in R$ and $m \in M$. This allows us, for example, to consider $(\mathbb{R} \cup \{\infty\}, min)$ as a left $\mathbb{R}^+$-semimodule with scalar multiplication defined by $am = m$ if $a > 0$ and $0m = \infty$.

Let M be a left R-semimodule and let $\{N_i \mid i \in \Omega\}$ be a family of subsemimodules of M. Then $\cap_{i\in\Omega} N_i$ is a subsemimodule of M which, indeed, is the largest subsemimodule of M contained in each of the N_i. In particular, if A is a subset of a left R-semimodule M then the intersection of all subsemimodules of M containing A is a subsemimodule of M, called the subsemimodule **generated by** A. This semimodule is just $RA = \{r_1a_1 + \cdots + r_na_n \mid r_i \in R, a_i \in A\}$. If A generates all of the semimodule M, then A is a **set of generators** for M. Any set of generators for M contains a minimal set of generators. A left R-semimodule having a finite set of generators is **finitely generated**. An element m of the subsemimodule generated by a subset A of a subsemimodule M is a **linear combination** of the elements of A. The **rank** of a left R-semimodule M is the smallest n for which there exists a set of generators of M having cardinality n. This rank need not be the same as the cardinality of a minimal set of generators for M, as the following example shows.

(14.14) EXAMPLE. [Cechlárová & Plávka, 1996] Let $R = (\mathbb{R} \cup \{-\infty, \infty\})$ and let $m > 1$ be a positive integer. For an arbitrary positive integer k, select elements $a_1 < a_2 < \cdots < a_{k+m-1}$ in R and consider the elements

$$v_1 = (a_1, a_2, \ldots, a_m), v_2 = (a_2, a_3, \ldots, a_{m+1}), \ldots v_k = (a_k, a_{k+1}, \ldots, a_{k+m-1})$$

in R^m. Then none of the elements v_h is a linear combination of the others.

(14.15) EXAMPLE. If R is a semiring and M is a left R-semimodule then two minimal sets of generators of M need not have the same cardinality, even if M is finitely-generated. For example, if S is a semiring and $R = S \times S$ then $\{(1_S, 1_S)\}$ and $\{(1_S, 0), (0, 1_S)\}$ are both minimal sets of generators for R, considered as a left semimodule over itself.

(14.16) EXAMPLE. [Dudnikov & Samborskiĭ, 1992] If R is the schedule algebra then $M = R^3$ is a left R-semimodule which is clearly finitely-generated over R. Let N be the subsemimodule of M generated by the elements of the two-dimensional disk of radius 1, which is orthogonal to the element $[1, 1, 1]$ of $\mathbb{R}^3$. Then N does not have a finite set of generators over R.

If M is a left R-semimodule then the set $\sum_{i\in\Omega} N_i$ of all finite sums of elements of $\cup_{i\in\Omega} N_i$ is a subsemimodule of M which is the smallest subsemimodule of M containing each of the N_i. If N is a subsemimodule of M and $I \in ideal(R)$, then, as we have already noted, IN is also a subsemimodule of M. Thus, in particular, if $m \in M$ we have the subsemimodule Rm of M defined by $Rm = \{rm \mid r \in R\}$. Surely $M = \sum_{m\in M} Rm$.

(14.17) EXAMPLE. If R is a ring and $R-fil$ is the set of all topologizing filters of left ideals of R then, as we noted in Example 1.6, $(R-fil,\cap,\cdot)$ is a semiring. If M is a left R-module, let $sub(M)$ be the family of all submodules of N and, for each $N \in sub(M)$ and $\kappa \in R-fil$, let $N\kappa$ be the κ-purification of N in M. That is to say, $N\kappa = \{m \in M \mid Im \subseteq N \text{ for some } I \in \kappa\}$. We claim that $(sub(M),\cap)$ is a right $(R-fil)$-semimodule. Indeed, if $N, N' \in sub(M)$ and $\kappa \in R-fil$ then surely $(N \cap N')\kappa \subseteq N\kappa \cap N'\kappa$. Conversely, if $m \in N\kappa \cap N'\kappa$ then there exist $I, H \in \kappa$ satisfying $Im \subseteq N$ and $Hm \subseteq N'$. But $I \cap H \in \kappa$ and $(I \cap H)m \subseteq N \cap N'$. Thus $m \in (N \cap N')\kappa$, proving that $N\kappa \cap N'\kappa = (N \cap N')\kappa$. If $N \in sub(M)$ and $\kappa, \kappa' \in R-fil$ then $(N\kappa\kappa') = (N\kappa)\kappa'$ by Proposition 4.5 of [Golan, 1987]. Also, we clearly have $N(\kappa \cap \kappa') \subseteq N\kappa \cap N\kappa'$. Conversely, if $m \in N\kappa \cap N\kappa'$ then there exist $I \in \kappa$ and $H \in \kappa'$ with $Im \subseteq N$ and $Hm \subseteq N$. Thus $(I+H)m \subseteq N$. But $I + H \in \kappa \cap \kappa'$, proving that $N(\kappa \cap \kappa') = N\kappa \cap N\kappa'$. Finally, it is clear that $N\eta[R] = N$ and $N\eta[0] = M$.

(14.18) EXAMPLE. In Example 1.19 we saw that if $(L, \vee, \wedge)$ is a frame then the set $\mathbf{PN}(L)$ of all prenuclei on L is a zerosumfree simple semiring with addition given by $\wedge$ and multiplication by composition of functions. Moreover, it is clear that the dual frame $(L, \vee, \wedge)$ is then a left $\mathbf{PN}(L)$-semimodule, with scalar multiplication defined by $z \cdot c = z(c)$. For the case of that L is the frame of all torsion theories on a module category $R-mod$ over a ring R, this semiring has been studied in [Golan & Simmons, 1988].

(14.19) EXAMPLE. Let R be a semiring and let M be a left R-semimodule. Then $ssm(M)$ is a left $ideal(R)$-semimodule, where scalar multiplication is defined as above. In particular, $lideal(R)$ is a left $ideal(R)$-semimodule. For the case that R is a ring, this situation has been studied in [Anderson, 1977].

A left R-semimodule M is **entire** if and only if $rm \neq 0_M$ whenever $0 \neq r \in R$ and $0_M \neq m \in M$. A left R-semimodule which is both entire and zerosumfree is an **information semimodule** over R. A complete classification of all cyclic information semimodules over $\mathbb{N}$ is given in [Takahashi, 1985]. Since the classes of zerosumfree and entire semimodules are both clearly closed under taking submodules, we see that subsemimodules of information semimodules are again information semimodules.

(14.20) PROPOSITION. *A semiring R is entire and zerosumfree if and only if there exists a nontrivial information semimodule over R.*

PROOF. If R is entire and zerosumfree then it is surely a nontrivial information semimodule over R. Conversely, assume that there exists a nontrivial information semimodule M over R and let $0_M \neq m \in M$. If $r, r' \in R \setminus \{0\}$ then $r'm \neq 0_M$ and so $(rr')m = r(r'm) \neq 0_M$. Therefore $rr' \neq 0$, proving that R is entire. Moreover, $rm \neq 0_M$ as well and so $(r+r')m = rm + r'm \neq 0_M$ and so $r + r' \neq 0$. Thus R is zerosumfree. □

As in the case of subsets of R, we say that a nonempty subset N of a left R-semimodule M is **subtractive** if and only if $m + m' \in N$ and $m \in N$ imply that $m' \in N$ for all $m, m' \in M$. Similarly, N is **strong** if and only if $m+m' \in N$ implies that $m, m' \in N$ for all $m, m' \in M$. Every submodule of a left R-semimodule is subtractive. Indeed, if N is a submodule of an R-semimodule M and $m \in M$, $n \in N$ are elements satisfying $m + n \in N$ then $m = (m + n) + (-n) \in N$. In particular, $V(M)$ is a subtractive subsemimodule of any R-semimodule M. If $N' \subseteq N$ are subsemimodules of a left R-semimodule M such that N' is a subtractive subsemimodule of N and N is a subtractive subsemimodule of M then one sees immediately that N' is a subtractive subsemimodule of M. If $\{M_i \mid i \in \Omega\}$ is a family of subtractive [resp. strong] subsemimodules of a left R-semimodule M then $\cap_{i \in \Omega} M_i$ is again subtractive [resp. strong]. Thus every subsemimodule of a left R-semimodule M is contained in a smallest subtractive [resp. strong] subsemimodule of M, called its **subtractive closure** [resp. **strong closure**] in M.

We now introduce a construction due to Takahashi [1996a]. Let R be a semiring and let M be a left R-semimodule. If N and N' are R-subsemimodules of M then set $E_N^M(N') = \{m \in M \mid m + n \in N' \text{ for some } n \in N\}$. It is easy to verify that for all such N and N' we have:

(1) $E_N^M(N')$ is an R-subsemimodule of M containing N';
(2) $E_N^M(E_N^M(N')) = E_N^M(N')$;
(3) If N is a submodule of M then $E_N^M(N') = N + N'$.
(4) If M' is a subsemimodule of M containing both N and N' then $E_N^M(N') \cap M' = E_N^{M'}(N')$;
(5) If $N \subseteq N'$ then $E_N^M(N) \cap N' = E_N^{N'}(N)$.

Moreover, we see that a subsemimodule N of a left R-semimodule M then $E_N^M(N)$ is precisely the subtractive closure of N in M, and so N is subtractive if and only if $N = E_N^M(N)$. Similarly, $E_M^M(N)$ is precisely the strong closure of N in M, and so N is strong if and only if $E_M^M(N) = N$.

Any left R-semimodule M has two subtractive subsemimodules: $\{0\}$ and M itself. If these are the only subtractive subsemimodules of M, then M is **austere**. That is to say, M is austere if and only if it is the subtractive closure of each of its nonzero subsemimodules.

(14.21) PROPOSITION. *If M is an austere left R-semimodule then $(0 : M) = (0 : m)$ for all $0 \neq m \in M$.*

PROOF. Clearly $(0 : M) \subseteq (0 : m)$ for all $m \in M$ and so we must prove the reverse inclusion for all $0 \neq m \in M$. Indeed, assume that $m \in M$ satisfies the condition $I = (0 : m) \not\subseteq (0 : M)$ and let $N = \{m' \in M \mid I \subseteq (0 : m')\}$. Then N is a subsemimodule of M properly contained in M since $I \not\subseteq (0 : M)$. If n and n' are elements of M such that both n and $n + n'$ belong to N then $\{0\} = I(n + n') = In + In' = In'$ and so $n' \in N$. Thus N is subtractive and so, by austerity, $N = \{0\}$. Thus $(0 : m) \subseteq (0 : M)$ for all $0 \neq m \in M$. □

If M is a left R-semimodule then the **zeroid** of M is defined to be $Z(M) = \{m \in M \mid m + n = n \text{ for some element } n \text{ of } M\}$. Clearly this is a subsemimodule of M containing $I(M)$ which we claim is subtractive. Indeed, if m and m' are elements of M satisfying $m \in Z(M)$ and $m + m' \in Z(M)$ then there exist elements n and n' of M satisfying $m + n = n$ and $(m + m') + n' = n'$. Therefore $m' + (n + n') = m' + m + (n + n') = n + n'$ and so $m' \in Z(M)$. Note that if $a \in Z(R)$

and $m \in M$ then there exists an element r of R satisfying $a + r = r$ and so $am + rm = (a + r)m = rm$. Thus $am \in Z(M)$. Since $Z(M)$ is a subsemimodule of M, this shows that $Z(R)M \subseteq M$. A left R-semimodule M is **zeroic** when $Z(M) = M$ and **nonzeroic** when $Z(M) \neq M$.

(14.22) PROPOSITION. *Let R be a semiring and let M be a left R-semimodule. If N, N', and N'' are subsemimodules of M satisfying the conditions that N is subtractive and $N' \subseteq N$, then $N \cap (N' + N'') = N' + (N \cap N'')$.*

PROOF. Let $x \in N \cap (N' + N'')$. Then we can write $x = y + z$, where $y \in N'$ and $z \in N''$. By (2), we have $y \in N$ and so, by (1), $z \in N \cap N''$. Thus $x \in N' + (N \cap N'')$, proving that $N \cap (N' + N'') \subseteq N' + (N \cap N'')$. The reverse containment is immediate. □

(14.23) PROPOSITION. *If I is an ideal of a semiring R and M is a left R-semimodule then $N = \{m \in M \mid Im = \{0_M\}\}$ is a subtractive submodule of M.*

PROOF. Clearly N is a submodule of M. If $m, m' \in M$ satisfy the condition that m and $m + m'$ belong to N then for each $r \in I$ we have $0 = r(m + m') = rm + rm' = rm'$ so $m' \in N$. Thus N is subtractive. □

(14.24) PROPOSITION. *If N is a subtractive subsemimodule of a left R-semimodule M and if A is a nonempty subset of M then $(N : A)$ is a subtractive left ideal of R.*

PROOF. Since the intersection of an arbitrary family of subtractive left ideals of R is again subtractive, it suffices to show that $(N : m)$ is subtractive for each element m of M. Let $a \in R$ and $b \in (N : m)$ satisfy the condition that $a + b \in (N : m)$. Then $am + bm \in N$ and $bm \in N$ so $am \in N$, since N is subtractive. Thus $a \in (N : m)$. □

(14.25) EXAMPLE. If R and S are semirings and if M is an (R, S)-bisemimodule then the set A of all matrices of the form $\left[\begin{smallmatrix} r & m \\ 0 & s \end{smallmatrix}\right]$ for $r \in R$, $m \in M$, and $s \in S$ is a semiring under the operations of addition and multiplication defined by

$$\begin{bmatrix} r & m \\ 0 & s \end{bmatrix} + \begin{bmatrix} r' & m' \\ 0 & s' \end{bmatrix} = \begin{bmatrix} r + r' & m + m' \\ 0 & s + s' \end{bmatrix}$$

and

$$\begin{bmatrix} r & m \\ 0 & s \end{bmatrix} \begin{bmatrix} r' & m' \\ 0 & s' \end{bmatrix} = \begin{bmatrix} rr' & rm' + ms' \\ 0 & ss' \end{bmatrix}.$$

Note that there is a morphism of semirings from $R \times S$ to A given by $(r, s) \mapsto \left[\begin{smallmatrix} r & 0 \\ 0 & s \end{smallmatrix}\right]$.

In the special case of $S = R$ we have a subsemiring of this semiring consisting of all matrices of the form $\left[\begin{smallmatrix} r & m \\ 0 & r \end{smallmatrix}\right]$ for $m \in M$ and $r \in R$. This is called the **trivial extension** of the semiring R by the (R, R)-bisemimodule M.

If R is a semiring and M and N are left R-semimodules then a function α from M to N is an **R-homomorphism** if and only if the following conditions are satisfied:

(1) $(m + m')\alpha = m\alpha + m'\alpha$ for all $m, m' \in M$;

(2) $(rm)\alpha = r(m\alpha)$ for all $m \in M$ and $r \in R$.

The **kernel** of α is $ker(\alpha) = \{0_N\}\alpha^{-1}$. This is a subtractive subsemimodule of M. The set $M\alpha = \{m\alpha \mid m \in M\}$ is a subsemimodule of N.

Homomorphisms of right semimodules and of bisemimodules are defined similarly but are written as acting on the left.

(14.26) EXAMPLE. If M is a left R-semimodule generated by a subset A then we have a surjective R-homomorphism $R^{(A)} \to M$ defined by $f \mapsto \sum_{m\in supp(f)} f(m)m$. In particular, we always have a surjective R-homomorphism from $R^{(M)}$ to M.

(14.27) EXAMPLE. If R is a semiring and A is a nonempty set then for each word w in A^* we have a function $\epsilon_w : R\langle\langle A\rangle\rangle \to R$ given by $\epsilon_w : f \mapsto f(w)$. This is clearly a homomorphism of (R, R)-bisemimodules.

(14.28) EXAMPLE. Let R be a semiring and let $\{M_i \mid i \in \Omega\}$ be a family of left R-semimodules. Then $\times_{i\in\Omega} M_i$ also has the structure of a left semimodule under componentwise addition and scalar multiplication. We denote this left semimodule by $\prod_{i\in\Omega} M_i$. Similarly,

$$\coprod_{i\in\Omega} M_i = \{\langle m_i\rangle \in \prod M_i \mid m_i = 0 \text{ for all but finitely-many indices } i\}$$

is a subsemimodule of $\prod_{i\in\Omega} M_i$. For each h in Ω we have canonical R-homomorphisms $\pi_h : \prod M_i \to M_h$ and $\lambda_h : M_h \to \coprod M_i$ defined respectively by $\pi_h : \langle m_i\rangle \mapsto m_h$ and $m_h\lambda = \langle u_i\rangle$, where

$$u_i = \begin{cases} 0 & \text{if } i \neq h \\ m_h & \text{if } i = h \end{cases}.$$

The R-semimodule $\prod M_i$ is the **direct product** of the R-semimodules M_i and the R-semimodule $\coprod M_i$ is their **coproduct**. It is easy to verify that if M is a left R-semimodule and if $\{M_i \mid i \in \Omega\}$ is a family of left R-semimodules such that, for each $i \in \Omega$, we are given an R-homomorphism $\alpha_i : M \to M_i$ then there exits a unique R-homomorphism $\alpha : M \to \prod_{i\in\Omega} M_i$ such that $\alpha_i = \alpha\pi_i$ for each $i \in \Omega$. Similarly, if we are given an R-homomorphism $\beta_i : M_i \to M$ for each $i \in \Omega$ then there exists a unique R-homomorphism $\beta : \coprod_{i\in\Omega} M_i \to M$ such that $\beta_i = \lambda_i\beta$ for each $i \in \Omega$.

Let M and N be left R-semimodules and let α and β be R-homomorphisms from M to N. Then $K = \{m \in M \mid m\alpha = m\beta\}$ is a subsemimodule of M. The inclusion map $\lambda : K \to M$ is the **equalizer** of (α, β) in the sense that if $\theta : M' \to M$ is an R-homomorphism satisfying $\theta\alpha = \theta\beta$ then there exists a unique R-homomorphism $\theta' : M' \to K$ satisfying $\theta = \theta'\lambda$. By a well-known result in category theory, we see that since the category of all left R-semimodules has products and equalizers, it has arbitrary limits.

(14.29) EXAMPLE. Let R be a semiring and let M be a left R-semimodule. If $\alpha : M \to R$ is an R-homomorphism then we can define an operation $\odot_\alpha$ on M by setting $m \odot_\alpha n = (m\alpha)n$. Then $(M, +, \odot_\alpha)$ is a hemiring which is not, in general, a semiring.

(14.30) EXAMPLE. In Example 14.17 we saw that if R is a ring and M is a left R-module then $(sub(M), \cap)$ is a right $(R-fil)$-semimodule. An R-homomorphism $\alpha: M \to M'$ of left R-modules induces a map $\alpha^* : sub(M') \to sub(M)$ defined by $\alpha^*(N') = N'\alpha^{-1}$. Clearly $\alpha^*(N' \cap N'') = \alpha^*(N') \cap \alpha^*(N'')$. If $\kappa \in R - fil$ and $N' \in sub(M')$ then

$$\begin{aligned}[\alpha^*(N')]\kappa &= \{m \in M \mid (N'\alpha^{-1} : m) \in \kappa\} \\ &= \{m \in M \mid (N' : m\alpha) \in \kappa\} = \alpha^*(N'\kappa)\end{aligned}$$

and so α^* is a homomorphism of right $(R - fil)$-semimodules.

(14.31) APPLICATION. As we saw in Example 14.13, $M = (\mathbb{R} \cup \{\infty\}, min)$ is a left $\mathbb{R}^+$-semimodule. Every n-tuple $x = (m_1, \ldots, m_n)$ of elements of M defines an $\mathbb{R}^+$-homomorphism $\gamma_x: (\mathbb{R}^+)^n \to M$ by

$$\gamma_x: (a_1, \ldots, a_n) \mapsto min\{a_i m_i \mid 1 \leq i \leq n\} = min\{m_i \mid a_i > 0\}.$$

This allows us to consider linear optimization problems in the context of homomorphisms of semimodules, as is done in detail in [Zimmermann, 1981].

Another application of semimodule theory to optimization is the following: let R be the semifield $(\mathbb{R} \cup \{\infty\}, min, +)$, on which we have a metric d, defined by $d(a, b) = |e^{-a} - e^{-b}|$. For a locally-compact topological space X, let $C_0(X)$ be the R-semimodule of all continuous functions $f \in R^X$ satisfying the condition for each $\epsilon > 0$ there exists a compact subset K of X such that $d(f(x), \infty) < \epsilon$ for all $x \in X \setminus K$. The study of R-homomorphisms of the form $C_0(X) \to C_0(Y)$ is significant in the analysis of a wide range of deterministic problems in optimal control theory, and is developed for this purpose in [Kolokol'tsov, 1992].

We have already noted that if $\gamma: R \to S$ is a morphism of semirings and if M is a left S-semimodule then M is also a left R-semimodule, with scalar multiplication defined by $r \cdot m = \gamma(r)m$. If $\alpha: M \to N$ is an S-homomorphism of left S-semimodules then it is immediate that it is also an R-homomorphism.

Here we should note an important point in which semimodules over semirings differ from modules over rings. Let R be a semiring and let $\alpha: M \to N$ be an R-homomorphism of left R-semimodules. Given an element n of N, we are often interested in finding $n\alpha^{-1} = \{m \in M \mid m\alpha = n\}$. If we know one element m_0 of $n\alpha^{-1}$ then clearly $m_0 + m' \in n\alpha^{-1}$ for each element m' of $ker(\alpha)$. All elements of $n\alpha^{-1}$ are of this form if $m_0 \in V(M)$, but this need not be true in general.

(14.32) EXAMPLE. Let M be the left $\mathbb{N}$-semimodule $(\mathbb{N}[t], +)$ and let N be the left $\mathbb{N}$-semimodule $\{\mathbb{N} \cup \{-\infty\}, max)$ in which scalar multiplication is defined by $i \cot n = -\infty$ if $i = 0$ and $i \cdot n = n$ otherwise. Then the function $\alpha: M \to N$ defined by $\alpha: p(t) \mapsto deg(p)$ is a surjective $\mathbb{N}$-homomorphism the kernel of which is $\{0\}$. Nonetheless, $h\alpha^{-1}$ is an infinite set for each $0 \leq h \in \mathbb{N}$.

(14.33) EXAMPLE. Let a and b be elements of a frame R and let $\alpha: R \to R$ be the R-homomorphism defined by $\alpha: r \mapsto r \wedge a$. Then $b\alpha^{-1} = \{r \in R \mid r \wedge a = b\}$. If $a \not\geq b$ then this set is clearly empty. Otherwise, it is $\{r \in R \mid b \leq r \leq (b : a)\}$.

We have already noted that if $\gamma: S \to R$ is a morphism of semirings then every left R-semimodule is canonically a left S-semimodule and every R-homomorphism of left R-semimodules is a homomorphism of left S-semimodules as well. This allows us occasionally to modify our choice of the semiring R to make the problem before us easier to solve, as we will see later on.

If M and N are left R-semimodules then we will denote the set of all R-homomorphisms from M to N by $Hom_R(M, N)$. If α and β belong to $Hom_R(M, N)$ then so does the map $\alpha + \beta$ from M to N which is defined by $m(\alpha + \beta) = m\alpha + m\beta$. It is easy to check that $(Hom_R(M, N), +)$ is an $\mathbb{N}$-semimodule (i.e. a commutative monoid), the identity element of which is given by the map which sends each element of M to 0_N. If M is an (R, S)-bisemimodule then $Hom_R(M, N)$ is a left S-semimodule, with scalar multiplication defined by $s\alpha: m \mapsto (ms)\alpha$. If N is an (R, S)-bisemimodule then $Hom_R(M, N)$ is a right S-semimodule with scalar multiplication defined by $\alpha s : m \mapsto (m\alpha)s$.

If M, N, and P are left R-semimodules and if $\varphi: M \to N$ is an R-homomorphism then we have induced $\mathbb{N}$-homomorphims $Hom(P, \varphi): Hom_R(P, M) \to Hom_R(P, N)$ and $Hom(\varphi, P): Hom_R(N, P) \to Hom_R(M, P)$ given respectively by $\alpha \mapsto \alpha\varphi$ and $\beta \mapsto \varphi\beta$.

An R-homomorphism from a left semimodule M to itself is called an R-**endomorphism** of M. The set of all R-endomorphisms of M will be denoted by $End_R(M)$. In addition to the operation of addition on $End_R(M)$ we have an operation of multiplication given by composition of functions: $\alpha\beta: m \mapsto (m\alpha)\beta$ for all m in M. If M is an (R, S)-bisemimodule and let $S[t]$ be the semiring of polynomials in the indeterminate t over S. If $\alpha \in End_R(M)$ then α induces on M the structure of a $(R, S[t])$-bisemimodule by setting $m(\sum s_i t^i) = \sum m s_i \alpha^i$.

(14.34) EXAMPLE. Let A be a nonempty set, let R be a semiring, and let M be a left R-semimodule. If $u: B \to A$ is a function from a subset B of A to A then u induces an R-endomorphism α_u of M^A defined by $(\alpha_u f)(i) = (fu)(i)$ if $i \in B$ and $(\alpha_u f)(i) = 0$ otherwise.

In particular, we can consider the case of $A = \mathbb{N}$. If $B = \mathbb{N}$ and u is defined by $u: i \mapsto i + 1$ then α_u is the **left shift endomorphism** on M^A. If $B = \mathbb{P}$ and u is defined by $u: i \mapsto i - 1$ then α_u is the **right shift endomorphism** on M^A.

If M is the left R-semimodule $\{0\}$ then $End_R(M) = \{\iota\}$, where ι is the identity map $m \mapsto m$. If $M \neq \{0\}$ then $End_R(M)$ has at least two elements: the identity map and the map $m \mapsto 0$.

(14.35) APPLICATION. Let S be the semiring of all formal languages on a nonempty set A. Following the terminology of Abramsky and Vickers [1993], we say that a **transition system** $(P, \to)$ over A consists of a nonempty set P together with a subset $\to$ of $P \times A \times P$, where we write $p \xrightarrow{a} q$ instead of $(p, a, q) \in \to$. The elements of A are the **atomic actions** of the transition system, while the elements of P are the **processes** of the system. Each atomic action $a \in A$ defines a function

θ_a from the set C of all subsets of P to itself given by

$$\theta_a: U \mapsto \{q \in P \mid p \xrightarrow{a} q \text{ for some } p \in U\}.$$

We can expand this notion by defining θ_w for each word $w \in A^*$ recursively as follows:

(1) If $w = \square$ then θ_w is the identity map;
(2) If $w = va$ for some $v \in A^*$ and some $a \in A$, then $\theta_w = \theta_v\theta_a$.

Furthermore, if $L \in S$ we can define the function $\theta_L: C \to C$ by setting $\theta_L: U \mapsto \cup\{U\theta_w \mid w \in L\}$.

Of course, $(C, \cup)$ is a left $\mathbb{B}$-semimodule. If $L \in S$ then θ_L is a $\mathbb{B}$-endomorphism of C and, indeed, $\{\theta_L \mid L \in S\}$ is a subsemiring of the semiring of all $\mathbb{B}$-endomorphisms of C, so that C becomes a right S-module in which, for each $U \in C$ and $L \in S$, we have $UL = \{q \in P \mid p \xrightarrow{w} q \text{ for some } p \in P \text{ and } w \in L\}$. This semimodule is zerosumfree but not necessarily entire. But $S' = \{L \in S \mid L = \varnothing \text{ or } \square \in L\}$ then S' is a subsemiring of S and C is an information semimodule over S'.

(14.36) PROPOSITION. *If R is a semiring and $M \neq \{0\}$ is a left R-semimodule then $S = End_R(M)$ is a semiring and M is an (R, S)-bisemimodule.*

PROOF. The proof that S is a semiring with additive identity given by $m \mapsto 0$ and multiplicative identity given by $m \mapsto m$ is straightforward. Similarly, it is straightforward to show that M is an (R, S)-bisemimodule. Refer to Example 1.13. $\square$

Note that if M is an additively-idempotent left R-semimodule and if α belongs to $End_R(M)$ then $m(\alpha + \alpha) = m\alpha + m\alpha = m\alpha$ and so $\alpha + \alpha = \alpha$. Thus the semiring $End_R(M)$ is additively idempotent.

(14.37) EXAMPLE. If M is a nonzero left R-semimodule, it is possible for the semiring $S = End_R(M)$ to be zeroic. For example, set $R = M = \mathbb{B}$. Then S consists precisely of the maps $m \mapsto 0$ and $m \mapsto m$ and hence is surely zeroic.

(14.38) EXAMPLE. [Cornish, 1971] To each semiring R we can associate the semiring $S = End_R(R_R) \times End_R({}_RR)$ consisting of pairs (φ, θ) of maps from R to itself, the first member of which is a homomorphism of right semimodules and the second member of which is a homomorphism of left semimodules. Addition and multiplication on S are defined componentwise. An element (φ, θ) of S is a **bimultiplication** of R if and only if $a[\varphi b] = [a\theta]b$ for all $a, b \in R$. The set $bim(R)$ of all bimultiplications of R forms a subsemiring of S. Moreover, we have a map $\gamma: a \mapsto (\varphi_a, \theta_a)$ from R to $bim(R)$ defined by $\varphi_a: r \mapsto ar$ and $\theta_a: r \mapsto ra$ for all $r \in R$. This is in fact a morphism of semirings. We claim that γ is in fact an isomorphism. Indeed, if $\gamma(a) = \gamma(b)$ then $a = \varphi_a(1) = \varphi_b(1) = b$ and so γ is injective. If $(\varphi, \theta) \in bim(R)$ and if $a = \varphi 1$ then $1\theta = (1\theta)1 = 1(\varphi 1) = 1a = a$. Moreover, for each $r \in R$ we have $\varphi(r) = \varphi(1r) = (\varphi 1)r = ar$ and similarly $r\theta = ra$. Thus $(\varphi, \theta) = \gamma(a)$, showing that γ is surjective.

(14.39) EXAMPLE. Let R be a semiring, let $M \neq \{0\}$ be a left R-semimodule, and let $\theta: R \to End_R(M)$ be a morphism of semirings. Then we can define the structure of a semiring on $R \times M$ by defining addition and multiplication as follows: $(r, m) + (r', m') = (r + r', m + m')$ and $(r, m) \cdot (r', m') = (rr', rm' + m\theta(r'))$ for all $r, r' \in R$ and $m, m' \in M$. We will denote this semiring by $R \times_\theta M$. Moreover, the map $r \mapsto (r, 0)$ is a morphism of semirings.

Let $M_1, \ldots, M_n$ be left R-semimodules and let $M = \times_{i=1}^n M_i$. Let S be the set of all rectangular arrays $[\alpha_{ij}]$ where, for each $1 \leq i, j \leq n$, α_{ij} is an R-homomorphism from M_i to M_j. Define addition and multiplication in S by setting $[\alpha_{ij}] + [\beta_{ij}] = [\alpha_{ij} + \beta_{ij}]$ and $[\alpha_{ij}][\beta_{ij}] = [\theta_{ij}]$, where θ_{ij} is defined to be $\sum_{i=1}^n \alpha_{ih}\beta_{hj}$. Then S is a semiring. Moreover, we have a morphism of semirings $\gamma: S \to End_R(M)$ which takes $[\alpha_{ij}]$ to the endomorphism α of M defined by

$$\alpha: (m_1, \ldots, m_n) \mapsto \left(\sum_{i=1}^n m_i\alpha_{i1}, \ldots, \sum_{i=1}^n m_i\alpha_{in}\right).$$

For each $1 \leq i \leq n$, let $\lambda_i: M_i \to M$ be the canonical embedding of M_i into the ith component of M and let $\pi_i: M \to M_i$ be the canonical projection of M onto its ith component. Then each R-endomorphism α of M is the image under γ of $[\lambda_i\alpha\pi_j]$ and so γ is surjective. Moreover, γ is also clearly injective and hence is an isomorphism of semirings.

If M is a left R-semimodule and $\alpha \in End_R(M)$ then, for a given element m of M it is still difficult, in general, to find all of the members of $m\alpha^{-1} = \{m' \in M \mid m'\alpha = m\}$.

(14.40) EXAMPLE. [Kim & Roush, 1980] let $R = \mathbb{B}$, let $M = R^2$, and let $\alpha \in End_R(M)$ be defined by $\alpha: (a, b) \mapsto (a+b, b)$. Then α is multiplicatively regular since it is multiplicatively idempotent. Moreover, $(0, 1) \in (1, 1)\alpha^{-1}$. However, $(0, 1) \neq (1, 1)\alpha\beta$ for any $\beta \in End_R(M)$ satisfying $\alpha\beta\alpha = \alpha$, as can be easily verified directly.

Let R be a semiring and let $\alpha: M' \to M$ and $\beta: M \to M''$ be R-homomorphisms of left R semimodules. If for each $m \in M$ there exists a unique element $x \in M\alpha$ satisfying $m\beta = x\beta$ then the function $\pi_{\alpha,\beta}: m \mapsto x$ is an R-endomorphism of M satisfying the following conditions:

(1) $\pi_{\alpha,\beta} = \pi_{\alpha,\beta}^2$;
(2) $\alpha\pi_{\alpha,\beta} = \alpha$; and
(3) $\pi_{\alpha,\beta}\beta = \beta$.

In other words, $\pi_{\alpha,\beta}$ is the **projector onto $M\alpha$ parallel to $ker(\beta)$**. Such maps were first studied by Cohen et al. [1996, 1997].

(14.41) EXAMPLE. Even in the apparently-simple case of R taken to be the schedule algebra and the semimodules M', M, M'' taken to be of the form R^k for suitable values of k, the map $\pi_{\alpha,\beta}$ may be difficult to calculate for specific α and β. See [Gunawardena, 1994] or [Cohen et al., 1996].

15. FACTOR SEMIMODULES

Congruence relations played an important role in the theory of semirings and we would expect them to play a similar role in the theory of semimodules. Let R be a semiring and let M be a left R-semimodule. An equivalence relation ρ on M is an R-**congruence relation** if and only if $m \; \rho \; m'$ and $n \; \rho \; n'$ in M imply that $(m+n) \; \rho \; (m'+n')$ and $rm \; \rho \; rm'$ for all $r \in R$. In other words, an R-congruence relation ρ on M is an equivalence relation satisfying the condition that ρ is also a subsemimodule of $M \times M$. Denote the set of all R-congruence relations on M by $R - cong(M)$. This set is nonempty since it contains the **trivial** R-**congruence** $\equiv_t$ defined by $m \equiv_t m'$ if and only if $m = m'$ and the **universal** R-**congruence** $\equiv_u$ defined by $m \equiv_u m'$ for all $m, m' \in M$. If $M \neq \{0_M\}$ and these are the only two elements of $R - cong(M)$, then the R-semimodule M is **simple**. Moreover, $R - cong(M)$ is partially-ordered by the relation $\leq$ defined by $\rho \leq \rho'$ if and only if $m \; \rho \; m'$ implies that $m \; \rho' \; m'$. Clearly $\equiv_t \; \leq \rho \leq \; \equiv_u$ for all R-congruence relations ρ in $R - cong(M)$.

If W is a nonempty subset of $R - cong(M)$ then the relation ρ on M defined by $m \; \rho \; m'$ if and only if $m \; \rho' \; m'$ for each $\rho' \in W$ is also an R-congruence relation on M and $\rho'' \leq \rho'$ for each ρ' in W if and only if $\rho'' \leq \rho$. Thus $R - cong(M)$ is a complete lattice. If $m, m' \in M$ we will denote the unique smallest element ρ of $R - cong(M)$ satisfying $m \; \rho \; m'$ by $\rho_{(m,m')}$.

If ρ belongs to $R - cong(M)$ for some left R-semimodule M and if $a \in C(R)$ then we can define a relation ${}^a\rho$ on M by setting $m \; {}^a\rho \; m'$ if and only if $am \; \rho \; am'$. It is easy to verify that this is an R-congruence relation which, indeed, turns $(R - cong(M), \vee)$ into a left $C(R)$-semimodule.

If N is a subsemimodule of a left R-semimodule M and if ρ belongs to $R - cong(M)$, then the restriction of ρ to N is an R-congruence relation on N. Thus we have a canonical map from $R - cong(M)$ to $R - cong(N)$ given by restriction. If N is a subsemimodule of a left R-semimodule M and if ζ is a given R-congruence relation on N then there exists a unique maximal R-congruence relation on M the restriction of which to N is ζ. We will denote this congruence relation by $\equiv_{\zeta|N}$. In particular, we will denote by $\equiv_{=|N}$ the unique maximal R-congruence relation on M the restriction of which to N is the trivial relation.

Let ρ be an R-congruence relation on M and, for each $m \in M$, let m/ρ be the equivalence class of m with respect to this relation. Set M/ρ equal to $\{m/\rho \mid m \in$

$M\}$ and define operations of addition and scalar multiplication on M/ρ by setting $(m/\rho)+(n/\rho) = (m+n)/\rho$ and $r(m/\rho) = (rm)/\rho$ for all $m, n \in M$ and $r \in R$. Then M/ρ is a left R-semimodule, called the **factor semimodule** of M by ρ. Moreover, we have a surjective R-homomorphism $M \to M/\rho$ defined by $m \mapsto m/\rho$. Clearly M/ρ equals M/ρ' if and only if ρ and ρ' are equal.

If N is a subsemimodule of a left R-semimodule M then we have a canonical map from $R-cong(M)$ to $R-cong(N)$ given by restriction. If ρ is an R-congruence relation on M the restriction of which to N is ρ', then there is a monic R-homomorphism $N/\rho' \to M/\rho$ defined by $n/\rho' \mapsto n/\rho$. In particular, if the restriction of ρ to N is trivial then the function $N \to M/\rho$ given by $n \mapsto n/\rho$ is monic.

If ζ is an R-congruence relation on M/ρ then ζ defines a relation ζ^* on M by $m\ \zeta^*\ m'$ if and only if $(m/\rho)\ \zeta\ (m'/\rho)$. Clearly ζ^* is an R-congruence relation on M satisfying $\zeta^* \geq \rho$. Moreover, the function $\zeta \mapsto \zeta^*$ is a morphism of complete lattices from $R - cong(M/\rho)$ to $R - cong(M)$. If $\rho \leq \rho'$ in $R - cong(M)$ then we have an R-congruence ρ'/ρ in $R - cong(M/\rho)$ defined by the condition that $(m/\rho)\ \rho'/\rho\ (m'/\rho)$ if and only if $m\ \rho'\ m'$. Clearly, ρ/ρ is the trivial R-congruence on M/ρ.

Let M be a left R-semimodule and let ρ belong to $R - cong(M)$. If N is a subsemimodule of M, then $N' = \{m \in M \mid m\ \rho\ n \text{ for some } n \in N\}$ is a subsemimodule of M containing N. Moreover, if ρ' is the restriction of ρ to N' then N'/ρ' and N/ρ are equal.

(15.1) EXAMPLE. If $\alpha: M \to N$ is an R-homomorphism of left R-semimodules then α defines an R-congruence relation $\equiv_\alpha$ on M by $m \equiv_\alpha m'$ if and only if $m\alpha = m'\alpha$. Thus, α is monic precisely when $\equiv_\alpha$ is trivial. Note that α also induces an R-homomorphism α' from $M/\equiv_\alpha$ to N defined by $(m/\equiv_\alpha)\alpha' = m\alpha$ and that this R-homomorphism is in fact monic. More generally, if $\alpha: M \to N$ is an R-homomorphism of left R-semimodules and if $\rho \in R - cong(M)$ satisfies the condition that $\rho \leq \equiv_\alpha$, then there exists a unique R-homomorphism $\beta: M/\rho \to N$ satisfying $\beta: m/\rho \mapsto m\alpha$ for all $m \in M$.

Also note that if $\rho \in R - cong(M)$ and if $\alpha: M \to M/\rho$ is the R-homomorphism given by $\alpha: m \mapsto m/\rho$ then $\equiv_\alpha$ and ρ coincide.

(15.2) EXAMPLE. If R is a semiring and A is a submonoid of $C(R)$ then A defines an R-congruence relation $\equiv_A$ on any left R-semimodule M by setting $m \equiv_A m'$ if and only if there exists an element a of A satisfying $am = am'$. As a special case of this, let R be a commutative semiring and let $A = K^\times(R)$, the set of all multiplicatively-cancellable elements of R. If the R-congruence relation on M defined by $K^\times(R)$ is trivial then M is **classically torsionfree**.

(15.3) EXAMPLE. If N is a subsemimodule of a left R-semimodule M, then N induces an R-congruence relation $\equiv_N$ on M, called the **Bourne relation**, defined by setting $m \equiv_N m'$ if and only if there exist elements n and n' of N such that $m+n = m'+n'$. Note that, using the notation introduced in the previous chapter, $\equiv_N$ is just $E^{M\times M}_{N\times N}(\equiv_t)$.

If $m \in M$ then we write m/N instead of $m/\equiv_N$. The factor semimodule $M/\equiv_N$ is denoted by M/N. Note that $n/N = 0/N$ for all $n \in N$ and so if $m \in M$ then $am/N = 0/N$ for all $a \in (N : m)$. A slight modification of the proof of Proposition

6.50 shows that if N is a subsemimodule of a left R-semimodule M then $0/N$ is a subtractive subsemimodule of M and, indeed, is the subtractive closure of N in M.

We have already noted that both $ssm(M)$ and $R - cong(M)$ are left $C(R)$-semimodules. It is a straightforward consequence of the definitions that the function $\theta\colon ssm(M) \to R{-}cong(M)$ defined by $\theta\colon N \mapsto \equiv_N$ is in fact a $C(R)$-homomorphism. If N and N' are subtractive subsemimodules of M satisfying $\theta(N) = \theta(N')$ then for each $n \in N$ we have $0 \equiv_N n$ and so $0 \equiv_{N'} n$. Thus there exist $n', n'' \in N'$ such that $n + n' = n''$. Since N' is subtractive, this implies that $n \in N'$ and so $N \subseteq N'$. The reverse containment is proven similarly and so $N = N'$.

(15.4) EXAMPLE. If N is a submodule of a left R-semimodule M then N induces an R-congruence relation $[\equiv]_N$ on M, called the **Iizuka relation**, defined by setting $m\ [\equiv]_N\ m'$ if and only if there exist elements n and n' of N and an element m'' of M such that $m + n + m'' = m' + n' + m''$. If $m \in M$ we write $m[/]N$ instead of $m/[\equiv]_N$.

(15.5) EXAMPLE. If M is a left R-semimodule and if ρ and ρ' belong to $R - cong(M)$ Then the relation ρ'' on M defined by $m\ \rho''\ m'$ if and only if there exist $x, x' \in M$ such that $x\ \rho'\ x$ and $(m + x)\ \rho\ (m' + x')$. Then ρ'' is an R-congruence relation on M and $\rho \leq \rho''$.

If $\rho \leq \rho'$ in $R - cong(M)$ then we can define the relation $\rho'/\rho \in R - cong(M/\rho)$ by setting $(m/\rho)\ \rho'/\rho\ (m'/\rho)$ if and only if $m\ \rho'\ m'$. Then clearly $\rho = \rho'$ if and only if ρ'/ρ is the trivial congruence in $R - cong(M/\rho)$.

A nonzero subsemimodule N of a left R-semimodule M is **absorbing** if and only if the following conditions are satisfied:

(1) If $0 \neq n \in N$ and $m \in M$ then $0 \neq n + m \in N$;
(2) If $0 \neq n \in N$ then $(0 : n) = \{0\}$.

If N is an absorbing submodule of M we will write $N \sqsubseteq M$.

(15.6) PROPOSITION. *Let R be a semiring and let $\alpha\colon M \to M'$ be an R-homomorphism of left R-semimodules satisfying the condition that $ker(\alpha) \sqsubseteq M$. If $N' \sqsubseteq M'$ then $N = N'\alpha^{-1} \sqsubseteq M$.*

PROOF. Let $0 \neq n \in N$ and let $m \in M$. If $n \in ker(\alpha)$ then $n + m \in ker(\alpha) \subseteq N$ by hypothesis. Otherwise, $0 \neq n\alpha \in N'$ and so, by hypothesis, $(n + m)\alpha = n\alpha + m\alpha \in N'$, proving that $n + m \in N$. Similarly, let $r \in R$ satisfy $rn = 0$. If $n \in ker(\alpha)$ then $r = 0$ by hypothesis. Otherwise, $r(n\alpha) = 0$, where $0 \neq n\alpha \in N'$, and so $r = 0$. □

Thus $\{0\} \cup \{i \in \mathbb{N} \mid i > n\} \sqsubseteq \mathbb{N}$ for all $n \in \mathbb{N}$. Any nonzero R-semimodule which is an absorbing subsemimodule of itself is zerosumfree. Moreover, a necessary and sufficient condition for there to exist an R-semimodule which is an absorbing subsemimodule of itself is that R be an absorbing subsemimodule of itself, which is equivalent to R being entire and zerosumfree. An element w of a left R-semimodule M is **infinite** if and only if $w + m = w$ for all $m \in M$; it is **strongly infinite** if and only if $\{0_M, w\} \sqsubseteq M$. In this case, $\{0_M, w\}$ is called the **crux** of M and denote by $cr(M)$. If M has no strongly infinite elements, we set the crux of $cr(M) = \{0_M\}$.

Thus the crux of a left R-semimodule M is always an information subsemimodule of M and surely $cr(M) \sqsubseteq M$. If it equals all of M, we say that M is **crucial**. Otherwise it is **noncrucial**. Note that if M is an information semimodule over an entire zerosumfree semiring R then one can always adjoin a strongly infinite element to M. Indeed, pick an element $w \notin M$ and define addition and scalar multiplication on $M \cup \{w\}$ by setting $m + w = w + m = w$ for all $m \in M$, $rw = w$ for all $0 \neq r \in R$, and $0w = 0_M$.

(15.7) EXAMPLE. Let $\{(M_i, +_i) \mid i \in \Omega\}$ be a family of left information semimodules over a semiring R, the underlying sets of which are disjoint. Further assume that each M_i has a strongly-infinite element w_i. Set

$$M = \{0, w\} \cup \bigcup_{i \in \Omega} [M_i \setminus cr(M_i)],$$

where 0 and w are elements not in $\cup_{i \in \Omega} M_i$. Define addition and scalar multiplication on M as follows:

(1) $0 + m = m + 0 = m$ and $r0 = 0$ for all $m \in M$ and $r \in R$;
(2) $w + m = m + w = w$ and $rw = w$ for all $m \in M$ and all $0 \neq r \in R$;
(3) If $m, m' \in M \setminus \{0, w\}$, then

$$m + m' = \begin{cases} m +_h m' & \text{if } m, m' \in M_h \setminus cr(M_h) \\ w & \text{otherwise} \end{cases}.$$

(4) If $m \in M_h \setminus cr(M_h)$ and $r \in R$ then rm is the same as the corresponding value in M_h. Then M is an information semimodule over R having strongly-infinite element w. Moreover, for each $i \in \Omega$ we have an monic R-homomorphism $\lambda_i \colon M_i \to M$ defined by

$$\lambda_i \colon x \mapsto \begin{cases} w & \text{if } x = w_i \\ x & \text{otherwise} \end{cases}.$$

We denote the semimodule M constructed in this way by $\sqcup_{i \in \Omega} M_i$.

Let R be an entire zerosumfree semiring and let $\alpha \colon M \to M'$ be an R-homomorphism of left R-semimodules. As an immediate consequence of Proposition 15.6, wee see that if w' is a strongly-infinite element of M' contained in $M\alpha$ then $N = w'\alpha^{-1} \cup \{0_M\}$ is an absorbing subsemimodule of M.

A semimodule can have at most one [strongly] infinite element. If R is antisimple then every infinite element of a left R-semimodule is strongly infinite. Indeed, in this situation, any nonzero element r of R is of the form $1 + r'$ and so $rw = (1 + r')w = w + r'w = w$. If w is a strongly-infinite element of a left R-semimodule M and if N is a subsemimodule of M, then $N \cup \{w\}$ is also an R-semimodule of M. By Proposition 14.20, we note that if R is not entire and zerosumfree than no left R-module can have a strongly-infinite element.

(15.8) EXAMPLE. [Takahashi, 1984a] Let R be an antisimple semiring and let M be a left R-semimodule which is not an R-module. Then $N = \{0_M\} \cup [M \setminus V(M)]$ is an absorbing subsemimodule of M, Indeed, if $n, n' \in N$ then clearly $n + n' \in N$. If $0_M \neq n \in N$ and $0 \neq r \in R$ then $r = 1 + s$ for some $s \in R$. Thus $rn + n' = 0_M$ implies that $n + (sn + n') = 0_M$ and so $n \in V(M)$, which is a contradiction. Thus $rn \in N$ as well. Thus N is a subsemimodule of M. Let $m \in M$ and $n \in N \setminus \{0_M\} = M \setminus V(M)$. If $m + n \notin N \setminus \{0_M\}$ then $m + n \in V(M)$ so there is an element $m' \in M$ satisfying $0_M = m + n + m' = n + (m + m')$, contradicting the assumption that $n \notin V(M)$. Thus $N \sqsubseteq M$.

We also note the converse, if $N \sqsubseteq M$ and N satisfies the condition that $M \setminus V(M) \subseteq N$ then $M \setminus V(M) = N$ for if $0_M \neq x \in V(M) \cap N$ then there exists an element $0_M \neq y \in V(M)$ satisfying $x + y = 0$, contradicting the fact that $M + [N \subseteq \{0_M\}] = N \subseteq \{0_M\}$.

An absorbing subsemimodule N of a left R-semimodule M defines a congruence relation $\sim_N$ on M by setting $m \sim_N m'$ if and only if $m = m'$ or both m and m' belong to $N \setminus \{0\}$. Note that $M/\sim_N = (M \setminus N) \cup \{0, w\}$, where w is a strongly-infinite element of $M/\sim_N$.

If $\alpha\colon M \to N$ is an R-homomorphism of left R-semimodules and if m, m' are elements of M satisfying $m \equiv_{ker(\alpha)} m'$ then surely $m \equiv_\alpha m'$, but the converse does not necessarily hold. If the relations $\equiv_\alpha$ and $\equiv_{ker(\alpha)}$ coincide, then the R-homomorphism α is **steady**. Thus, for example, a steady R-homomorphism $\alpha\colon M \to N$ is monic if and only if $ker(\alpha) = \{0\}$.

(15.9) PROPOSITION. *If α is a steady R-endomorphism of a left R-semimodule M then α^k is steady for each $k \geq 1$.*

PROOF. The proof will be by induction on k. For $k = 1$ the result is given. Assume therefore that α^k is steady and let m and m' be elements of M satisfying $m\alpha^{k+1} = m'\alpha^{k+1}$. Then $(m\alpha^k)\alpha = (m'\alpha^k)\alpha$ so, by steadiness, there exist elements x and x' of $ker(\alpha)$ such that $m\alpha^k + x = m'\alpha^k + x'$. But then $(m+x)\alpha^k = (m'+x')\alpha^k$ so there exist elements y and y' of $ker(\alpha^k)$ such that $m + x + y = m' + x' + y'$, where $x + y$ and $x' + y'$ belong to $ker(\alpha^{k+1})$. Thus α^{k+1} is steady. □

(15.10) PROPOSITION. *Let R be a semiring and let $N' \subseteq N$ be subsemimodules of a left R-semimodule M. Then the function $\alpha\colon M/N' \to M/N$ defined by $\alpha\colon m/N' \mapsto m/N$ is a steady surjective R-homomorphism.*

PROOF. That α is a surjective R-homomorphism is clear. Suppose that m/N' and m'/N' are elements of M/N' satisfying $(m/N')\alpha = (m'/N')\alpha$. Then $m/N = m'/N$ and so there exist elements n and n' of N satisfying $m + n = m' + n'$. But n/N' and n'/N' belong to $ker(\alpha)$ and so $m \equiv_{ker(\alpha)} m'$. Thus α is steady. □

(15.11) PROPOSITION. *Let R be a semiring and let $\alpha\colon M \to N$ be an R-homomorphism between left R-semimodules. Let $\beta\colon M \to P$ be a surjective steady R-homomorphism between left R-semimodules satisfying $ker(\beta) \subseteq ker(\alpha)$. Then:*

(1) *There exists a unique R-homomorphism $\theta\colon P \to N$ satisfying $\alpha = \beta\theta$;*
(2) *If α is monic so is θ;*
(3) *$ker(\theta) = (ker(\alpha))\beta$; and*
(4) *$P\theta = M\alpha$.*

PROOF. (1) Since β is surjective we know that if $p \in P$ then $p\beta^{-1} \neq \varnothing$. If $m, m' \in p\beta^{-1}$ then $m \equiv_\beta m'$ and so, by steadiness, $m \equiv_{ker(\beta)} m'$. Thus there exist elements $k, k' \in ker(\beta) \subseteq ker(\alpha)$ satisfying $m + k = m' + k'$ and so

$$m\alpha = m\alpha + k\alpha = (m + k)\alpha = (m' + k')\alpha = m'\alpha + k'\alpha = m'\alpha.$$

Define the function $\theta: P \to N$ by $\theta: p \mapsto m\alpha$, where m is any element of $p\beta^{-1}$. Then this function is well-defined, and it is easy to show that it is an R-homomorphism of semimodules satisfying $\alpha = \beta\theta$. Moreover, if $\theta': P \to N$ is an R-homomorphism satisfying $\alpha = \beta\theta'$ and if $p \in P$ then for any $m \in p\beta^{-1}$ we have $p\theta' = m\beta\theta' = m\beta\theta = p\theta$, proving that $\theta = \theta'$.

(2) Assume α is monic. If $p_1\theta = p_2\theta$ and if $m_i \in p_i\beta^{-1}$ for $i = 1, 2$, then $m_1\alpha = m_1\beta\theta = m_2\beta\theta = m_2\alpha$ and so $m_1 = m_2$. Therefore $p_1 = m_1\beta = m_2\beta = p_2$, proving that θ is monic.

(3) Clearly $(ker(\alpha))\beta \subseteq ker(\theta)$. Conversely, if $p \in ker(\theta)$ and if $m \in p\beta^{-1}$ then $m\alpha = p\theta = 0_N$ so $p = m\beta \in (ker(\alpha))\beta$, establising equality.

(4) This is immediate from the definition. □

We now prove a dual of this result.

(15.12) PROPOSITION. *Let R be a semiring and let $\alpha: M \to N$ be an R-homomorphism between left R-semimodules. Let $\beta: P \to N$ be a monic R-homomorphism of left R-semimodules satisfying the condition that $P\beta$ is a subtractive subsemimodule of P containing $M\alpha$. Then:*

1. *(1) There exists a unique R-homomorphism $\theta: M \to P$ satisfying $\alpha = \theta\beta$;*
2. *(2) $ker(\theta) = ker(\alpha)$;*
3. *(3) The subtractive closure of $M\theta$ in P is $N'\beta^{-1}$, where N' is the subtractive closure of $M\alpha$ in N; and*
4. *(4) θ is monic if and only if α is monic.*

PROOF. (1) If $m \in M$ then $m\alpha \in M\alpha \subseteq P\beta$. Since β is monic, this means that there exists a unique element p of P satisfying $p\theta = m\alpha$. Set $m\theta = p$. By uniqueness, it is easily seen that the function $\theta: M \to P$ thus defined is an R-homomorphism satisfying $\alpha = \theta\beta$, which is unique.

(2) If $m \in ker(\alpha)$ then $0_P\beta = 0_N = m\alpha$ so $m\theta = 0_P$, proving that $m \in ker(\theta)$. Conversely, if $m \in ker(\theta)$ then $m\alpha = m\theta\beta = 0_N$ so $m \in ker(\alpha)$.

(3) Since $P\beta$ is subtractive, we note that $N' \subseteq P\beta$. Let P' be the subtractive closure of $M\theta$ in P. Then $p \in P' \Leftrightarrow p' + m\theta = m'\theta$ for some $m, m' \in M \Leftrightarrow p'\beta + m\alpha = m'\alpha \Leftrightarrow p'\beta \in N' \Leftrightarrow p' \in N'\beta^{-1}$.

(4) This is an immediate consequence of the definition. □

If $\alpha: M \to N$ is an R-homomorphism of left R-semimodules then we define the **coimage** of α to be $coim(\alpha) = M/ker(\alpha)$ and the **cokernel** of α to be $N/M\alpha$.

(15.13) PROPOSITION. *If M is a simple left R-semimodule then M has no subtractive subsemimodules other than $\{0_M\}$ and itself. The converse is true if M is an R-module.*

PROOF. Let M be a simple left R-semimodule and let N be a subtractive subsemimodule of M. Then the R-congruence relation $\equiv_N$ is either trivial or universal.

If it is trivial and if $n \in N$ then $n \equiv_N 0_M$ so $n = 0_M$. Thus $N = \{0_N\}$. If $\equiv_N$ is universal and $m \in M$ then $m \equiv_N 0_M$ so there exists an element n of N satisfying $m + n \in N$. Since N is subtractive, this implies that $m \in N$ and so we must have $N = M$.

Conversely, let M be a left R-module satisfying the condition that $\{0_M\}$ and M are its only subtractive submodules and let ρ be an R-congruence relation on M. Set $N = \{m \in M \mid m \,\rho\, 0_M\}$. Then N is a subtractive submodule of M. If $N = M$ then ρ is universal; if $N = \{0_M\}$ then ρ is trivial. Therefore M is simple. □

As an immediate consequence of Proposition 15.13 and Proposition 14.23, we see that if M is a simple left R-semimodule and if I is an ideal of R then either $I \subseteq (0 : M)$ or $I \nsubseteq (0 : m)$ for all $0_M \neq m \in M$. Also, if M is simple and if $\alpha: M \to N$ is a nonzero R-homomorphism, then $ker(\alpha)$ must be equal to $\{0\}$.

(15.14) Proposition. *If N is a subsemimodule of a left R-semimodule M then the R-congruence relations $\equiv_N$ and $\equiv_{0/N}$ on M coincide.*

Proof. The proof is essentially the same as that of Proposition 6.54. □

If M and N are left R-semimodules then an R-homomorphism $\alpha: M \to N$ is an R-**monomorphism** if and only if whenever β and β' are distinct R-homomorphisms $M' \to M$ for some left R-semimodule M' then $\beta\alpha \neq \beta'\alpha$. Dually, α is an R-**epimorphism** if and only if whenever β and β' are distinct R-homomorphisms $N \to N'$ for some left R-semimodule N' then $\alpha\beta \neq \alpha\beta'$. A function which is both an R-epimorphism and an R-monomorphism is an R-**isomorphism**. If $\alpha: M \to N$ is an R-isomorphism then it is easily verified that so is $\alpha^{-1}: N \to M$. An R-isomorphism of left R-semimodules $\alpha: M \to N$ induces an isomorphism of semirings $End_R(M) \to End_R(N)$ defined by $\gamma \mapsto \alpha^{-1}\gamma\alpha$.

A surjective R-homomorphism having kernel $\{0\}$ is an R-**semiisomorphism**. Surely R-isomorphisms are R-semiisomorphisms, but the converse is not the case. If M is a simple left R-semimodule then any surjective R-endomorphism of M is an R-semiisomorphism. If $\alpha: M \to M'$ and $\beta: M' \to M''$ are R-semiisomorphisms then so is $\alpha\beta: M \to M''$.

(15.15) Proposition. *If $\alpha: M \to N$ is an R-homomorphism between left R-modules then:*

(1) *α is monic if and only if it is an R-monomorphism;*
(2) *α is surjective if and only if it is an R-epimorphism and $M\alpha$ is subtractive.*

Proof. (1) Let $\alpha: M \to N$ be an R-homomorphism of left R-semimodules. If α is monic, it is clearly an R-monomorphism. If it is not monic then there exist elements $m \neq m'$ of M satisfying $m\alpha = m'\alpha$. Define R-homomorphisms β and β' from R, considered as a left semimodule over itself, to M by setting $\beta: a \mapsto am$ and $\beta': a \mapsto am'$. Then $\beta \neq \beta'$ but $\beta\alpha = \beta'\alpha$, showing that α is not an R-monomorphism.

(2) If α is surjective then it is clearly an R-epimorphism and $M\alpha$ is subtractive. Conversely, assume that α is an R-epimorphism satisfying the condition that $M\alpha$ is subtractive, but that α is not surjective. Set $N' = N\alpha$. Then we have R-homomorphisms from N to N/N' given by $\beta: n \mapsto 0/N'$ and $\beta': n \mapsto n/N'$. Moreover, $\alpha\beta = \alpha\beta'$. Since α is not surjective, there exists an element $n \in N \setminus N'$.

Then $n/N' = 0/N'$ implies that there exist elements $m\alpha$ and $m'\alpha$ of N' with $n + m\alpha = 0 + m'\alpha \in N'$ which, by the subtractiveness of $M\alpha$, implies that $n \in N'$. This is a contradiction and so we conclude that $n\beta' = n/N' \neq 0/N' = n\beta$ and hence $\beta \neq \beta'$, contradicting the assumption that α is an R-epimorphism. □

(15.16) EXAMPLE. Let R be a semiring and M be a left $\mathbb{N}$-semimodule. Then $(R, +)$ is also a left $\mathbb{N}$-semimodule. Let $M^{\#}$ be the set of all $\mathbb{N}$-homomorphisms from R to M, written as acting on the right. If θ and φ are elements of $M^{\#}$ and if $r \in R$ then we define $\theta + \varphi$ and $r\theta$ by $a(\theta + \varphi) = a\theta + a\varphi$ and $(a)(r\theta) = (ar)\theta$ for all $a \in R$. It is straightforward to see that, under the given definitions, $M^{\#}$ is a left R-semimodule. Moreover, if M is a left R-semimodule then we have an R-homomorphism $\lambda: M \to M^{\#}$ defined by $(m\lambda): a \mapsto am$ for all $a \in R$ and all $m \in M$. Since $1(m\lambda) = m$ for all $m \in M$, we see that the function λ is monic and hence an R-monomorphism. If $\alpha: M \to N$ is an R-homomorphism of left R-semimodules then α defines an R-homomorphism $\alpha^{\#}: M^{\#} \to N^{\#}$ given by $\theta \mapsto \theta\alpha$.

(15.17) PROPOSITION. *Let M be a left R-semimodule and let α be a steady R-endomorphism of M. Then a sufficient condition for α to be an R-isomorphism is that it be monic and M satisfy the descending chain condition on subsemimodules or that it be surjective and M satisfy the ascending chain condition on subsemimodules.*

PROOF. Assume that α is monic and M satisfy the descending chain condition on subsemimodules. By Proposition 15.15, α is an R-monomorphism. Moreover, $M\alpha^n \subseteq M\alpha^{n-1}$ for all $n > 0$, where α^0 is taken to be the identity map on M, and so there exists a natural number t such that $M\alpha^t = M\alpha^{t-1}$. Pick the smallest such t. If $t = 1$ then $M\alpha = M$ and so α is surjective, hence an R-epimorphism, and hence an R-isomorphism. If $t > 1$ and $y \in M\alpha^{t-1}$ then $y\alpha \in M\alpha^t = M\alpha^{t-1}$ and so there exists an element m of M such that $y\alpha = m\alpha^{t-1}$. Since α is monic, this implies that $y = m\alpha^{t-2} \in M\alpha^{t-2}$ and so $M\alpha^{t-2} = M\alpha^{t-1}$, contradicting the minimality of t. Hence we must have $t = 1$ and so α is an R-isomorphism.

Now assume that α is surjective and that M satisfies the ascending chain condition on subsemimodules. By Proposition 15.15, α is an R-epimorphism. Set $K_0 = \{0\}$ and, for each $i > 0$, set $K_i = ker(\alpha^i)$. Then there exists an integer $n > 0$ such that $K_n = K_{n-1}$. Let t be the smallest such integer. If $t = 1$ then $ker(\alpha) = \{0\}$ and so, by steadiness, α is monic, hence an R-monomorphism, and hence an R-isomorphism. Assume therefore that $t > 1$. If $m \in K_{t-1}$ write $m = y\alpha$. Then $0 = m\alpha^{t-1} = (y\alpha)\alpha^{t-1} = y\alpha^t$. Hence $y \in K_t = K_{t-1}$ so $0 = y\alpha^{t-1} = m\alpha^{t-2}$. Therefore $m \in K_{t-2}$ and so $K_{t-1} = K_{t-2}$, contradicting the minimality of t. Thus we must have $t = 1$ and so α is an R-isomorphism. □

The following result is the analog of Proposition 10.11.

(15.18) PROPOSITION. *Let R be a semiring and let M be a left R-semimodule. Then a subset N of M is a subtractive subsemimodule if and only if there exists an R-homomorphism $\alpha: M \to M'$ satisfying $N = ker(\alpha)$.*

PROOF. We have already noted that kernels of R-homomorphisms $M \to M'$ are subtractive submodules of M. Conversely, assume that N is a subtractive subsemimodule of M and let $M' = M/N$. Let $\alpha: M \to M'$ be the surjective R-homomorphism of left R-semimodules defined by $\alpha: m \mapsto m/N$. Then $ker(\alpha) = \{m \in M \mid m \equiv_N 0\} = \{m \in M \mid$ there exist $n, n' \in N$ satisfying $m + n = n'\}$. But, since N is subtractive, this is just N. □

(15.19) Proposition. *Let R be a semiring and let $\alpha: M \to N$ be an R-homomorphism of left R-semimodules. If N' is a subtractive subsemimodule of N and if $M' = N'\alpha^{-1} \subseteq M$, then:*

(1) *M' is a subtractive subsemimodule of M containing $ker(\alpha)$; and*
(2) *α induces an R-homomorphism $\beta: M/M' \to N/N'$ having kernel $\{0\}$.*

Proof. (1) If $m', m'' \in M'$ and if $r \in R$ then $(m' + m'')\alpha = m'\alpha + m''\alpha \in N'$ and $(rm')\alpha = r(m'\alpha) \in N'$ and so M' is a subsemimodule of M. Since $0_N \in N'$, clearly $ker(\alpha) \subseteq M'$. Finally, if $m' + m'' \in M'$ and $m'' \in M'$ then $m'\alpha + m''\alpha$ and $m''\alpha$ both belong to N'. Since N' is subtractive, this implies that $m'\alpha \in N'$ and so $m' \in M'$. Thus M' is also subtractive.

(2) Define β by $\beta: m/M' \mapsto m\alpha/N'$. This map is well-defined for if $x/M' = y/M'$ then $x \equiv_{M'} y$ and so there exist elements m' and m'' in M' satisfying $x + m' = y + m''$. As a consequence, $x\alpha + m'\alpha = (x + m')\alpha = (y + m'')\alpha = y\alpha + m''\alpha$. But $m'\alpha$ and $m''\alpha$ both belong to N' and so $x\alpha \equiv_{N'} y\alpha$, whence $x\alpha/N' = y\alpha/N'$. Moreover, β is clearly an R-homomorphism. If $x/M' \in ker(\beta)$ then $x\alpha/N' = 0$ and so there exist elements n' and n'' of N' satisfying $x\alpha + n' = n''$. Since N' is subtractive, this implies that $x\alpha \in N'$ and so $x \in M'$. Thus $x/M' = 0/M'$, proving that $ker(\beta) = \{0\}$. □

(15.20) Corollary. *Let R be a semiring and let $\alpha: M \to N$ be a surjective R-homomorphism of left R-semimodules. Then there exists an R-semiisomorphism $M/ker(\alpha) \to N$.*

Proof. This is a direct consequence of Proposition 15.19. □

(15.21) Corollary. *If R is a semiring and if $N' \subseteq N$ are subsemimodules of a left R-semimodule M then M/N is R-isomorphic to $(M/N')/(N/N')$.*

Proof. Let $\alpha: M/N' \to M/N$ be the surjective R-homomorphism defined by $\alpha: m/N' \mapsto m/N$. Then $ker(\alpha) = N/N'$. By Corollary 15.20, the function α induces an R-semiisomorphism $\beta: (M/N')/(N/N') \to M/N$. Moreover, by Corollary 15.10, we see that β is steady and hence is monic. Therefore β is an R-isomorphism. □

The following result is the semimodule equivalent of Proposition 10.19.

(15.22) Proposition. *Let R be a semiring. If N and N' are subsemimodules of a left R-semimodule M then there exists a canonical surjective R-homomorphism $\alpha: N'/[N \cap N'] \to [N + N']/N$, which is an R-semiisomorphism if N is subtractive.*

Proof. Define the function $\alpha: N'/[N \cap N'] \to [N + N']/N$ by $\alpha: n'/[N \cap N'] \mapsto n'/N$. This is clearly a well-defined surjective R-homomorphism. Now suppose that N is subtractive and that $n'/[N \cap N'] \in ker(\alpha)$. Then there exist elements m and m' of N satisfying $n' + m = m'$ and so, by subtractiveness, $n' \in N$. This shows that $n' \in N \cap N'$ and so $n'/[N \cap N'] = 0/[N \cap N']$, proving that $ker(\alpha) = \{0/[N \cap N']\}$. Thus α is an R-semiisomorphism. □

(15.23) Proposition. *Let R be a semiring. If M and N are R-isomorphic left R-semimodules not equal to $\{0\}$ then $End_R(M)$ and $End_R(N)$ are isomorphic semirings.*

Proof. If $\alpha: M \to N$ is an R-isomorphism then it is easy to verify that the function from $End_R(M)$ to $End_R(N)$ given by $\gamma \mapsto \alpha^{-1}\gamma\alpha$ is a morphism of semirings which is in fact an isomorphism. □

An element m of a left R-semimodule M is **cancellable** if $m + m' = m + m''$ implies that $m' = m''$. The semimodule M is **cancellative** if and only if every element of M is cancellable. Clearly any R-module is cancellative. As in the case of ideals, a left R-semimodule M is cancellative if and only if the subsemimodule $D = \{(m, m) \mid m \in M\}$ of $M \times M$ is subtractive. It is also easy to verify that if N is a submodule of a left R-semimodule M such that $0[/]N \neq M$ then the Iizuka factor module $M[/]N$ is cancellative.

(15.24) Proposition. *If N is a subsemimodule of a cancellative left R-semimodule M then both N and M/N are cancellative.*

Proof. Clearly N is cancellative. If m, m', and m'' are elements of M satisfying $m/N + m'/N = m/N + m''/N$ then there exist elements n' and n'' of N such that $m + m' + n' = m + m'' + n''$. Since M is cancellative, this implies that $m' + n' = m'' + n''$ and so $m'/N = m''/N$. Thus M/N is cancellative. □

(15.25) Proposition. *For a family $\{M_i \mid i \in \Omega\}$ of left R-semimodules then following conditions are equivalent:*

(1) *$\prod_{i\in\Omega} M_i$ is cancellative;*
(2) *$\coprod_{i\in\Omega} M_i$ is cancellative;*
(3) *M_i is cancellative for each $i \in \Omega$.*

Proof. (1) $\Rightarrow$ (2) $\Rightarrow$ (3): This is a direct consequence of Proposition 15.24 since $\coprod_{i\in\Omega} M_i$ is R-isomorphic to a subsemimodule of $\prod_{i\in\Omega} M_i$ and each M_i is R-isomorphic to a subsemimodule of $\coprod_{i\in\Omega} M_i$.

(3) $\Rightarrow$ (1): Let $m = \langle m_i \rangle$, $m' = \langle m_i' \rangle$, and $m'' = \langle m_i'' \rangle$ be elements of $\prod_{i\in\Omega} M_i$ satisfying $m + m' = m + m''$. Then for each $i \in \Omega$ we have $m_i + m_i' = m_i + m_i''$ and so, by (3), $m_i' = m_i''$. Therefore $m' = m''$. □

(15.26) Example. [Takahashi, 1981] If M is a left R-semimodule and ζ is the R-congruence relation on M defined by $m \,\zeta\, m'$ if and only if there exists an element m'' of M satisfying $m+m'' = m'+m''$ then M/ζ is a cancellative left R-semimodule.

(15.27) Proposition. *If R is a semiring and M is a simple left R-semimodule not equal to $\{0_M\}$ then precisely one of the following conditions holds:*

(1) *M is additively idempotent; or*
(2) *M is cancellative.*

Proof. It is immediate that if a left R-semimodule is both additively idempotent and cancellative then it must equal $\{0_M\}$ since $m + m = m = m + 0$ implies in that $m = 0$. Therefore both of these conditions cannot hold simultaneously.

Let $\alpha: M \to M$ be the R-homomorphism defined by $\alpha: m \mapsto m+m$. Since M is simple, $\equiv_\alpha$ is either trivial or universal. If $\equiv_\alpha$ is universal then $m \equiv_\alpha 0$ for all $m \in M$ and so, in particular, $m+m=0$ for all such m. If m, m', and m'' are elements of M satisfying $m+m'=m+m''$ then $m'=0+m'=m+m+m'=m+m+m''=0+m''=m''$. Therefore, in this situation, M is cancellative.

We will assume, therefore, that $\equiv_\alpha$ is the trivial R-congruence relation on M. This means that α is monic. Moreover, $M\alpha$ is cancellative [resp. additively idempotent] if and only if M is and so, without loss of generality, we can in fact assume that α is surjective as well. Define a relation ρ on M by setting $x \ \rho \ x'$ if and only if there exist elements m and m' of M and an element $k \in \mathbb{N}$ such that $x=m+x'\alpha^{-k}$ and $x'=m'+x\alpha^{-k}$. Clearly ρ is an equivalence relation. We claim that it is an R-congruence relation as well. Indeed, if $x \ \rho \ x'$ and if m, m', and k are as above then for each $r \in R$ we have $rx=rm+(r'x)\alpha^{-k}$ and $rx'=rm'+(rx)\alpha^{-k}$ and hence $rx \ \rho \ rx'$. Moreover, if $y \ \rho \ y'$ as well, then we can write $y=y\alpha^{-k}\alpha^k$. For $k>0$ we have

$$
\begin{aligned}
y &= y\alpha^{-k}\alpha^{1-k}+y\alpha^{-k}\alpha^{1-k} \\
&= y\alpha^{-k}\alpha^{1-k}+y\alpha^{-k}\alpha^{2-k}+y\alpha^{-k}\alpha^{2-k} \\
&= \cdots = m''+y\alpha^{-k}
\end{aligned}
$$

for some element m'' of M. Therefore $x+y=(m+m'')+(x'+y)\alpha^{-k}$ and $x'+y=(m'+m'')+(x+y)\alpha^{-k}$ so $x+y \ \rho \ x'+y$. Similarly, $(x'+y) \ \rho \ (x'+y')$ and so $(x+y) \ \rho \ (x'+y')$, establishing the claim.

Since M is simple, this implies that ρ is either trivial or universal. If it is the trivial R-congruence relation then for each $m \in M$ we have $m=0+(m\alpha)\alpha^{-1}$ and $m\alpha=m+m=m+m\alpha^{-1}\alpha=(m+m\alpha^{-1})+m\alpha^{-1}$ so that $m \ \rho \ m\alpha$. By triviality, this implies that $m=m\alpha=m+m$ for each $m \in M$, proving that M is additively idempotent.

We are left to consider the case of ρ universal and we wish to show that, in this case, M is cancellative. Let m, m', and m'' be elements of M satisfying $m+m''=m'+m''$. Set $N=\{x \in M \mid x+m=x+m'\}$. This set is nonempty since $m'' \in N$. If $x \in N$ then $(m+x\alpha^{-1})\alpha=m+m+x=m'+m+x=m'+m'+x=(m'+x\alpha^{-1})\alpha$ and so, since α is monic, we must have $m+x\alpha^{-1}=m'+x\alpha^{-1}$. Thus $x \in N$ implies that $x\alpha^{-1} \in N$ and so $x\alpha^{-k} \in N$ for all $k \in \mathbb{N}$. If y is an arbitrary element of M then $y \ \rho \ m''$ since ρ is universal and so there exists an element y' of M and an element k of $\mathbb{N}$ such that $y=y'+m''\alpha^{-k}$. But $m'' \in N$ implies that $m''\alpha^{-k} \in N$ so $m+y=m+y'+m''\alpha^{-k}=m'+y'+m''\alpha^{-k}=m'+y$, proving that $y \in N$. Thus we have shown that $N=M$. In particular, this means that both m and m' belong to N and so we have $m\alpha=m+m=m+m'=m'+m'=m'\alpha$. Since α is monic, we conclude that $m=m'$, proving that M is cancellative. □

(15.28) PROPOSITION. *If R is a commutative semiring then any simple cancellative left R-semimodule is an R-module.*

PROOF. If M is a simple cancellative left R-module then the R-congruence relation $\equiv_{V(M)}$ is either trivial or universal. If it is universal, then for each $m \in M$ there exist elements v and v' of $V(M)$ satisfying $m+v=0+v'$ and so $m = v' + (-v) \in V(M)$. Hence, in this case, $M = V(M)$ and so M is an R-module. Thus we need only consider the case that $\equiv_{V(M)}$ is trivial. In particular, we have $V(M) \neq M$. By Proposition 15.13, this implies that $V(M) = \{0_M\}$. Let ρ be the relation on M defined by $m \rho m'$ if and only if both m and m' are either zero or nonzero. This is surely an equivalence relation. Moreover, if $m \rho m'$ and $n \rho n'$ in M and if $m + n = 0$ then $m = n = 0$ since $V(M) = \{0\}$ and so $(m+n) \rho (m'+n')$. Let $r \in R$. Since R is commutative, $M' = \{m \in M \mid rm = 0\}$ is a subtractive subsemimodule of M and so, by Proposition 15.13, we see that $M' = M$ or $M' = \{0\}$. This implies that whenever we have $m \rho m'$, we have $rm \rho rm'$, proving that the relation belongs to $R - cong(M)$. Since M is not $\{0\}$, the relation is not universal and so it must be trivial. Hence M has precisely two elements, say $M = \{0, x\}$. Since $V(M) = \{0\}$, we must have $x + x = x$ and so M is additively idempotent, contradicting Proposition 15.27. Thus this case cannot happen, proving that M must be an R-module. □

A complete classification of simple R-modules, for R a commutative semiring, is given in [Jezek & Kepka, 1983].

(15.29) PROPOSITION. *If M is a cancellative left R-semimodule then $Z(R) \subseteq (0 : M)$.*

PROOF. Let $a \in Z(R)$ and let $r \in R$ satisfy $a + r = r$. If $m \in M$ then $am + rm = (a + r)m = rm$ so, by cancellation, $am = 0$. Thus $a \in (0 : M)$. □

A cancellative left R-semimodule M satisfying the condition that $(0 : M) = Z(R)$ is **faithfully cancellative**.

If N is a nonzero subsemimodule of a left R-semimodule M, set $P(N, M) = \{m \in M \mid rm + n \neq 0_M$ for all $r \in R$ and $0_M \neq n \in N\}$. Clearly this is a subsemimodule of M. Moreover, $P(M, M)$ is the set of all those elements m of M satisfying the condition that no nonzero multiple of m has an additive inverse. Thus we surely have $V(M) \cap P(M, M) = \{0_M\}$. If $N \neq \{0_M\}$ and $P(N, M)$ has an infinite element, then this element must also belong to N. By convention, we set $P(\{0_M\}, M) = M$ for every left R-semimodule M. We also note that $N \sqsubseteq M$ if and only if N is entire and $M = P(N, M)$. This surely implies that N is zerosumfree and so we see that an entire left R-semimodule M is an information semimodule if and only if $M \sqsubseteq M$.

(15.30) PROPOSITION. *If N, N', M' are subsemimodules of a left R-semimodule M then:*

(1) *$N \subseteq P(N, M)$ if and only if N is zerosumfree;*
(2) *$P(N, P(N, M)) = P(N, M)$;*
(3) *If M' is a subsemimodule of M containing N then $P(N, M') = P(N, M) \cap M'$;*
(4) *If $\{M_i \mid i \in \Omega\}$ is a family of left R-semimodules satifying $M = \cap_{i\in\Omega} M_i$ then $P(N, M) = \cap_{i\in\Omega} P(N, M_i)$.*
(5) *If $\{N_i \mid i \in \Omega\}$ is a family of subsemimodules of M satisfying $\cap_{i\in\Omega} N_i = N$, then $\cap_{i\in\Omega} P(N_i, M) \subseteq P(N, M)$;*
(6) *If M' is a subsemimodule of $P(N, M)$ then $N + M' = N \cup M'$ and $N = \{0_M\} \cup \{n + m' \mid 0_M \in N, m' \in M'\}$;*

(7) *If* $N \sqsubseteq M$ *and* $N \subseteq M'$ *then* $N \sqsubseteq M'$;
(8) *If* $N \sqsubseteq M'$ *then* $N' \subseteq P(N, M)$;
(9) *If* N *is zerosumfree then* $N \sqsubseteq P(N, M)$;
(10) *If* $\{N_i \mid i \in \Omega\}$ *is a family of absorbing subsemimodules of* M *with* $\cap_{i\in\Omega} N_i = N$ *and* $\cup_{i\in\Omega} N_i = N'$ *then* $N, N' \sqsubseteq M$;
(11) *If* $N, N' \sqsubseteq M$ *then* $N \cap N' \neq \{0_M\}$ *and* $N + N' \sqsubseteq M$, *where in fact we have* $N + N' = N \cup N'$;
(12) *If* $N \sqsubseteq M$ *then* $N \cap M' \sqsubseteq M'$;
(13) *If* M *is entire and* $N, N' \sqsubseteq M$ *then* $\{n+n' \mid 0_M \neq n \in N; n' \in N'\} \subseteq N \cap N'$;
(14) *If* $N \sqsubseteq M$ *then* $N \cup N'$ *is a subsemimodule of* M.

PROOF. The proof follows directly from the definitions. □

In particular, we see that if N is an information subsemimodule of a left R-semimodule M then $P(N, M)$ is the largest subsemimodule of M containing N as an absorbing subsemimodule. Also, we note that the family of all absorbing subsemimodules of M is a sublattice of the lattice of all subsemimodules of M, and that this sublattice is in fact distributive and so forms a semiring.

If M and N are disjoint left R-semimodules then a **Takahashi extension** of M by N is a left R-semimodule T the underlying set of which is $M \cup [N \setminus \{0_N\}]$ and the operations of addition and multiplication on which are defined so that $N \sqsubseteq T$. These extensions are first considered in [Takahashi, 1984a]. By what we have already seen, a necessary condition for a Takahashi extension to exist is that N be an information semimodule.

A **translation** of a left R-semimodule M is a function ψ from $M \setminus \{0_M\}$ to itself satisfying the condition that $\psi(m + m') = \psi(m) + m' = m + \psi(m')$ for all $m, m' \in M \setminus \{0_M\}$. The $trans(M)$ set of all translations of M is nonempty, since it includes the identity map and closed under composition of functions. Indeed, it is easily seen to be a monoid under composition of functions. If T is a Takahashi extension of M by N, then each element m of M defines a translation $\varphi_m \in trans(N)$ given by $\varphi_m : n \mapsto m+n$. Thus we have a function $\varphi_T : M \to trans(N)$ given by $\varphi_T : m \mapsto \varphi_m$, and this is in fact as morphism of monoids since, clearly, $\varphi_{m+m'} = \varphi_m \varphi_{m'}$ for all $m, m' \in M$. Moreover, if $r \in R$, $m \in M$, and $0_N \neq n \in N$ then

$$r[\varphi_T(m)(n)] = \varphi_T(rm)(rn).$$

A morphism of monoids from M to $trans(N)$ with this property is **admissible**. Thus, for example, if M is any left R-semimodule and N is an information semimodule over R disjoint from M, then the morphism $\epsilon : M \to trans(N)$ defined by $\epsilon(m) : n \mapsto n$ is admissible. The set of all admissible morphisms from the monoid $(M, +)$ to $trans(N)$ is denoted by $Adm(M, N)$. If $\oplus$ is the operation on $Adm(M, N)$ defined by $(\alpha \oplus \beta)(m) = \alpha(m)\beta(m)$ for all $m \in M$ then $(Adm(M, N), \oplus)$ is a monoid with identity element ϵ. If N has a nonzero cancellable lelement then this monoid is abelian.

Let φ be an admissible morphism of monoids from a left R-semimodule M to $trans(N)$, where N is an information semimodule over R disjoint from M. Set $T = M \cup [N \setminus \{0_N\}]$ and define operations of addition and scalar multiplication on T as follows:

(1) If $m, m' \in M$ and $r \in R$ then $m + m'$ and rm are the same as in M;

(2) If $n, n' \in N$ and $r \in R$ then $n + n'$ and rn are the same as in N;

(3) If $m \in M$ and $n \in N$ then $m + n = n + m = \varphi(m)(n)$.

These operations turn T into a left R-semimodule having N as an absorbing subsemimodule and M as a subtractive subsemimodule; hence T is a Takahashi extension of M by N. Thus there exists a bijective correspondence between the set of all Takahashi extensions of M by N and the set of all admissible morphisms of monoids from M to $trans(N)$. If $\varphi: M \to trans(N)$ is an admissible morphism of monoids, we will denote by $M \bullet_\varphi N$ the Takahashi extension of M by N defined by φ.

Let N be a nontrivial information submodule of a left R-semimodule M (the existence of which, recall, implies that R is both zerosumfree and entire). Let M' be a subsemimodule of $P(N, M)$ properly containing N. Define a relation $\sim_N$ on M' by setting $m_1' \sim_N m_2'$ if and only if $m_1' = m_2'$ or $\{m_1', m_2'\} \subseteq N \setminus \{0_M\}$. It is straightforward to check that this is an R-congruence relation on M'. Moreover, we see that $M'/\sim_N$ is just $[M' \setminus N] \cup \{0_M, w\}$, where w is a strongly-infinite element of $M'/\sim_N$. If $\alpha \to M'/\sim_N$ is the canonical surjection, then $ker(\alpha) = \{0_M\}$, but α is not monic unless N has precisely two elements. For notational convenience, we will denote $M'/\sim_N$ by $M'//N$ and will write $m'//N$ instead of $m'/\sim_N$. Thus,

$$m'//N = \begin{cases} \{m'\} & \text{if } m' \in M' \setminus N \\ \{w\} & \text{otherwise} \end{cases}.$$

The semimodule $M'//N$ is the **Rees factor semimodule** of M' by N.

If N is an absorbing subsemimodule of a left R-semimodule M, then $N'//N$ is defined for any R-subsemimodule N' of M containing N, and it is straightforward to verify that the map $N' \mapsto N'//N$ induces a bijective order-preserving correspondence between the family of all subsemimoudles of M containing N and the family of all subsemimodules of $M//N$, which in turn restricts to a bijective correspondence between the family of all absorbing subsemimodules of M containing N and the family of all absorbing subsemimodules of $M//N$.

(15.31) PROPOSITION. [Poyatos, 1973a] *Let R be an entire zerosumfree semiring and let N and N' be absorbing subsemimodules of a left R-semimodule M. Then:*

(1) $N \sqsubseteq N \cup N'$;

(2) $N \cap N' \sqsubseteq N'$;

(3) $N'//(N \cap N')$ *is R-isomorphic to* $(N \cup N')//N$.

PROOF. (1) and (2) are immediate consequences of the definition. As for (3), the desired R-isomorphism is given by $n'//(N \cap N') \mapsto n'//N$. □

It is similarly straightforward to show the following.

(15.32) PROPOSITION. [Poyatos, 1973b] *Let R be an entire zerosumfree semiring and let M be a left R-semimodule. If N, N', W, W' are subsemimodules of M satisfying $N' \sqsubseteq N$ and $W' \sqsubseteq W$, and if*

(i) $U = N' \cup (N \cap W)$,

(ii) $U' = N' \cup (N \cap W')$,

(iii) $V = W' \cup (N \cap W)$, *and*
(iv) $V' = W' \cup (N' \cap W)$

then $U' \sqsubseteq U$, $V' \sqsubseteq V$, *and* $U//U'$ *is R-isomorphic to* $V//V'$.

PROOF. The proof is essentially the same as the corresponding proof for modules over a ring. □

(15.33) COROLLARY. *Let R be an entire zerosumfree semiring. If $N \subset N'$ are proper absorbing subsemimodules of a left R-semimodule M, then the R-semimodule $(M//N)//(M//N')$ is R-isomorphic to $M//N'$.*

PROOF. This is an immediate consequence of Proposition 15.31 and Proposition 15.32. □

A strongly-infinite element w of a left R-semimodule M is **primitive** in M if and only if $m + m' = w$ for all $m, m' \in M \setminus \{0_M\}$. Thus, for example, if M has a strongly-infinite element w then w is a primitive element of $cr(M)$.

(15.34) EXAMPLE. [Goldstern, 1985] Let $R = \{a_0, a_1, \ldots, b_0, b_1, \ldots\}$ on which we define operations of addition and multiplication as follows:

(1) $a_i + a_j = a_{i+j}$ for all $i, j \in \mathbb{N}$;
(2) $b_i + b_j = b_0$ for all $i, j \in \mathbb{N}$;
(3) $a_j + b_i = b_i + a_j = b_k$, where $k = i - j$ if $i > j$ and $k = 0$ otherwise;
(4) $a_i a_j = a_{ij}$ for all $i, j \in \mathbb{N}$;
(5) $b_i b_j = b_0$ for all $i, j \in \mathbb{N}$;
(6) $b_i a_j = a_j b_i = b_0$ for all $i \in \mathbb{N}$ and all $j > 1$;
(7) $b_i a_0 = a_0 b_i = a_0$ for all $i \in \mathbb{N}$;
(8) $b_i a_1 = a_1 b_i = b_i$ for all $i \in \mathbb{N}$.

These operations turn R into a semiring with additive identity a_0 and multiplicative identity a_1 having a strongly-infinite element b_0 which is not primitive.

An absorbing subsemimodule N of a left R-semimodule M is **quasiminimal** if and only if it properly contains $cr(M)$ and there is no absorbing subsemimodule of M properly containing $cr(M)$ and properly contained in N. Thus, M itself is quasiminimal if and only if either M has no proper absorbing subsemimodules or it has precisely one such subsemimodule, namely its crux. A nontrivial left R-semimodule which is quasiminimal and has no primitive elements is **quasisimple**. That is to say, a quasiminimal left R-semimodule is quasisimple if it either has no strongly-infinite elements or has one such element which is not primitive. If w is a strongly-infinite element of an information semimodule M over R which has no proper absorbing subsemimodules other than $cr(M)$, and if $N = M \setminus \{0_M\}$, then we know by Proposition 15.30(11) that $N + N$ must either equal N or equal $\{w\}$. In the first case, M is quasisimple. In the second case, w is a primitive element of M.

(15.35) EXAMPLE. [Poyatos, 1973a] Let T be a nonempty set and let z, w be distinct elements not in T. Define addition on $X = T \cup \{z, w\}$ by setting

(1) $t + t' = w$ for all $t, t' \in T$;
(2) $x + z = z + x = x$ for all $x \in X$;
(3) $x + w = w + x = w$ for all $x \in X$;

For each $k \in \mathbb{N}$ and each $x \in X$, define the element kx of X as follows:

(4) $0x = z$;
(5) $1x = x$;
(6) $kx = w$ if $k > 1$ and $x \neq z$;
(7) $kz = z$ for all $k \in \mathbb{N}$.

Then X belonges a left $\mathbb{N}$-semimodule with strongly-infinite element w. If T' is any subset of T then $T' \cup \{z, w\} \sqsubseteq X$. Moreover, w is primitive in X.

From the definitions, we see that if N is an absorbing subsemimodule of a left R-semimodule M and if N' is an absorbing subsemimodule of M properly containing N, then $N'//N$ is quasisimple if and only if there is no absorbing subsemimodule of M properly containing N and properly contained in N'.

By Proposition 15.30(10), we see that if M is a left R-semimodule having at least one absorbing subsemimodule, then M has a unique maximal absorbing subsemimodule, which we will denote by $abs(M)$. If $abs(M) \neq M$ then surely $M//abs(M)$ is quasisimple.

Let R be an entire zerosumfree semiring and let M be a left R-semimodule. For $0_M \neq m \in M$, set $T(m) = \{0_M\} \cup \{rm + m' \mid 0 \neq r \in R; m' \in M\}$. This is surely an R-subsemimodule of M containing Rm. Moreover, it is easily to see that $m \in abs(M)$ if and only if $T(m) \sqsubseteq M$ and that, in that case,

$$T(m) = \{0_M\} \cup \{rm + m' \mid 0 \neq r \in R; m' \in abs(M)\}.$$

Thus, if $0_M \neq m \in abs(M)$ then $T(m') \subseteq T(m)$ for all $0_M \neq m' \in T(m)$. Set $T'(m) = \{0_M\} \cup \{0_M \neq m' \in T(m) \mid T(m') \neq T(m)\}$.

(15.36) Proposition. *Iet R be an entire zerosumfree semiring and let M be a left R-semimodule having an absorbing subsemimodule. If $0_M \neq m \in abs(M)$ then $T'(m)$ is a maximal proper absorbing subsemimodule of $T(m)$.*

Proof. If $m_1, m_2 \in T(m)$ and $0 \neq r \in R$ then $T(m_1 + m_2) \subseteq T(m_1)$ and $T(rm_1) \subseteq T(m)$ so $T'(m)$ is an absorbing subsemimodule of $T(m)$, which is proper since $m \in T(m) \setminus T'(m)$. Finally, assume that $T'(m) \subseteq N \subset T(m)$, where N is an absorbing subsemimodule of M, and let $0_M \neq x \in N$. Then $T(x) \subseteq N \neq T(m)$ and so $x \in T'(m)$, establishing the maximality of $T'(m)$. □

If $w \in M$ is strongly infinite then $w \in abs(M)$ and $rm + w = w$ for all $0_M \neq m \in abs(M)$ and $r \in R$. Therefore $w \in T(m)$ and so we conclude that $cr(M) \subseteq T(m)$ for each $0_M \neq m \in abs(M)$.

Let R be an entire zerosumfree semiring and let M be an R-semimodule having a strongly-infinite element w. Then $cr(M) = T(w)$, from which it is easy to deduce that:

(1) $L(M) = \{0_M\} \cup \{m \in abs(M) \mid T(m) = cr(M)\}$ is an absorbing subsemimodule of M which is the unique maximal subsemimodule N of $abs(M)$ satisfying $n + m = w$ for all $0_M \neq n \in N$ and all $0_M \neq m \in abs(M)$; and
(2) If $m \in abs(M) \setminus cr(M)$ then $T(m) = cr(M)$ if and only if $T(m)$ is quasiminimal.

(15.37) PROPOSITION. *Let R be an entire zerosumfree semiring and let M be a noncrucial R-semimodule having a strongly-infinite element w. Then M is quasisimple if and only if $T(m) = M$ for all $m \in M \setminus cr(M)$.*

PROOF. Assume M is quasisimple. If $m \in M \setminus cr(M)$ then $T(m) \sqsubseteq M$ and so, by quasisimplicity, $T(m) = M$. Conversely, assume that $T(m) = M$ for all $m \in M \setminus cr(M)$. Since M is noncrucial, it is quasiminimal and, indeed, quasisimple. □

Thus we see that if R is an entire zerosumfree semiring and M is a noncrucial quasisimple information R-semimodule having a strongly-infinite element w. Then $T'(m) = cr(M)$ for all $m \in M \setminus cr(M)$. Indeeed, let R be an entire zerosumfree semiring and let M be a noncrucial R-semimodule having a strongly-infinite element w. If N is a quasiminimal absorbing subsemimodule of M then either N has a primitive element or is quasisimple. Thus, in particular, if R is an entire zerosumfree semiring and M is a left R-semimodule having an absorbing subsemimodule then, for each $m \in abs(M)$, the left R-semimodule $\bar{T}(m) = T(m)//T'(m)$ is the **principal factor** of M at m. By the comments above, $\bar{T}(m)$ has a strongly-infinite element. Indeed, if $0_M \neq m \in abs(M)$ then $\bar{T}(m)$ either has a primitive element or is quasisimple. Moreover, if R is an entire zerosumfree semiring, if M is a left R-semimodule and if N' is a maximal proper absorbing subsemimodule of a subsemimodule N of $abs(M)$ then $N//N' \cong \bar{T}(m)$ for any $m \in N \setminus N'$.

If M is a left R-semimodule then an **absorbing series** for M is a chain

$$cr(M) = N_t \sqsubseteq \cdots \sqsubseteq N_1 \sqsubseteq N_0 = M$$

of subsemimodules of M. An **absorbing quasiseries** for M is an absorbing series for $abs(M)$. Any chain obtained from a given absorbing series by inserting further terms is a **refinement** of that series. If new subsemimodules are actually inserted, such a refinement is **proper**. Two absorbing series

$$cr(M) = N_t \sqsubseteq \cdots \sqsubseteq N_1 \sqsubseteq N_0 = M$$

and

$$cr(M) = L_s \sqsubseteq \cdots \sqsubseteq L_1 \sqsubseteq L_0 = M$$

for M are **isomorphic** if and only if $t = s$ and there is a permutation σ of $\{1, \ldots, t\}$ such that $N_{i-1}//N_i \cong L_{\sigma(i)-1}//L_{\sigma(i)}$ for each $1 \leq i \leq t$. Given these notions, Poyatos [1972, 1973a, 1973b] has extended the Jordan-Hölder theorem for modules.

(15.38) PROPOSITION. *If R is an entire zerosumfree semiring and M is a left R-semimodule then any two absorbing quasiseries of M have isomorphic refinements.*

The proof is similar to the proof the the Jordan-Hölder theorem for modules. See [Poyatos 1973b] for details.

16. SOME CONSTRUCTIONS FOR SEMIMODULES

In this chapter we present three basic constructions associated with semimodules. The first of these, the construction of the module of differences of an R-semimodule, is based on the corresponding construction for semirings. The other two, decomposition of a semimodule into a direct sum of indecomposable summands and the construction of the tensor product of semimodules, are inspired by the corresponding constructions for modules over a ring. In each case, however, the results differ somewhat from those in module theory due to the allowances we have to make for being in a semimodule environment.

In Chapter 7 we defined the ring of differences of a nonzeroic semiring R. We now show how, given a semiring R, we can define, in an analogous manner, the module of differences of any left R-semimodule. Indeed, let R be a semiring. If M is a left R-semimodule and if W is the subsemimodule of $M \times M$ defined by $W = \{(m,m) \mid m \in M\}$, then $(M \times M)/W$ is a left R-semimodule which is in fact a left R-module since for all $(m,m') \in M \times M$ we have $(m,m')/W + (m',m)/W = (0,0)/W$. This left R-module, denoted by M^{Δ}, is called the **R-module of differences** of M. We have a canonical R-homomorphism ξ_M from M to M^{Δ} defined by $\xi_M \colon m \mapsto (m,0)/W$. This R-homomorphism is not necessarily monic. As in the case of semirings, it is straightforward to establish that the following conditions on M are equivalent:

(1) M is cancellative;
(2) W is subtractive;
(3) $\xi_M \colon M \to M^{\Delta}$ is monic.

(16.1) PROPOSITION. *Let M be a left R-semimodule and let N be a left R-module. If $\alpha \colon M \to N$ is an R-homomorphism then there exists a unique R-homomorphism $\beta \colon M^{\Delta} \to N$ satisfying $\xi_{M\beta} = \alpha$.*

PROOF. Define the function β from M^{Δ} to N as follows: $\beta \colon (m,m')/W \mapsto m\alpha + [-(m'\alpha)]$, where $-(m'\alpha)$ is the additive inverse of $m'\alpha$ in the R-module N. If m, m', u, and u' are elements of M satisfying $(m,m')/W = (u,u')/W$ then there exist elements v and v' of M such that $m+v = u+v'$ and $m'+v = u'+v'$. Therefore $m\alpha+[-(m'\alpha)] = (m+v)\alpha+[-(m'+v)\alpha] = (u+v')\alpha+[-(u'+v')\alpha] = u\alpha+[-(u'\alpha)]$.

Thus the function β is well-defined. It is straightforward to establish that β is an R-homomorphism having the desired property. □

(16.2) PROPOSITION. *If M is a left R-semimodule then there exists a canonical surjection from $R - cong(M^\Delta)$ to $R - cong(M)$, which is a bijection if M is cancellative.*

PROOF. Any R-congruence relation ρ on M^Δ induces a corresponding R-congruence relation ρ^* on M defined by the condition that $m \;\rho^*\; m'$ if and only if $m\xi_M \;\rho\; m'\xi_M$. We claim that every R-congruence relation on M is of this form. Indeed, let ζ be an R-congruence relation on M and consider the R-congruence relation ρ on M^Δ defined by $(m, m')/W \;\rho\; (n, n')/W$ if and only if $(m+n') \;\zeta\; (m'+n)$ in M. This relation is well-defined since for any $m, m', m'' \in M$ we have

$$(m, m')/W \;\rho\; (m + m'', m' + m'')/W.$$

Moreover, if $m, m' \in M$ then $m\xi_M \;\rho\; m'\xi_M$ if and only if $m \;\zeta\; m'$ so ζ is just ρ^*. Thus the map $\rho \mapsto \rho^*$ is a surjection from $R - cong(M^\Delta)$ to $R - cong(M)$.

Now assume that M is cancellative and that ρ_1 and ρ_2 are R-congruence relations on M^Δ satisfying $(\rho_1)^* = (\rho_2)^*$. That is to say, if m and m' are elements of M then $m\xi_M \;\rho_1\; m'\xi_M$ if and only if $m\xi_M \;\rho_2\; m'\xi_M$. If $(m, m')/W$ and $(n, n')/W$ are arbitrary elements of M^Δ then $(m, m')/W = x\xi_M + (-m'\xi_M)$ and $(n, n')/W = n\xi_M + (-n'\xi_M)$ and so

$$\begin{aligned}
&(m, m')/W \;\rho_1\; (n, n')/W \\
&\qquad \Leftrightarrow (m, m')/W + [m'\xi_M + n'\xi_M] \;\rho_1\; (n, n')/W + [m'\xi_M + n'\xi_M] \\
&\qquad \Leftrightarrow (m + n')\xi_M \;\rho_1\; (n + m')\xi_M \\
&\qquad \Leftrightarrow (m + n')\xi_M \;\rho_2\; (n + m')\xi_M \\
&\qquad \Leftrightarrow (m, m')/W \;\rho_2\; (n, n')/W
\end{aligned}$$

and so the relations ρ_1 and ρ_2 coincide. Thus, in this case, the map $\rho \mapsto \rho^*$ is bijective. □

As a direct consequence of this result, we see that a left R-semimodule M is simple whenever M^Δ is simple, and that the converse is also true if M is cancellative.

(16.3) PROPOSITION. *If $\alpha: M \to N$ is an R-homomorphism of R-semimodules then there exists a unique R-homomorphism $\alpha^\Delta: M^\Delta \to N^\Delta$ of R-modules satisfying $\xi_M\alpha^\Delta = \alpha\iota_N$. Moreover, if α is surjective or is an isomorphism then so is α^Δ.*

PROOF. Set $\beta = \alpha\iota_N$. By Proposition 16.1, there exists a unique R-homomorphism $\alpha^\Delta: M^\Delta \to N^\Delta$ extending β. Indeed, it is easy to see that if $W = \{(m, m) \mid m \in M\}$ and $W' = \{(n, n) \mid n \in N\}$ then α^Δ is defined by $(m, m')/W \mapsto (m\alpha, m'\alpha)/W'$. From this and uniqueness, the second assertion is immediate. □

If $\alpha: M \to M'$ and $\beta: M' \to M''$ are R-homomorphisms of left R-semimodules then, by uniqueness, we have $(\alpha\beta)^\Delta = \alpha^\Delta\beta^\Delta$. Also, if $\alpha: M \to M$ is the identity map then $\alpha^\Delta: M^\Delta \to M^\Delta$ is also the identity map. Thus we note that if R is a semiring then $(\_)^\Delta$ is a functor from the category of all left R-semimodules to the category of all left R-modules.

Let R be a nonzeroic semiring and let $S = R^\Delta$ be the ring of differences of R. For notational convenience, we will denote an element $(a, a')/D$ of S by $\langle a, a' \rangle$. If N is a left R-module then for each $\langle a, a' \rangle \in S$ and $n \in N$ set $\langle a, a' \rangle n = an + (-a'n)$. This is well-defined since if $\langle a, a' \rangle = \langle b, b' \rangle$ in S then there exists an element d of R satisfying $a + b' + d = a' + b + d$. Hence $an + b'n + dn = a'n + bn + dn$ and so, adding $-(a'n + b'n + dn)$ to each side, we see that $an + (-a'n) = bn + (-b'n)$. It is now easy to see that with the given addition in N and with the scalar multiplication defined as above, N is a left S-module. Moreover, if $\alpha: N \to N'$ is an R-homomorphism of left R-modules then it is also an S-homomorphism of left S-modules.

Combining both of the above constructions, we see that, given a nonzeroic semiring R, R, there exists a canonical functor from the category of all left R-semimodules to the category of all R^Δ-modules. The properties of this functor are worked out in detail in [Poyatos, 1971]. In particular, if M is a left R-semimodule then M^Δ is a left R^Δ-module with addition and scalar multiplication defined by $\langle m, m' \rangle + \langle n, n' \rangle = \langle m + n, m' + n' \rangle$ and $\langle a, b \rangle \langle m, m' \rangle = \langle am + bm', bm + am' \rangle$ where $\langle m, m' \rangle = (m, m')/W$ and $\langle n, n' \rangle = (n, n')/W$ are elements of M^Δ and $\langle a, b \rangle = (a, b)/D$ is an element of R^Δ.

Let R be a nonzeroic semiring and let M be a nonzero left R-semimodule having R-endomorphism semiring S. As above, we will denote elements of R^Δ by $\langle a, b \rangle$ and elements of M^Δ by $\langle m, m' \rangle$. To each pair (α, β) of elements of S define a function $\Phi(\alpha, \beta): M^\Delta \to M^\Delta$ by setting $\Phi(\alpha, \beta): \langle m, m' \rangle \mapsto \langle m\alpha + m'\beta, m\beta + m'\alpha \rangle$. This function is well-defined since if $\langle m, m' \rangle = \langle u, u' \rangle$ in M^Δ then there exist elements n and n' of M satisfying $(m + n, m' + n) = (u + n', u' + n')$. Therefore

$$\begin{aligned}
\langle m\alpha + m'\beta, m\beta + m'\alpha \rangle &= \langle m\alpha + m'\beta + n\alpha + n\beta, n\beta + m'\alpha + n\alpha + n\beta \rangle \\
&= \langle [m + n]\alpha + [m' + n]\beta, [m + n]\beta + [m' + n]\alpha \rangle \\
&= \langle [u + n']\alpha + [u' + n']\beta, [u + n']\beta + [u' + n']\alpha \rangle \\
&= \langle u\alpha + u'\beta + n'\alpha + n'\beta, u\beta + u'\beta + n'\alpha + n'\beta \rangle \\
&= \langle u\alpha + u'\beta, u\beta + u'\alpha \rangle.
\end{aligned}$$

Indeed, it is straighforward to show that $\Phi(\alpha, \beta)$ is in fact an R^Δ-endomorphism of M^Δ.

Now assume that, in addition, the semiring S is also nonzeroic. Let $G = \{(\alpha, \alpha) \mid \alpha \in S\}$ and let $S^\Delta = (S \times S)/G$. In line with our previous notation, we will denote the element $(\alpha, \beta)/G$ of S^Δ by $\langle \alpha, \beta \rangle$. If $\langle \alpha, \beta \rangle = \langle \alpha', \beta' \rangle$ in S^Δ then there exist elements θ and θ' in S such that $\langle \alpha + \theta, \beta + \theta \rangle = \langle \alpha' + \theta', \beta' + \theta' \rangle$. For each element $\langle m, m' \rangle$ of M^Δ we then have

$$\begin{aligned}
\langle m, m' \rangle \Phi(\alpha, \beta) &= \langle m\alpha + m'\beta, m\beta + m'\alpha \rangle \\
&= \langle m\alpha + m'\beta + m\theta + m'\theta, m\beta + m'\alpha + m\theta + m'\theta \rangle \\
&= \langle m[\alpha + \theta] + m'[\beta + \theta], m[\beta + \theta] + m'[\alpha + \theta] \rangle \\
&= \langle m[\alpha' + \theta'] + m'[\beta' + \theta'], m[\beta' + \theta'] + m'[\alpha' + \theta'] \rangle \\
&= \langle m, m' \rangle \Phi(\alpha', \beta').
\end{aligned}$$

This shows that Φ induces a function Φ' from the ring S^Δ to the ring of all R^Δ-endomorphisms of M^Δ defined by $\Phi': \langle\alpha,\beta\rangle \mapsto \Phi(\alpha,\beta)$. It is again straightforward (and rather tedious) to verify that Φ' is in fact a ring homomorphism.

(16.4) PROPOSITION. *Let R be a nonzeroic semiring and let M be a nonzero left R-semimodule having nonzeroic R-endomorphism semiring S. Then then kernel of the ring homomorphism Φ' from S^Δ to the ring of all R^Δ-endomorphisms of M^Δ is $\{\langle\alpha,\beta\rangle \mid$ for each $m \in M$ there exists an element $n \in M$ such that $m\alpha + n = m\beta + n\}$. In particular, a sufficient condition for Φ' to be monic is that M be cancellative.*

PROOF. Suppose that $\langle\alpha,\beta\rangle \in ker(\Phi')$. Then for each $m \in M$ we have $\langle 0,0\rangle = \langle m,0\rangle\Phi'(\langle\alpha,\beta\rangle) = \langle m\alpha, m\beta\rangle$ and so there exists an element n of M such that $m\alpha + n = m\beta + n$. Conversely, assume that this condition holds for $\langle\alpha,\beta\rangle \in S^\Delta$. Then if $\langle m,m'\rangle \in M^\Delta$ there exist elements n and n' of M such that $m\alpha+n = m\beta+n$ and $m'\alpha + n' = m'\beta + n'$. Therefore

$$\langle m\alpha + m'\beta, m'\alpha + m\beta\rangle = \langle m\alpha + m'\beta + n + n', m'\alpha + m\beta + n + n'\rangle = \langle 0,0\rangle$$

so $\langle\alpha,\beta\rangle \in ker(\Phi')$.

If M is cancellative and $\langle\alpha,\beta\rangle \in ker(\Phi')$ then the above result shows that $m\alpha = m\beta$ for all $m \in M$ and so $\alpha = \beta$. Thus $\langle\alpha,\beta\rangle = \langle 0,0\rangle$, proving that Φ' is monic. □

Let R be a semiring. If $\{M_i \mid i \in \Omega\}$ is a family of nonempty subsemimodules of a left R-semimodule M and if $\beta_i: M_i \to M$ is the inclusion map for each $i \in \Omega$, then we have a unique R-homomorphism $\beta: \coprod_{i\in\Omega} M_i \to M$ satisfying $\beta_i = \lambda_i\beta$ for each $i \in \Omega$. Indeed, β is defined by $\langle m_i\rangle \mapsto \sum_{i\in\Omega} m_i$ (where this sum is well-defined since only finitely-many of the m_i are nonzero). If β is an R-isomorphism, then M is the **direct sum** of the submodules M_i and we write $M = \oplus_{i\in\Omega}M_i$. As in the case of modules over a ring, it is straightforward to establish that $M = \oplus_{i\in\Omega}M_i$ if and only if each element m of M can be written in a unique way as $\sum m_i$, where $m_i \in M_i$ for each $i \in \Omega$ and only finitely-many of the m_i are nonzero.

A subsemimodule N of a left R-semimodule M is a **direct summand** of M if and only if there exists a subsemimodule N' of M satisfying $M = N \oplus N'$. In particular, every element m of M can be written in a unique manner as $n + n'$, where $n \in N$ and $n' \in N'$, and we have a surjective R-homomorphism $\pi_N: M \to N$ defined by $m \mapsto n$, called the **projection** of M onto N. Similarly, we have the projection π'_N of M onto N' the kernel of which is precisely N. Thus, in particular, any direct summand of M is subtractive.

The set of all nonzero direct summands of a left R-semimodule M will be denoted by $summ(M)$. This set is nonempty since $M \in summ(M)$.

(16.5) EXAMPLE. Let R be a zerosumfree entire semiring and let M be a left R-semimodule. If $\alpha: M \to R$ is an R-homomorphism having kernel K and if $m, m' \in M \setminus K$ then $(m + m')\alpha = m\alpha + m'\alpha \neq 0$ and so $m + m' \notin K$. Moreover, if $0 \neq r \in R$ then $(rm)\alpha = r(m\alpha) \neq 0$ and so $rm \notin K$. Thus $N = (M \setminus K) \cup \{0\}$ is an R-submodule of M and clearly $M = N \oplus K$.

(16.6) PROPOSITION. *Let R be a semiring and let M be a left R-semimodule having endomorphism semiring S. Then the following conditions on a subsemimodule N of M are equivalent:*

(1) *N is a direct summand of M;*
(2) *$N = M\alpha$ for some $\alpha \in comp(S)$;*
(3) *There is a subsemimodule N' of M such that $M = N + N'$ and such that the restriction of $\equiv'_N$ to N and the restriction of $\equiv_N$ to N' are trivial.*

PROOF. (1) $\Leftrightarrow$ (2): If N is a direct summand of M and if $\lambda: N \to M$ is the inclusion map then $N = M\pi_N$ where π_N is considered as an endomorphism of M. Moreover, if $M = N \oplus N'$ then $\pi_N + \pi_{N'} = 1_S$ and $\pi_N \pi_{N'} = \pi_{N'}\pi_N = 0_S$ so $\pi_N \in comp(S)$. Conversely, assume that $N = M\alpha$ for some $\alpha \in comp(S)$. If $N' = M\alpha^{\perp}$ then it is straightforward to verify that $M = N \oplus N'$.

(1) $\Rightarrow$ (3): Assume (1) and let N' be a subsemimodule of M satisfying $M = N \oplus N'$. Then surely $M = N + N'$. If $x, x' \in N$ satisfy $x \equiv_{N'} x'$ then there exist $y, y' \in N'$ such that $x + y = x' + y'$. By uniqueness of representation, this implies that $x = x'$ and $y = y'$ so, in particular, the restriction of $\equiv_{N'}$ to N is trivial. Similarly, the restriction of $\equiv_N$ to N' is trivial.

(3) $\Rightarrow$ (1): Assume (3). Then any element of M can be written as $x + y$, where $x \in N$ and $y \in N'$. Let $x, x' \in N$ and $y, y' \in N'$ satisfy $x + y = x' + y'$. Then $x \equiv_{N'} x'$ and so $x = x'$. Similarly $y \equiv_N y'$ and so $y = y'$. Thus such representations are unique, proving that $M = N \oplus N'$. □

(16.7) PROPOSITION. *Let R be a semiring and let M be a left R-semimodule. If $N_2 \subseteq N_1$ are direct summands of M then N_2 is a direct summand of N_1.*

PROOF. By assumption, there exist subsemimodules M_1 and M_2 of M satisfying $M = N_1 \oplus M_1 = N_2 \oplus M_2$. In particular, if $x \in N_1$ then we can write $x = n_2 + m_2$ for some elements n_2 of N_2 and m_2 of M_2. Furthermore, $m_2 = n_1 + m_1$ for some $n_1 \in N_1$ and $m_1 \in M_1$. Thus $x = (n_1 + n_2) + m_1$ for $n_1 + n_2 \in N_1$ and $m_1 \in M_1$. By the uniqueness of representation, we must have $m_1 = 0$ and $x = n_1 + n_2$, where $n_1 = m_2 \in N_1 \cap M_2$. Therefore $N_1 = N_2 + (N_1 \cap M_2)$. Moreover, any representation of an element of N_1 as a sum of an element of N_2 and an element of $N_1 \cap M_2$ is unique, and so N_2 is a direct summand of N_1. □

A nonzero left R-semimodule M is **indecomposable** if and only if there do not exist nonzero subsemimodules N and N' of M satisfying $M = N \oplus N'$. An R-semimodule which is not indecomposable is **decomposable**.

(16.8) EXAMPLE. Let R be a semiring considered as a left semimodule over itself. If $e \in comp(R) \setminus \{0, 1\}$ then $R = Re + Re^{\perp}$. If $ae + be^{\perp} = 0$ then

$$ae = ae^2 = ae^2 + be^{\perp}e = (ae + be^{\perp})e = 0$$

and similarly $be^{\perp} = 0$. Therefore this sum is direct. Thus we see that if R is indecomposable as a left semimodule over itself then it is integral.

(16.9) PROPOSITION. *If R is a semiring and M is a left R-semimodule satisfying the descending chain condition on subtractive submodules then M has a decomposition as a direct sum of finitely-many indecomposable subsemimodules.*

PROOF. Let M be a left R-semimodule. By hypothesis, $summ(M)$ has a minimal element N_1, which is surely indecomposable. Write $M = N_1 \oplus Y_1$. If Y_1 is indecomposable, we are done. If not, $summ(Y_1)$ has a minimal element N_2 and we can write $Y_1 = N_2 \oplus Y_2$ and so $M = N_1 \oplus N_2 \oplus Y_2$. Continue in this manner. Since $Y_1 \supset Y_2 \supset \dots$ is a properly descending chain of subtractive submodules of M, it must terminate after a finite number of steps, and so there exists some natural number t such that $M = N_1 \oplus \cdots \oplus N_t \oplus Y_t$, where all of the direct summands are indecomposable. □

(16.10) PROPOSITION. *Let R be a semiring. If M is a zerosumfree left R-semimodule and if $N_1, \dots, N_k, Y_1, \dots, Y_t$ are indecomposable submodules of M satisfying $M = N_1 \oplus \cdots \oplus N_k = Y_1 \oplus \cdots \oplus Y_t$ then $k = t$ and there exists a permutation σ of $\{1, \dots, k\}$ with $N_i = Y_{\sigma(i)}$ for all i.*

PROOF. If $x \in N_1$ we can write $x = y_1 + \cdots + y_t$, where the $y_i \in Y_i$ for each i. In turn, each y_i can be written as $y_i = n_{i1} + \cdots + n_{ik}$ with $n_{ij} \in N_j$ for each j. Therefore

$$x = \sum_{i=1}^{t} \sum_{j=1}^{k} n_{ij} = \sum_{j=1}^{k} \left[\sum_{i=1}^{t} n_{ij} \right],$$

where $n_j = \sum_{i=1}^{t} n_{ij} \in N_j$ for each $1 \leq j \leq k$. By uniqueness of representation, we have $n_j = 0$ for all $j > 1$ and so, by zerosumfreeness, $n_{ij} = 0$ for all i and all $j > 1$. Thus $y_i = n_{i1} \in Y_i \cap N_1$ for each i. Moreover, $x = n_{11} + \cdots + n_{t1}$ and so we see that $N_1 + (Y_1 \cap N_1) + \cdots + (Y_t \cap N_1)$, where this sum is in fact direct. Since N_1 is indecomposable, this means that $Y_i \cap N_1 = \{0\}$ for all i except one. Renumbering the Y_i if necessary, we can assume that $Y_1 \cap N_1 \neq \{0\}$. Thus $N = Y_1 \cap N_1 \subseteq Y_1$. A reversal of this argument shows that $Y_1 \subseteq N_1$ and thus we have equality. Thus $M = N_1 \oplus Y_2 \oplus \cdots \oplus Y_t$. Continuing in this manner, we show that we must have $k = t$ and that the Y_i are just a (possible) rearrangement of the N_j. □

(16.11) PROPOSITION. *[**Fitting's Lemma**] Let M be a cancellative left R-semimodule satisfying both the ascending chain condition and the descending chain condition on subsemimodules and let $\alpha: M \rightarrow M$ be a steady R-endomorphism of M satisfying the condition that $M = M\alpha^t + ker(\alpha^t)$ for some positive integer t. Then there exists a positive integer h for which $M = M\alpha^h \oplus ker(\alpha^h)$.*

PROOF. Since $ker(\alpha) \subseteq ker(a^2) \subseteq \dots$ and $M\alpha \supseteq M\alpha^2 \supseteq \dots$ we know that there exist positive intergers u and v such that $ker(\alpha^u) = ker(\alpha^{u+i})$ for all $i \in \mathbb{N}$ and $M\alpha^v = M\alpha^{v+i}$ for all $i \in \mathbb{N}$. Set $h = max\{t, u, v\}$ and let $\varphi = \alpha^t$. Then, by construction, $\varphi = \varphi^2$.

If $m \in M$ then there exist elements x of M and y of $ker(\alpha^t)$ satisfying $m = x\alpha^t + y$. Similarly, there exist elements x' of M and y' of $ker(\alpha^t)$ such that $x = x'\alpha^t + y'$. Hence $x\alpha^t = x'\alpha^{2t} + y\alpha^t = x'\alpha^{2t}$. Continuing in this manner, we can find an element x'' of M and an integer n such that $nt > h$ and $m = x''\alpha^{nt} + y = x\alpha^h + y$, where $y \in ker(\alpha^t) \subseteq ker(\alpha^h)$. Thus $M = M\varphi + ker(\varphi)$.

Suppose that m and m' are elements of M satisfying $m\varphi =_{ker(\varphi)} m'\varphi$. Then there exist elements x and x' of $ker(\varphi)$ satisfying $m\varphi + x = m'\varphi + x'$ so $m\varphi = m\varphi^2 = (m\varphi+x)\varphi = (m'\varphi+x')\varphi = m'\varphi^2 = m'\varphi$ and thus the restriction of $\equiv_{ker(\varphi)}$ to $M\varphi$ is trivial. Now s[/]{0}uppose that $x, x' \in ker(\varphi)$ satisfy $x \equiv_{M\varphi} x'$. Then there exist elements m and m' of M such that $x + m\varphi = x' + m'\varphi$. As above, this implies that $m\varphi = m'\varphi$ and so $x = x'$ since M is cancellative. Thus the restriction of $\equiv_{M\varphi}$ to $ker(\varphi)$ is trivial. The result now follows from Proposition 16.6 and its proof. □

The notion of the tensor product of semimodules over a semiring was defined in [Takahashi, 1982a]. For the application of this construction to the study of iterative nondeterministic algebras, see [Wechler, 1988].

Let R be a semiring, let M be a right R-semimodule, and let N be a left R-semimodule. Let A be the set $M \times N$ and let U be the $\mathbb{N}$-semimodule $R^{(A)} \times R^{(A)}$. Then every element of the R-semimodule $R^{(A)}$ can be written in a unique manner as a linear combination of the elements of the set $\{f[m,n] \mid (m,n) \in M \times N\}$, where $f[m,n]$ is the function from $M \times N$ to R defined by

$$f[m,n]\colon (m',n') \mapsto \begin{cases} 1 & \text{if } (m',n') = (m,n) \\ 0 & \text{otherwise} \end{cases}.$$

Let W be the subset of U consisting of all elements of the following forms:

(1) $(f[m+m',n], f[m,n]+f[m',n])$,
(2) $(f[m,n]+f[m',n], f[m+m',n])$,
(3) $(f[m,n+n'], f[m,n]+f[m,n'])$,
(4) $(f[m,n]+f[m,n'], f[m,n+n'])$,
(5) $(f[mr,n], f[m,rn])$,
(6) $(f[m,rn], f[mr,n])$

for $m, m' \in M$, $n, n' \in N$, and $r \in R$. Let U' be the $\mathbb{N}$-subsemimodule of U generated by W. Then every element of U' can be written (not necessarily uniquely) as a finite sum $\sum k_i(g_i,h_i) = (\sum k_i g_i, \sum k_i h_i)$ for $k_i \in \mathbb{N}$ and $g_i, h_i \in W$. We also note that $(g,g) \in U'$ for all $g \in R^{(A)}$. We can therefore define an R-congruence relation ρ on $R^{(A)}$ by setting $f \rho f'$ if and only if there exists an element (g,h) of U' such that $f + g = f' + h$. The factor $\mathbb{N}$-semimodule $R^{(A)}/\rho$ will be denoted by $M \otimes_R N$, and is called the **tensor product** of M and N over R. If $m \in M$ and $n \in N$ then the element $f[m,n]/\rho$ will be denoted by $m \otimes n$. Since $R^{(A)}$ is generated by the elements of the form $f[m,n]$, we see that $M \otimes_R N$ is generated by the elements of the form $m \otimes n$ and so every element of $M \otimes_R N$ can be written (not necessarily uniquely) as a finite sum $\sum (m_i \otimes n_i)$ for $m_i \in M$ and $n_i \in N$. Moreover, by the above construction we see that for all $m, m' \in M$, for all $n, n' \in N$, all $r \in R$, and all $k \in \mathbb{N}$, we have:

(1) $(m+m') \otimes n = m \otimes n + m' \otimes n$;
(2) $m \otimes (n+n') = m \otimes n + m \otimes n'$;
(3) $mr \otimes n = m \otimes rn$;
(4) $k(m \otimes n) = km \otimes n = m \otimes kn$;
(5) $0 \otimes n = m \otimes 0 = 0$.

Moreover, if S is a semiring then, as in the case of tensor products of modules over rings, it is straightforward to verify that if M is an (S, R)-bisemimodule then $M \otimes_R N$ is a left S-semimodule with scalar multiplication defined by $s(m \otimes n) = (sm) \otimes n$. Similarly, if N is an (R, S)-bisemimodule then $M \otimes_R N$ is a right S-bisemimodule with scalar multiplication defined by $(m \otimes n)s = m \otimes (ns)$.

(16.12) PROPOSITION. *Let R be a semiring. If M is a right R-semimodule and N is a left R-semimodule then the semimodule $M \otimes_R N$ is cancellative.*

PROOF. Let $A = M \times N$ and suppose that f, f', and f'' are elements of $R^{(A)}$ satisfying $f/\rho + f''/\rho = f'/\rho + f''/\rho$. Then there exists a pair $(g, h) \in U'$ satisfying $f + f'' + g = f + f'' + h$. But, by construction, (f'', f'') also belongs to U' and hence so does $(f'' + g, f'' + h)$. This implies that $f/\rho = f'/\rho$, proving that $M \otimes_R N$ is cancellative. □

Let R be a semiring. If M is a right R-semimodule, if N is a left R-semimodule, and if T is an $\mathbb{N}$-semimodule, then a function $\theta: M \times N \to T$ is R-**balanced** if and only if, for all $m, m' \in M$, for all $n, n' \in N$, and for all $r \in R$ we have:

(1) $\theta(m + m', n) = \theta(m, n) + \theta(m', n)$;
(2) $\theta(m, n + n') = \theta(m, n) + \theta(m, n')$;
(3) $\theta(mr, n) = \theta(m, rn)$.

(16.13) EXAMPLE. If N is a left R-semimodule and T is an $\mathbb{N}$-semimodule, then the set $Hom(N, T)$ of all $\mathbb{N}$-homomorphisms from N to T has the structure of a right R-semimodule, when we define $(\alpha + \beta)n = \alpha n + \beta n$ and $(\alpha r)n = \alpha(rn)$. If M is a right R-semimodule and $\varphi: M \to Hom(N, T)$ is an R-homomorphism then we have an R-balanced function $\theta: M \times N \to T$ defined by $\theta: (m, n) \mapsto (\varphi(m))(n)$. Conversely, if $\theta: M \times N \to T$ is an R-balanced function then we can define an R-homomorphism $\varphi: M \to Hom(N, T)$ by $(\varphi(m))(n) = \theta(m, n)$.

We now note the universal property of the tensor product. First, however, we recall from Example 15.4 that if R is a semiring and M is a left R-semimodule then we have a relation $[\equiv]_{\{0\}}$ on M defined by $m[\equiv]_{\{0\}}m'$ if and only if there exists an element m'' of M satisfying $m + m'' = m' + m''$. The equivalence class of an element m with respect to this relation is denoted by $m[/]\{0\}$.

(16.14) PROPOSITION. *Let R be a semiring, let M be a right R-semimodule, let N be a left R-semimodule, and let T be an $\mathbb{N}$-semimodule. If $\theta: M \times N \to T$ is an R-balanced function then there exists a unique $\mathbb{N}$-homomorphism $\psi: M \otimes_R N \to T[/]\{0\}$ satisfying the condition that $\psi(m \otimes n) = \theta(m, n)[/]\{0\}$ for all $m \in M$ and $n \in N$.*

PROOF. Set $A = M \times N$. The function $\theta: A \to T$ can be uniquely extended to an $\mathbb{N}$-homomorphism $\theta^*: R^{(A)} \to T$ satisfying $\theta^*(f[m, n]) = \theta(m, n)$ for all $m \in M$ and $n \in N$. Indeed, for each $g \in R^{(A)}$ we have $\theta^*(g) = \sum\{g(m, n)\theta(m, n) \mid (m, n) \in supp(g)\}$.

Let ρ be the equivalence relation used in defining $M \otimes_R N$, i.e. the relation such that $M \otimes_R N = R^{(A)}/\rho$. Similarly, let U' and W be as above. We define a function $\psi: M \otimes_R N \to T' = T[/]\{0\}$ by setting $\psi(f/\rho) = \theta^*(f)[/]\{0\}$. This function is well-defined since if $f \; \rho \; f'$ in $R^{(A)}$ then there exists a pair $(g, h) \in U'$ such that

$f + g = f' + h$. It follows that $\theta^*(f) + \theta^*(g) = \theta^*(f') + \theta^*(h)$. By definition of U', we know that (g, h) can be written as $(\sum k_i g_i, \sum k_i h_i)$, where $k_i \in \mathbb{N}$ and $g_i, h_i \in W$ for each i. Then $\theta^*(g) = \sum k_i \theta^*(g_i) = \sum k_i \theta^*(h_i) = \theta^*(h)$. Since $T[/]\{0\}$ is cancellative by Example 15.26, we see that $\theta^*(f)[/]\{0\} = \theta^*(f')[/]\{0\}$ and so $\psi(f/\rho) = \psi(f'/\rho)$.

We now claim that ψ is an $\mathbb{N}$-homomorphism. Indeed, if $f, g \in R^{(A)}$ then $\psi(f/\rho + g/\rho) = \psi([f+g]/\rho) = \theta^*(f+g)[/]\{0\} = \theta^*(f)[/]\{0\} + \theta^*(g)[/]\{0\} = \psi(f/\rho) + \psi(g/\rho)$. Now suppose that $f \in R^{(A)}$ satisfies the condition that $f \; \rho \; 0$. Then there exists a pair (g, h) in U' such that $f + g = h$ and so $\theta^*(f) + \theta^*(g) = \theta^*(h)$. Since $\theta^*(g) = \theta^*(h)$ and $T[/]\{0\}$ is cancellative, we have $\theta^*(f) = 0[/]\{0\}$. Thus $\psi(0/\rho) = 0[/]\{0\}$, proving that ψ is an $\mathbb{N}$-homomorphism.

Finally, we note that ψ clearly has the desired property, and uniqueness is straightforward to check. □

In particular, we note that if M is a right R-semimodule, N is a left R-semimodule, and if T is a cancellative $\mathbb{N}$-semimodule then for any R-balanced function $\theta: M \times N \to T$ there exists a unique $\mathbb{N}$-homomorphism $\psi: M \otimes_R N \to T$ satisfying $\psi(m \otimes n) = \theta(m, n)$ for all $m \in M$ and all $n \in N$.

(16.15) Proposition. *Let R be a semiring, left M be a right R-semimodule, left N be a left R-semimodule, and let T be a cancellative $\mathbb{N}$-semimodule. Then there exists a canonical isomorphism of $\mathbb{N}$-semimodules $\psi: Hom(M \otimes_R N, T) \to Hom_R(M, Hom(N, T))$.*

Proof. If α is an $\mathbb{N}$-homomorphism from $M \otimes_R N$ to T, let α^* be the function from M to $Hom(N, T)$ given by $\alpha^*: n \mapsto \alpha(m \otimes n)$. Then α^* is an R-homomorphism of right R-semimodules. That is to say, $\alpha^* \in Hom_R(M, Hom(N, T))$. Let

$$\psi: Hom(M \otimes_R N, T) \to Hom_R(M, Hom(N, T))$$

be the function defined by $\psi: \alpha \mapsto \alpha^*$. It is straightforward to verify that this is an $\mathbb{N}$-homomorphism.

If $\alpha^* = \beta^*$ then $\alpha(m \otimes n) = \beta(m \otimes n)$ for all $m \in M$ and $n \in N$ and so we must have $\alpha = \beta$. Thus ψ is monic. To show that it is surjective as well, let δ be an R-homomorphism from M to $Hom(N, T)$ and let θ be the function from $M \times N$ to T defined by $\theta(m, n) = \delta(m)(n)$. Then θ is R-balanced and so, by the remark after Proposition 16.14, there exists a unique $\mathbb{N}$-homomorphism $\alpha: M \otimes_R N \to T$ satisfying $\theta(m, n) = \alpha(m \otimes n)$ for all $m \in M$ and all $n \in N$. By definition, $\alpha^* = \delta$, proving that ψ is surjective and hence an $\mathbb{N}$-isomorphism. □

(16.16) Proposition. *If M is a left R-semimodule then $R \otimes_R M$ is isomorphic to $M[/]\{0\}$.*

Proof. Let $\theta: R \times M \to M[/]\{0\}$ be the function defined by $\theta: (r, m) \mapsto rm[/]\{0\}$. Then θ is R-balanced and so, by Proposition 16.14, there exists a unique $\mathbb{N}$-homomorphism $\psi: R \otimes_R M \to M[/]\{0\}$ satisfying $\psi(r \otimes m) = \theta(rm)$ for all $r \in R$ an $m \in M$. Indeed, ψ is an R-homomorphism of left R-modules. On the other hand, we have a function $\varphi: M[/]\{0\} \to R \otimes_R M$ given by $\varphi: m[/]\{0\} \mapsto 1 \times m$. It is easy to verify that this function is indeed well-defined and that it in fact is an R-homomorphism. Since $\varphi\psi(r \otimes m) = \varphi(rm[/]0) = 1 \otimes rm = r \otimes m$ and

$\psi\varphi(m[/]0) = \psi(1 \otimes m) = m[/]\{0\}$ for all $r \in R$ and $m \in M$, we see that ψ must be both surjective and monic and so it is an R-isomorphism. □

17. FREE, PROJECTIVE, AND INJECTIVE SEMIMODULES

Let R be a semiring and let M be a left R-semimodule. If A is a nonempty subset of M then there exists an R-homomorphism $\alpha\colon R^{(A)} \to M$ defined by $\alpha\colon f \mapsto \sum_{m\in A} f(m)m$. The set A is a set of generators for M precisely when this R-homomorphism is surjective. Moreover, α induces an R-congruence relation $\equiv_\alpha$ on $R^{(A)}$ as defined in Example 15.1. The set A is **linearly independent** if and only if $\equiv_{\alpha}$ is the trivial relation, i.e. if and only if $\sum_{m\in A} f(m)m = \sum_{m\in A} g(m)m$ implies that $f = g$. If A is not linearly independent then it is **linearly dependent**. A linearly-independent set of generators for M is a **basis** of M over R. We note that if A is linearly dependent and if $B \subset A$ then the subsemimodules of M generated by B and $A \setminus B$ have no nozero element in common.

The set A is **weakly linearly independent** if and only if $ker(\alpha) = \{0\}$. Linearly independent subsets of M are surely weakly linearly independent. If A is not weakly linearly independent then it is **weakly linearly dependent**. Any subset of M containing a [weakly] linearly dependent set is again [weakly] linearly dependent. If m is an element of a semimodule M which is a linear combination of a subset A of M then $A \cup \{m\}$ is linearly dependent but not necessarily weakly linearly dependent. A weakly linearly-independent set of generators for M is a **weak basis** of M over R. A nonempty subset A of a left R-semimodule M is **linearly attached** if and only if there exists a partition $A = B \cup C$ of A into a union of disjoint subsets, together with nonzero functions $f \in R^{(B)}$ and $g \in R^{(C)}$ such that

$$\sum_{m'\in B} f(m')m' = \sum_{m''\in C} g(m'')m''.$$

(By convention, the sum taken over an empty set equals 0_M.) Every linearly attached subset of M is linearly dependent, but the converse need not be true if M is not an R-module.

(17.1) EXAMPLE. Let R be a semiring and let A be a nonempty set. For each $a \in A$, let $f_a \in R^{(A)}$ be the characteristic function on $\{a\}$. Clearly $\{f_a \mid a \in A\}$ is a basis for $R^{(A)}$. In particular, if R is a semiring and n is a positive integer, then R^n, on which we have componentwise addition and scalar multipliction, has a basis.

(17.2) EXAMPLE. [Kim & Roush, 1980] It is easy to verify that, for each positive integer n, every finitely-generated subsemimodule of $\mathbb{N}^n$ has a unique basis and every finitely-generated subsemimodule of $(\mathbb{R}^+)^n$ has a basis unique up to nonzero multiples.

(17.2h) EXAMPLE. [Dudnikov & Samborskiĭ, 1991] Let R be an entire zerosumfree semiring and let n be a positive integer. Then clearly

$$B = \{[1, 0, \ldots, 0], [0, 1, 0, \ldots, 0], \ldots, [0, \ldots, 0, 1]\}$$

is a basis for the left R-semimodule $M = R^n$. Conversely, assume that $B' = \{v_1, \ldots, v_n\}$ be a set of generators for M having n elements. If $a_1, \ldots, a_n$ are elements of R satisfying $[1, 0, \ldots, 0] = \sum_{i=1}^{n} a_i v_i$ then, by zerosumfreeness, we see that $a_i = 0$ unless v_i is of the form $[c, 0, \ldots, 0]$. A similar argument can be made for each element of B and so we see that B' must be of the form

$$\{[c_1, 0, \ldots, 0], [0, c_2, 0, \ldots, 0], \ldots, [0, \ldots, 0, c_n]\},$$

where the c_i are nonzero elements of R. Moreover, this is then a basis for M.

(17.3) EXAMPLE. [Zhao, 1990] Let $R = \{0, a, b, 1\}$ be partially-ordered by the relations $0 \leq a \leq 1$ and $0 \leq b \leq 1$. Then R is a bounded distributive lattice and hence a semiring. We see that $\{1\}$ and $\{a, b\}$ are both bases for R, considered as a left semimodule over itself.

(17.4) EXAMPLE. A nonempty subset M of $\mathbb{Z}^n$ is a **polyhedral monoid** if $[0, \ldots, 0] \in M$ and there exists an $m \times n$ matrix A over $\mathbb{Z}$ such that

$$M = \{x \in \mathbb{Z}^n \mid Ax^T \leq [0, \ldots, 0]^T\}.$$

Every polyhedral monoid is surely a left $\mathbb{N}$-semimodule. Hilbert [1890] showed that every polyhedral monoid has a finite basis. For a constructive method of finding such a basis, see [Bachem, 1978].

(17.5) PROPOSITION. *Let R be a semiring and let M be a left R-semimodule. Any basis for M over R is a minimal set of generators for M over R.*

PROOF. Let A be a basis for M over R and suppose that A properly contains a set of generators B for M over R. Pick $x \in A \setminus B$ and let $f \in R^{(B)}$ satisfy the condition that $x = \sum_{m \in B} f(m)m$. Extend f to a function $f' \in R^{(A)}$ by setting $f'(m) = 0$ if $m \in A \setminus B$. Then surely $x = \sum_{m \in B} f'(m)m$. But on the other hand, $x = \sum_{m \in B} g(m)m$, where $g \in R^{(A)}$ is defined by

$$g \colon m \mapsto \begin{cases} 1 & \text{if } m = x \\ 0 & \text{otherwise} \end{cases}.$$

Since $g \neq f$, this contradicts the assumption that A is a basis for M over R. □

(17.6) PROPOSITION. *Let R be an entire zerosumfree simple semiring satisfying the condition that $a+b \neq 1$ unless $a=1$ or $b=1$. If M is a left R-semimodule having a finite basis then that basis is unique.*

PROOF. Let $B = \{x_1, \ldots, x_k\}$ and $B' = \{y_1, \ldots, y_n\}$ be bases for M. Without loss of generality, we can assume that $k \geq n$. Then there exist elements a_{ij} and b_{ji} $(1 \leq i \leq k; 1 \leq j \leq n)$ of R such that $x_i = \sum_{j=1}^{n} a_{ij} y_j$ for all $i \leq i \leq k$ and $y_j = \sum_{i=1}^{k} b_{ji} x_i$ for all $1 \leq j \leq n$. Thus $x_i = \sum_{h=1}^{k} (\sum_{j=1}^{n} a_{ij} b_{jh}) x_h$. Since B is a basis, this implies that $\sum_{j=1}^{n} a_{ij} b_{ji} = 1$ for all $1 \leq i \leq k$ and $\sum_{j=1}^{n} a_{ij} b_{jh} = 0$ when $1 \leq i \neq h \leq k$. Since R is entire and zerosumfree, this means that $a_{ij} = 0$ or $b_{jh} = 0$ if $i \neq h$ and $1 \leq j \leq n$ and that there is an index $1 \leq j \leq n$ for which $a_{ij} b_{ji} = 1$, which, by the simplicity of R, implies that $a_{ij} = b_{ji} = 1$. Thus $b_{jh} = 0$ for $h \neq i$ and so for each $1 \leq i \leq k$ there exists a $1 \leq j \leq n$ such that $x_i = y_j$. Thus $B \subseteq B'$ and, since we assumed that $k \geq n$, we must in fact have equality. □

If A is set of generators for a left R-semimodule M and if $x \in A$, it does not necessarily follow that $x = \sum_{m \in A} f(m)m$ implies that $f(x)x = x$. A set of generators having this property is **standard**.

(17.7) EXAMPLE. If $M = \mathbb{I}^3$ then $\{(\frac{1}{2}, \frac{1}{2}, \frac{1}{2}), (0, 1, \frac{1}{2}), (0, \frac{1}{2}, 1)\}$ is a standard basis for M over $\mathbb{I}$ but $\{\frac{1}{2}, 1, \frac{1}{2}), (0, 1, \frac{1}{2}), (0, \frac{1}{2}, 1)\}$ is a basis which is not standard since $(\frac{1}{2}, 1, \frac{1}{2}) = \frac{1}{2}(\frac{1}{2}, 1, \frac{1}{2}) + (0, 1, \frac{1}{2})$ in M. For each $n \in \mathbb{P}$, the left $\mathbb{I}$-semimodule $\mathbb{I}^n$ has a unique standard basis. See [Kim & Roush, 1980] for details.

Note that $B = \{(\frac{3}{5}, \frac{3}{10}, \frac{3}{5}), (\frac{2}{5}, \frac{1}{2}, \frac{1}{2}), (\frac{1}{2}, \frac{3}{5}, \frac{1}{2}), (\frac{4}{5}, \frac{3}{5}, \frac{7}{10})\}$ is a linearly-independent subset of M having more than three elements while, for any element $\frac{4}{5} \leq b \leq 1$, the set $\{(b, \frac{3}{10}, \frac{7}{10}), (\frac{2}{5}, \frac{3}{5}, \frac{1}{2})\}$ is a basis for the subsemimodule of M generated by B. Refer to [Guo et al., 1988].

(17.8) EXAMPLE. [Takahashi, 1985] If M is a left $\mathbb{N}$-semimodule and $m \in M$ then $\{m\}$ is weakly linearly independent if and only if $am \neq 0$ for all $a \in \mathbb{P}$. It is linearly dependent if and only if there exist integers $0 < a < b$ in $\mathbb{N}$ such that $am = bm$. In this case, there exists a unique pair of integers (a', b') in $\mathbb{N}$ such that $0 < a' < b'$, $a'm = b'm$, and $\mathbb{N}m = \{0, m, \ldots, (b'-1)m\}$.

(17.9) PROPOSITION. *Let R be a semiring and let M be a left R-semimodule having a finite set of generators $A = \{x_1, \ldots, x_n\}$ satisfying the condition that for each $1 \leq h \leq n$ there exist $a_{h1}, \ldots, a_{hn}$ in R with a_{hh} is left absorbing and $x_h = \sum_{i=1}^{n} a_{hi} x_i$. Then $\{a_{11}x_1, \ldots, a_{nn}x_n\}$ is a standard set of generators for M.*

PROOF. Set $y_1 = a_{11}x_1$. If $m \in M$ then there exist $b_1, \ldots, b_n \in R$ such that

$$\begin{aligned} m &= b_1 x_1 + \cdots + b_n x_n \\ &= b_1 \left(y_1 + \sum_{h=2}^{n} a_{1h} x_h \right) + \sum_{h=2}^{n} b_h x_h \\ &= b_1 y_1 + \sum_{h=2}^{n} (b_1 a_{1h} + b_h) x_h \end{aligned}$$

and so $\{y_1, x_2, \ldots, x_n\}$ is a set of generators for M. Moreover, if $y_1 = c_1y_1 + c_2x_2 + \cdots + c_nx_n$ then, by assumption, $c_1y_1 = c_1a_{11}x_1 = a_{11}x_1 = y_1$. Now repeat this procedure for $y_2 = a_{22}x_2$, etc. □

(17.10) PROPOSITION. *Let R be an entire zerosumfree semiring. Then a left R-semimodule M is an information semimodule over R if and only if it has a weak basis.*

PROOF. Assume that M is an information semimodule over R. Set $A = M \setminus \{0_M\}$ and let $\alpha: R^{(A)} \to M$ be defined by $\alpha: f \mapsto \sum_{m\in A} f(m)m$. If f is not the 0-map then $f(m)m \neq 0_M$ whenever $f(m) \neq 0$ since M is entire, and so $f(m) \neq 0_M$ since M is zerosumfree. Therefore $ker(\alpha) = \{0\}$, showing that A is a weak basis for M. Conversely, assume that M has a weak basis A and let α be as before. If $m, m' \in M \setminus \{0_M\}$ then there exists nonzero functions $f, g \in R^{(A)}$ satisfying $m = \alpha(f)$ and $m' = \alpha(g)$. Since R is entire and zerosumfree, we know that $f + g \neq 0$ and so $m + m' = \alpha(f + g) \neq 0_M$. Similarly, if $0 \neq r \in R$ then $rf \neq 0$ and so $rm = \alpha(rf) \neq 0_M$. Thus M is an information semimodule. □

A left R-semimodule having a basis over R is called a **free** R-semimodule. If R is a ring and M is a left R-module, this reduces to the usual definition of a free module. Since not every module over a ring is free, certainly not every semimodule over a semiring is free. As a consequence of the definitions, we note that for any nonempty set A the left R-semimodule $R^{(A)}$ is free, and that every free left R-semimodule is R-isomorphic to $R^{(A)}$ for some suitable nonempty set A. For the use of free semimodules in defining automata over semirings, refer to [Peeva, 1991].

(17.11) PROPOSITION. *If R is a semiring and M is a left R-semimodule then there exists a free R-semimodule N and a surjective R-homomorphism from N to M.*

PROOF. Let M be a left R-module. Since the result is trivial for the case of $M = \{0\}$, we can assume that $M \neq \{0\}$. Let $M' = M \setminus \{0\}$ and let $N = R^{(M')}$. Let $\alpha: N \to M$ be defined by $\alpha: f \mapsto \sum_{m\in supp(f)} f(m)$. This is clearly a surjective R-homomorphism. □

(17.12) PROPOSITION. *Let M be a free left R-semimodule having a basis U and let N be an arbitrary left R-semimodule. For each function $g \in N^U$ there is a unique R-homomorphism $\alpha: M \to N$ satisfying $u\alpha = g(u)$ for all $u \in U$.*

PROOF. We know that each element m of M can be written uniquely in the form $\sum_{u\in U} r_uu$, where the r_u are elements of R only finitely-many of which are nonzero. Define the function $\alpha: M \to N$ by $\sum r_uu \mapsto \sum r_ug(u)$. It is straightforward to verify that α is indeed an R-homomorphism having the desired property. Moreover, if $\beta: M \to N$ is an R-homomorphism satisfying $u\beta = g(u)$ for all $u \in U$ then $(\sum r_uu)\beta = \sum r_u(u\beta) = \sum r_ug(u) = \sum r_uu\alpha = (\sum r_uu)\alpha$ and so $\beta = \alpha$. This shows that α is unique. □

Note that we have already implicitly made use of this result in the proof of Proposition 16.14.

By combining Example 17.1 and Proposition 17.12, we see that if M is a left R-semimodule, if A is a nonempty set, and if g is a function from A to M, then there exists a unique R-homomorphism $\alpha: R^{(A)} \to M$ satisfying $\alpha: f_a \mapsto g(a)$ for all $a \in A$.

Let M be a free left R-semimodule with basis U and let N be a free left R-semimodule with basis V. If $\alpha: M \to N$ is an R-homomorphism then, by Proposition 17.12, the action of α is completely determined by its action on U. For each $u \in U$ we have $u\alpha = \sum_{v \in V} a_{uv}v$, where the a_{uv} are elements of R only finitely-many of which are nonzero. Thus α is effectively represented by the column-finite matrix $[a_{uv}] \in R^{U \times V}$. Note that if M and N are free left R-semimodules and if α is represented by a matrix A in terms of given fixed bases of M and N, then in trying to determine $n\alpha^{-1}$ we are trying to solve the equation $XA = B$, where B is the vector of coefficients of the representation of n in terms of the given basis for N.

(17.13) EXAMPLE. Solution of equations of the form $XA = B$, where B is an element of a finitely-generated free $\mathbb{I}$-semimodule was first considered in [Sanchez, 1976]. This was extended to consideration of equations of this form for free semimodules over a totally-ordered lattice in [Di Nola, 1985], over a complete and completely distributive lattice in [Zhao, 1987, 1990], and over a frame in [Di Nola & Lettieri, 1989].

A left R-semimodule P is **projective** if and only if the following condition holds: if $\varphi: M \to N$ is a surjective R-homomorphism of left R-semimodules and if $\alpha: P \to N$ is an R-homomorphism then there exists an R-homomorphism $\beta: P \to M$ satisfying $\beta\varphi = \alpha$. In other words, P is projective if and only if

$$Hom(P, \varphi): Hom_R(P, M) \to Hom_R(P, N)$$

is a surjective $\mathbb{N}$-homomorphism for every surjective R-homomorphism $\varphi: M \to N$.

(17.14) PROPOSITION. *Every free left R-semimodule is projective.*

PROOF. Let P be a free left R-semimodule with basis A. Let $\varphi: M \to N$ be a surjective R-homomorphism of left R-semimodules and let $\alpha: P \to N$ be an R-homomorphism. Since φ is surjective, we see that for each element a of A there exists an element m_a of M such that $m_a\varphi = a\alpha$. By Proposition 17.12, we see that there is a unique R-homomorphism $\beta: P \to M$ satisfying $a\beta = m_a$. Then $a\beta\varphi = m_a\varphi = a\alpha$ for all $a \in A$ and so, by the uniqueness part of Proposition 17.12, we must have $\alpha = \beta\varphi$. □

A left R-semimodule N is a **retract** of a left R-semimodule M if and only if there exist a surjective R-homomorphism $\theta: M \to N$ and an R-homomorphism $\psi: N \to M$ satisfying the condition that $\psi\theta$ is the identity map on N. If N is a direct summand of a left R-semimodule M then surely N is a retract of M. Also, if N is a retract of M and M is a retract of M' then N is immediately seen to be a retract of M'.

(17.15) EXAMPLE. If R is a semiring and n is a positive integer then any matrix $A = [a_{ij}]$ in $\mathcal{M}_n(R)$ defines an R-endomorphism α of the left R-semimodule R^n given by $\alpha: (r_1, \ldots, r_n) \mapsto (s_1, \ldots, s_n)$ where, for each $1 \leq h \leq n$, $s_h = \sum_{i=1}^{n} r_i a_{ih}$. If the matrix A is multiplicatively regular then there exists a matrix B in $\mathcal{M}_n(R)$

satisfying $ABA = A$ and this matrix similarly defines an R-endomorphism β of R^n. Let $N = R^n\alpha$ and let β' be the restriction of β to N. Then for each $m \in R^n$ we have $m\alpha = m\alpha\beta\alpha = (m\alpha)\beta'\alpha$ and so $\beta'\alpha$ is the identity map on N. Thus N is a retract of R^n.

(17.16) PROPOSITION. *A left R-semimodule is projective if and only if it is a retract of a free left R-semimodule.*

PROOF. If P is a projective left R-semimodule then, by Proposition 17.11, there exists a free R-semimodule F and a surjective R-homomorphism $\theta: F \to P$. By definition of projectivity, there exists an R-homomorphism $\psi: P \to F$ such that $\psi\theta$ is the identity map on P.

Conversely, assume that P is a retract of a free left R-semimodule$F$ and let $\theta: F \to P$ and $\psi: P \to F$ be R-homomorphisms such that θ is surjective and $\psi\theta$ is the identity map on P. Let $\varphi: M \to N$ be a surjective R-homomorphism of left R-semimodules and let $\alpha: P \to N$ be an R-homomorphism. Since F is projective by Proposition 17.14, there exists an R-homomorphism $\beta: F \to M$ such that $\beta\varphi = \theta\alpha$. Therefore $\psi\beta\varphi = \psi\theta\alpha = \alpha$, and so $\psi\beta: P \to M$ is a map having the property we seek in order to prove projectivity. □

(17.17) COROLLARY. *Any retract of a projective left R-semimodule is projective.*

PROOF. This is a direct consequence of Proposition 17.16. □

(17.18) EXAMPLE. If R is a semiring then $I^+(R)$ is an idempotent subsemimodule of R, considered as a left semimodule over itself. If the semiring R is additively regular then, from remarks in Chapter 12, we note that the function $\alpha: R \to I^+(R)$ defined by $\alpha: a \mapsto a^\circ = a + a^\#$ is a surjective R-homomorphism of left R-semimodules. Furthermore, the restriction of α to $I^+(R)$ is the identity map. Therefore $I^+(R)$ is a retract of R. Since R is projective as a left semimodule over itself by Proposition 17.14, we see that $I^+(R)$ is also projective.

(17.19) PROPOSITION. *If $\{P_i \mid i \in \Omega\}$ is a family of left R-semimodules then $P = \coprod_{i\in\Omega} P_i$ is projective if and only if each P_i is projective.*

PROOF. If P is projective then each P_i is a retract of P and hence is projective by Corollary 17.17. Conversely, assume that each P_i is projective. For each $i \in \Omega$, let $\theta_i: P \to P_i$ be the surjective R-homomorphism $\langle p_h\rangle \mapsto p_i$ and let $\psi_i: P_i \to P$ be the inclusion map.

Let $\varphi: M \to N$ be a surjective R-homomorphism of left R-modules and let $\alpha: P \to N$ be an R-homomorphism. Then, by projectivity, for each $i \in \Omega$ there exists an R-homomorphism $\beta_i: P_i \to M$ satisfying $\beta_i\varphi = \psi_i\alpha$. Define the R-homomorphism $\beta: P \to M$ by $\beta: p \mapsto \sum_{i\in\Omega} p\theta_i\beta_i$. Then for $p \in P$ we have

$$p\beta\varphi = \sum_{i\in\Omega} p\theta_i\beta_i\varphi = \sum_{i\in\Omega} p\theta_i\psi_i\alpha = p\alpha$$

and so $\beta\varphi = \alpha$. □

(17.20) Application. The algebraic formulation of linear systems theory over a field is given in [Kalman, Falb & Arbib, 1969] and was extended to systems over rings in [Eilenberg, 1974], [Sontag, 1976], and [Naudé & Nolte, 1982]. It is easily extended to the case of systems over semirings. This is often desirable, since it is useful to consider systems over $\mathbb{N}$ or over the schedule algebra. If R is a semiring then a (**discrete-time, constant, linear dynamical**) **system** over R is a sextuple $(U, X, Y, \varphi, \psi, \theta)$ where U is a left R-semimodule, called the **input semimodule** of the system, X is a left R-semimodule, called the **state semimodule** of the system, Y is a left R-semimodule, called the **output semimodule** of the system, $\varphi: U \to X$ is an R-homomorphism called the **input homomorphism** of the system, ψ is an R-endomorphism of X called the **state updating homomorphism**, and $\theta: X \to Y$ is an R-homomorphism called the **output homomorphism** of the system. The R-semimodules U, X, and Y are often taken to be free. If $(U, X, Y, \varphi, \psi, \theta)$ and $(U', X', Y', \varphi', \psi', \theta')$ are systems, then a **system morphism** from the first to the second consists of a triple (α, β, γ) of R-homomorphisms $\alpha: U \to U'$, $\beta: X \to X'$, and $\gamma: Y \to Y'$ such that the diagram

$$\begin{array}{ccccccc} U & \xrightarrow{\varphi} & X & \xrightarrow{\psi} & X & \xrightarrow{\theta} & Y \\ \downarrow \alpha & & \downarrow \beta & & \downarrow \beta & & \downarrow \gamma \\ U' & \xrightarrow{\varphi'} & X' & \xrightarrow{\psi'} & X' & \xrightarrow{\theta'} & Y' \end{array}$$

commutes. Many of the results in [Sontag, 1976] are easily transferable to this context.

Let $R[t]$ be the semiring of polynomials in an indeterminate t over a semiring R and let S be the subsemiring of $\mathcal{M}_3(R[t])$ consisting of all matrices $[p_{ij}(t)]$ satisfying the following conditions:

(1) $p_{12}(t) = p_{13}(t) = p_{23}(t) = 0$;

(2) $p_{11}(t)$ and $p_{33}(t)$ have degree at most 0 (i.e. they are elements of R).

If U and Y are (R, R)-bisemimodules (and so, in particular, if they are free or if R is commutative) then we can consider the left R-semimodule $U \times X \times Y$ as a right S-semimodule by defining

$$(u, x, y)[p_{ij}(t)] = (up_{11}, u\varphi p_{21}(\psi) + xp_{22}(\psi), u\alpha p_{31}(\psi)\theta + xp_{32}(\psi)\theta + yp_{33}).$$

Moreover, any system morphism canonically becomes an S-homomorphism of these semimodules. Conversely, each such S-semimodule defines an system and each S-homomorphism defines a system morphism between such systems.

Sometimes it is interesting to consider a weaker version of projectivity. A left R-semimodule P is **steady projective** if and only if the following condition holds: if $\varphi: M \to N$ is a surjective R-homomorphism of left R-semimodules and if $\alpha: P \to N$ is an R-homomorphism then there exists an R-homomorphism $\beta: P \to M$ satisfying $\beta\varphi = \alpha$. For the properties of such semimodules, see [Al-Thani, 1995, 1996].

Let R be a semiring. A left R-semimodule E is **injective** if and only if, given a left R-semimodule M and a subsemimodule N, any R-homomorphism from N to E can be extended to an R-homomorphism from M to E. In other words, E is injective if and only if

$$Hom(\varphi, E): Hom_R(M, E) \to Hom_R(N, E)$$

is a surjective $\mathbb{N}$-homomorphism for all monic R-homomorphisms $\varphi: N \to M$.

We know that if R is a ring then any left R-module is contained in an injective left R-module. However, for arbitrary semirings this may be far from the case. The following result was communicated privately to the author by B. Banaschewski.

(17.21) PROPOSITION. *If R is an entire, cancellative, zerosumfree semiring then the only injective left R-semimodule is $\{0\}$.*

PROOF. Let E be an injective left R-semimodule and, for $e \in E$, let $\alpha_e: R \to E$ be the R-homomorphism defined by $r \mapsto re$. By injectivity, there exists an R-homomorphism $\beta_e: R^{\Delta} \to E$ extending α_e. Then $e + (-1)\beta_e = 1\beta_e + (-1)\beta_e = 0\beta_e = 0$ and so e has an additive inverse in E. Thus E is an R-module.

Now let $E' = E\{\infty\}$ as defined in Example 14.8. Then the identity map on E can be extended to an R-homomorphism β from E' to E. Set $u = \infty\beta$. For each $e \in E$ we have $e + u = e\beta + \infty\beta = (e + \infty)\beta = \infty\beta = u$. This contradicts the fact that u, like every element of E, has an additive inverse unless $u = 0$. But in that case we must have $E = \{0\}$. □

In particular, there are no nonzero injective $\mathbb{N}$-semimodules. This does not mean, of course, that the situation is similarly bleak for semimodules over all semirings, even those far from being rings. For example, if R is a frame then every left R-semimodule can be embedded in an injective R-semimodule [Joyal & Tierney, 1984].

(17.22) EXAMPLE. [H. Wang, 1994] We claim that $\mathbb{B}$ is injective as a left semimodule over itself. To see this, let N be a subsemimodule of a left $\mathbb{B}$-semimodule M and let $\alpha: N \to \mathbb{B}$ be a $\mathbb{B}$-homomorphism. Define the function $\beta: M \to \mathbb{B}$ by setting $m\beta = 0$ if there exists an element $x \in M$ such that $m + x \in ker(\alpha)$ and $x\beta = 1$ otherwise. We claim that β is a $\mathbb{B}$-homomorphism. Indeed, suppose that $m_1, m_2 \in M$. If $(m_1 + m_2)\beta = 0$ then there exists an element x of M such that $m_1 + m_2 + x \in ker(\alpha)$. Then $(m_1 + [m_2 + x])\alpha = 0 = (m_2 + [m_1 + x])\alpha$ and so $m_1\beta + m_2\beta = 0 + 0 = 0 = (m_1 + m_2)\beta$. If $(m_1 + m_2)\beta = 1$ then either $m_1\beta = 1$ or $m_2\beta = 1$ for otherwise, if both of these were equal to 0, there would exist elements x_1 and x_2 of M satisfying $(m_1 + x_1)\alpha = 0 = (m_2 + x_2)\alpha$ and so $(m_1 + m_2 + [x_1 + x_2])\alpha = 0$, implying that $(m_1 + m_2)\beta = 0$. Therefore, again, $m_1\beta + m_1\beta = (m_1 + m_2)\beta$.

(17.23) PROPOSITION. *Let R be a semiring and let E be an injective left R-semimodule. Then*

(1) *E^A is an injective left R-semimodule for every nonempty set A;*
(2) *Any direct summand of E is injective.*

PROOF. (1) By Example 14.4, E^A is then a left R-semimodule. If N is a subsemimodule of a left R-semimodule M and if α is an R-homomorphism from N to E^A then for each a in A we have an R-homomorphism $\alpha_a: N \to E$ defined by $n\alpha_a = (n\alpha)(a)$. Since E is injective, we know that for each $a \in A$ there exists an R-homomorphism $\beta_a: M \to E$ extending α_a. Define a function $\beta: M \to E^A$ by $(m\beta)(a) = m\beta_a$ for all $m \in M$ and all $a \in A$. Then β is an R-homomorphism extending α. Thus E^A is injective.

(2) Let E' be a direct summand of E and let E'' be a subsemimodule of E satisfying $E = E' \oplus E''$. Then there exists a surjective R-homomorphism $\pi: E \rightarrow E'$, the kernel of which is precisely E''. Let $\lambda: E' \rightarrow E$ be the inclusion map. If N is a subsemimodule of a left R-semimodule M and if $\alpha: N \rightarrow E'$ is an R-homomorphism then, by injectivity, there exists an R-homomorphism $\beta: M \rightarrow E$ extending $\alpha\lambda$. In particular, if $x \in N$ then $x\beta \in E'$ and so $x\beta\pi = x\alpha\lambda\pi = x\alpha$. Therefore $\beta\pi: M \rightarrow E'$ extends α, proving that E' is injective. □

(17.24) COROLLARY. *If A is a nonempty set the $\mathbb{B}^A$ is injective as a left $\mathbb{B}$-semimodule.*

PROOF. This is an immediate consequence of Example 17.22 and Proposition 17.23. □

Let $\gamma: S \rightarrow R$ be a semiring homomorphism. If R is canonically a left S-semimodule if we define scalar multplication by $s \cdot r = \gamma(s)r$ for all $s \in S$ and $r \in R$. Let M be a left S-semimodule. Then $Hom_S(R, M)$ is a left R-semimodule with respect to componentwise addition and scalar multiplication given by $r'\alpha: r \mapsto (rr')\alpha$ for all $\alpha \in Hom_S(R, M)$ and $r, r' \in R$.

(17.25) PROPOSITION. *Let $\gamma: S \rightarrow R$ be a semiring homomorphism. If M is an injective left S-semimodule then $Hom_S(R, M)$ is injective as a left R-semimodule.*

PROOF. Set M be an injective left S-semimodule and set $Y = Hom_S(R, M)$. Let N' be a subsemimodule of a left R-semimodule N and let $\alpha: N' \rightarrow Y$ be an R-homomorphism. Note that N is also a left S-semimodule, with scalar multiplication defined by $s \cdot x = \gamma(s)x$ for all $s \in S$ and $x \in N$. Moreover, N' is an S-subsemimodule of N. Define a function $\varphi: N' \rightarrow M$ by setting $\varphi: n \mapsto (1)(n\alpha)$. Then φ is an S-homomorphism, as can easily be verified. Therefore, by injectivity, there exists an S-homomorphism $\theta: N \rightarrow M$ extending φ. We claim that the function $\beta: N \rightarrow Y$ defined by $n\beta: r \mapsto (rn)\theta$ is an R-homomorphism. Indeed, for all $n_1, n_2 \in N$ and all $r \in R$ we have

$$\begin{aligned}(r)[(n_1 + n_2)\beta] &= (r[n_1 + n_2])\theta = (rn_1 + rn_2)\theta \\ &= (rn_1)\theta(rn_2)\theta = (r)[\beta(n_1)] + (r)[\beta(n_2)] \\ &= (r)[\beta(n_1) + \beta(n_2)]\end{aligned}$$

and for all $n \in N$ and $r, r' \in R$ we have $(r)[\beta(r'n)] = (r[r'n])\theta = ([rr']n)\theta = (rr')\beta(n) = (r)[r'\beta(n)]$. This establishes the claim. Moreover, β extends α since for any $n' \in N'$ and $r \in R$ we have $(r)[\beta(n)] = (rn)\theta = (rn)\varphi = (1)[(rn)\alpha] = (1)((n)r\alpha) = (r)[n\alpha]$. □

In particular, if M is a left R-semimodule which is injective as a left $\mathbb{N}$-semimodule then the semimodule $M^{\#}$ defined in Example 15.16 is injective as a left R-semimodule.

An R-monomorphism $\alpha: M \rightarrow N$ of left R-semimodules is **essential** if and only if, for any R-homomorphism $\beta: N \rightarrow N'$, the map $\alpha\beta$ is an R-monomorphism only when β is an R-monomorphism. A subsemimodule M' of a left R-semimodule M is **large** in M if and only if the inclusion map $M' \rightarrow M$ is an essential R-homomorphism. Equivalently, $\alpha: M \rightarrow N$ is an essential R-homomorphism if and only if $M\alpha$ is a large subsemimodule of N. It is immediately evident that a subsemimodule M' of a left R-semimodule M is large in M if and only if every subsemimodule of M containing M' is large in M.

(17.26) PROPOSITION. *If N is a subsemimodule of a left R-semimodule M then the following conditions are equivalent:*

(1) *N is large in M;*

(2) *If ρ is a nontrivial R-congruence relation on M then the restriction of ρ to N is also nontrivial;*

(3) *If m and m' are distinct elements of M then there exist distinct elements n and n' of N satisfying $n\ \rho_{(m,m')}\ n'$.*

PROOF. (1) $\Rightarrow$ (2): Let ρ be a nontrivial R-congruence relation on M and let $\beta: M \to M/\rho$ be the R-homomorphism defined by $m \mapsto m/\rho$. Then β is not an R-monomorphism and hence, by (1), neither is its restriction to N. This implies that there are elements $n \neq n'$ of N satisfying $n\ \rho\ n'$, proving that the restriction of ρ to N is nontrivial. (2) $\Rightarrow$ (3): This is immediate. (3) $\Rightarrow$ (1): Let $\beta: M \to M'$ be an R-homomorphism, the restriction of which to N is an R-monomorphism. If β is not injective then there exist distinct elements m and m' of M satisfying $m\ \rho\ m'\beta$. By (3), there exist distinct elements n and n' of N satisfying $n\rho_{(m,m')}n'$ and hence $n\ \rho\ n'\beta$, which is a contradiction. Thus β must be an R-monomorphism, proving (1). □

(17.27) EXAMPLE. Let R be the semiring $(\mathbb{I}, max, min)$ and let $H = R \setminus \{1\}$, which is a left ideal of R and so is a left R-semimodule. If ρ is a nontrivial R-congruence relation on R which restricts to the trivial R-congruence relation on H then there must exist an element $a \in H$ satisfying $a\ \rho\ 1$. If $b \in H$ satisfies $a < b < 1$ then $a = ba\ \rho\ b1 = b$, which is a contradiction. Thus we see that every nontrivial R-congruence relation on R restricts to a nontrivial relation on H, proving that H is large in R by Proposition 17.26.

(17.28) PROPOSITION. *Let N be a subsemimodule of a left R-semimodule M and let ρ be the largest R-congruence relation on M the restriction of which to N is trivial. Then the canonical R-monomorphism $\alpha: N \to M/\rho$ is essential.*

PROOF. Let ζ be a nontrivial R-congruence relation on M/ρ and let ζ^* be the R-congruence relation on M defined by the condition that $m\ \zeta^*\ m'$ if and only if $m/\rho\ \zeta\ m'/\rho$. Then $\zeta^* \geq \rho$ and, indeed, ζ^* and ρ are not equal. By the definition of ρ, this means that there exist elements $n \neq n'$ of N satisfying $n\ \zeta^*\ n'$ and so $n\alpha\ \zeta\ n'\alpha$. Thus ζ restricts to a nontrivial relation on $N\alpha$, proving that $N\alpha$ is large in M/ρ. □

(17.29) PROPOSITION. *If E is an injective left R-semimodule then every essential R-monomorphism $\alpha: E \to E'$ is an R-isomorphism.*

PROOF. Let E be injective and let $\alpha: E \to E'$ be an essential R-monomorphism. Then there exists an R-homomorphism $\beta: E' \to E$ such that $\alpha\beta$ is the identity map on E. Assume that $x \in E' \setminus E\alpha$. Then $x\beta \in E$ so $x\beta\alpha \neq x$. Since $x\beta\alpha \equiv_\beta x$, this means that $\equiv_\beta$ is a nontrivial R-congruence relation on E' and so, by Proposition 17.26, it restricts to a nontrivial R-congruence relation on $E\alpha$. In particular, there exist elements $e \neq e'$ of E satisfying $e\alpha\beta = e'\alpha\beta$, contradicting the choice of β. Therefore we must have $E' = E\alpha$ and so α is an R-isomorphism. □

Let M be a left R-semimodule. If there exists an injective left R-semimodule E and an essential R-monomorphism $\alpha: M \to E$ then E is an **injective hull** of M. As we have seen in Proposition 17.21, injective hulls of nonzero R-semimodules need not exist for every semiring R.

(17.30) EXAMPLE. If R is an additively-regular semiring then any left R-semimodule has an injective hull. See [Katsov, 1997] for details. In particular, any semilattice (i.e. left $\mathbb{B}$-semimodule) has an injective hull. These hulls are completely characterized in [Bruns & Lakser, 1975].

It is well-known, however, that they do always exist if R is a ring. If injective hulls exist, they are unique, as the next result shows.

(17.31) PROPOSITION. *If $\alpha: M \to E$ and $\alpha': M \to E'$ are injective hulls of a left R-semimodule M then there exists an R-isomorphism from E to E'.*

PROOF. By injectivity, there exists an R-homomorphism $\theta: E \to E'$ satisfying $\alpha\theta = \alpha'$. We claim that this is the isomorphism we seek. Indeed, since $\alpha' = \alpha\theta$ is an R-monomorphism, we see by essentiality that θ is also an R-monomorphism. Assume that $\beta: E' \to N$ is an R-homomorphism satisfying the condition that $\theta\beta$ is an R-monomorphism. Then $\alpha\theta\beta = \alpha'\beta$ is also an R-monomorphism. But α' is essential and so β is an R-monomorphism. Thus θ is essential and so, by Proposition 17.29, it is an R-isomorphism, as claimed. □

Let R be an an additively-idempotent semiring. Then $B(R) = \{0_R, 1_R\}$ and so, as we have already noted, R can be considered as a left $\mathbb{B}$-semimodule. For any nonempty set A, we can then set $I(A) = Hom_{\mathbb{B}}(R, \mathbb{B}^A)$ and it is straightforward to show that $I(A)$ is a left R-semimodule with addition and scalar multiplication defined by $(\eta + \eta')(r)(a) = \eta(r)(a) + \eta'(r)(a)$ and $(sf)(r)(a) = f(rs)(a)$ for all $\eta, \eta' \in I(A)$, $r, s \in R$, and $a \in A$. As a consequence of Proposition 17.25, we see that if R is an additively-idempotent semiring and M is a left R-semimodule then $I(M)$ is an injective left R-semimodule.

We can now extend Joyal and Tierney's result.

(17.33) PROPOSITION. [Wang, 1994] *If R is an additively-idempotent semiring then every left R-semimodule can be embedded in an injective left R-semimodule.*

PROOF. Let M be a left R-module. For $m \in M$ set $U(m) = M \setminus \{m+m' \mid m' \in M\}$. Define the function $\theta: M \to I(M)$ as follows: if $m \in M$ and $r \in R$ then $(r)[m\theta]$ is the characteristic function on $U(m)$. Thus, in particular, $(0_M)\theta: r \mapsto 0$ for all $r \in R$. If $r_1, r_2 \in R$ then $U((r_1 + r_2)m) = U(r_1m + r_2m) = [U(r_1m) \cap U(r_2m)] = U(r_1m) \cup U(r_2m)$ we see that $(r_1+r_2)(m\theta) = (r_1)(m\theta)+(r_2)(m\theta)$ for all $r_1, r_2 \in R$. Thus $m\theta$ is a $\mathbb{B}$-homomorphism and so belongs to $I(M)$.

Suppose that $m\theta = m'\theta$. Then $U(rm) = U(rm')$ for all $r \in R$ and so, in particular, $U(m) = U(m')$ Since $m \notin U(m)$, this means that $m \notin U(m')$ and so there exists an element x of M satisfying $m = m' + x$. Similarly there exists an element y of M satisfying $m' = m + y$. Since M is additively idempotent, we have $m = m+m = m+m'+x = m+m'+m'+x = m+m+y+m'+x = m+m'+x+y$.

Similarly, $m' = m + m' + x + y$ and so $m = m'$. Thus θ is monic. The proof that θ is an R-homomorphism is straightforward. □

(17.34) COROLLARY. *If R is an additively-idempotent semiring then every left R-semimodule has an injective hull.*

PROOF. By Proposition 17.32 we know that every left R-semimodule can be embedded in an injective one. We are left to show that in this case every such semimodule has maximal essential extension which too is injective. This is a consequence of general results in universal algebra. See page 261 of [Cohn, 1965]. □

Let R be as semiring and let $\mathcal{C}$ is a nonempty class of left R-semimodules. A left R-semimodule E is **$\mathcal{C}$-injective** if and only if, given an R-subsemimodule M of a left R-semimodule N such that both M and N are in $\mathcal{C}$, then any R-homomorphism from M to E can be extended to an R-homomorphism from N to E. It is easily seen that if $\{E_i \mid i \in \Omega\}$ is a family of left R-semimodules, then $\prod_{i \in \Omega} E_i$ is $\mathcal{C}$-injective if and only if each E_i is $\mathcal{C}$-injective. If $\mathcal{C}$ is the class of all cancellative left R-semimodules, this situation has been studied by Hall and Pianskool [1996]. We will say that a left R-semimodule is **c-injective** if and only if it is $\mathcal{C}$-injective , where $\mathcal{C}$ is the class of all cancellative left R-semimodules.

(17.35) EXAMPLE. [Hall and Pianskool, 1996] If $(D, +)$ is a divisible abelian group, for example $\mathbb{Q}/\mathbb{Z}$, then $D^{\#} = Hom_{\mathbb{N}}(R, D)$ is a left R-module which, by the same reasoning as above, can be shown to be c-injective. Using variants of the above arguments and the standard arguments from the theory of modules over a ring, one can show that if R is a semiring then any cancellative left R-semimodule can be embedded in a c-injective left R-module.

18. LOCALIZATION OF SEMIMODULES

In Chapter 10 we constructed semirings of fractions of certain semirings. We now subsume that construction in the more general construction of localizations of semimodules over semirings. Our method follows the method for modules over rings given in [Golan, 1986]. If R is a semiring then a nonempty subset κ of $lideal(R)$ is a **topologizing filter** if and only if the following conditions are satisfied:

(1) If $I \subseteq H$ are left ideals of R with $I \in \kappa$ then $H \in \kappa$;
(2) If $I, H \in \kappa$ then $I \cap H \in \kappa$;
(3) If $I \in \kappa$ and $a \in R$ then $(I : a) \in \kappa$.

The family of all topologizing filters of left ideals of R will be denoted by $R - fil$. Note that $R \in \kappa$ for all $\kappa \in R - fil$.

(18.1) Example. If $I \in ideal(R)$ then $\eta[I] = \{H \in lideal(R) \mid I \subseteq H\}$ belongs to $R - fil$.

It is clear that the intersection of an arbitrary family of elements of $R - fil$ again belongs to $R - fil$. Thus $R - fil$ is a complete lattice. Moreover, if κ and κ' are elements of $R - fil$ then we set $\kappa\kappa'$ equal to the set of all those elements I of $lideal(R)$ satisfying the condition that there exists an element H in κ' containing I for which $(I : a) \in \kappa$ for all $a \in H$. It is straightforward to verify that this is again an element of $R - fil$.

(18.2) Example. If $\kappa \in R - fil$ and if I is an ideal of R, then $\kappa\eta[I]$ equals $\{I' \in lideal(R) \mid$ there exists an element H of κ satisfying $IH \subseteq I' \subseteq H\}$.

By a straightforward translation of the proofs in Chapter 3 of [Golan, 1987] one can show that $(R - fil, \cap, \cdot)$ is a zerosumfree simple semiring for any semiring R, thus extending Example 1.7.

If R is a semiring, then a nonempty subset κ of $lideal(R)$ is a **Gabriel filter** if and only if the following conditions are satisfied:

(1) If $H, I \in \kappa$ and if $\alpha \in Hom_R(H, R)$ then $I\alpha^{-1} \in \kappa$;
(2) If K is a left ideal of R and $I \in \kappa$ satisfies the condition that for each $a \in I$ there exists an element H_a of κ with $H_a a \subseteq K$, then $K \in \kappa$.

The set of all Gabriel filters of left ideals of R will be denoted by $R - gab$. Note that if $\kappa \in R - gab$ and if I, H are elements of κ then for each $a \in H$ we have $(IH : a) = \{r \in R \mid ra \in IH\} \supseteq I$ and so $(IH : a) \in \kappa$ for each $a \in H$. Moreover, $(IH : a)a \subseteq IH$ for each such a. Therefore $IH \in \kappa$.

Clearly $\eta[\{0\}] = lideal(R)$ belongs to $R - gab$ and, indeed, this filter contains all other elements of $R - gab$.

Gabriel filters of ideals of bounded distributive lattices are considered in [Georgescu, 1988].

(18.3) Example. [Huq, 1983] An ideal I of a commutative semiring R is **separating** if and only if for each pair $r \neq r'$ of distinct elements of R there exists an element a of I such that $ar \neq ar'$. Then the family of all separating ideals of R is a Gabriel filter of ideals of R. This is the filter used in [Schmid, 1983] for the construction of lattices of fractions of bounded distributive lattices.

(18.4) Proposition. *If R is a semiring then $R - gab = I^{\times}(R - fil)$.*

Proof. Let $\kappa \in R - gab$. Assume that $I \subseteq H$ are left ideals of R with $I \in \kappa$. If $a \in I$ then $Ia \subseteq I \subseteq H$ and so, by condition (2) of the definition of a Gabriel filter, we have $H \in \kappa$. Let I and H be elements of κ and let $\alpha : I \rightarrow R$ be the inclusion map, which is an R-homomorphism. Then $I \cap H = H\alpha^{-1} \in \kappa$. Let $I \in \kappa$, let $a \in R$, and let $\alpha : R \rightarrow R$ be the R-homomorphism of left R-semimodules defined by $\alpha : r \mapsto ra$. Then $(I : a) = I\alpha^{-1} \in \kappa$. Thus $\kappa \in R - fil$.

It is easy to see that $\kappa \subseteq \kappa^2$ for any $\kappa \in R - fil$. If $\kappa \in R - gab$, then condition (2) in the definition of a Gabriel filter implies that the reverse containment is also true and so we have equality. Therefore $R - gab \subseteq I^{\times}(R - fil)$. Conversely, assume that $\kappa \in I^{\times}(R - fil)$. If $H, I \in \kappa$ and $\alpha \in Hom_R(H, R)$ then for each $a \in H$ we have $(I\alpha^{-1} : a) = \{r \in R \mid ra \in I\alpha^{-1}\} = \{r \in R \mid ra\alpha \in I\} = (I : a\alpha) \in \kappa$ and so $I\alpha^{-1} \in \kappa^2 = \kappa$. Furthermore, if K is a left ideal of R and $I \in \kappa$ satisfies the condition that for each $a \in I$ there exists an element H_a of κ satisfying $H_a a \subseteq K$ then, as an immediate consequence of the definition, $K \in \kappa^2 = \kappa$. Thus $\kappa \in R - gab$, proving that $R - gab = I^{\times}(R - fil)$. □

(18.5) Proposition. *Let R be a semiring and let M be a left R-semimodule. Then any $\kappa \in R - fil$ defines an R-congruence relation $\equiv_{\kappa}$ on M by setting $m \equiv_{\kappa} m'$ if and only if there exists an element I of κ such that $am = am'$ for all $a \in I$.*

Proof. Clearly $m \equiv_{\kappa} m$ for each $m \in M$ and if $m \equiv_{\kappa} m'$ then surely $m' \equiv_{\kappa} m$. If $m \equiv_{\kappa} m'$ and $m' \equiv_{\kappa} m''$ then there exist elements I and H of κ such that $am = am'$ for all $a \in I$ and $bm' = bm''$ for all $b \in H$. Therefore $cm = cm''$ for all $c \in I \cap H \in \kappa$ and so $m \equiv_{\kappa} m''$. Similarly, if $m \equiv_{\kappa} m'$ and $n \equiv_{\kappa} n'$ then there exist elements I and H of κ such that $am = am'$ for all $a \in I$ and $bn = bn'$ for all $b \in H$. Hence $c(m+n) = c(m'+n')$ for all $c \in I \cap H \in \kappa$ and so $m+n \equiv_{\kappa} m'+n'$. Finally, if $m \equiv_{\kappa} m'$ and $r \in R$ then there exists an ideal I in κ such that $am = am'$ for all $a \in I$ and so $b(rm) = b(rm')$ for all $b \in (I : r) \in \kappa$. Thus $rm \equiv_{\kappa} rm'$, proving that $\equiv_{\kappa}$ is an R-congruence relation on M. □

The relation $\equiv_{\kappa}$ defined in Proposition 18.5 is called the κ-**torsion congruence** on M. A left R-semimodule M is κ-**torsion** if and only if the R-congruence relation $\equiv_{\kappa}$ is universal on M. To show that this holds, it is necessary and sufficient to show that $m \equiv_\kappa 0_M$ for every element m of M. The left R-semimodule M is **strongly κ-torsionfree** if and only if then R-congruence $\equiv_\kappa$ on M is trivial.

(18.6) PROPOSITION. *Let R be a semiring and let $\kappa \in R - fil$. Then R, considered as a left semimodule over itself, is κ-torsion if and only if $\kappa = \eta[\{0\}]$.*

PROOF. If $\kappa = \eta[\{0\}]$ then surely R is κ-torsion since $0a = 0b$ for all $a, b \in R$. Conversely, assume that R is κ-torsion. Then $0 \equiv_\kappa 1$ and so there exists an element I of κ satisfying $0 = 0a = 1a = a$ for all $a \in I$. But this means that $I = \{0\}$ and so $\{0\} \in \kappa$, proving that $\kappa = \eta[\{0\}]$. □

(18.7) PROPOSITION. *Let $\kappa \in R - fil$ for some semiring R.*

(1) *If $\alpha: M \to N$ is an R-homomorphism of R-semimodules and if $m \equiv_\kappa m'$ in M then $m\alpha \equiv_\kappa m'\alpha$ in N.*
(2) *If N is a submodule of a left R-module M then $n \equiv_\kappa n'$ in N if and only if $n \equiv_\kappa n'$ in M.*
(3) *If M is a left R-semimodule then $N = M/\equiv_\kappa$ is strongly κ-torsionfree.*
(4) *If $\alpha: M \to N$ is an essential R-monomorphism of left R-semimodules then N is strongly κ-torsionfree if M is.*

PROOF. (1) - (3) are immediate consequences of the definitions; (4) is an immediate consequence of the definitions and of Proposition 17.26. □

In particular, we see that any subsemimodule of a κ-torsion left R-semimodule is again κ-torsion and a subsemimodule of a strongly κ-torsionfree R-semimodule is again strongly κ-torsionfree.

Let κ be a topologizing filter of left ideals of a semiring R. A subsemimodule N of a left R-semimodule M is **κ-dense** in M if and only if $(N : m) \in \kappa$ for all $m \in M$.

(18.8) PROPOSITION. *If R is a semiring and $\kappa \in R - gab$ then the following conditions on a subtractive left ideal I of R are equivalent:*

(1) *$I \in \kappa$;*
(2) *R/I is a κ-torsion left R-semimodule;*
(3) *I is κ-dense in R.*

PROOF. (1) ⇒ (2): If $a \in R$ then $(I : a) \in \kappa$ and $(I : a)a \subseteq I$. Therefore $a/I \equiv_\kappa 0/I$, whence R/I is a κ-torsion left R-semimodule.

(2) ⇒ (3): By (2) we know that for each a in R there exists a left ideal H in κ such that for each $h \in H$ there is an element b of I with $ha + b \in I$. Since I is subtractive, this means that $ha \in I$ for each $h \in H$ and so $H \subseteq (I : a)$. Hence $(I : a) \in \kappa$ for all $a \in R$, proving (3).

(3) ⇒ (1): If $a \in R$ then $H_a = (I : a) \in \kappa$ satisfies $H_a a \subseteq I$. Since $R \in \kappa$, this implies that $I \in \kappa$. □

(18.9) PROPOSITION. *Let R be a semiring and let $\kappa \in R - gab$. If N is a subsemimodule of a left R-semimodule M then N is κ-dense in M if and only if M/N is κ-torsion.*

PROOF. Assume that N is κ-dense in M. If $m \in M$ then $(N : m) \in \kappa$ and $am/N = 0/N$ for all $a \in (N : m)$. Therefore $m/N \equiv_\kappa 0/N$ for each $m \in M$, proving that M/N is κ-torsion. Conversely, if M/N is κ-torsion and if $m \in M$ then $m/N \equiv_\kappa 0/N$ and so there exists an element I of κ satisfying $am/N = 0/N$ for all $a \in I$. Therefore $I \subseteq (N : m)$ and so $(N : m) \in \kappa$, proving that N is κ-dense in M. □

(18.10) Corollary. *Let $\kappa \in R - gab$ for some semiring R. If $\alpha: M \to N$ is an R-homomorphism of left R-semimodules then the following conditions are equivalent:*

(1) *A subsemimodule N' of N is κ-dense in N if and only if $N'\alpha^{-1}$ is κ-dense in M;*

(2) *If N is κ-torsion and $ker(\alpha) = \{0\}$ then M is κ-torsion.*

Proof. (1) $\Rightarrow$ (2): If N is κ-torsion then, by Proposition 18.9, we see that $\{0_N\}$ is κ-dense in M and so $\{0_M\} = \{0_N\}\alpha^{-1}$ is κ-dense in M by (1). Therefore, by Proposition 18.9, we conclude that M is κ-torsion.

(2) $\Rightarrow$ (1): Let N' be a κ-dense submodule of N. Then, by Proposition 18.9, N/N' is κ-torsion. Moreover, by Proposition 15.19, the map α induces an R-homomorphism $M/N'\alpha^{-1} \to N/N'$ with kernel $\{0\}$ and so $M/N'\alpha^{-1}$ is κ-torsion. Hence, by Proposition 18.9, $N'\alpha^{-1}$ is κ-dense in M. □

(18.11) Proposition. *Let $\kappa \in R - gab$ for some semiring R and let M be a strongly κ-torsionfree left R-semimodule. If N' is a κ-dense subsemimodule of a left R-semimodule N and if $\alpha: N' \to M$ is an R-homomorphism then there is at most one R-homomorphism $\beta: N \to M$ extending α.*

Proof. Assume that $\beta \neq \beta'$ are R-homomorphisms from N to M extending α and let N be an element of N satisfying $n\beta \neq n\beta'$. The left ideal $I = (N' : n)$ of R belongs to κ and $a(n\beta) = (an)\alpha = a(n\beta')$ for all $a \in I$. Therefore $n\beta =_\kappa n\beta'$ in M and so, by strong κ-torsionfreeness, $n\beta = n\beta'$, which is a contradiction. □

If $\kappa \in R - gab$ for some semiring R and if M is a left R-semimodule, set $T_\kappa(M) = \{m \in M \mid (0 : m) \in \kappa\}$. We call $T_\kappa(M)$ the **κ-torsion subsemimodule** of M. This terminology is justified by the following result.

(18.12) Proposition. *If $\kappa \in R - gab$ for some semiring R and if M is a left R-semimodule then:*

(1) *$T_\kappa(M)$ is a subsemimodule of M;*

(2) *$T_\kappa(M)$ is κ-torsion;*

(3) *$T_\kappa(T_\kappa(M)) = T_\kappa(M)$;*

(4) *$T_\kappa(M/T_\kappa(M)) = \{0\}$;*

(5) *If $\alpha: M \to N$ is an R-homomorphism of left R-semimodules then $T_\kappa(M)\alpha \subseteq T_\kappa(N)$;*

(6) *If M' is a subsemimodule of M then $T_\kappa(M') = T_\kappa(M) \cap M'$.*

Proof. (1) If $m, m' \in M$ and if $r \in R$ then $(0 : m + m') \supseteq (0 : m) \cap (0 : m')$ and so, by Proposition 18.4, $(0 : m + m') \in \kappa$. Moreover, $(0 : rm) = ((0 : m) : r)$ and so, by Proposition 18.4, $(0 : rm) \in \kappa$. Thus $m + m'$ and rm both belong to $T_\kappa(M)$ and so $T_\kappa(M)$ is a subsemimodule of M.

(2) If $m \in T_\kappa(M)$ then $am = a0$ for all $a \in (0 : m) \in \kappa$ and so $m \equiv_\kappa 0$. Thus $\equiv_\kappa$ is universal on $T_\kappa(M)$ and so $T_\kappa(M)$ is κ-torsion.

(3) This is an immediate consequence of the definition.

(4) Assume that $m/T_\kappa(M) \in T_\kappa(M/T_\kappa(M))$. Then $(T_\kappa(M) : m) \in \kappa$. For each $a \in (T_\kappa(M) : m)$ there exists an element H_a of κ with $H_a am = \{0\}$. Therefore, by the definition of a Gabriel filter, $K = \cup\{H_a a \mid a \in (T_\kappa(M) : m)\} \in \kappa$ and $Km = \{0\}$ so $m \in T_\kappa(M)$. Hence $m/T_\kappa(M) = 0$.

(5) If $x \in T_\kappa(M)$ then $(0 : m\alpha) \supseteq (0 : m)$ so $(0 : m\alpha) \in \kappa$. Thus $m\alpha \in T_\kappa(N)$.

(6) This is an immediate consequence of the definitions. □

If $\kappa \in R-gab$ and if M is a left R-semimodule satisfying $T_\kappa(M) = \{0\}$ then M is **κ-torsionfree**. Strongly κ-torsionfree left R-semimodules are surely κ-torsionfree.

/medskip If $\kappa \in R - gab$ for some semiring R then a left R-semimodule E is **κ-injective** if and only if, given a left R-semimodule M and a κ-dense submodule N of M, any R-homomorphism from N to E can be extended to an R-homomorphism from M to E. Such an extension may be not be unique. However, we do have the following result.

(18.13) Proposition. *Let R be a semiring. The following conditions on a left R-semimodule E are equivalent:*

(1) *Given a left R-semimodule and a κ-dense subsemimodule N of M, any R-homomorphism from N to E can be uniquely extended to an R-homomorphism from M to E;*

(2) *E is κ-injective and strongly κ-torsionfree.*

Proof. (1) $\Rightarrow$ (2): Surely (1) implies that E is κ-injective. If x and y are elements of E satisfying $x \equiv_\kappa y$ then there exists an element I of κ satisfying $ax = ay$ for all $a \in I$. By Proposition 18.8, I is κ-dense in R. Let $\alpha: I \to E$ be then R-homomorphism of left R-semimodules defined by $\alpha. a \mapsto ax = ay$. Then the R-homomorphisms from R to E defined by $r \mapsto rx$ and $r \mapsto ry$ both extend α and so, by (1), they must be equal. Hence $x = 1x = 1y = y$, proving that E is strongly κ-torsionfree.

(2) $\Rightarrow$ (1): The existence of such an extension follows from the κ-injectivity of E, and its uniqueness follows from strong κ-torsionfreeness by Proposition 18.11. □

(18.14) Proposition. *Let $\kappa \in R - gab$ for some semiring R and let E be a κ-injective left R-semimodule. Let E' be a subsemimodule of E satisfying the condition that E/E' is κ-torsionfree. Then E' is also κ-injective.*

Proof. Let N be a κ-dense subsemimodule of a left R-semimodule M and let $\alpha: N \to E'$ be an R-homomorphism. Then there exists an R-homomorphism $\beta: M \to E$ extending α. If $m \in M$ then $I = (N : m) \in \kappa$ so $I(m\beta) = (Im)\beta \subseteq N\beta \subseteq E'$. If $m\beta \notin E'$ then $m\beta/E'$ is a nonzero element of $T_\kappa(E/E')$, which is a contradiction. Thus we must have $m\beta \in E'$, proving that E' is also κ-injective. □

We would now like to construct "semimodule of quotients" of a left R-semimodule with respect to a Gabriel filter κ. We are hampered in emulating the construction in module theory [Golan, 1986] by the possible lack of sufficiently-many injective semimodules, as demonstrated in Proposition 17.21.

(18.15) PROPOSITION. *Let $\kappa \in R - gab$ for some semiring R and let M be a strongly κ-torsionfree left R-semimodule. Then there exists a strongly κ-torsionfree left R-semimodule $Q_\kappa(M)$ and an R-monomorphism $\varphi_M: M \rightarrow Q_\kappa(M)$ such that:*

(1) $M\varphi_M$ is a κ-dense subsemimodule of $Q_\kappa(M)$;

(2) If N is a strongly κ-torsionfree left R-semimodule and $\alpha: M \rightarrow N$ is an R-monomorphism such that $M\alpha$ is a κ-dense subsemimodule of N, then there exists a unique R-monomorphism $\beta: N \rightarrow Q_\kappa(M)$ extending φ_M.

PROOF. Let M be a strongly κ-torsionfree left R-semimodule and let $W_\kappa(M) = \{(I, \alpha) \mid I \in \kappa \text{ and } \alpha \in Hom_R(I, M)\}$. Define a relation ζ on $W_\kappa(M)$ by setting $(I, \alpha) \ \zeta \ (I', \alpha')$ if and only if there exists an $I'' \in \kappa$ satisfying $I'' \subseteq I \cap I'$ while the restrictions of α and α' to I'' coincide. Then ζ is clearly an equivalence relation. Let $Q_\kappa(M) = W_\kappa(M)/\zeta$ and denote the equivalence class $(I, \alpha)/\zeta$ by $I//\alpha$. Define an operation $+$ on $Q_\kappa(M)$ by setting $I//\alpha + H//\beta = (I \cap H)//(\alpha + \beta)$. This is well-defined since if $I//\alpha = I'//\alpha'$ and $H//\beta = H'//\beta'$ then there exist elements I'' and H'' of κ satisfying $I'' \subseteq I \cap I'$ and $H'' \subseteq H \cap H'$ such that α and α' agree on I'' while β and β' agree on H''. Hence $I'' \cap H'' \in \kappa$ and $I'' \cap H'' \subseteq (I \cap H) \cap (I' \cap H')$ while $\alpha + \beta$ and $\alpha' + \beta'$ agree on $I'' \cap H''$, proving that $(I \cap H)//(\alpha + \beta) = (I' \cap H')//(\alpha' + \beta')$. Moreover, as an immediate consequence of the definition we see that $(Q_\kappa(M), +)$ is a commutative monoid with additive identity $R//\theta$, where $\theta: R \rightarrow M$ is the R-homomorphism defined by $\theta: r \mapsto 0_M$ for all $r \in R$.

If $I//\alpha \in Q_\kappa(M)$ and $r \in R$, let $r(I//\alpha) = (I : r)//r\alpha$, where $r\alpha: (I : r) \rightarrow M$ is the R-homomorphism defined by $a \mapsto (ar)\alpha$. Again, this is well-defined since if $I//\alpha = I'//\alpha'$ then there exists an element I'' of κ such that $I'' \subseteq I \cap I'$ while α and α' coincide on I''. But then $(I'' : r) \subseteq (I : r) \cap (I' : r)$ while $r\alpha$ and $r\alpha'$ coincide on $(I'' : r)$. Hence $r(I//\alpha) = r(I'//\alpha')$.

Given these definitions, it is a straightforward computation to show that $Q_\kappa(M)$ is in fact a left R-semimodule. Moreover, for each $m \in M$ we have an R-homomorphism $\theta_m: R \rightarrow M$ defined by $\theta_m: r \mapsto rm$ and so $R//\theta_m$ is an element of $Q_\kappa(M)$. If $m, m' \in M$ then $\theta_{m+m'} = \theta_m + \theta'_m$ while if $m \in M$ and $a \in R$ then $\theta_{am} = a\theta_m$. Thus, if $\varphi_M: M \rightarrow Q_\kappa(M)$ is the function given by $m \mapsto R//\theta_m$ then φ_M is an R-homomorphism of left R-semimodules. Indeed, it is injective since if $m\varphi_M = m'\varphi_M$ then $R//\theta_m = R//\theta'_m$ and so there exists an element I of κ such that $am = am'$ for all $a \in I$. But M was assumed to be strongly κ-torsionfree, and so this implies that $m = m'$. Thus, by Proposition 15.15, φ_M is an R-monomorphism. Next, we claim that $M\varphi_M$ is κ-dense in $Q_\kappa(M)$. Indeed, if $I//\alpha$ belongs to $Q_\kappa(M)$ then for each $a \in I$ we have $a(I//\alpha) = R//\theta_m$, where $m = (a)\alpha \in M$. Thus $(M\varphi_M: I//\alpha) \in \kappa$. If $I//\alpha$ and $I'//\alpha'$ are elements of $Q_\kappa(M)$ for which there exists an element I'' of κ such that $a(I//\alpha) = a(I'//\alpha')$ for all $a \in I''$ then for each $a \in I''$ there exists an element H_a of κ contained in $(I : a) \cap (I' : a)$ such that $a\alpha$ and $a\alpha'$ coincide on H_a. Set $H = \sum_{a \in I''} H_a$. Then $H \in \kappa$, $H \subseteq I \cap I'$, while α and α' coincide on H. Therefore $I//\alpha = I'//\alpha'$. This shows that $Q_\kappa(M)$ is strongly κ-torsionfree.

Now assume that $\alpha: M \rightarrow N$ is an R-monomorphism such that N is strongly κ-torsionfree and such that $M\alpha$ is κ-dense in N. If $n \in N$ then $(M\alpha : n) \in \kappa$. Define the function $\beta: N \rightarrow Q_\kappa(M)$ by $\beta: n \mapsto (M\alpha : n)//\psi_n$, where ψ_n is the R-homomorphism defined by $a \mapsto (an)\alpha^{-1}$. If $n = m\alpha$ for some $m \in M$, then $(M\alpha : n) = R$ and $\psi_n: r \mapsto rm$. In other words, $\psi_n = \theta_m$, proving that $\alpha\beta = \varphi_M$.

We are left to show that β is in fact an R-monomorphism. If $n, n' \in N$ then $H = (M\alpha : n) \cap (M\alpha : n')$ belongs to κ, while the restrictions of $\psi_{n+n'}$ and $\psi_n + \psi'_n$ coincide on H. Therefore $n\beta + n'\beta = (n+n')\beta$. Similarly $r(n\beta) = (rn)\beta$ for all $r \in R$ and $n \in N$. Thus β is an R-homomorphism. If $n, n' \in N$ satisfy $n\beta = n'\beta$ then there exists an element H of κ satisfying $Hn\alpha^{-1} = Hn'\alpha^{-1}$ and so $Hn = Hn'$ since α is injective. But M is strongly κ-torsionfree, and so this implies that $n = n'$. Thus β is an R-monomorphism. It is unique by Proposition 18.11. □

(18.16) PROPOSITION. *Let $\kappa \in R - gab$ for some semiring R and let M be a strongly κ-torsionfree left R-semimodule having an injective hull $\alpha\colon M \to E$. Then $Q_\kappa(M)$ is κ-injective.*

PROOF. Let E' be the submodule of E containing $M\alpha$ and defined by $E'/M\alpha = T_\kappa(E/M\alpha)$. Then $M\alpha$ is κ-dense in E' and E/E' is κ-torsionfree by Proposition 18.12(4). By Proposition 18.14, E' is κ-injective. By Proposition 18.7(4), E is strongly κ-torsionfree and so E' is also strongly κ-torsionfree.

Since E' is κ-injective, there exists an R-homomorphism $\theta\colon Q_\kappa(M) \to E'$ satisfying $\alpha = \varphi_M\theta$. On the other hand, by Proposition 18.15, there exists an R-homomorphism $\psi\colon E' \to Q_\kappa(M)$ satisfying $\varphi_M = \alpha\psi$. Therefore $\alpha = \alpha(\psi\theta)$ and $\varphi_M = \varphi_M(\theta\psi)$. By Proposition 18.11, we see that $\theta\psi$ must be the identity map on $Q_\kappa(M)$ and $\psi\theta$ must be the identity map on E'. Therefore E' is R-isomorphic to $Q_\kappa(M)$, proving that $Q_\kappa(M)$ is κ-injective. □

Let $\kappa \in R - gab$ for some semiring R. If M is a left R-semimodule then $M/\equiv_\kappa$ is strongly κ-torsionfree by Proposition 18.7(3). We then set $Q_\kappa(M) = Q_\kappa(M/\equiv_\kappa)$. This strongly κ-torsionfree left R-semimodule is called the **semimodule of κ-quotients** of M.

Now let us look at R, considered as a left semimodule over itself. If $\eta[\{0\}] \neq \kappa \in R - gab$ then, by Proposition 14.36, $R_\kappa = End_R(Q_\kappa(R))$ is a semiring, called the **semiring of κ-quotients** of R. This terminology is justified by the following result.

(18.17) PROPOSITION. *If R is a semiring and $\eta[\{0\}] \neq \kappa \in R - gab$ then R_κ is canonically a left R-semimodule R-isomorphic to $Q_\kappa(R)$.*

PROOF. Each element q of $Q_\kappa(R)$ defines an R-homomorphism β_q from R, considered as a left semimodule over itself, to $Q_\kappa(R)$, given by $a \mapsto aq$. If $a \equiv_\kappa b$ in R then $aq \equiv_\kappa bq$. But $Q_\kappa(R)$ is strongly κ-torsionfree and so we must have $aq = bq$. Therefore β_q induces an R-homomorphism from $R' = R/\equiv_\kappa$ to $Q_\kappa(R)$ which, by Proposition 18.15, can be uniquely extended to an R-endomorphism θ_q of the left R-semimodule $Q_\kappa(R)$. For $r \in R$, set $r' = r/\equiv_\kappa \in R'$. Then we can define the structure of a left R-semimodule on R_κ by setting $r\alpha = \theta'_r\alpha$ for each $r \in R$ and $\alpha \in R_\kappa$. (The proof that this indeed does turn R_κ into a left R-semimodule is straightforward, relying on the uniqueness of θ_q.) Let $\theta\colon Q_\kappa(R) \to R_\kappa$ be the function defined by $q \mapsto \theta_q$. We claim that this is an R-homomorphism of left R-semimodules. Indeed, if $x, y \in Q_\kappa(R)$ then $x\theta + y\theta$ and $(x+y)\theta$ both extend the map from R' to $Q_\kappa(R)$ induced by $\beta_x + \beta_y$ and so must be equal. Similarly, if $r \in R$ and $x \in Q_\kappa(R)$ then $r(x\theta)$ and $(rx)\theta$ are equal. This establishes the claim. We note that, in fact, θ is surjective since if $\alpha \in R_\kappa$ then α is the image of $(1/\equiv_\kappa)\alpha$ under θ. Similarly, θ is injective since if $x \neq y$ in $Q_\kappa(R)$ then $(1/\equiv_\kappa)\theta_x = x \neq y = (1/\equiv_\kappa)\theta_y$

and so $x\theta \neq y\theta$. Thus, by Proposition 15.15, θ is an R-monomorphism and so is the R-isomorphism we seek. □

The isomorphism θ defined in Proposition 18.17, composed with the R-homomorphism from R to $Q_\kappa(R)$ defines an R-homomorphism λ^κ from R to R_κ. Indeed, λ^κ is a morphism of rings for if $a, b \in R$ then $\lambda^\kappa(ab)$ and $\lambda^\kappa(a)\lambda^\kappa(b)$ are both R-endomorphisms of $Q_\kappa(R)$ which extend the R-homomorphism from $R/\equiv_\kappa$ to $Q_\kappa(R)$ defined by $r/\equiv_\kappa \mapsto rab/\equiv_\kappa$ and so, by Proposition 18.15, they must be equal.

(18.18) PROPOSITION. *If R is a semiring and $\eta[\{0\}] \neq \kappa \in R-gab$ then any strongly κ-torsionfree left R-semimodule has the structure of a left R_κ-semimodule which naturally extends its structure as an R-semimodule.*

PROOF. If N is a strongly κ-torsionfree left R-semimodule then any element x of N defines an R-homomorphism from R to N given by $a \mapsto ax$. If $a \equiv_\kappa b$ in R then $ax \equiv_\kappa bx$ in N and so $ax = bx$. Therefore this R-homomorphism induces an R-homomorphism from $R/\equiv_\kappa$ to N which, by Proposition 18.15, can be uniquely extened to an R-homomorphism $\psi_x: Q_\kappa(R) \to N$. If $\alpha \in R_\kappa$ define $\alpha \cdot x$ to be $(1)\alpha\psi_x \in N$. This defines on N the structure of a left R_κ-semimodule. Furthermore, if $\alpha = \lambda^\kappa(r)$ for some element r of R then $\alpha \cdot x = rx$ and so the R_κ-semimodule structure on N naturally extends its R-semimodule structure. □

Further results on the nature of semimodules of quotients can be developed along the lines of the corresponding results for modules presented in [Golan, 1986].

19. LINEAR ALGEBRA OVER A SEMIRING

The techniques of linear algebra over a semiring have important applications in optimization theory, models of discrete event networks, and graph theory, particularly if the semiring is in fact a semifield. For further examples, see [Baccelli & Mairesse, 1998] and [Gaujal & Jean-Marie, 1998].

If A and B are nonempty sets and if R is a semiring then $R^{A\times B}$ can be turned into an (R, R)-bisemimodule by defining addition and scalar multiplication componentwise. We denote this bisemimodule by $\mathcal{M}_{A\times B}(R)$. If the set A is either finite or countably-infinite then $\mathcal{M}_{A\times B}(R)$ can be turned into a left $\mathcal{M}_{A,r}(R)$-semimodule by defining addition componentwise and scalar multiplication as follows: if $u \in \mathcal{M}_{A\times B}(R)$ and $f \in \mathcal{M}_{A,r}(R)$ then $fu\colon (i,j) \mapsto \sum_{k\in A} f(i,k)u(k,j)$ for all $(i,j) \in A \times B$. (Note that the sum is well-defined since, for each $i \in A$, only finitely-many values of $f(i,k)$ are nonzero.) As in the case of elements of $\mathcal{M}_A(R)$, we often use matrix notation rather than functional notation to denote the elements of $\mathcal{M}_{A\times B}(R)$. If the set A [resp. B] is finite and has order n, we will sometimes write $\mathcal{M}_{n\times B}(R)$ [resp. $\mathcal{M}_{A\times n}(R)$] instead of $\mathcal{M}_{A\times B}(R)$.

If $u \in \mathcal{M}_{A\times B}(R)$ and $v \in \mathcal{M}_{B\times C}(R)$ satisfy the condition that either u is row finite or v is column finite, then we can define the **matrix product** uv to be the element of $\mathcal{M}_{A\times C}(R)$ defined by $uv\colon (i,j) \mapsto \sum_{k\in B} u(i,k)v(k,j)$ for all $(i,j) \in A \times C$. If v is a fixed column-finite element of $\mathcal{M}_{B\times C}(R)$, then it is straightforward to see that for all $u, u' \in \mathcal{M}_{A\times B}(R)$ and for all $f \in \mathcal{M}_{A,r}(R)$ we have $(u+u')v = uv+u'v$ and $(fu)v = f(uv)$. Thus we see that the function $u \mapsto uv$ is an $\mathcal{M}_{A,r}(R)$-homomorphism from $\mathcal{M}_{A\times B}(R)$ to $\mathcal{M}_{A\times C}(R)$. Moreover, if A and B are nonempty sets and if $n \in \mathbb{P}$ then we have a function

$$\theta_n\colon \mathcal{M}_{A\times n}(R) \times \mathcal{M}_{n\times B}(R) \to \mathcal{M}_{A\times B}(R)$$

defined by $\theta_n\colon (u,v) \mapsto uv$. If $S = \mathcal{M}_n(R)$ then $\mathcal{M}_{A\times n}(R)$ is a right S-semimodule and $\mathcal{M}_{n\times B}(R)$ is a left S-semimodule. Moreover, the map θ_n is S-balanced. Therefore, by Proposition 16.14, there exists a unique $\mathbb{N}$-homomorphism

$$\psi\colon \mathcal{M}_{A\times n}(R) \otimes_S \mathcal{M}_{n\times B}(R) \to \mathcal{M}_{A\times B}(R)[/]\{0\}$$

satisfying the condition that $\psi(u \otimes v) = uv[/]\{0\}$.

A matrix $u \in M_{A\times B}(R)$ defines a function $u_{i*} \in R^B$ for each $i \in A$ by setting $u_{i*}: j \mapsto u(u, j)$. Similarly, for each $j \in B$ the matrix u defines a function $u_{*j} \in R^A$ by setting $u_{*j}: i \mapsto u(i, j)$. The **row semimodule** of u is the subsemimodule of R^B generated by $\{u_{i*} \mid i \in A\}$ and the **column semimodule** of u is the subsemimodule of R^A generated by $\{u_{*j} \mid j \in B\}$. The **row rank** [resp. **column rank**] of u is the rank of its row [resp. column] semimodule. If these two values coincide, and they may not, their common value is the **rank** of u.

Other definitions of rank abound in the literature. Following a defintion originally given for semiring theory [Schein, 1976], one can define the **Schein rank** of a matrix u to be the cardinality of the smallest set of matrices having rank 1 the sum of which is u.

(19.1) EXAMPLE. [Kim & Roush, 1980] Let $R = (\mathbb{I}, \vee, \wedge)$ and let

$$u = \begin{bmatrix} 1 & 0.8 & 0 \\ 0.8 & 0.7 & 0 \\ 0.7 & 0.6 & 0 \end{bmatrix}.$$

Then the column rank and Schein rank of u equal 2, while its row rank is 3.

Another notion of rank is the following: an element of $M_{A\times B}(R)$ belonging to the image of θ_n but not to the image of θ_k for any $k < n$ has **factor rank** equal to n.

(19.2) EXAMPLE. [Beasley & Pullman, 1988b] The matrix

$$u = \begin{bmatrix} 1 & 0 & 1 & 1 \\ 1 & 1 & 0 & 0 \\ 0 & 1 & 1 & 0 \end{bmatrix}$$

in $\mathcal{M}_{3\times 4}(\mathbb{B})$ has column rank 4 but factor rank 3.

Matrices of factor rank 1 over subsemirings of $\mathbb{I}$ are discussed by Scully [1991, 1993].

If $u \in M_{A\times B}(R)$ then the **transpose** of u is the matrix $u^T \in M_{B\times A}(R)$ defined by $u^T: (j, i) \mapsto u(i, j)$ for all $i \in A$ and $j \in B$. Clearly $(u + v)^T = u^T + v^T$ and $u^{TT} = u$ for all matrices u and v in $\mathcal{M}_{A\times B}$.

Let A be a nonempty set which is either finite or countably-infinite and let B be an arbitrary nonempty set. Each morphism of semirings $\gamma: R \to S$ defines a function $\mathcal{M}_{A\times B}(\gamma): \mathcal{M}_{A\times B}(R) \to \mathcal{M}_{A\times B}(S)$ by $u \mapsto \gamma u$. This function is surjective when γ is. Moreover, if $u \in \mathcal{M}_{A\times B}(R)$ and $v \in \mathcal{M}_{B\times C}(R)$ satisfy the condition that either u is row finite or v is column finite, then $\gamma(uv) = (\gamma u)(\gamma v)$.

We begin by looking at some more properties of semirings of matrices over certain semirings.

(19.3) PROPOSITION. *Let R be a zerosumfree semiring and n a positive integer. An element $A = [a_{ij}]$ of $\mathcal{M}_n(R)$ is a unit if and only if the following conditions are satisfied:*

(1) *For each $1 \leq i, j \leq n$ there exist elements b_{ij}, x_{ij}, and y_{ij} of R satisfying $a_{ij}b_{ji} + x_{ij} = b_{ji}a_{ij} + y_{ij} = 1$ and $b_{ji}x_{ij} = a_{ij}y_{ij} = x_{ij}a_{ij} = y_{ij}b_{ji} = 0$;*

(2) *For each* $1 \leq i \leq n$ *we have* $a_{i1}b_{1i} + \cdots + a_{in}b_{ni} = 1$ *and* $a_{ij}b_{ji}a_{ik}b_{ki} = 0$ *for all* $1 \leq j \neq k \leq n$;

(3) *For each* $1 \leq j \leq n$ *we have* $b_{j1}a_{1j} + \cdots + b_{jn}a_{nj} = 1$ *and* $b_{ji}a_{ij}b_{jk}a_{kj} = 0$ *for all* $1 \leq i \neq k \leq n$.

PROOF. Since R is zerosumfree, A is a unit of $\mathcal{M}_n(R)$ if and only if there exists a matrix $[b_{ij}] \in \mathcal{M}_n(R)$ such that the conditions

(*) $\sum_{k=1}^{n} a_{ik}b_{ki} = 1 = \sum_{k=1}^{n} b_{ik}a_{ki}$ for all $1 \leq i \leq n$; and

(**) $a_{ik}b_{kj} = 0 = b_{ik}a_{kj}$ for all $1 \leq i \neq j \leq n$ and all $1 \leq k \leq n$

are satisfied.

If these conditions are satisfied, then for each $1 \leq i, j \leq n$ define $x_{ij} = \sum_{k \neq j} a_{ik}b_{kj}$ and $y_{ij} = \sum_{k \neq i} b_{jk}a_{kj}$, and it is straightforward to verify that conditions (1) – (3) hold.

Conversely, assume that these conditions hold. Then (*) is satisfied. By (2) and (3) we see that for all $1 \leq i \neq j \leq n$ and all $1 \leq k \leq n$ we have

$$a_{ik}b_{kj} = (a_{ik}b_{ki}a_{ik})(b_{kj}a_{jk}b_{kj}) = a_{ik}(b_{ki}a_{ik}b_{kj}a_{jk})b_{kj} = 0$$

and similarly $b_{ik}a_{kj} = 0$, showing that (**) is satisfied as well. □

(19.4) PROPOSITION. *Let R be a commutative zerosumfree semiring and n a positive integer. Let $A = [a_{ij}]$ be an element of $\mathcal{M}_n$ and let $B = [b_{ij}]$ be a matrix satisfying $AB = I$, where I is the multiplicative identity of $\mathcal{M}_n(R)$. Then:*

(1) $a_{ij}a_{ik} = a_{ji}a_{ki} = b_{ij}b_{ik} = b_{ji}b_{ki} = 0$ *for all* $1 \leq i \leq n$ *and all* $1 \leq j \neq k \leq n$;

(2) $a_{ik}b_{kj} = a_{ki}b_{jk} = 0$ *for all* $1 \leq k \leq n$ *and all* $1 \leq i \neq j \leq n$;

(3) $[\sum_{k=1}^{n} a_{ik}][\sum_{m=1}^{n} b_{mj}] = [\sum_{k=1}^{n} a_{ki}][\sum_{m=1}^{n} b_{jm}] = 1$ *for all* $1 \leq i, j \leq n$;

(4) $a_{ij}b_{ji} \in I^{\times}(R)$ *for all* $1 \leq i, j \leq n$;

(5) *If* $e_{\sigma} = \prod_{i=1}^{n} a_{i,\sigma(i)}b_{\sigma(i),i}$ *for each permutation* σ *of* $\{1, \ldots, n\}$ *then* $\{e_{\sigma}\}$ *is a complete set of orthogonal central idempotents of* R.

PROOF. (1) Since $AB = I$ and since R is zerosumfree, we have $a_{ik}b_{kj} = 0$ for all $1 \leq i \neq j \leq n$ and $1 = \prod_{i=1}^{n}[\sum_{k=1}^{n} a_{ik}b_{kj}]$. By commutativity, all terms of the form $a_{ik}b_{ki}a_{jk}b_{kj}$ with $i \neq j$ are equal to 0 and so this product reduces to $\sum_{\sigma}[\prod a_{i,\sigma(i)}b_{\sigma(i),i}]$, where the product ranges over all permutations σ of $\{1, \ldots, n\}$. If $j \neq k$ then $a_{ik}a_{jk} = a_{ik}a_{jk}(\sum_{\sigma}[\prod a_{i,\sigma(i)}b_{\sigma(i),i}])$ and this is 0, since each summand contains a factor of the form $a_{ik}a_{jk}b_{k,\sigma(k)}$ with either $\sigma(k) \neq i$ or $\sigma(k) \neq j$. By similar arguments we obtain the rest of (1).

(2) We have already noted that $a_{ik}b_{kj} = 0$ if $i \neq j$. Also,

$$a_{ki}b_{jk} = a_{ki}b_{jk}\left(\sum_{\sigma}\left[\prod a_{i,\sigma(i)}b_{\sigma(i),i}\right]\right) = 0$$

since each term has a factor of the form $a_{ji}b_{jk}b_{i,\sigma(i)}a_{\sigma(j),j} = a_{ki}b_{i,\sigma(i)}a_{\sigma(j),j}b_{jk}$ and either $k \neq \sigma(i)$ or $k \neq \sigma(j)$.

(3) By the above,

$$\left[\sum_{k=1}^{n} a_{ik}\right]\left[\sum_{m=1}^{n} b_{mi}\right] = \sum_{k=1}^{n} a_{ik}b_{ki} = 1.$$

The other equality is proven analogously.

(4) If $1 \leq i, j \leq n$ then

$$a_{ij}b_{ji}a_{ij}b_{ji} = a_{ij}b_{ji}\left[\sum_{k=1}^{n} a_{ik}b_{ki}\right] = a_{ij}b_{ji}1 = a_{ij}b_{ji}.$$

(5) By (4) and the commutativity of R, we see that each e_σ is a central idempotent. Moreover, $\sum_\sigma e_\sigma = \sum_\sigma \prod_{i=1}^{n} a_{i,\sigma(i)}b_{\sigma(i),i} = 1$. If $\sigma \neq \tau$ then there exist $1 \leq i \neq j \leq n$ such that $\sigma(i) = \tau(j)$ and so $e_\sigma e_\tau = 0$ since it has a factor $a_{i,\sigma(i)}b_{\sigma(i),j} = 0$. $\square$

If n is a positive integer then $\mathcal{S}_n$ denotes the symmetric group on $\{1, \ldots, n\}$ and $\mathcal{A}_n$ denotes the alternating group on $\{1, \ldots, n\}$. If R is a semiring, if n is a positive integer and if $A \in \mathcal{M}_n(R)$, we define the **positive determinant** $|A|^+$ and the **negative determinant** $|A|^-$ of A as follows:

$$|A|^+ = \sum\{a_{1,\sigma(1)} \cdot \ldots \cdot a_{n,\sigma(n)} \mid \sigma \in \mathcal{A}_n\}$$

and

$$|A|^- = \sum\{a_{1,\sigma(1)} \cdot \ldots \cdot a_{n,\sigma(n)} \mid \sigma \in \mathcal{S}_n \setminus \mathcal{A}_n\}.$$

The pair $|A|^{\pm} = (|A|^-, |A|^+)$ is the **bideterminant** of A. If R is a nonzeroic semiring with ring of differences R^Δ and if $\nu: R \to R^\Delta$ is the canonical morphism, then the **determinant** of A is defined to be $|A| = \nu(|A|^+)\nu(|A|^-)$. For an application of bideterminants in graph theory, see [Kuntzmann 1972].

Let $A \in \mathcal{M}_n(R)$. If the matrix $B \in \mathcal{M}_n(R)$ is formed by multiplying all of the entries of one row or one column of A by an element r of R, then $|B|^{\pm} = r|A|^{\pm}$. In particular, if one column or one row of A consists entirely of 0's then $|A|^{\pm} = (0,0)$. If one column [resp. row] of A is a linear combination of the other columns [resp. rows] of A then $|A|^+ = |A|^-$. The converse of this statement is false, as the following example shows.

(19.5) EXAMPLE. [Gondran & Minoux, 1984b] Let R be the semiring $(\mathbb{R}^+ \cup \{\infty\}, max, min)$ and let A be the matrix

$$\begin{bmatrix} 12 & 14 & 3 \\ 11 & 15 & 7 \\ 6 & 8 & 10 \end{bmatrix}.$$

Then $|A|^+ = |A|^- = 10$ but no column [resp. row] is a linear combination of the other columns [resp. rows].

If $\gamma: R \to S$ is a morphism of semirings then for each positive integer n we have an induced morphism of semirings $\gamma_n: \mathcal{M}_n(R) \to \mathcal{M}_n(S)$ defined by $\gamma_n: [r_{ij}] \mapsto [\gamma(r_{ij})]$. As an immediate consequence of the definitions, we see that for each $A \in \mathcal{M}_n(R)$ we have $\gamma(|A|^+) = |\gamma_n(A)|^+$ and similarly $\gamma(|A|^-) = |\gamma_n(A)|^-$.

(19.6) PROPOSITION. *If n is a positive integer and if R is a commutative semiring then for $A = [a_{ij}]$ and $B = [b_{ij}]$ in $\mathcal{M}_n(R)$ we have:*

(1) *There exists an element r of R satisfying $|AB|^+ = |A|^+|B|^+ + |A|^-|B|^- + r$ and $|AB|^- = |A|^+|B|^- + |A|^-|B|^+ + r$.*

(2) $|AB|^+ + |A|^+|B|^- + |A|^-|B|^+ = |AB|^- + |A|^+|B|^+ + |A|^-|B|^-$.

PROOF. (1) By Proposition 9.42, we know that there exists a cancellative semiring S and a surjective morphism $\gamma: S \to R$. By the above remarks, it therefore suffices to assume that R is cancellative and thus is a subsemiring of R^Δ. Since R is commutative, so is R^Δ.

By definition, $|AB|^+ = \sum\{\prod_{i=1}^n a_{i1}b_{i,\sigma(1)} + \cdots + a_{in}b_{n,\sigma(n)} \mid \sigma \in \mathcal{A}_n\}$. If we expand this product of a sum of monomials, we obtain $n!$ terms of the form $\prod_{i=1}^n a_{i,\tau(i)}b_{\tau(i),\sigma(i)}$, for $\tau \in \mathcal{S}_n$, as well as various other terms, the sum of which we will denote by r'. For fixed $\sigma \in \mathcal{A}_n$, let $\tau' = \sigma\tau^{-1}$. Then $\tau' \in \mathcal{A}_n$ if and only if $\tau \in \mathcal{A}_n$. Moreover, we have

$$\prod_{i=1}^n a_{i,\tau(i)}b_{\tau(i),\sigma(i)} = \left(\prod_{i=1}^n a_{i,\tau(i)}\right)\left(\prod_{i=1}^n b_{i,\tau'(i)}\right).$$

Thus

$$\begin{aligned}|AB|^+ &= \sum\left\{\left(\prod_{i=1}^n a_{i,\tau(i)}\right)\left(\prod_{i=1}^n b_{i,\tau'(i)}\right) \,\middle|\, \sigma \in \mathcal{A}_n, \tau \in \mathcal{S}_n; \tau' = \sigma\tau^{-1}\right\} + r' \\ &= \sum\left\{\left(\prod_{i=1}^n a_{i,\tau(i)}\right)\left(\prod_{i=1}^n b_{i,\rho(i)}\right) \,\middle|\, \tau, \rho \in \mathcal{A}_n\right\} \\ &\quad + \sum\left\{\left(\prod_{i=1}^n a_{i,\tau(i)}\right)\left(\prod_{i=1}^n b_{i,\rho(i)}\right) \,\middle|\, \tau, \rho \in \mathcal{S}_n \setminus \mathcal{A}_n\right\} + r' \\ &= |A|^+|B|^+ + |A|^-|B|^- + r'.\end{aligned}$$

A similar computation shows that there exists an element r'' of R such that $|AB|^- = |A|^+|B|^- + |A|^-|B|^+ + r''$. We are left to show that $r' = r''$. But R is a subsemiring of the commutative ring R^Δ and there we have

$$\begin{aligned}&[|A|^+ - |A|^-][|B|^+ - |B|^-] \\ &\quad = |A||B| = |AB| = |AB|^+ - |AB|^- \\ &\quad = [|A|^+|B|^+ + |A|^-|B|^- + r'] - [|A|^+|B|^- + |A|^-|B|^+ + r''] \\ &\quad = [|A|^+ - |A|^-][|B|^+ - |B|^-] + (r' - r''),\end{aligned}$$

implying that $r' = r''$.

(2) This is an immediate consequence of (1). □

If R is a semiring, if n is an integer greater than 1, and if $A \in \mathcal{M}_n(R)$ then for each $1 \leq i, j \leq n$ we can define the (i, j)-**positive minor** $pm_{i,j}(A)$ of A to be $|A'|^+$, where A' is the matrix in $\mathcal{M}_{n-1}(R)$ obtained by deleting the ith row and

jth column of A. Similarly, we define the **(i,j)-negative minor** $nm_{ij}(A)$ of A to be $|A'|^-$. We then define the **positive comatrix** A^+ of A to be the matrix $[b_{ij}]$ defined by the condition that $b_{ij} = pm_{ij}(A)$ when $i+j$ is even and $b_{ij} = nm_{ij}(A)$ when $i+j$ is odd. Similarly, we define the **negative comatrix** A^- of A to be the matrix $[c_{ij}]$ defined by the condition that $c_{ij} = nm_{ij}(A)$ when $i+j$ is even and $c_{ij} = pm_{ij}(A)$ when $i+j$ is odd.

(19.7) PROPOSITION. *If n is an integer greater than 1, if R is a commutative semiring, and if $A = [a_{ij}] \in \mathcal{M}_n(R)$ then for each $1 \leq j \leq n$ we have:*

(1) $|A|^+ = a_{1j}pm_{ij}(A) + a_{2j}nm_{2j}(A) + a_{3j}pm_{3j}(A) + \ldots$ *and*
$|A|^- = a_{1j}nm_{1j}(A) + a_{2j}pm_{2j}(A) + a_{3j}nm_{3j}(A) + \ldots$ *whenever j is odd;*

(2) $|A|^+ = a_{1j}nm_{1j}(A) + a_{2j}pm_{2j}(A) + a_{3j}nm_{3j}(A) + \ldots$ *and*
$|A|^- = a_{1j}pm_{1j}(A) + a_{2j}nm_{2j}(A) + a_{3j}pm_{3j}(A) + \ldots$ *whenever j is even.*

PROOF. Let S be the cancellative semiring $\mathbb{N}[\{x_{ij} \mid 1 \leq i,j \leq n\}]$ in n^2 commuting indeterminates x_{ij}. Then there exists a canonical morphism of semirings $\gamma: \mathcal{M}_n(S) \to \mathcal{M}_n(R)$ which takes the matrix $X = [x_{ij}]$ to A. Since, by the above remarks, $|A|^+ = \gamma(|X|^+)$ and $|A|^- = \gamma(|X|^-)$, it suffices to prove then proposition for the case $R = S$ and $A = X$. We see that R is a subsemiring of the ring $R^\Delta = \mathbb{Z}[x_{ij}]$. There, we have

$$|A|^+ - |A|^- = |A| = \sum_{i=1}^{n}(-1)^{i+j}x_{ij}[pm_{ij}(A) - nm_{ij}(A)].$$

If we assume that j is odd, we obtain the equation

$$|A|^+|A|^- = [x_{1j}pm_{ij}(A) + x_{2j}nm_{2j}(A) + \ldots] - [x_{1j}nm_{ij}(A) + x_{2j}pm_{2j}(A)\ldots].$$

Note that the left-hand side of this equation is the sum of n monomials in the x_{ij}. There is no cancellation of terms in $|A|^+$ and $|A|^-$ since these are all distinct. Since each of the two sums on the right-hand side of the equation is the sum of $n!/2$ monomials, there can be no cancellation here either. Thus we can identity the positive and negative parts of the two sides of the equation, and (1) follows. The proof of (2) is analogous. □

(19.8) PROPOSITION. *Let $n > 1$ be an integer, let R be a commutative semiring, and let $A = [a_{ij}] \in \mathcal{M}_n(R)$. Then*

(1) $A^+A = [c_{ij}]$, *where*

$$c_{ij} = \begin{cases} |A|^+ & \text{if } i = j \\ |C_{ij}|^+ & \text{if } i \neq j \end{cases},$$

where C_{ij} is the matrix obtained from A by replacing the ith column of A by its jth column.

(2) $A^-A = [d_{ij}]$ *where*

$$d_{ij} = \begin{cases} |A|^- & \text{if } i = j \\ c_{ij} & \text{if } i \neq j \end{cases}.$$

PROOF. (1) If i is odd then

$$c_{ij} = pm_{1i}(A)a_{1j} + nm_{2i}(A)a_{2j} + pm_{3i}(A)a_{3j} + \dots .$$

If $i = j$, this is just $|A|^+$, by Proposition 19.7. If $i \neq j$ then, by Proposition 19.7, it equals $|C_{ij}|^+$. (Since C_{ij} has two equal columns, this is the same as $|C_{ij}|^-$.) The reasoning is analogous of i is even.

(2) The proof of this is similar. □

(19.9) PROPOSITION. *Let $n > 1$ be an integer, let R be a commutative semiring, and let $A = [a_{ij}]$ and $B = [b_{ij}]$ be elements of $\mathcal{M}_n(R)$ satisfying $AB = I$, where I is the multiplicative identity of $\mathcal{M}_n(R)$. Let $C = [c_{ij}]$ be the matrix defined by the condition that $c_{ij} = 0$ if $i = j$ while $c_{ij} = |C_{ij}|^+$ otherwise, where C_{ij} is the matrix obtained from A by replacing the ith column by the jth column. Then the matrices $|B|^+$ and $|B|^-C$ have additive inverses in $\mathcal{M}_n(R)$.*

PROOF. Since $AB = I$, we have $\sum_{k=1}^{n} a_{ik}b_{kj} = 0$ for all $i \neq j$. Therefore $a_{ik}b_{kj}$ has an additive inverse in R for all $1 \leq i, j, k \leq n$ with $i \neq j$.

If $i = j$, then the (i, j)-entry in $|B|^+$ is 0, which certainly has an additive inverse. If $i \neq j$ then the (i, j)-entry is

$$\begin{aligned} |B|^+ c_{ij} &= \left[\sum_{\sigma \in \mathcal{A}_n} b_{1,\sigma(1)} \cdot \ldots \cdot b_{n,\sigma(n)}\right] \left[\sum_{\tau \in \mathcal{A}_n} u_{\tau(1),1} \cdot \ldots \cdot u_{\tau(n),n}\right] \\ &= \sum \left\{ \prod_{k=1}^{n} b_{k,\sigma(k)} u_{\tau(k),k} \;\middle|\; \sigma, \tau \in \mathcal{A}_n \right\}, \end{aligned}$$

where

$$u_{km} = \begin{cases} a_{km} & \text{if } m \neq i \\ a_{kj} & \text{if } m = i \end{cases}$$

for all $1 \leq k \leq n$. For each of the permutations σ and τ in $\mathcal{A}_n$, either $\sigma(j) \neq \tau(j)$ or $\sigma(j) \neq \tau(i)$. In either case, each term in the above sum contains a factor of the form $a_{rs}b_{st}$, with $r \neq t$. Thus each term has an additive inverse in R and so $|B|^+ c_{ij}$ has an additive inverse in R. Thus $|B|^+$ has an additive inverse in $\mathcal{M}_n(R)$. The proof that the same is true for $|B|^-C$ is similar. □

(19.10) PROPOSITION. *If n is a positive integer, if R is a commutative semiring, and if $A = [a_{ij}]$ and $B = [b_{ij}]$ are elements of $\mathcal{M}_n(R)$ satisfying $AB = I$, where I is the multiplicative identity of $\mathcal{M}_n(R)$, then $BA = I$.*

PROOF. If $n = 1$ the result follows directly from the commutativity of R. Hence we can assume that $n > 1$. Let C be the matrix obtained from A as in the statement of Proposition 19.9. Then by Proposition 19.8 we see that $AB = I$ implies that $A^+A = A^+(AB)A = |A|^+BA + CBA$ and $A^-A = A^-(AB)A = |A|^-BA + CBA$. If we multiply the first of these equations by $|B|^+$ and the second by $|B|^-$ and then add, we obtain $|B|^+A^+A + |B|^-A^-A = [|A|^+|B|^+ + |A|^-|B|^-]BA + |B|^+CBA + |B|^-CBA$.

We now apply Proposition 19.8 again and obtain $|A|^+|B|^+ + |A|^-|B|^- + |B|^+ + |B|^-C = [|A|^+|B|^+ + |A|^-|B|^-]BA + |B|^+CBA + |B|^-CBA$. Since $AB = I$, we know by Proposition 19.6(2) that

$$1 + |A|^+|B|^- + |A|^-|B|^+ = |A|^+|B|^+ + |A|^-|B|^-.$$

We apply this and Proposition 19.8 to the above equation, obtaining

$$\begin{aligned}
1 + |A|^+|B|^- &+ |A|^-|B|^+ + |B|^+ + |B|^-C \\
&= BA + [|A|^+|B|^- + |A|^-|B|^+]BA + |B|^+CBA + |B|^-CBA \\
&= BA + |B|^+[|A|^-I + C]BA + |B|^-[|A|^+I + C]BA \\
&= BA + |B|^+A^-ABA + |B|^-A^+ABA \\
&= BA + |B|^+A^-A + |B|^-A^+A \\
&= BA + |B|^+[|A|^-I + C] + |B|^-[|A|^+I + C] \\
&= BA + |A|^+|B|^- + |A|^-|B|^+ + |B|^+ + |B|^-C.
\end{aligned}$$

By Proposition 19.9, $|B|^+$ and $|B|^-C$ have additive inverses in $\mathcal{M}_n(R)$. From Proposition 19.6(1) and the fact that $|AB|^- = |I|^- = 0$, we conclude that $r = |A|^+|B|^- + |A|^-|B|^+$ has an additive inverse in R and so R has an additive inverse in $\mathcal{M}_n(R)$. Hence we can conclude that $I = BA$, as desired. □

If R is a semiring, n a positive integer, and $A \in \mathcal{M}_n(R)$ then $perm(A) = |A|^+ + |A|^-$ is the **permanent** of A. This is an element of R and so we do not need to avail ourselves of the map $\nu: R \to R^{\Delta}$. For the theory of permanents over rings, see [Minc, 1978]. Permanents of matrices in $\mathcal{M}_n(\mathbb{B})$ play an important role in the analysis of switching circuits and have been extensively studied. Permanents of matrices in $\mathcal{M}_n(R)$, where R is a bounded totally-ordered set on which addition is *max* asnd multiplication is *min*, are studied in [Cechlárová & Plávka, 1996]. For the special case of permanents of matrices in $\mathcal{M}_n(\mathbb{I})$, see [J. B. Kim, 1984] and [Kim, Baartmans & Sahadin, 1989]. For permanents of matrices over the schedule algebra, see [Olsder & Roos, 1988]. For permanents of matrices of bounded distributive lattices, see [Zhang, 1994].

(19.11) EXAMPLE. Permanents do not have the nice properties of determinants. For example, $perm(AB)$ is not necessarily equal to $perm(A)perm(B)$. Thus, for example, in $\mathcal{M}_2(\mathbb{I})$, we note that if $A = \begin{bmatrix} 0.14 & 0.25 \\ 0.12 & 0.14 \end{bmatrix}$ and $B = \begin{bmatrix} 0.5 & 0.3 \\ 0.2 & 0.16 \end{bmatrix}$ then $perm(A)perm(B) = 0.14 \neq 0.16 = perm(AB)$. See [Kim, Baartmans & Sahadin, 1989].

(19.12) EXAMPLE. The fact that $perm(A) \neq 0$ does not necessarily imply that A is invertible or even multiplicatively cancellable. For example, if R is the schedule algebra $(\mathbb{R} \cup \{-\infty\}, max, +)$ and if $A \in \mathcal{M}_3(R)$ is the matrix

$$\begin{bmatrix} 1 & 3 & 5 \\ 2 & 0 & 1 \\ 0 & 4 & 2 \end{bmatrix}$$

then $perm(A) \neq -\infty$ but

$$A\begin{bmatrix} 0 & -\infty & -\infty \\ 2 & 0 & -\infty \\ 3 & -\infty & 0 \end{bmatrix} = A\begin{bmatrix} 2 & -\infty & -\infty \\ 2 & 0 & -\infty \\ 3 & -\infty & 0 \end{bmatrix}.$$

If A is a matrix in $\mathcal{M}_{m\times n}(R)$ and if $h \in \mathbb{N}$ then we define $r_h(A)$ by setting $r_0(A) = 1$ and, for $h > 0$, setting $r_h(A)$ as the sum of all submatrices of A in $\mathcal{M}_h(R)$. The polynomial $r_A(t) = \sum r_h(A)t^h$ in $R[t]$ is the **rook polynomial** of A. For the importance of this polynomial in the analysis of matrices over certain semirings, see [Beasley & Pullman, 1988a].

(19.13) EXAMPLE. If R is a simple semiring and $A = [a_{ij}] \in \mathcal{M}_n(R)$ is a matrix satisfying $a_{ii} = 1$ for all $1 \leq i \leq n$ then $perm(A) = 1$.

Let R and S be semirings and let M be an (R,S)-bisemimodule. If $s \in S$, then an element m of M is an R-**eigenelement** of s in M if and only if there exists an element r of R satisfying $rm = ms$. Such an element r is called an R-**eigenvalue** of s associated with the element m. Given an element r of R, we denote the (possibly empty!) set of all elements m of M for which R is an associated eigenvalue of s by $eig_M(r,s)$. If this set is nonempty, it is a subsemigroup of $(M,+)$. If M is also a right R'-semimodule for some semiring R', then $eig_M(r,s)$ is a right R'-semimodule of M.

In particular, if R is a semiring and $S = \mathcal{M}_n(R)$, then R^n is an (R,S)-bisemimodule and we will consider the eigenelements of matrices in S as being in R^n and the associated eigenvalues as being in R. If M is a left R-semimodule and α is an R-endomorphism of M then the eigenelements of α belong to M and the associated eigenvalues are in R.

(19.14) EXAMPLE. In the "classical" example, $M = R^n$ for some positive integer n and $S = \mathcal{M}_n(R)$. For the case of R being the semiring $(\mathbb{R}\cup\{\infty,-\infty\}, max, +)$, this situation is considered in [Cuninghame-Green, 1979], [Cuninghame-Green & Burkard, 1984] and [Cuninghame-Green & Huisman, 1982]. If $A = [a_{ij}] \in S$ and $r \in R$. then $m = (b_1,\ldots,b_n) \in eig_M(r,s)$ if and only if

$$r + b_i = max\{a_{ij} + b_j \mid 1 \leq j \leq n\}$$

for each $1 \leq i \leq n$. In the case of R being the semiring $(\mathbb{R}^+, max, \cdot)$ or the semiring $(\mathbb{R}^+ \cup \{\infty\}, min, \cdot)$, this situation is considered in [Vorobjev, 1963]. Refer also to [Gondran & Minoux, 1978], especially for the case of R a division semiring. For the case of R being the schedule algebra $(\mathbb{R}\cup\{-\infty\}, max, +)$, refer to [Cochet-Terrason et al., 1998] for specific computational algorithms.

For the general relation between eigenvalue problems and problems in graph theory and combinatorics, see [Zimmermann, 1981].

(19.15) EXAMPLE. [Gaubert, 1996a; Olsder, 1992] Consider the schedule algebra $R = (\mathbb{R}\cup\{-\infty\}, max, +)$. A matrix $A = [a_{ij}] \in \mathcal{M}_n(R)$ is **irreducible** if and only if no permutation matrix P exists such that $P^{-1}AP$ has upper-triangular block structure. Any such matrix admits a unique eigenvalue $\rho(A) = \sum_{k=1}^{n} tr(A^k)^{1/k}$, where $tr(B)$ denotes the trace of a matrix M. Moreover, $\rho(A) \leq \sum_{i=1}^{n} \sum_{j=1}^{n} a_{ij}$, with equality in the case A is symmetric. Furthermore, if $c \in R$ then there exists an element $u \in R^n$ satisfying $Au^T + cu^T = cu^T$ if and only if $\rho(A) \leq c$ and there exists an element $u \in R^n$ satisfying $Au^T + cu^T = Au^T$ if and only if $c \leq \rho(A)$.

If the matrix A is not irreducible, it may have more than one eigenvalue. Thus, for example, the matrix $\begin{bmatrix} 1 & -\infty \\ -\infty & 2 \end{bmatrix}$ has eigenvalues 1 and 2, associated respectively with eigenelements $[0, -\infty]^T$ and $[-\infty, 0]^T$. On the other hand, it is straightforward to check that the matix $\begin{bmatrix} 1 & 0 \\ -\infty & 0 \end{bmatrix}$ has a unique eigenvalue, 1, even though it is not irreducible. See [Baccelli et al., 1992] for details. Also refer to [Mairesse, 1997].

The matrix $A = \begin{bmatrix} 0 & -\infty \\ 2 & 2 \end{bmatrix} \in \mathcal{M}_2(R)$ has two distinct eigenvalues, 0 and 2, corresponding to eigenelements $[0, -\infty]^T$ and $[0, 0]^T$. Nonetheless, A is not a unit of $\mathcal{M}_2(R)$. See [Wagneur, 1991].

We can consider a matrix $A = [a_{ij}] \in \mathcal{M}_n(R)$ as representing a weighted directed graph with nodes $\{1, \ldots, n\}$, where a_{ij} is the weight on the arc $i \to j$ (if this weight is $-\infty$ then the arc doesn't exist). The condition that the matrix A be irreducible corresponds to the condition that this graph be strongly connected. In that case, the unique eigenvalue is equal to the maximum cycle mean of the graph.

For the importance of eigenvalue problems over the schedule algebra in graph theory and the theory of discrete-event dynamical systems, refer also to [Baccelli et al., 1992], [Braker & Olsder, 1993], and [Braker & Resing, 1993].

(19.15h) APPLICATION. [Gaubert & Max Plus, 1997] In statistical physics one considers the asymptotics, as $h \to 0^+$, of the spectrum of $n \times n$ matrices with nonnegative real entries of the form $A_h = [exp(h^{-1}a_{ij})]$, where $a_{ij} \in \mathbb{R}$ for $1 \leq i, j \leq n$. The **Perron eigenvalue** $\rho(B)$ of a matrix B with nonnegative entries is the maximal eigenvalue associated with a nonnegative eigenvector, which is equal to the spectral radius of B. Then one can show that $lim_{h \to 0^+} log\ \rho(A_h)$ is just the maximal eigenvalue of $A = [a_{ij}]$ over the schedule algebra. For related results, see [Akian, Bapat & Gaubert, 1998].

(19.15m) EXAMPLE. [Lesin & Sambourskiĭ, 1992] Let S be a closed subsemiring of the schedule algebra $R = (\mathbb{R} \cup \{-\infty\}, max, +)$ and let $\varphi: \mathbb{R} \to \mathbb{R}$ be a concave differentiable function satisfying the condition that $lim_{r \to \infty} \varphi(r) = 0$. (For example, we can take $\varphi: x \mapsto e^x$.) Then φ defines a metric on R given by

$$d(a, b) = |\varphi(a) - \varphi(b)|,$$

which in turn restricts to a metric on S. Let X be a totally-bounded metric space and let $C(X, S)$ be the S-semimodule of all continuous bounded functions from X to S. Each S-valued relation $\theta: X \times X \to S$ on X defines a function $\alpha_\theta: C(X, S) \to C(X, S)$ given by

$$(\alpha_\theta)(f): x \mapsto \bigvee_{y \in X} \theta(x, y) + f(y).$$

If θ is a bounded function uniformly continuous in the first argument and equicontinuous in the second, then α_θ is an S-endomorphism of $C(X,S)$. Moreover, there exists a nonzero subsemimodule N of $C(X,S)$ and an element s of S such that s is an eigenvalue of α_θ and $N \subseteq eig_{C(X,S)}(s, \theta_\alpha)$. This result has important applications in dynamic programming.

It is natural at this stage to try to define the notion of the characteristic polynomial of a matrix over a commutative semiring. If t is an indeterminate over $\mathcal{M}_n(R)$ and if I is the multiplicative identity of $\mathcal{M}_n(R)$ then we can consider the polynomial $perm(A+tI) \in R[t]$. Such polynomials over the schedule algebra are studied in [Cuninghame-Green, 1983] in connection with their application to problems in optimization theory.

Another approach is given in [Straubing, 1983a]. If R is a semiring, n is a positive integer, and t is an indeterminate over $\mathcal{M}_n(R)$, then for each matrix A in $\mathcal{M}_n(R)$ we can construct the matrix $A^*(t)$ in $\mathcal{M}_{2n}(R[t])$ defined by $A^*(t) = \begin{bmatrix} A & tI \\ I & I \end{bmatrix}$, where I is the multiplicative identity of $\mathcal{M}_n(R)$. Note that an element R of R is an R-eigenvalue of A associated with some element m of R^n if and only if the columns of this matrix are linearly dependent over R. The **positive characteristic polynomial** $p^+(t)$ of A is now defined to be $p_A^+(t) = |A^*(t)|^+$ and the **negative characteristic polynomial** of A is defined to be $p_A^-(t) = |A^*(t)|^-$.

If R is a commutative semiring, if A is a matrix in $\mathcal{M}_n(R)$ and if $g \in R[t]$ is a polynomial over R in an indeterminate t then we can define $g(A)$ to be $\sum_{i\in\mathbb{N}} g(i)A^i$. This is well-defined since $\mathcal{M}_n(R)$ is a left R-semimodule and since only finitely-many of the values $g(i)$ are nonzero.

(19.16) Proposition. **(Cayley-Hamilton Theorem)** *If R is a commutative semiring and if $A \in \mathcal{M}_n(R)$ for some positive integer n then $p_A^+(A) = p_A^-(A)$.*

Proof. Let $A = [a_{ij}]$. If $X = \{t_{11}, \ldots, t_{nn}\}$ is a set of n^2 distinct elements then we have a function $\varphi: X \to R$ defined by $\varphi: t_{ij} \mapsto a_{ij}$. As in Example 9.19, this function induces a φ-evaluation morphism $\gamma = \epsilon_\varphi: \mathbb{N}\langle X\rangle \to R$ which, in turn, induces a morphism $\gamma_n: \mathcal{M}_n(\mathbb{N}\langle X\rangle) \to \mathcal{M}_n(R)$ having the property that $\gamma_n(T) = A$, where T is the matrix $[t_{ij}]$. By the remarks before Proposition 19.6, it therefore suffices to prove the result for the case of $R = \mathbb{N}\langle X\rangle$. But $\mathbb{N}\langle X\rangle$ is a commutative cancellative semiring and so can be embedded in a commutative ring of differences. Here the result follows by the usual Cayley-Hamilton Theorem for commutative rings. □

(19.17) Example. [Baccelli et al., 1992] Let $R = (\mathbb{R} \cup \{-\infty\}, max, +)$. If

$$A = \begin{bmatrix} 1 & 2 & 3 \\ 4 & 1 & -\infty \\ 0 & 5 & 3 \end{bmatrix}$$

then $p_A^+(t) = max\{3t, 4+t, 9\}$ and $p_A^-(t) = max\{3+2t, 6+t, 12\}$ and, indeed,

$$p_A^+(A) = p_A^-(A) = \begin{bmatrix} 12 & 11 & 9 \\ 10 & 12 & 10 \\ 12 & 11 & 12 \end{bmatrix}.$$

20. PARTIALLY-ORDERED SEMIRINGS

Many of the semirings originally studied, such as $\mathbb{N}$ and $ideal(R)$, have a partial-order structure in addition to their algebraic structure and, indeed, the most interesting theorems concerning them make use of the interplay between these two structures. In is therefore natural for us to study semirings, and semimodules over them, on which a partial order is defined. A hemiring $(R, +, \cdot)$ is **partially-ordered** if and only if there exists a partial order relation $\leq$ on R satisfying the following conditions for elements r, r', and r'' of R:

(1) If $r \leq r'$ then $r + r'' \leq r' + r''$;
(2) If $r \leq r'$ and $r'' \geq 0$ then $rr'' \leq r'r''$ and $r''r \leq r''r'$.

If the relation $\leq$ is in fact a total order, then R is **totally-ordered**.

An element a of a partially-ordered hemiring R is **positive** if and only if $a \geq 0$. Clearly, this condition is equivalent to the condition that $a+r \geq r$ for all $r \in R$. The set R^+ of all positive elements of R, called the **positive cone** of R, is nonempty since $0 \in R^+$. It is also easy to verify that this set is closed under finite sums and products. Thus it is a subhemiring of R, which is a subsemiring if R is a semiring satisfying $1 > 0$. The partially-ordered hemiring R is **positive** when $R^+ = R$. The nonzero elements of R^+ are said to be **strictly positive**.

(20.1) EXAMPLE. [Acharyya, Chattopadhyay & Ray, 1993] The positive cone of a partially-ordered semiring R does not determine R, even if R is a ring. Indeed, let define a partial order on $\mathbb{R}$ by setting $a \leq b$ if and only if $b - a \in \mathbb{N}$. Then $\mathbb{R}^+ = \mathbb{N}$. But $\mathbb{Z}^+$ is also $\mathbb{N}$, where $\mathbb{Z}$ is ordered with the usual order.

Note that if a is a positive infinite element of a semiring R then $a = a + r \geq r$ so a is the unique maximal element of R^+. In particular, if R is a positive simple semiring then $1 \geq r \geq 0$ for all $r \in R$.

(20.2) EXAMPLE. Any frame $(R, \vee, \wedge)$ is a partially-ordered semiring.

(20.3) EXAMPLE. The semiring $\mathbb{N}$ with the usual order is a totally-ordered semiring. Similarly, the schedule algebra $(\mathbb{R} \cup \{-\infty\}, max, +)$ with the usual order is a totally-ordered semiring, as is $(\mathbb{R}^+ \cup \{\infty\}, min, +)$, with the reverse of the usual order. If $R = (\mathbb{R} \cup \{-\infty\}, max, +)$ then we can define the structure of a totally-ordered semiring on R^2 by setting

$$(1)\ (a,b)+(c,d)=\begin{cases}(a,b) & \text{if } a>c \text{ or } a=c \text{ and } b\geq d\\ (c,d) & \text{otherwise}\end{cases}.$$

$$(2)\ (a,b)\cdot(c,d)=(ac,bd).$$

for all $(a,b),(c,d) \in R^2$. The order on this semiring is just the **lexicographic order**. Obviously, this construction can be iterated.

Wagneur [1991] has considered the problems of linear independence and bases for semimodules over certain classes of totally-ordered semirings.

We know that if R is a hemiring then R can be embedded in its Dorroh extension $S = R \times \mathbb{N}$, which is a semiring. If R is partially-ordered by a relation $\leq$, then we can define a relation $\leq'$ on S by setting $(a,n) \leq' (a',n')$ if and only if $n \leq n'$ and $a \leq a'$. This relation restricts to $\leq$ on the image $R \times \{0\}$ of R in S and it is straightforward to check that S is a partially-ordered semiring under $\leq'$. Georg Karner, in a private communication, has pointed out that the lexicographic order on S defined by setting $(a,n) \leq' (a'n')$ if and only if $n < n'$ or $n = n'$ and $a \leq a'$ does not work. Indeed, if $R = 2\mathbb{N}$ then we would have $(10,1) <' (1,2)$ and $(0,0) <' (2,0)$ but $(10,1)\cdot(2,0) = (22,0) \not\leq' (6,0) = (1,2)\cdot(2,0)$.

(20.4) EXAMPLE. [Janowitz, 1976] A ring R having no nonzero nilpotent elements is a **Rickart ring** [Maeda, 1960] if and only if for each $a \in R$ there exists a (necessarily unique) idempotent e_a of R satisfying $ab = 0$ if and only $b = e_a b$. Thus, for example, any integral domain R is a Rickart ring with $e_a = 0$ for $a \neq 0$ and $e_0 = 1$. On a Rickart ring there is a natural partial order defined by $a \leq b$ if and only if $ab = a^2$ (or, equivalently, if and only if $ba = a^2$). Note that $0 \leq a$ for each $a \in R$. If we set $a \wedge b = e_{a-b}a$ for all $a, b \in R$ then $(R, \wedge)$ is a meet semilattice in which multiplication distributes over meets from either side. Moreover, $a \leq b$ if and only if $a = eb$ for some idempotent e of R.

Let R be a Rickart ring and let ∞ be an element not in R. Set $S = R \cup \{\infty\}$ and extend the definitions of $\cdot$ and $\wedge$ to S by setting $s \cdot \infty = \infty \cdot s = \infty$ and $s \wedge \infty = \infty \wedge s = s$ for each $s \in S$. Then $(S, \wedge, \cdot)$ is a partially-ordered semiring in which $0_S = \infty$ and $1_S = 1_R$. Moreover, $S^+ = \{\infty\}$.

(20.5) EXAMPLE. If A is a nonempty set and $R = sub(A)^A$ is the semiring defined in Example 6.3, then we can define a partial order on R by setting $f \leq g$ if and only if $f(a) \subseteq g(a)$ for all $a \in A$. This partial order turns R into a positive partially-ordered semiring.

(20.6) EXAMPLE. A subsemiring of a partially-ordered semiring is itself partially-ordered under the induced partial order. However, a given semiring may be a subsemiring of many other semirings, and inherit different partial orders from each of them. Consider the following example [Karner, 1992]: Let $R = (\mathbb{R} \cup \{\infty\}, +, \cdot)$. This semiring is surely totally-ordered under the restriction of the usual relation $\leq$. For each $1 \leq a \in R$, the subset $R_a = \mathbb{N} \cup \{r \in R \mid r \geq a\}$ of R is in fact a subsemiring and it has a partial order $\sqsubseteq_a$ defined by $r \sqsubseteq_a r'$ if and only if there exists an $s \in R_a$ satisfying $r + s = r'$. If $a \leq b$, then R_b is a subsemiring of R_a

and the restriction of $\sqsubseteq_a$ to R_b is also a partial order on R_b, which is different from $\sqsubseteq_b$. Therefore, for each $a > 1$, there are uncountably-many different partial orders definable on R_a.

(20.7) EXAMPLE. If R is a cancellative partially-ordered semiring having ring of differences R^Δ and if $\nu: R \to R^\Delta$ is the canonical morphism, then the partial order on R can be extended to a partial order on R^Δ by setting $\nu(a)\nu(b) \leq \nu(c)\nu(d)$ if and only if $a + d \leq b + c$. Indeed, this is the only possible way of extending the partial order on R to one on R^Δ.

(20.8) EXAMPLE. If $\{R_i \mid i \in \Omega\}$ is a family of partially-ordered semirings then the product semiring $R = \times_{i\in\Omega} R_i$ is also partially-ordered when we define $\langle a_i \rangle \leq \langle b_i \rangle$ if and only if $a_i \leq b_i$ for all $i \in \Omega$. Indeed, R is just the product of the R_i in the category of all partially-ordered semirings. Conversely, if $R = \times_{i\in\Omega} R_i$ is partially-ordered then each R_i is partially-ordered by restriction.

If each R_i is positive then so is R and conversely. In particular, a semiring R is partially-ordered if and only if R^A is partially-ordered for some (and hence every) nonzero set A, and it is positive if and only if some (and hence every) R^A is positive. Thus, for example, since $\mathbb{N}\{\infty\}$ is a partially-ordered semiring, so is the semiring of multisets on any nonempty set A. Similarly, R is partially-ordered if and only if $\mathcal{M}_n(R)$ is partially-ordered for some (and hence every) natural number n.

(20.9) EXAMPLE. [Wechler, 1977] Let $\{R_i \mid i \in \Omega\}$ be a family of disjoint positive partially-ordered hemirings and, for each $i \in \Omega$, let 0_i be the additive identity of R_i. Denote the operations on R_i by $+_i$ and $\cdot_i$ and the order relation on R_i by $\leq_i$. Let u, z be elements not in any of the R_i and let $S = \cup\{[R_i \backslash \{0_i\}] \mid i \in \Omega\}$. Define addition, multiplication, and order on $R = S \cup \{u, z\}$ as follows:

(1) If $a, b \neq z$ then $a + b$ equals $a +_i b$ if a and b both belong to R_i, and equals u otherwise;
(2) If $a, b \neq z$ then $a \cdot b$ equals $a \cdot_i b$ if a and b both belong to R_i, and equals u otherwise;
(3) If $a, b \in S$ then $a \leq b$ if and only if $a \leq_i b$ for some i;
(4) $z + a = a + z = a$ and $z \cdot a = a \cdot z = z$ for all $a \in R$, while $z \leq a$ for all $a \in R$;
(5) $u + a = a + u = u$ and $u \cdot a = a \cdot u = u$ for all $z \neq a \in R$, while $a \leq z$ for all $a \in R$.

Then $(R, +, \cdot)$ is a hemiring and, in fact, is the coproduct of the R_i in the category of all hemirings.

A subsemiring S of a partially-ordered semiring R is **full** in R if and only if $a \leq b \in S$ implies that $a \in S$.

(20.10) EXAMPLE. [Wongseelashote, 1979] Let $R = \mathbb{N}\{\infty\}^A$ be the semiring of multisets of elements of a nonempty set A. Then the subsemirings $\{f \in R \mid \sum_{a\in A} f(a) \neq \infty\}$ and $\{f \in R \mid supp(f) \text{ is countable}\}$ of R are full.

(20.11) EXAMPLE. If n is a positive integer and R is a partially-ordered semiring then the matrix semiring $\mathcal{M}_n(R)$ can be partially ordered by $[a_{ij}] \leq [b_{ij}]$ if and only if $a_{ij} \leq b_{ij}$ for all $1 \leq i, j \leq n$. If $n \geq 2$, this order cannot be total. Similarly, if A is a nonempty set and R is a partially-ordered semiring then the semirings $R\langle\langle A\rangle\rangle$ and $R\langle A\rangle$ are partially-ordered by the order $f \leq g$ if and only if $f(w) \leq g(w)$ for all $w \in A^*$. If R is positive then so are $R\langle\langle A\rangle\rangle$ and $R\langle A\rangle$.

(20.12) EXAMPLE. [Bleicher & Bourne, 1965] Let $1 < c \in \mathbb{N}$ and define a new partial order $\preceq$ on $\mathbb{N}$ by setting $a \preceq b$ if and only if $a \leq b$ and $a \equiv b \pmod{c}$. Call this new semiring R. Then $R^+ = \{a \in \mathbb{N} \mid a \equiv 0 \pmod{c}\}$ so R is not positive.

(20.13) EXAMPLE. Let $\gamma: R \to S$ be a morphism of 3 semirings and assume that S is a partially-ordered semiring with respect to the relation $\leq$. Define a relation $\preceq$ on R by setting $r \preceq r'$ if and only if $\gamma(r) \leq \gamma(r')$. Then it is straightforward to verify that R is partially ordered by the relation $\preceq$.

(20.14) EXAMPLE. Let R be a semiring and let $\{\leq_i \mid i \in \Omega\}$ be a family of partial-order relations on R each of which turns R into a partially-ordered semiring. Then R is a partially-ordered semiring with respect to the relation $\leq$ defined by $r \leq r'$ if and only if $r \leq_i r'$ for all $i \in \Omega$.

(20.15) PROPOSITION. *If a is an element of a partially-ordered semiring R satisfying $a \leq b$ for all $b \in R$ then $a \in I^+(R)$.*

PROOF. By hypothesis we have $a \leq 0$ and so $a \leq a + a \leq a + 0 = a$. Thus $a = a + a$. □

(20.16) PROPOSITION. *Positive partially-ordered semirings are zerosumfree.*

PROOF. Let R be a positive partially-ordered semiring. If $a, b \in R$ then $b \geq 0$ so $a + b \geq a + 0 = a \geq 0$. Hence $a + b = 0$ implies that $0 \geq a \geq 0$ and thus $a = 0$. Similarly $b = 0$. □

An element a of a partially-ordered semiring R is **transitive** if and only if $a^2 \leq a$. Clearly 0 is transitive, as is every element a of R satisfying $0 \leq a \leq 1$. If R is a commutative semiring then the set of all transitive elements of R is closed under taking products. The transitive elements of semirings of the form $\mathcal{M}_n(R)$, where R is an additively idempotent semiring, are studied in [Hashimoto, 1985].

(20.17) PROPOSITION. *If R is a positive partially-ordered Gel'fand semiring then for each $a, b \in R$ there exist units $u, v \in U(R)$ such that $ab \leq au$ and $ab \leq vb$.*

PROOF. Since R is Gel'fand we know that if $a, b \in R$ then $u = 1+b$ and $v = 1+a$ are units of R. Moreover, since R is positive we have $a \leq a + 1$ and $b \leq b + 1$. The result then follows from the definition of a partially-ordered semiring. □

Note that if R is a simple semiring then $ab \leq a$ and $ba \leq a$ for all $a, b \in R$, by Proposition 4.3.

(20.18) PROPOSITION. *If $a_1, \ldots, a_n$ are elements of a positive simple semiring R and if $1 \leq h < k \leq n$ are indices such that $a_h a_k = 0$ then $a_1 \cdot \ldots \cdot a_n = 0$.*

PROOF. Let $b = a_1 \cdot \ldots \cdot a_h$ and let $c = a_{h+1} \cdot \ldots \cdot a_k$. Then, by Proposition 4.3, $0 \leq b \leq a_h$ and $0 \leq c \leq a_k$. This implies that $0 \leq bc \leq a_h a_k = 0$ and so $bc = 0$. Hence $a_1 \cdot \ldots \cdot a_n = bca_{k+1} \cdot \ldots \cdot a_n = 0$. □

(20.19) PROPOSITION. *If R is an additively-idempotent semiring then R is partially-ordered by the relation $a \leq b$ if and only if $a + b = b$. Under this relation, R is positive and, indeed, R is a join semilattice with $a \vee b = a + b$. Moreover, if $a, b \in U(R)$ then $a \geq b$ if and only if $a^{-1} \leq b^{-1}$.*

PROOF. That R is partially-ordered by the above-defined relation is an immediate consequence of the definition and of the additive idempotence of R. Clearly $0 \leq a$ for every element a of R, so R is positive. Moreover, since R is additively idempotent we have $a, b \leq a+b$ for all $a, b \in R$. Now let c be an element of R satisfying $a, b \leq c$. Then $a+c = c$ and $b+c = c$ so $(a+b)+c = a+(b+c) = a+c = c$ and hence $a+b \leq c$. Thus we have $a+b = a \vee b$. Finally, we note that if $a, b \in U(R)$ then $a \leq b \Leftrightarrow a+b = b \Leftrightarrow a^{-1} = a^{-1}bb^{-1} = a^{-1}(a+b)b^{-1} = b^{-1}+a^{-1} \Leftrightarrow b^{-1} \leq a^{-1}$. □

From now on, whenever we consider additively idempotent semirings we will assume that they are partially-ordered by the relation given in Proposition 20.19. In particular, this will be true of simple semirings.

(20.20) PROPOSITION. *If R is an multiplicatively-cancellative additively-idempotent commutative semiring and if $a_1, \ldots, a_n \in R$ then $a_1 \cdot \ldots \cdot a_n \leq \sum_{i=1}^{n} a_i^n$.*

PROOF. By the remark after Proposition 4.43 we see that

$$\sum_{i=1}^{n} a_i^n = \left(\sum_{i=1}^{n} a_i\right)^n = a_1 \cdot \ldots \cdot a_n + b$$

for some $b \in R$, and so the result follows. □

(20.21) EXAMPLE. [Shubin, 1992] Proposition 20.20 is not true if multiplicative cancellation is not assumed. For example, let $M = (\mathbb{N}^2, +)$ and let $R = sub(M)$, the operations on which are given in Example 1.10. Then R is additively-idempotent but not multiplicatively-cancellative. If $a = \{[0,0],[0,1]\}$ and $b = \{[0,0],[1,0]\}$ then $ab = \{[0,0],[1,0],[0,1],[1,1]\}$ while $a^2 + b^2 = \{[0,0],[1,0],[0,1]\}$.

(20.23) PROPOSITION. *If R is a simple ring then*

(1) *$a \leq b \lhd c \leq d$ in R implies that $a \lhd d$.*

Moreover, if R is positive then

(2) *$a \lhd b$ and $c \lhd d$ in R imply that $ac + ca \lhd bd$; and*
(3) *$a, c \lhd b$ implies $a + c \lhd b$.*

In particular, if R is positive and $a \in R$ then $I = \{r \in R \mid r \lhd a\}$ is an ideal of R.

PROOF. (1) By hypothesis, there exists an element e of R satisfying $be = 0 = eb$ and $e+c = 1$. Since $a \leq b$ we have $a+b = b$. By Proposition 4.3, we have $b = ab+b$ and so $a+b = ab+b$. Thus $ae = ae+be = abe+be = 0$. A similar argument shows that $ea = 0$. Since $c \leq d$ we have $c + d = d$. Hence $e + d = e + c + d = 1 + d = 1$ by simplicity. Therefore $a \lhd d$.

(2) By hypothesis there exist elements r and s of R satisfying $ar = ra = 0 = cs = sc$ and $r + b = 1 = s + d$. Set $e = bs + r$. Then $e(ac + ca) = bsac + bsca + rac + rca$ and this equals 0 by Proposition 20.18. Similarly, $(ac + ca)e = 0$. Moreover, $e + bd = b(d + s) + r = b + r = 1$. Therefore $ac + ca \lhd bd$.

(3) Since $a, c \lhd b$ we know there exist elements r and s of R satisfying $ar = ra = 0 = cs = sc$ and $r + b = 1 = s + b$. Set $d = rs$. Then, since R is positive, we have $0 \leq d(a + c) = rsa + rsc \leq ra + sc = 0$, implying that $d(a + c) = 0$. Similarly, $(a + c)d = 0$. Moreover, $d + b = d + 1b = d + (r + 1)b = rs + rb + b = r(s + b) + b = r1 + b = r + b = 1$. Therefore $a + c \lhd b$.

Finally, if $r \in I$ and $r' \in R$ then $r'r \leq r \lhd a$ and so, by (1), $r'r \lhd a$. Similarly $rr' \lhd a$. Thus $r'r, rr' \in I$. If $r, r' \lhd a$ then $r + r' \lhd a$ by (3). Thus $r + r' \in I$. □

A partially-ordered semiring R is [**uniquely**] **difference ordered** if and only if $a \leq b$ in R when and only when there exists an element [resp. a unique element] c of R such that $a + c = b$. Difference-ordered semirings are clearly positive and hence zerosumfree.

(20.23) PROPOSITION. *A difference-ordered semiring R is uniquely difference-ordered if and only if it is cancellative.*

PROOF. Cancellative difference-ordered semirings are surely uniquely difference-ordered. Conversely, assume that R is uniquely difference-ordered and that $a + b = a + c$ for elements a, b, and c of R. If this common value is d then we have $d \geq a$ and so, by uniqueness, we must have $b = c$. □

A difference-ordered semiring is totally ordered precisely when it is a yoked semiring. A semiring R is **extremal** if and only if $a + b \in \{a, b\}$ for all $a, b \in R$. The boolean semiring $\mathbb{B}$ and the semirings in Example 1.8 and Example 1.22 are extremal. The extremal semirings R are precisely the additively-idempotent yoked difference-ordered semirings. Simple extremal semirings are called **Dijkstra semirings**. The semirings $\mathbb{B}$, $(\mathbb{I}, max, \cdot)$, $(\mathbb{N} \cup \{\infty\}, min, +)$, and $(\mathbb{R} \cup \{\infty\}, max, min)$ are examples of Dijkstra semirings. Determinants of matrices over extremal semirings are studied in [Gondran & Minoux, 1978]. In particular, it is shown there that if R is an extremal division semiring and if $A \in \mathcal{M}_n(R)$ satisfies the condition that $|A|^+ = |A|^-$ then the columns (and rows) of A are linearly dependent.

(20.24) EXAMPLE. The semiring $\mathbb{N}$ is uniquely difference ordered in its usual ordering.

(20.25) EXAMPLE. If R is a semiring then $ideal(R)$ is a simple difference-ordered semiring with infinite element R.

(20.26) EXAMPLE. The order on an additively-idempotent semiring defined in Proposition 20.19 is just the difference order. Indeed, if $a + b = b$ then surely $a \leq b$ in the difference order. Conversely, assume that $a \leq b$ in the difference order. Then there exists an element c of R such that $a + c = b$ so $b = a + c = a + c + c = b + c$. This implies that $a + b = a + b + c = b + b = b$.

(20.27) EXAMPLE. If R is a difference-ordered semiring and A is a nonempty set then the partially-ordered semiring R^A (see Example 20.8) is difference-ordered. Indeed, if $f, g \in R^A$ satisfy $f \leq g$ then $f(a) \leq g(a)$ for each $a \in A$ and so for each $a \in A$ there exists an element $h(a) \in R$ such that $f(a) + h(a) = g(a)$. Therefore $f + h = g$ in R^A. Similarly, the partially-ordered semirings $\mathcal{M}_n(R)$, $R\langle\langle A\rangle\rangle$ and $R\langle A\rangle$ (see Example 20.11), are difference ordered.

A difference-ordered semiring R is a **weakly uniquely difference-ordered semiring (WUDO-semiring)** if and only if $a < b$ implies that there is a unique element c of R satisfying $a + c = b$. We will denote this unique element c by $b \ominus a$. We also set $a \ominus a = 0_R$ for all $a \in R$. WUDO-semirings were first introduced and studied in [Wu, 1998].

(20.28) EXAMPLE. Surely every uniquely difference-ordered semiring is a WUDO-semiring. The following are examples of totally-ordered semirings which are weakly uniquely difference ordered but not uniquely difference ordered:

(1) $\mathbb{B}$;
(2) $(\mathbb{I}, max, \sqcap)$, where $\sqcap$ is any triangular norm on $\mathbb{I}$;
(3) $(\mathbb{I}, min, \sqcup)$, where $\sqcup$ is any triangular conorm on $\mathbb{I}$;
(4) $(\mathbb{R} \cup \{-\infty\}, max, +)$;
(5) $(\mathbb{R} \cup \{\infty\}, min, +)$

In particular, if R is a WUDO-semiring and $a \leq b$ are elements of R then $a \ominus 0_R = a$ and $b = (b \ominus a) + a$. Since R is positive, one also sees that $b \geq b \ominus a$ for all $a \leq b$ in R and that if $a \leq b \leq c$ in R then $c \ominus a \leq c \ominus b$.

(20.29) PROPOSITION. *Let R be a WUDO-semiring*

(1) *If $a \leq c$ and $b \leq c$ then $c \ominus a \geq b$ and $c \ominus b \geq a$ imply that $(c \ominus b) \ominus a = (c \ominus a) \ominus b$;*
(2) *$a + b = a + c > a$ then $b = c$.*
(3) *If $b \geq a$ and $c > 0_R$ then $(b + c) \ominus a = (b \ominus a) + c$ if and only if $b + c > a$.*

PROOF. (1) The result is obvious if $a = 0_R$ or $b = 0_R$, so we may assume that that is not the case. By definition, we then see that $(c \ominus b) \ominus a = 0_R$ if $c \ominus b = a$ and, otherwise, $(c \ominus b) \ominus a$ is the unique element d of R satisfying $c \ominus b = a + d$. On the other hand, if $c \ominus b = a$ then $c = b + a$, which implies that $c \ominus a = b$ (since $c \ominus a \geq b > 0_R$ implies that $c > a$). Hence $c \ominus a) \ominus b = 0_R$. If $c \ominus b > a$ then $c = (c \ominus b) + b = a + d + b$. Again, since $c \ominus a \geq b > 0_R$ we have $c > a$ and so $c \ominus a = b + d$. If $c \ominus a = b$ then $c = a + b$ and $c \ominus b = a$, which contradicts the fact that $c \ominus b > a$. Thus we must have $c \ominus a > b$ and $(c \ominus a) \ominus b = d$. Therefore $(c \ominus a) \ominus b = 0_R$ if $c \ominus b = a$ and $(c \ominus a) \ominus b = d$ if $c \ominus b > a$, proving that $(c \ominus b) \ominus a = (c \ominus a) \ominus b$.

(2) Let $d = a + b = a + c$. Then both $d > a$ and so, by uniqueness, $b = c$.

(3) Assume $b + c > a$ then

$$((b + c) \ominus a) + a = b + c = ((b \ominus a) + a) + c = ((b \ominus a) + c) + a$$

and so $(b+c) \ominus a = (b \ominus a) + c$ by (2). Conversely, assume that $(b+c) \ominus a = (b \ominus a) + c$. Then $(b \ominus a) + c \geq c > 0_R$ and so $(b+c) \ominus a > 0_R$, which implies that $b + c > a$. □

If R is uniquely difference-ordered then it is straightforward to check that the following conditions are equivalent for $a, b, c \in R$:

(1) $a \geq b + c$;
(2) $a \ominus b \geq c$;
(3) $a \ominus c \geq b$.

Also, if $a, b, c \in R$ and $a \geq b$ then $a \ominus (a \ominus b) = b$ and $(a \ominus b) + c = (a + c) \ominus b$.

A nonempty subset A of a partially-ordered semiring R is **convex** if and only if $a, b \in A$ and $a \leq r \leq b$ imply that $r \in A$.

(20.30) PROPOSITION. *Every zerosumfree division semiring R is difference ordered. Moreover, if $\gamma: R \to S$ is a morphism of semirings then $mker(\gamma)$ is convex with respect to this order.*

PROOF. If a and b are elements of a division semiring R we will write $a \leq b$ if and only if there exists an element c of R satisfying $a + c = b$. Clearly $a \leq a$ for all $a \in R$ and $a \leq b \leq c$ implies that $a \leq c$. Moreover, it is also clear that if we complete showing that $\leq$ is indeed a partial order relation on R then, relative to that relation, R is a partially-ordered semiring. What we are left to show is that if $a \leq b$ and $b \leq a$ in R then $a = b$.

First, however, we will prove the second part of the proposition. Let $K = mker(\gamma)$ for some morphism of semirings $\gamma: R \to S$ and let a and b be elements of K. Let r be an element of R satisfying $a \leq r \leq b$. We can clearly assume that $r \neq a, b$. Then there exist nonzero elements u and v of R satisfying $a + u = r$ and $r + v = b$. Since R is zerosumfree, we also see that $u + v \neq 0$. Set $w = (u + v)^{-1}$. Then $wv + wu = 1$ and so, by Proposition 10.24, $awv + bwu \in K$. But

$$awv + bwu = awv + (a + u + v)wu = aw(u + v) + (u + v)wu = a + u = r,$$

proving that $r \in K$.

Now suppose that $a \leq b$ and $b \leq a$ in R. If both a and b equal 0, we are done. Hence we can assume that $a \neq 0$. Then $1 \leq a^{-1}b \leq 1$. Since $\{1\} = mker(\iota)$, where ι is the identity morphism from R to itself, we see that $a^{-1}b = 1$ and so $a = b$, as required. □

(20.31) PROPOSITION. *If R is an additively-idempotent semifield then every pair of elements of R has an infimum in R.*

PROOF. Let $a, b \in R$. If $a = 0$ or $b = 0$ then 0 is the infimum of $\{a, b\}$ since R is difference ordered. Hence we can assume that $a, b \in R \setminus \{0\}$. Since R is additively idempotent and hence zerosumfree, this implies that $a + b \neq 0$. Set $c = ab(a+b)^{-1}$. Then $(c + a)(a + b) = ab + a^2 + ab = ab + a^2 = a(a + b)$ and so $c + a = a$, proving that $c \leq a$. Similarly $c \leq b$. Now suppose that $d \leq a$ and $d \leq b$. Then $d + a = a$ and $d + b = b$ so

$$\begin{aligned} d(a + b) + ab &= da + db + (d + a)(d + b) = da + db + d^2 + da + db + ab \\ &= d^2 + da + db + ab = (d + a)(d + b) = ab \end{aligned}$$

and hence $d(a + b) \leq ab$, implying $d \leq c$. Therefore c is the infimum of $\{a, b\}$ in R. □

As a consequence of Proposition 20.30, we can characterize the morphisms from $\mathbb{Q}^+$ to other semirings.

(20.32) PROPOSITION. *If $\gamma: \mathbb{Q}^+ \to R$ is a morphism of semirings then either $im(\gamma)$ is isomorphic to $\mathbb{B}$ or γ is monic.*

PROOF. If $mker(\gamma) = \mathbb{Q}^+ \setminus \{0\}$ then γ is the morphism which sends 0 to 0_R and every nonzero element of $\mathbb{Q}^+$ to 1_R so that $im(\gamma)$ is isomorphic to $\mathbb{B}$. Otherwise, we can assume that $mker(\gamma)$ is properly contained in $\mathbb{Q}^+ \setminus \{0\}$. By Proposition 10.25, γ is injective if we can show that $mker(\gamma) = \{1\}$. Assume that this is not the case and let a be an element of $mker(\gamma)$ not equal to 1. Then a^{-1} also belongs to $mker(\gamma)$ and so, without loss of generality, we can assume that $0 < a < 1$. Similarly, let $0 \neq b \in \mathbb{Q}^+ \setminus mker(\gamma)$. Then $b^{-1} \notin mker(\gamma)$ and so we can assume that $0 < b < 1$ as well. But then there exists a natural number k such that $a^k < b < 1$. Since $a^k \in mker(\gamma)$, this contradicts the convexity of $mker(\gamma)$. Thus γ is injective. □

(20.33) COROLLARY. *If there exists a morphism from $\mathbb{Q}^+$ to a semiring R then it is unique.*

PROOF. Assume there exists a morphism $\gamma: \mathbb{Q}^+ \to R$. If $im(\gamma) = \{0, 1\}$ then R is additively idempotent since $1_R = \gamma(1) = \gamma(2) = \gamma(1) + \gamma(1) = 1_R + 1_R$. Hence it cannot have a subsemiring isomorphic to $\mathbb{Q}^+$ and so, by Proposition 20.32, there is no other morphism from $\mathbb{Q}^+$ to R. Therefore we can assume that γ is injective. If δ is another morphism from $\mathbb{Q}^+$ to R then it too must be injective by the above reasoning. For each $n \in \mathbb{N}$ we therefore have $\delta(n) = n\delta(1) = n \cdot 1_R = n\gamma(1) = \gamma(n)$ and so for each $p/q \in \mathbb{Q}^+$ we have $\delta(q)\delta(p/q) = \delta(p) = \gamma(p) = \gamma(q)\gamma(p/q)$ so $\delta(p/q) = \gamma(p/q)$. Thus $\delta = \gamma$. □

If R is a zerosumfree division semiring then there does exist indeed exist a morphism of semirings from $\mathbb{Q}^+$ to R, namely the map γ_0 defined by $\gamma_0: p/q \mapsto (p1_R)(q1_R)^{-1}$. For such a semiring R, let $L(R)$ be the set of all those elements $r \in R$ for which there exist elements a and b of $\mathbb{Q}^+$ satisfying $\gamma_0(a) \leq r \leq g_0(b)$.

(20.34) PROPOSITION. *If R is a zerosumfree division semiring then $L(R) = mker(\delta)$ for some morphism $\delta: R \to S$.*

PROOF. Let $r, r' \in L(R)$. Then there exist $a, a', b, b' \in \mathbb{Q}^+$ satisfying $\gamma_0(a) \leq r \leq \gamma_0(b)$ and $\gamma_0(a') \leq r' \leq \gamma_0(b')$. As an immediate consequence of the definition of a partial order we then have $\gamma_0(aa') \leq rr' \leq \gamma_0(bb')$, $\gamma_0(b)^{-1} \leq r^{-1} \leq \gamma_0(a)^{-1}$, and $\gamma_0(a) = u^{-1}\gamma_0(a)u \leq u^{-1}ru \leq u^{-1}\gamma_0(b)u = \gamma_0(b)$ for any $0 \neq u \in R$. Thus $L(R)$ is a normal divisor of R. Moreover, if u and v are elements of R satisfying $u + v = 1$, if $a'' = min\{a, a'\}$, and if $b'' = max\{b, b'\}$ then

$$\gamma_0(a'') = \gamma_0(a'')u + \gamma_0(a'')v \leq ru + r'v \leq \gamma_0(b'')u + \gamma_0(b'')v = \gamma_0(b'')$$

so that $ru + r'v \in L(R)$. The result now follows from Proposition 10.24. □

(20.35) PROPOSITION. *If R is an additively-idempotent partially-ordered semiring satisfying $0 < 1$ then $S = \{r \in R \mid 0 \leq r \leq 1\}$ is a subsemiring of R.*

PROOF. Clearly $\{0, 1\} \subseteq S$. If $s, s' \in S$ then $0 \leq s + s' \leq 1 + 1 = 1$ and $0 = s0 \leq ss' \leq s1 = s \leq 1$ so both $s + s'$ and ss' belong to S. □

Note that if R and S are difference-ordered semirings and if $\gamma: R \to S$ is a morphism of semirings then γ preserves partial order. Indeed, if $a \leq b$ in R then there exists an element c of R satisfying $a + c = b$ so $\gamma(a) + \gamma(c) = \gamma(b)$, which implies that $\gamma(a) \leq \gamma(b)$.

(20.36) PROPOSITION. *If R is an arbitrary semiring then there exists a difference-ordered semiring S and a surjective morphism of semirings $S \to R$.*

PROOF. Let $A = \{a_r \mid r \in R\}$ be a set bijectively corresponding to R and let $\lambda: A \to R$ be this bijection. Then λ extends to a map λ^* from A^* to R given by $\lambda^*(\square) = 1$ and $\lambda^*(w) = \lambda(a_1) \cdot \ldots \lambda(a_n)$ for each nonempty word $w = a_1 \cdot \ldots \cdot a_n$ in A^*. Let $S = \mathbb{N}\langle A\rangle$. Then there exists a morphism of semirings $\gamma: S \to R$, which is clearly surjective, defined by $\gamma: f \mapsto \sum f(w)\lambda^*(w)$. Moreover, the semiring $\mathbb{N}$ is surely difference-ordered and hence so is $\mathbb{N}\langle A\rangle$ by Example 20.27. $\square$

(20.37) PROPOSITION. *The following conditions on a semiring R are equivalent:*

(1) *R is difference ordered;*
(2) *If a, b, c are elements of R satisfying $a = a + b + c$, then $a = a + b$.*

PROOF. Assume that R is difference ordered and let $b, c \in R$. If $a = a + b + c$ then $a \leq a + b \leq a$, proving that $a = a + b$. Conversely, assume that (2) holds and define the relation $\leq$ on R by setting $r \leq r'$ if and only if there exists an element r'' of R satisfying $r + r'' = r'$. Then clearly $a \leq a$ for all $a \in R$, while $a \leq b$, and $b \leq c$ imply $a \leq c$ for all $a, b, c \in R$. If a and b are elements of R for which there exist elements c and d satisfying $a + c = b$ and $b + d = a$ then $a + c + d = a$ so $a + c = a$. This implies that $b = a + c = a$, proving that $\leq$ is a partial order on R which turns R into a partially-ordered semiring. Thus R is difference ordered. $\square$

(20.38) COROLLARY. *If R is a nonzeroic difference-ordered semiring then $Z(R)$ is a strong ideal of R.*

PROOF. If $b + c \in Z(R)$ then, by definition, there exists an element a of R such that $a+b+c = a$. By Proposition 20.37, this implies that $a+b = a$ and so $b \in Z(R)$. Similarly $c \in Z(R)$. $\square$

(20.39) PROPOSITION. *An ideal I of a difference-ordered semiring R is strong if and only if $a \leq b$ and $b \in I$ imply that $a \in I$.*

PROOF. Assume that I is strong. If $a \leq b$ and $b \in I$ then there exists an element c such that $a+c \in I$. By assumption, $a \in I$. Conversely, assume the given condition holds. If a and b are elements of R satisfying $a + b \in I$ then $a \leq a + b \in I$ and so $a \in I$. Similarly $b \in I$ and so I is strong. $\square$

(20.40) PROPOSITION. *If R is a difference-ordered semiring and A is a subset of R then $(0 : A)$ is a strong ideal of R.*

PROOF. If $c \leq b \in (0 : A)$ and $a \in A$ then $0 \leq ca \leq ba = 0$ so $ca = 0$. Thus $c \in (0 : A)$ and so, by Proposition 20.39, $(0 : A)$ is a strong ideal of R. $\square$

(20.41) PROPOSITION. *If R is a difference-ordered Gel'fand semiring and $d \geq c \in U(R)$ in R then $d \in U(R)$.*

PROOF. Since R is difference-ordered, we know that there exists an element r of R satisfying $d = c + r$. Then $d \in U(R)$ by Proposition 4.50. $\square$

An element a of a partially-ordered semiring R is **prime** if and only if a is not a unit and $bc \leq a$ in R implies $b \leq a$ or $c \leq a$. An element a of R is **semiprime** if and only if a is not a unit and $b^2 \leq a$ in R implies that $b \leq a$. If R is multiplicatively idempotent then clearly every nonunit of R is semiprime.

(20.42) EXAMPLE. If R is a semiring then the prime elements of the semiring *ideal*(R) are precisely the prime ideals of R and the semiprime elements of *ideal*(R) are precisely the semiprime ideals of R.

(20.43) EXAMPLE. [Solian & Viswanathan, 1988] If A is a set having at least two elements then. $f \in \mathbb{I}^A$ is prime if and only if there is an element a of A such that $f(a) < 1$ and $f(b) = 1$ for all $b \in A \setminus \{a\}$.

(20.44) EXAMPLE. If R is the semiring of all open subsets of a topological space X then the prime elements of R are precisely the complements of closed irreducible sets. If the space X is sober (namely, if every closed irreducible set is the closure of a point) then the primes of R are the complements of closures of points in X.

A **maximal nonunit** of a partially-ordered semiring R is an element a of $R \setminus U(R)$ satisfying the condition that $\{r \in R \mid r > a\}$ is a nonempty subset of $U(R)$. Note that if R is simple then the maximal nonunits of R are precisely the coatoms of R. Frames R for which the prime elements are coatoms have been studied in detail in [Rosický & Šmarda, 1985].

Since *ideal*(R) is certainly a positive difference-ordered Gel'fand semiring, we see that the following result generalizes Corollary 7.13.

(20.45) PROPOSITION. *If R is a positive difference-ordered Gel'fand semiring then every maximal nonunit of R is prime.*

PROOF. Let a be a maximal nonunit of R; let b and c be elements of R satisfying $b \not\leq a$, $c \not\leq a$, and $bc \leq a$. Then $a, b \leq a + b$ and so $a < a + b$, proving that $a + b \in U(R)$. Similarly $a + c \in U(R)$ and so $d = (a + b)(a + c) \in U(R)$. But

$$\begin{aligned} d &= (a+b)a + ac + bc \leq (a+b)a + ac + a \\ &= (a+b)a + a(c+1) \leq (a+b+1)c(c+1), \end{aligned}$$

where $a + b + 1$ and $c + 1$ are units of R, since R is a Gel'fand semiring. Therefore, since R is positive, $(a + b + 1)^{-1}d(c + 1)^{-1} \leq a$ and so, by Proposition 20.41, we have $a \in U(R)$. This is a contradiction, proving that a must be prime. □

(20.46) COROLLARY. *Any coatom of a simple difference-ordered semiring is prime.*

PROOF. This is an immediate consequence of Proposition 20.45. □

(20.47) PROPOSITION. *Let R be a simple difference-ordered semiring and let A be a nonempty subset of R satisfying the condition that if $a, a' \in A$ then there exists an element $a'' \in A$ with $a'' \leq aa'$. Let $B = \{r \in R \mid a \not\leq r$ for all $a \in A\}$. Then every additively-idempotent maximal element of B is prime.*

PROOF. Let b be an additively-idempotent maximal element of B and let $r, r' \in R$ satisfy $r, r' \not\leq b$. Then $r + b, r' + b > b$. By the choice of b, this means that $r + b$ and $r' + b$ do not belong to B and so there exist elements a, a', and a'' of A satisfying $a \leq r + b$, $a' \leq r' + b$, and $a'' \leq aa'$. Hence $a'' \leq (r + b)(r' + b) = rr' + br' + rb + b^2$. By Proposition 4.3, $br' + rb + b^2 \leq b + b + b = b$ and so $a'' \leq rr' + b$. If $rr' \leq b$ then $a'' \leq b + b = b$, contradicting the fact that $b \in B$. Thus $rr' \not\leq b$, proving that b is prime. □

(20.48) PROPOSITION. *Let R be a simple difference-ordered additively-idempotent semiring and let a be an element of R which is not a unit. Then a is prime if and only if there exists a character γ_a on R satisfying $ker(\gamma_a) = \{r \in R \mid r \leq a\}$. Moreover, if a and b are distinct prime elements of R then γ_a and γ_b are also distinct.*

PROOF. Assume that a is prime and define the function $\gamma_a : R \to \mathbb{B}$ by $\gamma_a(r) = 0$ if and only if $r \leq a$. Then $\gamma_a(0) = 0$ since R is difference ordered and $\gamma_a(1) = 1$ since $1 \not\leq a$. Moreover, if $r, r' \in R$ then $\gamma_a(r + r') = 0 \Leftrightarrow r + r' \leq a \Leftrightarrow r \leq a$ and $r' \leq a \Leftrightarrow \gamma_a(r) = 0 = \gamma_a(r')$ so $\gamma_a(r) + \gamma_a(r') = \gamma_a(r + r')$. Similarly, by primeness, $\gamma_a(rr') = 0 \Leftrightarrow rr' \leq a \Leftrightarrow r \leq a$ or $r' \leq a \Leftrightarrow \gamma_a(r)\gamma_a(r') = 0$ and so $\gamma_a(rr') = \gamma_a(r)\gamma_a(r')$.

Conversely, assume that there exists a morphism $\gamma_a: R \to \mathbb{B}$ satisfying $ker(\gamma_a) = \{r \in R \mid r \leq a\}$. If $rr' \in R$ satisfy $rr' \leq a$ then $\gamma_a(r)\gamma_a(r') = \gamma_a(rr') = 0$ so $\gamma_a(r) = 0$ or $\gamma_a(r') = 0$. Hence either $r \leq a$ or $r' \leq a$, proving that a is prime.

Finally, if a and b are prime elements of R satisfying $\gamma_a = \gamma_b$ then $\gamma_b(a) = 0$ and so $a \leq b$. Similarly $b \leq a$ and so $a = b$. □

If R is a partially-ordered semiring then a function $\nu: R \to R$ is a **nucleus** if and only if the following conditions are satisfied:

(1) If $r \leq r'$ in R then $\nu(r) \leq \nu(r')$;
(2) If $r \in R$ then $\nu^2(r) = \nu(r) \geq r$;
(3) If $r, r' \in R$ then $\nu(rr') \geq \nu(r)\nu(r')$.

Note that if ν is a nucleus then $r \leq \nu(r')$ if and only if $\nu(r) \leq \nu(r')$.

(20.49) PROPOSITION. *Let R be a partially-ordered semiring and let $\nu: R \to R$ be a nucleus on R.*

(1) *If R is positive then $\nu(rr') = \nu(\nu(r)r') = \nu(\nu(r)\nu(r')) = \nu(r\nu(r'))$ for all $r, r' \in R$;*
(2) *If R is additively idempotent then $\nu(\nu(r)+\nu(r')) = \nu(r+r')$ for all $r, r' \in R$.*

PROOF. (1) If R is positive and if $r, r' \in R$ then $\nu(r) \geq r$ and $\nu(r') \geq r'$ so $\nu(r)\nu(r') \geq \nu(r)r' \geq rr'$. This implies that $\nu(\nu(r)r') \geq \nu(rr') = \nu(\nu(rr')) \geq \nu(\nu(r)\nu(r')) \geq \nu(\nu(r)r')$, proving that $\nu(rr') = \nu(\nu(r)r') = \nu(\nu(r)\nu(r'))$. A similar proof shows that $\nu(rr') = \nu(r\nu(r')) = \nu(\nu(r)\nu(r'))$.

(2) If $r, r' \in R$ then $r + r' \geq r, r'$ and so $\nu(r + r') \geq \nu(r), \nu(r')$. Since R is additively idempotent, this implies that $\nu(r + r') \geq \nu(r) + \nu(r')$. Thus $\nu(r + r') = \nu^2(r + r') \geq \nu(\nu(r) + \nu(r'))$. Conversely, $\nu(r) \geq r$ and $\nu(r') \geq r'$ so $\nu(r) + \nu(r') \geq r + r'$. Therefore $\nu(\nu(r) + \nu(r')) \geq \nu(r + r')$ and hence we have equality. □

A nucleus ν on R is **strict** if and only if condition (3) can be replaced by:

(3') If $r, r' \in R$ then $\nu(rr') = \nu(r)\nu(r')$.

The notion of a nucleus is usually defined for frames. Refer to [Johnstone, 1982]. Examples of several nuclei on the frame of torsion theories are given in [Golan & Simmons, 1988].

(20.50) EXAMPLE. [Kirby, 1969] If R is a commutative ring R then the map $I \mapsto \sqrt{I}$ is a nucleus on the semiring $ideal(R)$.

(20.51) EXAMPLE. [Banaschewski & Harting, 1985] Let R be a ring and let ν be the function which assigns to each ideal I of R the ideal $\nu(I)$ defined by the condition that $\nu(I)/I$ is the Levitzki radical of R/I. Then ν is a nucleus on the semiring $ideal(R)$. The same is true if we take, instead, the Jacobson radical or the Brown-McCoy radical.

(20.52) EXAMPLE. If R is a C^*-algebra, then the map which takes each element of $ideal(R)$ to its closure is a nucleus on the semiring $ideal(R)$.

(20.53) EXAMPLE. If R is a commutative integral domain then the semiring $fract(R)$ of all fractional ideals of R is partially-ordered by inclusion. A function ν from $fract(R)$ to itself is a **$*$-operation** on R if and only if the following conditions are satisfied:

(1) $\nu(aR) = aR$ and $\nu(aI) = a\nu(I)$ for all $a \in R$ and all $I \in fract(R)$;
(2) $I \subseteq \nu(I) = \nu^2(I)$ for all $I \in fract(R)$;
(3) $\nu(I) \subseteq \nu(H)$ whenever $\subseteq H$ in $fract(R)$.

See [Gilmer, 1972] for details. Examples of such functions include the function ν_v which assigns to each $I \in fract(R)$ the intersection of all principal fractional ideals of R containing I, and the function ν_t defined by

$$\nu_t(I) = \sum\{\nu_v(H) \mid H \text{ a finitely-generated subideal of } I\}.$$

For further examples, see [Anderson & Anderson, 1991]. It is straightforward to verify that a $*$-operation on $fract(R)$ is a nucleus. Commutative integral domains R satisfying the condition that the nuclei ν_v and ν_t coincide are studied in [Houston & Zafrullah, 1988]. Noetherian domains, Krull domains, and Mori domains are of this type.

(20.54) EXAMPLE. If R is a difference-ordered semiring then any morphism of semirings $\nu: R \to R$ satisfies conditions (1) and (3') of the definition of a nucleus. Thus such a morphism is a strict nucleus if and only if $\nu^2(r) = \nu(r) \geq r$ for all and only if ν is a closure operator on R.

(20.55) EXAMPLE. Let R be a semiring and let the function $\nu: ideal(R) \to ideal(R)$ be defined by $\nu: I \mapsto 0/I$. Then ν is a nucleus which, in general, is not strict.

If ν is a nucleus defined on an additively-idempotent semiring and if ρ is the relation on R defined by $r \;\rho\; r'$ if and only if $\nu(r) = \nu(r')$, then it is immediate that ρ is an equivalence relation. Moreover, as a straightforward consequence of Proposition 20.49, we see that it is in fact a congruence relation.

If R is a partially-ordered semiring then a function $\mu: R \to R$ is a **modality** if and only if the following conditions are satisfied:

(1) If $r \leq r'$ in R then $\mu(r) \leq \mu(r')$;
(2) If $r \in R$ then $\mu^2(r) = \mu(r) \leq r$;
(3) If $r, r' \in R$ then $\mu(\mu(r)\mu(r')) = \mu(r)\mu(r')$.

(20.56) EXAMPLE. If H is an ideal of a semiring R then the function $I \mapsto I \cap H$ is a modality on the semiring *ideal*(R).

Let R be a partially-ordered semiring. A left R-semimodule M is **partially-ordered** if and only if there exists a partial order relation $\leq$ defined on M satisfying the following conditions:

(1) If $m \leq m'$ in M and if $m'' \in M$ then $m + m'' \leq m' + m''$;
(2) If $m \leq m'$ in M and if $r \geq 0$ in R then $rm \leq rm'$;
(3) If $r \leq r'$ in R and if $m \geq 0$ in M then $rm \leq r'm$.

Partially-ordered right R-semimodules are defined analogously. If the relation $\leq$ is in fact a total order, then we say that the semimodule is **totally-ordered**. Totally-ordered $\mathbb{N}$-semimodules are studied in [Clifford, 1958] and [Lugowski, 1964a, 1964b].

(20.57) EXAMPLE. Clearly $\mathbb{Q}$ is a totally-ordered left (and right) semimodule over $\mathbb{N}$.

(20.58) EXAMPLE. If R is a partially-ordered semiring then R^+ is a partially-ordered left and right R-semimodule.

(20.59) EXAMPLE. Let R be a partially-ordered semiring and let A be a nonempty set. Define a relation $\leq$ on R^A by setting $f \leq g$ if and only if $f(a) \leq g(a)$ for all $a \in A$. Then R^A is a partially-ordered left R-semimodule. A function $f \in R^A$ is **bounded** if and only if there exists an element $r_f \in R$ such that $f(a) \leq b_f$ for all $a \in A$. The set $B(A, R)$ of all bounded elements of R^A is clearly a subsemimodule of R^A. For an analysis of such R-semimodules of bounded functions, when R is an additively-idempotent semifield, refer to [Dudnikov, 1992] and [Dudnikov and Sambourskiĭ, 1989].

If R is a commutative additively-idempotent semiring and if $B(A, R) \cong M_1 \coprod M_2$ for nonzero R-semimodules M_1 and M_2, then A can be partitioned into a disjoint union $A = A_1 \cup A_2$ such that $M_1 \cong B(A_i, R)$ for $i = 1, 2$. See [Dudnikov and Sambourskiĭ, 1992] for details.

(20.60) EXAMPLE. Let R be a partially-ordered semiring and let m and n be positive integers. Then the set $\mathcal{M}_{m,n}(R)$ of all $m \times n$ matrices with entries from R can be partially-ordered by setting $[a_{ij}] \leq [b_{ij}]$ if and only if $a_{ij} \leq b_{ij}$ for all $1 \leq i \leq m$ and $1 \leq j \leq n$. Moreover, as we have seen in Example 20.11, $\mathcal{M}_m(R)$ is a partially-ordered semiring and one easily checks that $\mathcal{M}_{m,n}(R)$ is a partially-ordered left $\mathcal{M}_m(R)$-semimodule, under the usual definition of matrix multiplication and addition.

(20.61) EXAMPLE. A **Riesz space** is a vector space L over $\mathbb{R}$ which is partially-ordered in such a manner that the positive cone $\{f \in L \mid f \geq 0\}$ is a partially-ordered $\mathbb{R}^+$-semimodule. Such spaces play an important part in functional analysis, and especially spectral analysis. They were studied in [Freudenthal, 1936] and in detail by H. Nakano in an unpublished manuscript partially printed many years later in [Nakano, 1966]. For a detailed work on Riesz spaces and their place in functional analysis, refer to [Luxemburg & Zaanen, 1971] and [Zaanen, 1983]. For topological Riesz spaces, see Fremlin, 1974.

As with semirings, a partially-ordered left R-semimodule M is **difference ordered** if and only if $m \geq m'$ in M when and only when there exists an element $m'' \in M$ satisfying $m = m' + m''$. Additively-idempotent semimodules are canonically difference-ordered under the partial order defined by $m \geq m'$ if and only if $m = m + m'$.

(20.62) PROPOSITION. *If R is a positive semiring and M is a difference-ordered left R-semimodule then:*

(1) *$M' = \{m \in M \mid m \leq 0\}$ is a submodule of M;*
(2) *$M'' = \{m \in M \mid m \geq 0\}$ is a subsemimodule of M;*
(3) *If $0 \neq m \in M'$ then $m + m < m$;*
(4) *If $0 \neq m \in M''$ then $m + m > m$.*

PROOF. (1) If $m, m' \in M'$ then $m + m' \leq 0 + 0 = 0$ so $m + m' \in M'$. If $m \in M'$ and $r \in R$ then $r \geq 0$ and so $rm \leq r0 = 0$. Thus M' is a subsemimodule of M. If $m \in M$ then $m \leq 0$ and so there exists an element m'' of M such that $m + m'' = 0$. Hence $V(M') = M'$.

(2) The proof that M'' is a subsemimodule of M follows as in (1).

(3) If $m < 0$ in M then there exists an element m'' of M satisfying $m + m'' = 0$ and so $m + m + m'' = m$. Thus $m + m \leq m$. If $m + m = m$ then $m = m + (m + m'') = (m + m) + m'' = m + m'' = 0$, which contradicts the choice of m.

(4) This is proven in a manner similar to the proof of (3). □

Note that if R is a positive semiring and M is a difference-ordered left R-semimodule then either one of the sets M' and M'' defined in Proposition 20.62 equals $\{0\}$ or $M \neq M' \cup M''$. Indeed, if $0 \neq m' \in M'$ and $0 \neq m'' \in M''$, consider $m = m' + m''$. If $m \in M'$ then there is an element u of M satisfying $0 = m + u = m'' + (m' + u)$ and so $m'' \in M' \cap M''$, which is a contradiction since $m'' \neq 0$. A similar contradiction is obtained if we assume that $m \in M''$. Thus $m \notin M' \cup M''$.

21. LATTICE-ORDERED SEMIRINGS

A semiring R is **lattice-ordered** if and only if it also has the structure of a lattice such that, for all a and b in R:

(1) $a + b = a \vee b$; and

(2) $ab \leq a \wedge b$,

where partial order here is the one induced naturally by the lattice structure on R. If R, as a lattice, is distributive, then R is a **distributive lattice-ordered semiring (DLO-semiring)**. Clearly any lattice-ordered semiring is a partially-ordered semiring in the sense of Chapter 18, with respect to the partial order induced by the lattice structure. (Note in passing that some authors replace (1)by a weaker condition; see, for example, [Ranga Rao, 1981].)

As an immediate consequence of the definition we see that lattice-ordered semirings are additively idempotent. Also, if a and b are elements of a lattice-ordered semiring satisfying $ab = a$ or $ba = a$ then $a \leq b$. Therefore, if a is an element of a lattice-ordered semiring R then $a = a1 \leq 1$. Any element a of a lattice-ordered semiring R defines a nucleus ν_a on R given by $\nu_a\colon r \mapsto r + a$.

(21.1) EXAMPLE. Any bounded distributive lattice R is clearly a DLO-semiring if we define $a+b = a \vee b$ and $ab = a \wedge b$ for all $a, b \in R$. The set R' of all complemented elements of R is a subsemiring of R which is a ring (in fact, it is a boolean algebra).

(21.2) EXAMPLE. The semiring of all ideals of a semiring is lattice ordered but is not, in general, a DLO-semiring. Indeed, as we have seen in Example 6.36, the lattice $ideal(R)$ need not even be modular.

If d is a derivation on a semiring R then a d-**differential ideal** of R is an ideal I satisfying $d(a) \in I$ for all $a \in I$. If I and H are d-differential ideals of R then clearly so are $I + H$ and $I \cap H$. Moreover, if $\{a_1, \ldots, a_n\} \subseteq I$ and $\{b_1, \ldots, b_n\} \subseteq H$ then

$$d\left(\sum_{i=1}^{n} a_i b_i\right) = \sum_{i=1}^{n} a_i d(b_i) + \sum_{i=1}^{n} d(a_i) b_i \in IH$$

so IH is again a d-differential ideal of R. Thus the set of all d-differential ideals of R forms a subsemiring of $ideal(R)$ which is also a lattice-ordered semiring.

(21.3) EXAMPLE. One does not even need a semiring. If $(M, *)$ is a commutative groupoid, i.e. a set on which we have a binary operation, which may not even be associative, then a nonempty subset I of M is an **ideal** of M if and only if $a * m \in I$ for all $a \in I$ and $m \in M$. Denote the set of all ideals of M by $ideal(M)$ and set $I * H = \{a * b \mid a \in I, b \in H\}$ for all $I, H \in ideal(M)$. Then $ideal(M), \cup, \cap)$ is a lattice and $(ideal(M), \cup, *)$ is a lattice-ordered semiring.

(21.4) EXAMPLE. [Alarcón & Anderson, 1994a] If R is a division semiring which is not a ring and which does not have characteristic 0 then R is a lattice-ordered semiring of the form $(R, \vee, \cdot)$, where $(R \setminus \{0\}, \cdot)$ is a lattice-ordered group.

(21.5) EXAMPLE. F. A. Smith [1966] defines operations $+$, $\cdot$, and $\wedge$ on $R = \{0, 1, a, b, c, d\}$ which turn R into a lattice-ordered semiring which is not a DLO-semiring.

(21.6) EXAMPLE. In Example 1.7 we noted that the set $R - fil$ of all topologizing filters of left ideals of a ring R is a semiring in which addition is given by intersection and multiplication is given by the Gabriel product. It is easy to see that this semiring is zerosumfree and hence additively idempotent. If we consider $R - fil$ as a partially-ordered set with the order being the *reverse* of the usual order, we see that $R - fil$ has the structure of a lattice in which join is taken to be $\cap$. This lattice is not necessarily distributive. See [Golan, 1987] for details.

(21.7) EXAMPLE. [Arnold, 1951] Let A and B be bounded distributive lattices. Then $R = A \times B$ is again a bounded distributive lattice on which the operations of join and meet are defined by $(a, b) \vee (a', b') = (a \vee a', b \vee b')$ and $(a, b) \wedge (a', b') = (a \wedge a', b \wedge b')$. We can define an operation $*$ on R which is equal to neither of these by setting $(a, b) * (a', b') = (a \wedge a', b \vee b')$. Then $(R, \vee, *)$ is a commutative additively-idempotent semiring which is not a DLO-semiring since $(a, b) * (a', b') \geq (a, b) \wedge (a', b')$.

(21.8) EXAMPLE. Let a be a prime element of a lattice-ordered semiring R. Then $R_a = \{0\} \cup \{r \in R \mid r \not\leq a\}$ is a subsemiring of R.

(21.9) EXAMPLE. Define a lattice structure on $\mathbb{N}$ by setting $a \sqcap b$ equal to the least common multiple of a and b and $a \sqcup b$ equal to the greatest common divisor of a and b. (Note that this is the reverse of the usual definitions.) Thus, in this lattice $a \preceq b$ if and only if b divides a. If $\cdot$ is ordinary multiplication in $\mathbb{N}$ then $(\mathbb{N}, \sqcup, \cdot)$ is a lattice-ordered semiring.

(21.10) EXAMPLE. Let R be the set of all functions from $\mathbb{I}$ to itself and let $\oplus$, $\otimes$, and $\sqcap$ be the operations on R defined by:

(1) $(f \oplus g)(a) = min\{f(a), g(a)\}$ for all $a \in \mathbb{I}$;
(2) $(f \otimes g)(a) = f(a)g(a)$ for all $a \in \mathbb{I}$;
(3) $(f \sqcap g)(a) = max\{f(a), g(a)\}$ for all $a \in \mathbb{I}$.

Then $(R, \oplus, \sqcap)$ is a lattice and $(R, \oplus, \otimes)$ is a LO-semiring.

In Proposition 20.19 we saw that any additively-idempotent semiring is naturally partially ordered. We now consider a condition for such a semiring to be a DLO-semiring.

(21.11) PROPOSITION. *Let R be an additively-idempotent semiring satisfying the condition that $a \in bR \cap Rb$ whenever $a \leq b$ in R. Then we can define an operation $\wedge$ on R such that $(R, +, \wedge)$ is a distributive lattice; and multiplication distributes over $\wedge$ from either side. If R is simple it is a DLO-semiring.*

PROOF. We begin by noting a consequence of the condition in the hypothesis. If $r \in R$ then surely $r \leq r$ and so there exists an element r^* of R satisfying $r * r = r$. If $r' \leq r$ in R then there is an element r'' of R satisfying $r' = rr''$ and so $r * r' = r * rr'' = rr'' = r'$.

By hypothesis, we know that if $a, b \in R$ then there exist elements a', a'', b', b'' of R satisfying $a = (a + b)a' = a''(a + b)$ and $b = (a + b)b' = b''(a + b)$. Moreover, $ab' = a''(a + b)b' = a''b$, while $ba' = b''(a + b)a' = b''a$. Then

$$b = ab' + bb' \geq ab' = a''b \leq a''a + a''b = a$$

so $ab' \leq a, b$. Similarly, $ba' \leq a, b$. Therefore $ab' + ba' \leq a + a = a$ and similarly $ab' + ba' \leq b$. Suppose that r is an element of R satisfying $r \leq a, b$. Then there exists an element r' of R such that $r = r'(a + b)$ and hence $r = r'a + r'b = r'(a+b)a' + r'(a+b)b' = ra' + rb' \leq ba' + ab' = a''b + b''a$. Thus $ba' + ab' = a''b + b''a$ is a well-defined infimum of a and b in R, which is independent of the choice of a', b', a'', and b'', and which we will denote by $a \wedge b$.

If c is another element of R then $ca = (ca + cb)a'$ and $cb = (ca + cb)b'$ and so $ca \wedge cb = (ca)b' + (cb)a' = c(ab' + ba') = c(a \wedge b)$. Similarly, $(a \wedge b)c = ac \wedge bc$. Therefore multiplication distributes over $\wedge$ from either side.

Since $b = aa' + ba'$, we see that $ba' \leq b$ and so there exists an element d of R satisfying $ba' = db$. Set $e = (a + b)^* \wedge d$. Then

$$eb = (a + b)^*b \wedge db = b \wedge db = b \wedge ba' = ba' \leq a$$

while, similarly, $ea = a \wedge da \leq a$. Therefore $ba' = eb = e(a+b)b' = (ea+eb)b' \leq ab'$. Similarly, $ab' \leq ba'$ and so we conclude that in fact $a \wedge b = ab' = ba' = a''b = b''a$.

To complete the proof that $(R, +, \wedge)$ is a distributive lattice, we must show that if a, b, and c are elements of R then $c \wedge (a + b) = (c \wedge a) + (c \wedge b)$. It is trivial that $c \wedge (a + b) \geq (c \wedge a) + (c \wedge b)$ and so all we need to establish is the reverse inequality. Since $c \leq c + a + b$, we know by hypothesis that there exists an element r of R satisfying $c = r(c + a + b)$. Set $r' = (c + a + b)^* \wedge r$. Then, as above, $c = r'(c + a + b)$, $r'a \leq a$, and $r'b \leq b$, from which we conclude that $c + r'a = r'(c + a + b + a) = c$. Hence $r'a \leq c$ and similarly $r'b \leq c$. Hence $c \wedge (a + b) = r'(a + b) = r'a + r'b \leq (c \wedge a) + (c \wedge b)$, as desired.

Finally, if R is simple then for $a, b \in R$ we have $ab \leq a$ and $ab \leq b$ by Proposition 4.3 and so $ab \leq a \wedge b$. Hence, in this case, R is a DLO-semiring. □

(21.12) PROPOSITION. *If a, b, and c are elements of a lattice-ordered semiring R then:*

(1) $a + ab = a$;
(2) $ab + c = (a + c)b + c$;
(3) $a \leq b$ *implies that* $ca \leq cb$ *and* $ac \leq bc$;
(4) $a \leq b$ *implies that* $a^2 \leq ab \leq b^2$;
(5) $ab \wedge ac \geq a(b \wedge c)$ *and* $ba \wedge ca \geq (b \wedge c)a$;
(6) $(a \wedge b)(a + b) \leq ba + ab$;
(7) *If* $a + b = 1$ *then* $a \wedge b = ab + ba$;
(8) *If* $a + b = 1$ *then* $ac \leq b$ *or* $ca \leq b$ *implies* $c \leq b$;
(9) *If* $a + b = a + c = 1$ *then* $a + bc = a + (b \wedge c) = 1$.

PROOF. (1) By definition, $ab \leq a$ and so $ab + a = ab \vee a = a$.

(2) By (1), $(a + c)b + c = ab + cb + c = ab + c$.

(3) By definition, $a \leq b$ implies that $a + b = a \vee b = b$ and so $cb = c(a + b) = ca + cb = ca \vee cb$. Thus $ca \leq cb$. Similarly, $ac \leq bc$.

(4) This is an immediate consequence of (3).

(5) Since $b \wedge c \leq b$, we have $a(b \wedge c) \leq ab$. Similarly, $a(b \wedge c) \leq ac$ and so $a(b \wedge c) \leq ab \wedge ac$. The second inequality is proven similarly.

(6) By (3) we have $(a \wedge b)(a + b) = (a \wedge b)a + (a \wedge b)b \leq ba + ab$.

(7) From the definition of a lattice-ordered semiring we know that $ab + ba = ab \vee ba \leq a \wedge b$. The converse follows from (6).

(8) Assume that $ac \leq b$. Then $c = (a + b)c = ac + bc \leq b + b = b$. If $ca \leq b$ the proof is similar.

(9) If $a + b = a + c = 1$ then $1 = (a + b)(a + c) = a^2 + ab + ac + bc \leq a + bc \leq 1$ so $a + bc = 1$. Since $bc \leq b \wedge c$, we immediately have $a + (b \wedge c) = 1$ as well. □

(21.13) COROLLARY. *If R is a commutative lattice-ordered semiring and a, b, and c are elements of R satisfying $a + b = a + c = 1$ then $a(b \wedge c) = ab \wedge ac$.*

PROOF. By Proposition 21.12(9) we have $a + (b \wedge c) = 1$ and so by Proposition 21.12(7) we have $a(b \wedge c) = (a \wedge b \wedge c)(a + [b \wedge c]) = a \wedge b \wedge c = (a \wedge b)(a \wedge c) = ab \wedge ac$. □

Note too that, as a direct consequence of Proposition 21.12(9), we see that if a and b are elements of a lattice-ordered semiring satisfying $a + b = 1$ then $a^h + b^k = 1$ for all positive integers h and k.

(21.14) PROPOSITION. *A lattice-ordered semiring R is multiplicatively idempotent if and only if $ab = a \wedge b$ for all $a, b \in R$.*

PROOF. If R is multiplicatively idempotent and $a, b \in R$ then $a \wedge b = (a \wedge b)^2 \leq ab \leq a \wedge b$ and so $ab = a \wedge b$. The converse is trivial. □

(21.15) PROPOSITION. *Every lattice-ordered semiring R is simple and positive, having 1 as its sole unit.*

PROOF. We note that R is simple by Proposition 21.12(1) and Proposition 4.3. Also, for any element a of R we have $0 = a0 \leq a \wedge 0 \leq a$ and so 0 is the unique smallest element of R. This proves that R is positive. Finally, we have already noted that $U(R) = \{1\}$ for any simple semiring R. □

(21.16) Proposition. *If r is a semiprime element of a lattice-ordered semiring R and if a and b are elements of R then the following conditions are equivalent:*

(1) $ab \leq r$;
(2) $ba \leq r$;
(3) $a \wedge b \leq r$.

Proof. By the definition of a lattice-ordered semiring, (3) implies (1) and (2). Conversely, assume (1). Then, by Proposition 21.12(3) we have $(a \wedge b)^2 \leq ab \leq r$ and so, by semiprimeness, $a \wedge b \leq r$, proving (3). Similarly (2) implies (3). □

(21.17) Corollary. *If r and r' are semiprime elements of a lattice-ordered semiring then $r \wedge r'$ is semiprime as well.*

Proof. By Proposition 21.15, we note that $r \wedge r'$ is not a unit. If $a^2 \leq r \wedge r'$ then $a^2 \leq r, r'$ and so, by semiprimeness, $a \leq r, r'$. Therefore $a \leq r \wedge r'$, proving that $r \wedge r'$ is semiprime. □

If m and m' are elements of a partially-ordered monoid $(M, *)$, we define the **interval** $[m, m']$ to be $\{m'' \in M \mid m \leq m'' \leq m'\}$. (Note that this set may be empty!) We will denote the set of all such intervals by $int(M)$ and define the operation $[*]$ on $int(M)$ by $[m, m'][*][n, n'] = [m * n, m' * n']$. It is easy to see that if $(M, *)$ is a partially-ordered monoid with identity element e then $(int(M), [*])$ is a monoid with identity element $[e, e]$. In particular, we note that if R is a lattice-ordered semiring then $(R, +)$ and $(R, \cdot)$ are partially-ordered monoids, with the partial order being that coming from the lattice-structure of R. As an immediate consequence of the definitions, we see that if R is a lattice-ordered semiring then $(int(R), [\vee], [\wedge])$ is a lattice and $(int(R), [+], [\cdot])$ is a lattice-ordered semiring. Moreover, we have a morphism of semirings $R \to int(R)$ given by $r \mapsto [r, r]$ for all $r \in R$.

(21.18) Proposition. *If R is a DLO-semiring then in $int(R)$ we have:*

(1) $[a, b][+][c, d] = \{u + v \mid u \in [a, b], v \in [c, d]\}$;
(2) $[a, b][\cdot][c, d] \supseteq \{uv \mid u \in [a, b], v \in [c, d]\}$.

Proof. (1) If $u \in [a, b]$ and $v \in [c, d]$ then $a + c \leq u + v \leq b + d$ and so $u + v \in [a, b][+][c, d]$. Conversely, if $w \in [a, b][+][c, d]$ then $a \leq w \wedge b \leq b$ and $c \leq w \wedge d \leq d$. Moreover, $(w \wedge b) + (w \wedge d) = w \wedge (b + d) = w$. Thus we have equality.

(2) If $u \in [a, b]$ and $v \in [c, d]$ then $ac \leq av \leq uv \leq bv \leq bd$ and so $uv \in [a, b][\cdot][c, d]$. □

(21.19) Example. If $(R, \vee, \wedge)$ is a lattice which is not distributive, then $[a, b][\vee][c, d]$ is not necessarily equal to $\{u \vee v \mid u \in [a, b], v \in [c, d]\}$ and, indeed, the latter may not be an interval at all. To see this, consider the lattice $R = \{0, a_1, a_2, a_3, 1\}$ in which $0 \leq a_i \leq 1$ for $1 \leq i \leq 3$ but the a_i are incomparable among themselves. Then $R = [0, a_1][\vee][0, a_2]$ but $[0, a_3] \notin \{u \vee v \mid u \in [0, a_1], v \in [0, a_2]\}$.

If we have equality in Proposition 21.18(2), then we say that the lattice-ordered semiring R is **divisory**.

(21.20) APPLICATION. The lattice-ordered semiring $int(\mathbb{R})$ is introduced and studied in detail in [Moore, 1966] as a tool in numerical analysis and the treatment of floating-point computations in computers with a fixed word length. Refer also to [Alefeld & Herzberger, 1983] and [Moore, 1979]. For the use of $int(\mathbb{R})$ and $int(\mathbb{R} \cup \{-\infty, \infty\})$ in global optimization theory, see [Hansen, 1992].

A subset A of a lattice L is a **lattice ideal** if and only if $a \in A$ and $b \in L$ imply that $a \wedge b \in A$. In particular, if $a \in A$ and $b \leq a$ then $b \in A$. Thus every subset A of L is contained in a unique smallest lattice ideal of L, namely $(A] = \{b \in L \mid b \leq a \text{ for some } a \in A\}$. If $a \in L$ we write $(a]$ instead of $(\{a\}]$.

(21.21) EXAMPLE. Let $(R, +, \cdot)$ be a lattice-ordered semiring and let $1 \neq a \in R$. Set $[a) = \{r \in R \mid r \geq a\}$ and define an operation $*$ on $[a)$ by $r * r' = rr' + a$. Then, using Proposition 21.12, it is straightforward to verify that $([a), +, *)$ is also a lattice-ordered semiring with additive identity a and multiplicative identity 1. Moreover, the function $\gamma_a \colon R \to [a)$ given by $r \mapsto r + a$ is a surjective morphism of semirings. Note that $[a)$ is a lattice ideal of R but is not an ideal of R.

(21.22) PROPOSITION. *The following conditions on an ideal I of a lattice-ordered semiring R are equivalent:*

(1) *I is a lattice ideal;*
(2) *I is a strong ideal;*
(3) *I is a subtractive ideal.*

PROOF. (1) $\Rightarrow$ (2): Let a and b be elements of R satisfying $a + b \in I$. Then $a = a \wedge (a \vee b) = a \wedge (a + b)$ and so, by (1), $a \in I$. Similarly, $b \in I$ and so I is a strong ideal of R.

(2) $\Rightarrow$ (3): This is trivial.

(3) $\Rightarrow$ (1): Let $a \in I$ and let $r \in R$. Then $a = a \vee (a \wedge r) = a + (a \wedge r) \in I$ and so, by (3), $a \wedge r \in I$. Thus I is a lattice ideal of R. □

In Chapter 5 we noted that the sum of subtractive ideals of a semiring need not be subtractive. However, this condition does hold in the case of lattice-ordered semirings.

(21.23) COROLLARY. *If $\{I_k \mid k \in \Omega\}$ is a family of subtractive ideals of a lattice-ordered semiring R then $\sum_{j \in \Omega} I_j$ is subtractive.*

PROOF. Let $a \in \sum_{j \in \Omega} I_j$ and let $b \in R$. Then there exists a finite subset Λ of Ω and elements $a_k \in I_k$ for all $k \in \Lambda$ such that $a = \sum_{k \in \Lambda} a_k$. Thus, by Proposition 21.22, $a \wedge b = (\sum_{k \in \Lambda} a_k) \wedge b = \sum_{k \in \Lambda} (a_k \wedge b) \in \sum_{j \in \Omega} I_j$ and so $\sum I_j$ is a lattice ideal of R and hence, by Proposition 21.22, a subtractive ideal of R. □

In particular, if R is a lattice-ordered semiring then the family consisting of R and all of its subtractive ideals is a sublattice of $ideal(R)$.

(21.24) PROPOSITION. *If I is an ideal of a lattice-ordered semiring R which is also a lattice ideal and if a and b are elements of R satisfying $ab \in I$ then $(a)(b) \subseteq I$.*

PROOF. If r, r', and r'' are elements of R, then $rar' \leq a$ and $br'' \leq b$ so $rar'br'' \leq ab$. Thus $rar'br'' \in I$. Since every element of $(a)(b)$ is a finite sum of elements of R of this form, it follows that $(a)(b) \subseteq I$. □

(21.25) PROPOSITION. *The following conditions on an ideal I of a lattice-ordered semiring R which is also a lattice ideal are equivalent:*

(1) *I is prime.*

(2) *If a and b are elements of R satisfying $ab \in I$ then either $a \in I$ or $b \in I$.*

PROOF. Assume (1) and suppose that I is prime and that a and b are elements of R satisfying $ab \in I$. By Proposition 21.24 we see that $(a)(b) \subseteq I$ and so either $(a) \subseteq I$ or $(b) \subseteq I$. This implies that either $a \in I$ or $b \in I$. Thus (1) implies (2). Conversely, (2) implies (1) by Proposition 7.4. □

Let R and S be lattice-ordered semirings. If $\gamma: R \to S$ is a morphism of semirings then γ is a morphism between the semigroups $(R, +)$ and $(S, +)$, and so is order-preserving. Indeed, if $r \geq r'$ in R then $r' + r = r' \vee r = r$ and so $\gamma(r') \vee \gamma(r) = \gamma(r') + \gamma(r) = \gamma(r' + r) = \gamma(r)$ and so $\gamma(r) \geq \gamma(r')$.

(21.26) PROPOSITION. *Let R and S be lattice-ordered semirings, let $\gamma: R \to S$ be a morphism of semirings, and let $\delta: R \to S$ be a morphism of lattices. Then:*

(1) *$\gamma(r \wedge r') \leq \gamma(r) \wedge \gamma(r')$ for all $r, r' \in R$;*

(2) *$\delta(r + r') \geq \delta(r) + \delta(r')$ for all $r, r' \in R$.*

Moreover, if either γ or δ is bijective then the corresponding inequality becomes an equality.

PROOF. Since, as already noted, morphisms of semirings between lattice-ordered semirings are order-preserving, we have $\gamma(r \wedge r') \leq \gamma(r)$ and similarly $\gamma(r \wedge r') \leq \gamma(r')$. This suffices to establish (1). If γ is bijective then it has an inverse γ^{-1}. Note that $\gamma^{-1}(\gamma(r) + \gamma(r')) = \gamma^{-1}\gamma(r + r') = r + r' = \gamma^{-1}\gamma(r) + \gamma^{-1}\gamma(r')$ and so γ^{-1} is a morphism between the semigroups $(S, +)$ and $(R, +)$. This implies that γ^{-1} is order-preserving and so, as before, $\gamma^{-1}(\gamma(r) \wedge \gamma(r')) \leq \gamma^{-1}\gamma(r) \wedge \gamma^{-1}\gamma(r') = r \wedge r'$ so, applying γ, we get $\gamma(r) \wedge \gamma(r') \leq \gamma(r \wedge r')$, and thus we have equality, proving (1). The proof of (2) is similar. □

If R is a lattice-ordered semiring and $\nu: R \to R$ is a nucleus then $\nu(ab) \leq \nu(a) \wedge \nu(b)$ for all $a, b \in R$, since ν is order-preserving. The following result gives necessary and sufficient conditions for equality.

(21.27) PROPOSITION. *If R is a lattice-ordered semiring and $\nu: R \to R$ is a nucleus then $\nu(ab) = \nu(a) \wedge \nu(b)$ for all $a, b \in R$ if and only if the following conditions are satisfied:*

(1) *$\nu(ab) = \nu(ba)$ for all $a, b \in R$;*

(2) *$\nu(a^2) = \nu(a)$ for all $a \in R$.*

PROOF. If $\nu(ab) = \nu(a) \wedge \nu(b)$ for all $a, b \in R$ then surely (1) and (2) are satisfied. Conversely, assume that they are satisfied. Then $a \wedge b \leq \nu(a) \wedge \nu(b)$ and so $\nu(a \wedge b) \leq \nu(\nu(a) \wedge \nu(b)) = \nu([\nu(a) \wedge \nu(b)]^2) \leq \nu(\nu(a)\nu(b)) \leq \nu(\nu(ab)) = \nu(ab)$. The reverse inequality, as we have noted above, is always true and so we have equality. □

Note that the nucleus given in Example 20.50 satisfies the conditions of Proposition 21.27.

A **multiplicative filter** on a partially-ordered semiring R is a nonempty subset F of R satisfying $r, r' \in F \Rightarrow rr' \in F$ and $r \geq r' \in F \Rightarrow r \in F$.

(21.28) EXAMPLE. If a is an element of a semiring R then $F = \{I \in ideal(R) \mid a^n \in I\}$ for some $n \geq 1$ is a multiplicative filter on the partially-ordered semiring $ideal(R)$.

(21.29) EXAMPLE. Let ρ be a congruence relation an a difference-ordered semiring R having a multiplicatively-idempotent infinite element a. Then $\{r \in R \mid r \rho a\}$ is a multiplicative filter on R.

If $\{F_i \mid i \in \Omega\}$ is a family of multiplicative filters on a partially-ordered semiring R then $\cap_{i \in \Omega} F_i$ is again a multiplicative filter on R, so the set of all multiplicative filters on R is a complete lattice. If $a \in R$ then the smallest multiplicative filter on R containing A is $\{r \in R \mid r \geq a^k\}$ for some $k \in \mathbb{N}$.

If F is a multiplicative filter on an additively-idempotent simple semiring R and a is an element of R, set $F_a = \{r \in R \mid r + a \in F\}$. Then clearly $F_0 = F$. Define a relation $\equiv_F$ on R by setting $a \equiv_F b$ if and only if $F_a = F_b$. This is clearly an equivalence relation. We claim that it is a congruence relation as well. Indeed, assume that $a \equiv_F b$ and $c \equiv_F d$ in R. Then

$$\begin{aligned} F_{a+c} &= \{r \in R \mid r + a + c \in F\} \\ &= \{r \in R \mid r + a \in F_c\} \\ &= \{r \in R \mid r + a \in F_d\} \\ &= \{r \in R \mid r + a + d \in F\} \\ &= \{r \in R \mid r + d \in F_a\} \\ &= \{r \in R \mid r + d \in F_b\} \\ &= \{r \in R \mid r + b + d \in F\} \\ &= F_{b+d} \end{aligned}$$

Thus $a+c \equiv_F b+d$. Moreover, $r \in F_{ac} \Rightarrow r+ac \in F \Rightarrow r+a \in F$ and $r+c \in F \Rightarrow r+b \in F$ and $r+d \in F \Rightarrow (r+b)(r+d) \in F \Rightarrow r^2+br+rd+bd \in F \Rightarrow r+bd \in F$ since $r \geq r^2 + br + rd \Rightarrow r \in F_bd$ and so $F_{ac} \subseteq F_{bd}$. A similar argument shows the reverse containment, and so we have $F_{ac} = F_{bd}$. Thus $ac \equiv_F bd$.

The same argument used above can make another point: let R be a lattice-ordered semiring and let F be a multiplicative filter on R. If $a, b \in R$ and $r \in F_{ab}$ then $r + ab \in F$ and so $r + (a \wedge b) \in F$ since $ab \leq a \wedge b$. Thus $r \in F_{a \wedge b}$. Conversely, if $r \in F_{a \wedge b}$ then $r + (a \wedge b) \in F$, whence $r + a \in F$ and $r + b \in F$. Thus $r^2 + rb + ar + ab = (r + a)(r + b) \in F$ and so $r + ab \in F$, whence $r \in F_{ab}$. This shows that $ab \equiv_F a \wedge b$. As an immediate consequence we see that $ab \equiv_F ba$ for all $a, b \in R$.

If R is an additively-idempotent simple semiring and F is a multiplicative filter on R, we can therefore form the factor semiring $R_F = R/\equiv_F$. The elements of R_F are just the sets of the form F_a for $a \in R$. The operations on R_F are defined by $F_a + F_b = F_{a+b}$ and $F_a \cdot F_b = F_{ab}$. In particular, we note that if R is a lattice-ordered semiring then $F_a \cdot F_b = F_b \cdot F_a$ and $(F_a)^2 = F_a$ for all $a, b \in R$. Thus R_F is commutative and multiplicatively idempotent. Moreover, it is simple and additively idempotent since R is. Therefore, by the result cited in Example 1.5, R_F is a bounded distributive lattice.

22. COMPLETE SEMIRINGS

We now want to consider the possibility of infinite sums in semirings. Semirings having infinite sums, such as *ideal*(R) for any ring R, are well-known and the ability to take infinite (or at least countably-infinite) sums is, as we shall see, very important in certain applications.

Let R be a semiring. A family $\mathcal{A}$ of functions of the form $\theta: \Omega \to R$, where Ω is a set, is **admissible** if and only if to each θ in $\mathcal{A}$ we can assign a value $\sum \theta$ in R such that the following conditions are satisfied:

(1) If $\Omega = \varnothing$ then $\sum \theta = 0$.

(2) If $\Omega = \{h_1, \ldots, h_n\}$ is a finite set, then any function $\theta: \Omega \to R$ belongs to $\mathcal{A}$ and $\sum \theta = \theta(h_1) + \cdots + \theta(h_n)$.

(3) A function $\theta: \Omega \to R$ belongs to $\mathcal{A}$ if and only if, for each $r \in R$, the functions $r\theta: \Omega \to R$ and $\theta r: \Omega \to R$ defined by $r\theta: i \mapsto r\theta(i)$ and $\theta r: i \mapsto \theta(i)r$ belong to $\mathcal{A}$. Moreover, in this situation, $\sum[r\theta] = r[\sum \theta]$ and $\sum[\theta r] = [\sum \theta]r$.

(4) If $\Omega = \cup_{j \in \Lambda} \Omega_j$ is a partition of Ω then $\theta: \Omega \to R$ belongs to $\mathcal{A}$ if and only if the restriction θ_j of θ to each Ω_j belongs to $\mathcal{A}$ and the function $\varphi: \Lambda \to R$ defined by $\varphi: j \mapsto \sum \theta_j$ belongs to $\mathcal{A}$ as well. Moreover, in this situation, $\sum \theta = \sum \varphi$.

The assignment $\theta \mapsto \sum \theta$ is called a **summation** on $\mathcal{A}$. A semiring R is $\mathcal{A}$-**complete** if $\mathcal{A}$ is an admissible family of functions with values in R with a specified summation. In particular, R is **countably complete** if $\mathcal{A}$ is the family of all functions from countable sets to R and R is **complete** if and only if $\mathcal{A}$ is the family of all functions with values in R.

Complete semirings have been studied in [Eilenberg, 1974], [Goldstern, 1985], and [Krob, 1987], based on ideas first presented in [Conway, 1971], all in connection with automata theory, where the problem of infinite summation is central. Semirings which are $\mathcal{A}$-complete were also studied in [Mahr, 1984]. See also [Higgs, 1980]. For an application of such semirings to quantum statistics, refer to [Belavkin, 1987]. On the face of it, the notion of a complete semiring seems to run into foundational difficulties since the family of all functions with values in R is clearly a proper class. However, Goldstern [1985] has shown that if R is a complete semiring then there exists a cardinal number c such that for each function $\theta: \Omega \to R$ there exists a subset Λ of Ω having cardinality at most c such that $\sum \theta = \sum \theta'$, where θ' is the restriction of θ to Λ.

Note: to simplify notation, we will often identify a function $\theta: \Omega \to R$ with its image in R and write $\sum A$, where A is an indexed family of (not necessarily distinct) elements of R.

(22.1) EXAMPLE. The boolean semiring $\mathbb{B}$ is complete if, for each function $\theta: \Omega \to \mathbb{B}$ we define $\sum \theta$ to equal 0 if $im(\theta) = \{0\}$ and to equal 1 otherwise.

(22.2) EXAMPLE. If $(R, \vee, \wedge)\}$ is a frame, then R is a complete additively-idempotent semiring in which, for each function $\theta: \Omega \to R$, we define $\sum \theta = \vee_{i \in \Omega}\theta(i)$. In particular, the semiring of open subsets of a topological space is complete.

(22.3) EXAMPLE. Let R be the semiring $(\mathbb{R}^+ \cup \{-\infty, \infty\}, max, +)$. Then R is complete since, for each function $\theta: \Omega \to R$ we can define $\sum \theta$ to be $sup\{\theta(h) \mid h \in \Omega\}$.

(22.4) EXAMPLE. In Example 1.10 we considered the semiring $(sub(A^*), \cup, \cdot)$ of all languages on an alphabet A. This semiring is surely complete.

(22.5) EXAMPLE. If R is a complete semiring and A is a nonempty set then the semiring R^A is complete if, for each function $\theta: \Omega \to R^A$, we define $(\sum \theta)(a) = \sum_{a \in A} \theta(a)$.

(22.6) EXAMPLE. Countably-complete semirings have important applications in the analysis of iteration theories [Bloom & Ésik, 1993] and in automata theory. Such semirings are not necessarily complete [Krob, 1987]. Indeed, let P be the set of all countable subsets of $\mathbb{R}$ and let $R = P \cup \{\mathbb{R}\}$. Then $(R, \cup, \cap)$ is an additively-idempotent semiring. Moreover, if $\theta: \mathbb{N} \to R$ then we can define $\sum \theta$ to be $\cup_{i \in \mathbb{N}}\theta(i)$. Thus R is countably complete. We claim that R is not complete. Indeed, let $\Omega = \mathbb{R} \setminus \{0\}$ and let $\theta: \Omega \to R$ be the function defined by $\theta: r \mapsto \{r\}$. Set $b = \sum \theta$ Suppose $b \in P$. Since Ω is uncountable, there exists an element c in $\Omega \setminus b$. If θ' is the restriction of θ to $\Omega \setminus b$, then $b = \{c\} \cup \sum \theta'$ and hence $c \in b$, which is a contradiction. Therefore we must have $b = \mathbb{R}$. Then $b \cap \{0\} = \{0\} = \sum[\theta \cap \{0\}] = \sum \varnothing = \varnothing$, which is again a contradiction. Thus $\sum \theta$ cannot exist, and so R is not complete.

(22.7) EXAMPLE. [Goldstern, 1985] Let $R = \mathbb{N} \cup \{z, z'\}$, where z and z' are elements not in $\mathbb{N}$. Define operations of addition and multiplication on R to be the usual operations of addition and multiplication on $\mathbb{N}$ augmented as follows:

(1) $n + z = nz = z + z = zz = z$ for all $0 \neq n \in \mathbb{N}$;
(2) $n + z' = nz' = z + z' = z' + z' = z'z' = z'z = zz' = z'$ for all $0 \neq n \in \mathbb{N}$;
(3) $0 + z = z$;
(4) $0 + z' = z'$;
(5) $0z = 0z' = z0 = z'0 = 0$.

If $\theta: \Omega \to R$ then we set

$$\sum \theta = \begin{cases} \sum\{\theta(i) \mid \theta(i) \neq 0\} & \text{if } \Lambda = supp(\theta) \text{ is finite} \\ z' & \text{if } \Lambda \text{ is uncountable or if there exists a } j \in \Omega \\ & \text{such that } \theta(j) = z' \\ z & \text{otherwise} \end{cases}$$

Then R is a complete semiring.

(22.8) APPLICATION. Let D be a nonempty set and let $S = sub(D \times D)$ be the set of all relations on D. Define addition and multiplication on S by setting $r + s = r \cup s$ and $rs = \{(d, d'') \in D \times D \mid$ there exists an element $d' \in D$ with $(d, d') \in r$ and $(d', d'') \in s\}$. Then S is a complete positive zerosumfree semiring partially ordered by inclusion, having additive identity $\varnothing$ and multiplicative identity $\Delta = \{(d, d) \mid d \in D\}$. Assume that D has a distinguished element $\perp$ and let $R = \{\varnothing\} \cup \{r \in S \mid (\perp, d) \in r$ if and only if $d = \perp\}$. Then R is a subsemiring of S. As a partially-ordered set, R has a unique atom $z = \{(\perp, \perp)\}$. If $\varnothing \neq r, r' \in R$ then $z = z^2 \subseteq rr'$ and so R is entire. This semiring R was used in [Main & Black, 1993] as a model for computations in an abstract "computer" completely determined by being in one of a set D of states, among them the distinguished state $\perp$ of being in an unending loop. The nonempty elements of R are **nondeterministic procedures** which the computer is to execute.

Let $\{R_i \mid i \in \Lambda\}$ be a family of semirings and let $R = \times_{i \in \Lambda} R_i$. For each i in Λ, let $\gamma_i \colon R \to R_i$ be the canonical projection onto the ith component. Let $\mathcal{A}$ be a family of functions into R and, for each $i \in \Lambda$, let $\mathcal{A}_i = \{\gamma_i \theta \mid \theta \in \mathcal{A}\}$. Then it is easy to verify that:

(1) If $\mathcal{A}$ is admissible then so is $\mathcal{A}_i$ for each $i \in \Lambda$. In this case, summation on $\mathcal{A}_i$ is defined by $\sum \gamma_i(\theta) = \gamma_i(\sum \theta)$ for each $\theta \in \mathcal{A}$.

(2) If $\mathcal{A}_i$ is admissible for each $i \in \Lambda$ then $\mathcal{A}$ is admissible. In this case, the summation on $\mathcal{A}$ is defined by the condition that $\gamma_i(\sum \theta) = \sum \gamma_i \theta$ for all $i \in \Lambda$.

The following examples show how a complete semiring generates new semirings.

(22.9) EXAMPLE. [Goldstern, 1985] If R is a complete semiring and if A is a nonempty set then the semiring of formal power series $R\langle\langle A \rangle\rangle$ in A over R is also complete. Indeed, if $\theta \colon \Omega \to R\langle\langle A \rangle\rangle$ is a function, then for each word $w \in A^*$ we have a function $\theta_w \colon \Omega \to R$ defined by $\theta_w(i) = [\theta(i)](w)$. Then define $(\sum \theta)(w) = \sum \theta_w$ for each $w \in A^*$.

(22.10) EXAMPLE. If R is an $\mathcal{A}$-complete semiring and Ω is a nonempty set such that every function from Ω to R belongs to $\mathcal{A}$, then we can define the **semiring $\mathcal{M}_\Omega(R)$ of $(\Omega \times \Omega)$-matrices** on R. This is the set $R^{\Omega \times \Omega}$ on which addition is defined componentwise. If $f, g \in R^{\Omega \times \Omega}$ and if $(i, j) \in \Omega \times \Omega$, then $(fg)(i, j)$ is defined to be $\sum \theta$, where $\theta \colon \Omega \to R$ is the function defined by $\theta(k) = f(i, k)g(k, j)$.

(22.11) EXAMPLE. If R is a complete semiring and if $(M, *)$ is a monoid, then any favorable family $\mathcal{C}$ of subsets of M is R-favorable and so we can define the convolution algebra $(R[\mathcal{C}], +, \langle * \rangle)$.

(22.12) APPLICATION. By Example 1.5, we know that $(\mathbb{I}, max, min)$ is a simple semiring, which is complete, idempotent, and commutative. A function $f \colon \mathbb{R}^+ \to \mathbb{I}$ is called a **fuzzy nonnegative real number** if and only if, for each $\{r \in \mathbb{R}^+ \mid f(r) \geq h\}$ is a nonempty closed interval in $\mathbb{R}^+$. In particular, we see that this condition implies that there exists a real number r_0 for which $f(r_0) = 1$. We will denote the set of all fuzzy real numbers by $fuzz(\mathbb{R}^+)$. For details, see [Kaufmann & Gupta, 1985]. In particular, it is shown there that $fuzz(\mathbb{R}^+)$ is closed under the convolutions $\langle + \rangle$, defined by

$$(f\langle + \rangle g)(r) = sup\{min\{f(r'), g(r'')\} \mid r = r' + r''\}$$

and $\langle \cdot \rangle$ defined by

$$(f\langle \cdot \rangle g)(r) = sup\{min\{f(r'), g(r'')\} \mid r = r'r''\}.$$

Moreover, $(fuzz(\mathbb{R}^+), \langle + \rangle, \langle \cdot \rangle)$ is a commutative semiring, the additive identity f_0 and the multiplicative identity f_1 in which are defined by $f_k(r) = k$ if $r = k$ and $f_k(r) = 0$ otherwise. This semiring has important applications in the computation under conditions of uncertainty.

The notion of a fuzzy nonnegative real number can be generalized in many directions to fit different applications. Many such generalizations are considered in [Kaufmann & Gupta, 1985]. Thus, for example, we can define the notion of the set $fuzz(\mathbb{R}^{+n})$ of all **fuzzy nonnegative real numbers of dimension n**, for n some positive integer, which, in a manner analogous to the above, can be turned into a semiring by using the convolutions $\langle + \rangle$ and $\langle \cdot \rangle$ appropriately defined.

(22.13) APPLICATION. Let R be a complete simple semiring. A **constraint system** over R consists of a pair (D, V), where D is a finite set and V is an ordered set of variables. A **constraint** is a pair (K, δ), where $K \subseteq V$ is the **type** of the constraint and $\delta: D^{|K|} \to R$ is the **value** of the constraint. Problems involving constraints and constraint satisfaction play an important role in optimization theory and the modeling of optimization schemes. By considering the general framework of constraints over semirings, one can model not only classical constraint problems but also fuzzy constraint problems and weighed constraint problems. Refer to [Bistarelli et al., 1997a, 1997b] and [Georget & Codognet, 1998] for an introduction to this theory.

(22.14) PROPOSITION. *Let R be an entire positive totally-ordered semiring and let ∞ be an element not in R. Extend the order on R to one on $R\{\infty\}$ by setting $r < \infty$ for all $r \in R$. For each function $\theta: \Omega \to R\{\infty\}$ define $\sum \theta$ to be $sup\{\sum \theta' \mid \theta'$ the restriction of θ to a finite subset of $\Omega\}$. Then $R\{\infty\}$ is a complete semiring.*

PROOF. First note that R is zerosumfree by Proposition 20.16 and so, the semiring $R\{\infty\}$ is well-defined. It is easily seen that it is complete. □

If particular, we note that $\mathbb{N}\{\infty\}$ and $\mathbb{R}^+\{\infty\}$ are complete semirings, as is $R\{\infty\}$, where R is the schedule algebra.

(22.15) EXAMPLE. [Goldstern, 1985] The above construction can be generalized as follows: Let R be a uniquely difference-ordered semiring (e.g. $\mathbb{N}$) having additive identity z and multiplicative identity e and let $\bar{R} = \{\bar{r} \mid r \in R\}$ be a set bijectively corresponding to R and disjoint from it. Set $S = R \cup \bar{R}$ and define operations $\oplus$ and $\odot$ on S as follows:

(1) If $a, b \in R$ then $a \oplus b$ and $a \odot b$ are just the sum and product of these elements in R;

(2) If $\bar{a}, \bar{b} \in \bar{R}$ then $\bar{a} \oplus \bar{b} = \bar{z} = \bar{a} \odot \bar{b}$;
(3) If $a \in R$ and $\bar{b} \in \bar{R}$ then $a \oplus \bar{b} = \bar{b} \oplus a = \bar{c}$, where $c = z$ if $a \nleq b$ and c is the unique element of R satisfying $a + c = b$ otherwise.
(4) $e \odot \bar{a} = \bar{a} = \bar{a} \odot e$ for all $\bar{a} \in \bar{R}$;
(5) $z \odot \bar{a} = z = \bar{a} \odot z$ for all $\bar{a} \in \bar{R}$;
(6) $a \odot \bar{b} = \bar{z} = \bar{b} \odot a$ for all $a \in R \setminus \{e, z\}$ and $\bar{b} \in \bar{R}$.

Extend the partial order on R to a partial order on S by setting $a \leq \bar{b}$ for all $a \in R$ and $\bar{b} \in \bar{R}$, and $\bar{a} \leq \bar{b}$ in $\bar{R}$ if and only if $a \geq b$ in R.

For each function $\theta: \Omega \to S$, define $\sum \theta$ to be equal to $\theta(h_1) \oplus \cdots \oplus \theta(h_n)$ if θ has finite support $\{h_1, \ldots, h_n\}$, and equal to $\bar{z}$ otherwise. Then the semiring S is complete.

The following result is extremely important.

(22.16) PROPOSITION. *Let R be a $\mathcal{A}$-complete semiring and let Ω and Λ be sets between which there exists a bijective correspondence $\tau: \Omega \to \Lambda$. If $\theta: \Omega \to R$ and $\psi: \Lambda \to R$ are functions in $\mathcal{A}$ satisfying $\psi\tau = \theta$ then $\sum \theta(\Omega) = \sum \psi(\Lambda)$.*

PROOF. Set $a = \sum \theta(\Omega)$. Define a partition $\{\Omega_j \mid j \in \Lambda\}$ of Ω by setting $\Omega_j = \tau^{-1}(j)$. Then $\theta(\Omega_j) = \psi(j)$ for each $j \in \Lambda$ and so, by the definition of a complete semiring, $a = \sum \psi(\Lambda)$. □

In particular, Proposition 22.16 implies that if R is an $\mathcal{A}$-complete semiring and if θ is a function in $\mathcal{A}$ then the sum $\sum \theta$ is independent of any ordering of Ω. This is a generalization of the condition that in a semiring the operation of addition is commutative. Next, we note that summations are not necessarily uniquely determined by the addition in R.

(22.17) EXAMPLE. [Goldstern, 1985; Kuich, 1987] Let ∞ be an element not in $\mathbb{B}$ and let $R = \mathbb{B}\{\infty\}$. Then R is commutative and additively idempotent. Furthermore, it is totally ordered by the relation $0 \leq 1 \leq \infty$. Then, for each function $\theta: \Omega \to R$, we can define $\sum \theta$ in two different ways:

(1) $\sum \theta = sup\{\theta(i) \mid i \in \Omega\}$; and
(2) $\sum \theta = \infty$ if and only if θ does not have finite support.

We will say that a summation on an admissible family $\mathcal{A}$ of functions is **necessary** if and only if, for functions $\theta, \theta': \Omega \to R$ in $\mathcal{A}$ satisfying the condition that each finite subset Λ of Ω is contained in a finite subset Λ' of Ω such that $\sum\{\theta(i) \mid i \in \Lambda'\} = \sum\{\theta'(i) \mid i \in \Lambda'\}$ we have $\sum \theta = \sum \theta'$. In particular, a countably-complete semiring R has a necessary summation if and only if it satisfies the condition that for any $\theta, \theta': \mathbb{N} \to R$ we have $\sum \theta = \sum \theta'$ whenever $\theta(0) + \cdots + \theta(n) = \theta'(0) + \cdots + \theta'(n)$ for each natural number n, for all n greater than or equal to some $n_0 \in \mathbb{N}$.

(22.18) EXAMPLE. The summation given in Example 22.17(1) and in Proposition 22.14 is necessary that given in Example 22.17(2) is not necessary.

(22.19) EXAMPLE. [Goldstern, 1985] If R is a countably complete semiring having necessary summation and if $\gamma: R \to S$ is a surjective morphism of semirings then the summation in S need not be necessary. For example, let y be an element not in $\mathbb{N}$ and let $R = \mathbb{N}\{y\}$. For any countable set Ω and any function $\theta: \Omega \to R$, define $\sum \theta$ to be y if and only if θ does not have finite support. This summation is easily seen to be necessary. If $S = \mathbb{B}\{\infty\}$ for some element ∞ not in $\mathbb{B}$ and if for each function $\theta: \Omega \to S$, with Ω again a countable set, we define $\sum \theta = \infty$ if and only if θ does not have finite support then this summation is, as we have already noted, not necessary. However, there exists a surjective morphism of semirings $\gamma: R \to S$ defined by setting $\gamma(y) = \infty$, $\gamma(0) = 0$, and $\gamma(n) = 1$ for all $0 < n \in \mathbb{N}$.

(22.20) EXAMPLE. In Chapter 5 we noted that if R is a semiring then it is possible to define infinite summation in the semiring *ideal*(R) of all ideals of R. This summation is clearly necessary.

Let R be a positive partially-ordered semiring and let $\mathcal{A}$ be an admissible family of functions of the form $\theta: \Omega \to R$ such that R is $\mathcal{A}$-complete. Then R is **finitarily $\mathcal{A}$-complete** if and only if, for each function $\theta: \Omega \to R$ in $\mathcal{A}$ and for each element a of R satisfying the condition:

(*) If Λ is a finite subset of Ω then $\sum_{i\in\Lambda} \theta(i) \leq a$

we have $\sum \theta \leq a$. The semiring R is said to be **finitarily** [resp. **countably**] **complete** if and only if it is finitarily $\mathcal{A}$-complete, where $\mathcal{A}$ is the family of all functions [resp. from countable sets] with values in R.

(22.21) EXAMPLE. The semiring $\mathbb{N}\{\infty\}$, as defined in Proposition 22.14, is finitarily complete.

(22.22) EXAMPLE. Let R be the semiring $\mathbb{B}\{\infty\}$ with sums of the form $\sum \theta$ defined as in Example 22.17(2). Then R is not finitarily complete.

(22.23) EXAMPLE. The semiring S defined in Example 22.15 is not finitarily countably complete since if $\theta: \mathbb{N} \to S$ is defined by $\theta(n) = e$ for all $n \in \mathbb{N}$ then $\sum \theta = \bar{z}$, while $\sum_{i\in\Lambda} \theta(i) \leq \bar{e} < \bar{z}$ for all finite subsets Λ of $\mathbb{N}$.

(22.24) PROPOSITION. *Let R be a finitarily countably-complete semiring, let $a \in R$, and let $\theta: \mathbb{N} \to R$ be a function satisfying the condition that $\theta(0)+\cdots+\theta(n) = a$ for each natural number n. Then $\sum \theta = a$.*

PROOF. If Λ is a finite subset of $\mathbb{N}$ having maximal element n then $\sum_{i\in L} \theta(i) \leq \theta(0) + \cdots + \theta(n) = a$. Since R is finitarily countably-complete, this implies that $\sum \theta \leq a$. Since R is positive, $a \leq \sum \theta$ and so we have equality. □

(22.25) PROPOSITION. *Let R be a finitarily countably-complete semiring. Then $a + b + c = a$ implies $a + b = a$ for all elements a, b, and c of R.*

PROOF. Let $a + b + c = A$ and define a function $\theta: \mathbb{N} \to R$ by

$$\theta: n \mapsto \begin{cases} a & \text{if } n = 0 \\ b & \text{if } n = 2i-1 \text{ for some } i \in \mathbb{P} \\ c & \text{if } n = 2i \text{ for some } i \in \mathbb{P} \end{cases} .$$

Let $d = \sum \theta$. Then

$$d = \sum_{n\geq 1} [\theta(2n-1) + \theta(2n)] = a + (b+c) + (b+c) + \ldots.$$

Since $a+(b+c)=a$, we see that all of the finite partial sums are equal to a and so, by Proposition 22.24, we have $d=a$. On the other hand,

$$d=\sum_{n\geq 0}[\theta(2n)+\theta(2n+1)]=(a+b)+(c+b)+(c+b)+\ldots.$$

Since $(a+b)+(c+b)=a+b$, we see that all of the finite partial sums are equal to $a+b$ and so, by Proposition 22.24, we have $d=a+b$. Therefore $a=a+b$. □

Thus, combining Propositions 20.37 and 22.25, we see that every finitarily countably-complete semiring is difference ordered. Goldstern [1985] has proven the converse as well: any difference-ordered semiring is isomorphic to a subsemiring of a finitarily countably-complete semiring.

Note that if R is a $\mathcal{A}$-complete semiring and if $\theta\in\mathcal{A}$ is the 0-function then $\sum\theta=\sum_{i\in\Omega}0\theta(i)=0(\sum\theta)=0$.

(22.26) Proposition. *Let R be an $\mathcal{A}$-complete semiring and let $\theta:\Omega\to R$ and $\varphi:\Lambda\to R$ be elements of $\mathcal{A}$. Then the function $\psi:\Omega\times\Lambda\to R$ defined by $\psi:(i,j)\mapsto\theta(i)\varphi(j)$ belongs to $\mathcal{A}$ and $\sum\psi=(\sum\theta)(\sum\varphi)$.*

Proof. For each $j\in\Lambda$, the restriction of φ to $\Omega_j=\{(i,j)\mid i\in\Omega\}$ is just θb for $b=\varphi(j)$, and this function belongs to $\mathcal{A}$. Since $\Omega\times\Lambda=\cup\{\Omega_j\mid j\in\Lambda\}$ is a partition of $\Omega\times\Lambda$, the function ψ belongs to $\mathcal{A}$. For each $j\in\Lambda$, let $\theta_j:\Omega\to R$ be the function defined by $\theta_j:i\mapsto\theta(i)\varphi(j)$. Then $\sum\theta_j=(\sum\theta)\varphi(j)$ and so $\sum\psi=\sum\{\sum_{j\in\Lambda}\theta_j\}=\sum_{j\in\Lambda}(\theta)\varphi(j)=(\sum\theta)(\sum\varphi)$. □

Let us note one consequence of this result. A (possibly infinite) subset A of a complete semiring R is a **cover** of R if and only if $\sum A=1$. For example, if R is the set of all open subsets of $\mathbb{R}^n$ (in the usual topology) then $(R,\cup,\cap)$ is a complete semiring. For each $\epsilon>0$, the set of all elements of R having Lebesgue measure less than ϵ is a cover of R. If A and B are covers of R, Proposition 22.26 immediately implies that $A\cap B$ is also a cover of R.

(22.27) Proposition. *Every complete semiring R has an infinite element.*

Proof. Let $\Omega=R\times\mathbb{N}$ and define the function $\theta:\Omega\to R$ by $\theta:(r,i)\mapsto r$. For each $r\in R$, let $\Omega_r=\{(r,i)\in\Omega\mid i\in\mathbb{N}\}$ and let θ_r be the restriction of θ to Ω_r. Then $\Omega=\cup_{r\in R}\Omega_r$ is a partition of Ω and so if $a=\sum\theta$ then $a=\sum\{\sum_{r\in R}\theta_r\}$.

If $b\in R$ then $R=\{b\}\cup(R\setminus\{b\})$ and so $a=c+\sum\theta_b$, where

$$c=\sum\left\{\sum\theta_r\,\middle|\,b\neq r\in R\right\}.$$

Then $a+b=c+\sum\theta_b+b$. Now let $\tau:\mathbb{N}\to\mathbb{N}$ be the function defined by $\tau:i\mapsto i+1$. Then $\theta(b,i)=\theta(b,\tau(i))$ for all $i\in\mathbb{N}$. Moreover, $\mathbb{N}=\{0\}\cup im(\tau)$ so $\sum\theta_b=\theta(b,0)+\sum\theta_b'$, where θ_b' is the restriction of θ_b to $im(\tau)$. Thus $\sum\theta_b=b+\sum\theta_b'=b+\sum\theta_b$. This shows that $a+b=c+\sum\theta_b=a$ for all b in R, proving that a is an infinite element of R. □

(22.28) PROPOSITION. *Every countably complete semiring is zerosumfree.*

PROOF. Let a and b be elements of a countably-complete semiring R satisfying $a+b=0$. Define the function θ from $\mathbb{N}$ to R by setting $\theta(i)=a$ if i is even and $\theta(i)=b$ if i is odd. Set $c=\sum\theta$, $d=\sum\{\theta(i)\mid i \text{ odd}\}$, and $e=\sum\{\theta(i)\mid i \text{ even}\}$. Then $c=d+e$. On the other hand, $c=\sum\{\theta(i)+\theta(i+1)\mid i \text{ even}\}=0$.

If $\tau\colon\mathbb{N}\to\mathbb{N}$ is the function defined by $\tau\colon i\mapsto i+2$ then $\theta(i)=\theta\tau(i)$ for all $i\in\mathbb{N}$. Moreover,

$$\begin{aligned} e &= \theta(0)+\sum\{\theta(i+2)\mid i \text{ even}\} \\ &= \theta(0)+\sum\{\theta\tau(i)\mid i \text{ even}\} \\ &= \theta(0)+\sum\{\theta(i)\mid i \text{ even}\} \\ &= \theta(0)+e=a+e. \end{aligned}$$

Therefore $0=d+e=d+a+e=a$ and $0=a+b=0+b=b$. □

If we have a countably-complete semiring with necessary summation then we can improve the result in Proposition 22.28.

(22.29) PROPOSITION. *If R is a countably-complete semiring with necessary summation then R is difference ordered.*

PROOF. Define the relation $\leq$ on R by setting $a\leq b$ if and only if there exists an element c of R such that $a+c=b$. Then for elements a, a', and a'' of R we clearly have $a\leq a$ and $a\leq a''$ whenever $a\leq a'$, and $a'\leq a''$.

Assume that a and b are elements of R satisfying $a\leq b$ and $b\leq a$. Then there exist elements c and d of R such that $a+c=b$ and $b+d=a$. Set $e=c+d$. Then $a+e=a$ and $b+e=b$. Indeed, for each positive integer n we have $b+ne=b$. Let $\theta\colon\mathbb{N}\to R$ be the function defined by $\theta(i)=e$ for each i. Since R is countably complete, $u=\sum\theta$ exists. If $\theta_0\colon\mathbb{N}\to R$ is the function defined by $\theta_0(i)=0$ for all i then $b+\theta(0)+\cdots+\theta(n)=b+\theta_0(0)+\cdots+\theta_0(n)$ for each natural number n and so, since the summation on R is necessary, we see that $b+u=b+\sum\theta_0=b$. On the other hand if φ and φ' are the functions from $\mathbb{N}$ to R defined by $\varphi(i)=c$ and $\varphi'(i)=d$ for each natural number i then, by Proposition 22.16, we have $b=b+u=b+\sum\varphi+\sum\varphi'=b+d+\sum\varphi+\sum\varphi'=d+b+u=d+b=a$. Thus $\leq$ is a partial order relation on R, and so R is difference ordered. □

(22.30) EXAMPLE. Complete semirings need not be entire. For example, let R be the boolean algebra $(sub(\mathbb{N}),\cup,\cap)$. This is a complete semiring. The additive identity of R is $\varnothing$. If A is the set of all even elements of $\mathbb{N}$ and B is the set of all odd elements of $\mathbb{N}$ then both sets are nonzero but $A\cap B=0_R$.

(22.31) PROPOSITION. *If R is a complete simple additively-idempotent semiring with necessary summation then there exists a bijective correspondence between the set of all prime elements of R and the set of characters of R.*

PROOF. By Proposition 22.29 we see that R is difference ordered and so, by Proposition 20.48, we know that each prime element a of R defines a character γ_a

of R by $\gamma_a(r) = 0$ if and only if $r \leq a$. Moreover, if $a \neq b$ are prime elements of R then $\gamma_a \neq \gamma_b$. Conversely, let γ be a character of R and let $a = \sum ker(\gamma)$. Since R has necessary summation, we see that $\gamma(a) = 0$. If a is a unit, we obtain $\gamma(1) = \gamma(a)\gamma(a^{-1}) = 0$, which is a contradiction. Thus a cannot be a unit of R. If $r \leq a$ then there exists an element r' of R satisfying $r + r' = a$ and so $\gamma(r) + \gamma(r') = \gamma(a) = 0$. Since $\mathbb{B}$ is zerosumfree, this implies that $\gamma(r) = 0$. Thus $ker(\gamma) = \{r \in R \mid r \leq a\}$ and so, by Proposition 20.48, a is prime and $\gamma = \gamma_a$. This proves that the map $a \mapsto \gamma_a$ is the bijection we seek. □

(22.32) PROPOSITION. *If R is an entire zerosumfree semiring and $\infty \notin R$ then $R\{\infty\}$ is a complete semiring.*

PROOF. In Chapter 2 we saw how the semiring structure of a zerosumfree semiring R can be extended to $R\{\infty\}$. Consider a function $\theta: \Omega \to R\{\infty\}$, and let $\Lambda = \{i \in \Omega \mid \theta(i) \neq 0\}$. If Λ is finite, set $\sum \theta$ to be the sum of the elements of R of the form $\{\theta(j) \mid j \in \Lambda\}$. Otherwise, set $\sum \theta = \infty$. Then $R\{\infty\}$ is complete. □

(22.33) PROPOSITION. *The following conditions on a complete commutative semiring $(R, +, \cdot)$ are equivalent:*

(1) *$(R, +, \cdot)$ is a frame in which arbitrary joins are given by $\sum$;*
(2) *R is simple and idempotent.*

PROOF. (1) $\Rightarrow$ (2): This is immediate.

(2) $\Rightarrow$ (1): By Proposition 20.19 we see that R is partially ordered by the relation $a \leq b$ if and only if $a + b = b$ and, indeed, it is a meet semilattice with $a \vee b = a + b$ for all $a, b \in R$. If $a, b \in R$ then $ab + a = a(b+1) = a1 = a$ so $ab \leq a$. Similarly $ab \leq b$. On the other hand, if $c \leq a, b$ then $c + a = a$ and $c + b = b$ so $ab = (c+a)(c+b) = c + cb + ac + ab$ so $c \leq c + cb + ac \leq ab$. Thus R is a lattice, with $a \wedge b = ab$.

If we are given a function $\theta: \Omega \to R$ then $\sum \theta \geq \theta(i)$ for each $i \in \Omega$. Suppose that $b \geq \theta(i)$ for each $i \in \Omega$. Then $b\theta(i) = \theta(i)$ so $b(\sum \theta) = \sum_{i \in \Omega} b\theta(i) = \sum \theta$, whence $b \geq \sum \theta$. Therefore $\vee[\theta(\Omega)] = \sum \theta$. The distributivity of meet over arbitrary joins follows from the definition of a complete semiring. □

(22.34) EXAMPLE. If R is an entire zerosumfree semiring and $\infty \notin R$ then $R\{\infty\}$ is a complete semiring by Proposition 22.32. However, $(R\{\infty\}, +, \cdot)$ is not a frame since, if it were a frame, then by the uniqueness of infinite elements, we would have $a = a1 = a\infty = \infty$ for all $a \in R$, which is a contradiction.

If R is a complete semiring and if A, B, C are nonempty sets (or if R is an arbitrary semiring and the set B is finite) and if $h \in R^{A \times B}$ and $k \in R^{B \times C}$ are R-valued relations then we can define the R-valued relation $k \circ h \in R^{A \times C}$ by

$$k \circ h: (a, c) \mapsto \sum_{b \in B} h(a, b)k(b, c).$$

It is straightforward to show that $\circ$ is associative and distributes over addition from either side. Also, if $h = h' + h''$ in $R^{A \times B}$ or $k = k' + k''$ in $R^{B \times C}$ then $k \circ h = k \circ h' + k \circ h''$ and $k \circ h = k' \circ h + k'' \circ h$. If k and h are R-valued functions then $k \circ h$ is also an R-valued function. Indeed, if $a_0 \in A$ then, if there exists an element $b_0 \in B$ for which $f(a_0, b_0) \neq 0$ then that element must be unique. Similarly, if there exists an element $c_0 \in C$ for which $g(b_0, c_0) \neq 0$ then that element must be unique. On the other hand, if $(k \circ h)(a_0, c_1) \neq 0$ for some $c_1 \in C$, then there exists an element $b_1 \in B$ such that $h(a_0, b_1)k(b_1, c_1) \neq 0$, which, by the uniqueness of b_0 and c_0, implies that $b_1 = b_0$ and $c_1 = c_0$. However, we do note that $h \circ k$ could be the 0-map even if k and h are not. This would not be so if the semiring R is entire, for then we would have $(k \circ h)(a_0, c_0) = h(a_0, b_0)k(b_0, c_0) \neq 0$. If R is a complete difference-ordered semiring and A is a nonempty set, then $h \circ h = h$ for each R-valued equivalence relation h on A,

(22.35) PROPOSITION. *Let R be a difference-ordered complete semiring and let A be a nonempty set. Let $f, g, h \in R^{A \times A}$ be R-valued relations on A satisfying the following conditions:*

(1) *$im(f) \subseteq I^{\times}(R)$;*
(2) *The elements of $im(f)$ and $im(g)$ commute.*

Let $f' \in R^{A \times A}$ be defined by $f' \colon (a, b) \mapsto f(b, a)$. Then $(f \circ g)h \leq f \circ [g(f' \circ h)]$.

PROOF. If $a, b \in A$ then

$$\begin{aligned}
[(f \circ g)h](a, b) &= \left[\sum_{c \in A} f(a, c)g(c, b)\right] h(a, b) \\
&= \sum_{c \in A} f(a, c)g(c, b)h(a, b) \\
&= \sum_{c \in A} f(a, c)g(c, b)f'(c, a)h(a, b) \\
&\leq \sum_{c \in A} f(a, c)g(c, b)\,[(f' \circ h)(c, b)] \\
&= \sum_{c \in A} f(a, c)\,[g(f' \circ h)(c, b)] \\
&= (f \circ [g(f' \circ h)])\,(a, b).
\end{aligned}$$

from which the result follows. □

Compositions can, of course, be iterated. In particular, if $h \in R^{A \times A}$ we can define $h^{\circ k}$ for all $k \geq 0$ by setting $h^{\circ 0} = e_0$ and then setting $h^{\circ k} = h^{\circ (k-1)} \circ h$ for all $k > 0$. Moreover, if R is complete we can further define $h^{\circ *} = \sum_{k=0}^{\infty} h^{\circ k}$ to be the reflexive and transitive closure of h. These definitions lead to the operational semantics of R-valued computations, as studied in [Wechler, 1986a].

If R and S are [countably-] complete semirings then a function $\gamma \colon R \to S$ is a **morphism** of **[countably-] complete semirings** if and only if the following conditions are satisfied:

(1) $\gamma(0_R) = 0_S$;
(2) $\gamma(1_R) = 1_S$;
(3) $\gamma(rr') = \gamma(r) \cdot \gamma(r')$ for all $r, r' \in R$;
(4) $\gamma(\sum \theta) = \sum(\gamma\theta)$ for all functions θ from a [countable] set Ω to R.

(22.36) EXAMPLE. [Goldstern, 1985] We extend Example 9.4. If R is a countably-complete semiring then we define a morphism of countably-complete semirings $\gamma_R: \mathbb{N}\{\infty\} \to R$ by setting $\gamma_R(n) = n1_R$ if $n \in \mathbb{N}$ and $\gamma_R(\infty) = \sum \theta$, where $\theta: \mathbb{N} \to R$ is the function defined by $\theta(i) = 1$ for all $i \in \mathbb{N}$. Note that if $\delta: R \to S$ is a morphism of countably-complete semirings, then for each $a \in \mathbb{N}\{\infty\}$ and each $r \in R$ we have $\delta(\gamma_R(a)r) = \gamma_S(a)\delta(r) = \delta(r)\gamma_S(a)$.

(22.37) EXAMPLE. We now extend Example 9.19. Let R be a complete semiring, let A be a nonempty set, and let φ be a function from A to the center $C(R)$ of R. Then φ defines a morphism of complete semirings $\epsilon_\varphi: R\langle\langle A\rangle\rangle \to R$, called the **$\varphi$-evaluation morphism**, given by

$$\epsilon_\varphi: f \mapsto \sum \left\{ f(a_1a_2\cdot\ldots\cdot a_n)\varphi(a_1)\cdot\ldots\cdot(a_n) \mid a_1a_2\cdot\ldots\cdot a_n \in A^* \right\}.$$

Let R be an $\mathcal{A}$-complete semiring for some admissible family $\mathcal{A}$ of functions with values in R. An element b of R is **$\mathcal{A}$-compact** if and only if for each function $\theta: \Omega \to R$ in $\mathcal{A}$ satisfying $\sum\theta = \sum\theta + b$ there exists a finite subset Λ of Ω such that the restriction θ' of θ to Λ satisfies $\sum\theta' = \sum\theta' + b$. The element is **compact** if it is $\mathcal{A}$-compact for the family $\mathcal{A}$ of all functions with values in R. We note that compact elements play an important part in the study of frames; we will also make significant use of this concept in the next chapter. From the definition it follows that 1 is a compact element of R if and only if every cover of R contains a finite subcover. We will denote the set of all compact [resp. $\mathcal{A}$-compact] elements of a semiring R by $comp(R)$ [resp. $\mathcal{A} - comp(R)$].

(22.38) EXAMPLE. If R is a semiring, then any finitely-generated ideal of R is a compact element of the semiring $ideal(R)$. In particular, R is a compact element of $ideal(R)$.

Let R be a complete semiring and let $\mathcal{K}_R$ be the family of all subsets V of R satisfying the condition that if $\theta: \Omega \to R$ is a function and if $\sum\theta \in V$ then there exists a finite subset Λ_0 of Ω satisfying the condition that $\sum_{i\in\Lambda}\theta(i) \in V$ for all finite subsets $\Lambda \supseteq \Lambda_0$ of Ω. Then $\mathcal{K}_R$ is a topology on R, which we will call the **Karner topology**, since it was first introduced in [Karner, 1994].

(22.39) EXAMPLE. If $R = (\mathbb{Q}^+ \cup \{\infty\}, +, \cdot)$ then the Karner topology $\mathcal{K}_R$ is characterized as follows:

(1) Every $a \in \mathbb{Q}^+$ is isolated;
(2) A subset V of R containing ∞ belongs to $\mathcal{K}_R$ iff and only if $R \setminus V$ is well-ordered by the reverse of the usual order.

(22.40) EXAMPLE. If $R = (\mathbb{R}^+ \cup \{\infty\}, +, \cdot)$ then a base for the Karner topology $\mathcal{K}_R$ is given by $\{0\}$ and all subsets of R of the form $\{r \in R \mid a < r \leq b\}$ for some $a < b$ in R.

It is straightforward to verify that if R and S are complete semirings and if $\alpha: R \to S$ is a function satisfying the condition that $\alpha(\sum\theta) = \sum(\alpha\theta)$ for all functions $\theta: \Omega \to R$, then α is a continuous function from $(R, \mathcal{K}_R)$ to $S, \mathcal{K}_S)$.

If $\{R_i \mid i \in \Omega\}$ is a family of complete semirings and $R = \times_{i\in\Omega} R_i$, then projection maps $R \to R_h$ are continuous when R and R_h are endowed with the Karner topologies and so the Karner topology contains the product topology on R. However, these topologies need not be equal. Thus, for example, Karner [1994] points out that if $R_1 = R_2 = (\mathbb{Q}^+ \cup \{\infty\}, +, \cdot)$ then the map $R_1 \times R_2 \to R_1$ given by $(a, b) \mapsto a + b$ is not continuous when R_1 is endowed with the Karner topoogy and $R_1 \times R_2$ is endowed with the product topology but is continuous when both semirings are endowed with the Karner topology.

As expected, if $\alpha: R \to S$ is a surjective morphism of complete semirings then $\mathcal{K}_S$ coincides with the final topology on S generated by α.

23. COMPLETE SEMIMODULES

In a manner analogous to that in the preceeding chapter, we can also define the notion of a [countably-] complete semimodule over a [countably-] complete semiring R. Refer to [R. Lee, 1979]. Note that if $\{M_h \mid h \in \Gamma\}$ is a family of [countably-] complete left R-semimodules then the left R-semimodule $\prod_{h\in\Gamma} M_h$ is also [countably-] complete. Indeed, if $\{f_i \mid i \in \Omega\}$ is a (countable) family of elements of $\prod_{h\in\Gamma} M_h$, we define $\sum_{i\in\Omega} f_i$ to be the function from Γ to $\cup_{h\in\Gamma} M_h$ given by

$$\sum_{i\in\Omega} f_i : h \mapsto \sum_{i\in\Omega} f_i(h),$$

where $f_i(h) \in M_h$ for all $h \in \Gamma$ and all $i \in \Omega$. In particular, we note that a direct product of an arbitrary number of copies of a [countably-] complete semimodule is again [countably-] complete.

(23.1) Example. If R is a complete semiring and X is a nonempty set such that R^X is a complete left R-semimodule, then summation in R^X can be consdidered as a form of integration. An important example of this is the case in which we take R to be the schedule algebra and X to be a locally-compact space [Litvinov & Maslov, 1998]. Indeed, let us write $\int_X f(x) = \vee_{x\in X} f(x)$ for all $f \in R^X$. If f is continuous or upper-continuous, then we can define a function $m_f : sub(X) \to R$ by setting $m_f : B \mapsto \vee_{x\in B} f(x)$. Note that $m_f(\cup_{\in\Omega} B_j) = \vee_{i\in\Omega} m_f(B_i)$ for any index set Ω. The function m_f is the R-**measure** defined by f and we can define an analog of the Lebesgue integral with respect to this measure by setting

$$\int_X g(x)dm_f = \int_X [g(x) + f(x)] = \bigvee_{x\in X} [g(x) + f(x)].$$

(Recall that $+$ is the multiplication in R!). These considerations lead to the analogy between such functions and probability measures, which has been exploited so fruitfully in "idempotent analysis". Refer also to [Del Moral & Doisy, 1999a, 1999b].

Many other notions from functional analysis can be transfered to this context. For example, if X is an appropriate space we can use the above formula to define the notion of an R-**valued inner product** on R^X by setting

$$\langle f, g\rangle = \int_X [f(x) + g(x)] = \bigvee_{x\in X} [g(x) + f(x)].$$

Also note that we clearly have the analog of Fubini's Theorem: for a suitable function $f: X \times Y \to R$ we can easily reverse the order of integration:

$$\int_X \int_Y f(x,y) = \int_Y \int_X f(x,y).$$

Loreti and Pedicini [1998] in fact give a somewhat more general form of this: if F is a function from X to $sub(Y)$ and if, for each $y \in Y$, we write $F^{-1}(y) = \{x \in X \mid y \in F(x)\}$, then

$$\int_X \int_{F(x)} f(x,y) = \int_{F(X)} \int_{F^{-1}(y)} f(x,y).$$

If M and N are [countably-] complete left R-semimodules, then an R-homomorphism $\alpha: M \to N$ is **[countably-] complete** if and only if it satisfies the aditional condition that $\left(\sum_{i\in\Omega} m_i\right)\alpha = \sum_{i\in\Omega} m_i\alpha$ for all [countable] index sets Ω. We will denote the set of all [countably-] complete R-homomorphisms from M to N by $CHom_R(M,N)$ [resp. $CCHom_R(M,N)$]. Similary, we denote the set of all [countably-] complete endomorphisms from M by $Cend_R(M)$ [resp. $CCEnd_R(M)$].

If M is a countably-complete left R-semimodule then the function $\theta: M \to M$ which assigns to each $m \in M$ the sum $m\theta$ of a countably-infinite number of copies of m is an idempotent member of $CCEnd_R(M)$ satisfying the condition that $m+m\theta = m\theta$ for all $m \in M$. We also note that every element of $M\theta$ is idempotent. More generally, if M is a complete left R-semimodule and if c is an infinite cardinal, then the function $\theta_c: M \to M$ which assigns to each $m \in M$ the sum $m\theta_c$ of c copies of itself is a member of $Cend_R(M)$ satisfying the condition that $m\theta_c + m\theta_d = m\theta_c$ for all cardinals $d \leq c$ and all $m \in M$. Moreover, that every element of $M\theta_c$ is idempotent.

Let R be a partially-ordered semiring. A partially-ordered [countably-] complete left R-semimodule M is **finitary** if and only if for every (countable) family $\{m_i \mid i \in \Omega\}$ of elements of M and every element m' of M satisfying the condition that $\sum_{j\in\Lambda} m_j \leq m'$ for any finite subset Λ of Ω, we have $\sum_{i\in\Omega} m_j \leq m'$. As an immediate consequence of this definition, w see that a direct product of finitary left R-semimodules is again finitary.

(23.2) PROPOSITION. *The following conditions on a [countably-] complete positive left R-semimodule M over an additively-idempotent semiring R are equivalent:*

(1) *M is finitary;*
(2) *M is a [countably-] complete lattice, if we set $\vee_{i\in\Omega} m_i = \sum_{i\in\Omega} m_i$ for any [countable] family $\{m_i \mid i \in \Omega\}$ of elements of M;*
(3) *$m\theta_c = m$ for all infinite cardinals c [resp. $m\theta = m$] and all $m \in M$.*

PROOF. (1) $\Rightarrow$ (2): If $\{m_i \mid i \in \Omega\}$ is a [countable] family of elements of M then, by hypothesis, we know that $m' = \sum_{i\in\Omega} m_i$ exists and that $m_i \leq m'$ for all $i \in \Omega$. Now let $m'' \in M$ satisfy the condition that that $m_i \leq m''$ for all $i \in \Omega$. Then $m' = \sum_{i\in\Omega} m_i \leq \sum_{i\in\Omega}(m_i + m'')$. But $m_i + m'' = m''$ for all $i \in \Omega$ and so $\sum_{j\in\Lambda}(m_j + m'') = m''$ for each finite subset Λ of Ω. By (1), we then have $m' \leq \sum_{i\in\Omega}(m_i + m'') \leq m''$, proving (2).

(2) $\Rightarrow$ (3): If $m \in M$ and if $\{m_i \mid i \in \Omega\}$ is a [countably-] infinite family of elements of M with $m_i = m$ for all $i \in \Omega$ then, by (2), $\sum_{i\in\Omega} m_i = \vee_{i\in\Omega} m_i = m$, establishing (3).

(3) $\Rightarrow$ (1): Let $\{m_i \mid i \in \Omega\}$ is a [countably-] infinite family of elements of M. If $m' \in M$ satisfies the condition that $\sum_{j\in\Lambda} m_j \leq m'$ for each finite subset Λ of Ω then, in particular, $m_i \leq m'$ for all $i \in \Omega\}$. By (3), $\sum_{i\in\Omega} m'_i = m'$, where $m'_i = m'$ for all $i \in \Omega\}$. Hence

$$\sum_{i\in\Omega} m_i + m' = \sum_{i\in\Omega} m_i + \sum_{i\in\Omega} m'_i = \sum_{i\in\Omega}(m_i + m'_i) = \sum_{i\in\Omega} m'_i = m'$$

and so $\sum_{i\in\Omega} m_i \leq m'$, proving (1). □

(23.3) Proposition. *If R is an additively-idempotent semiring, then every left R-semimodule can be embedded in a finitary complete R-semimodule.*

Proof. Let M be a left R-semimodule. Since R is additively-idempotent, $B(R)$ is isomorphic to $\mathbb{B}$ and so R is a left $\mathbb{B}$-semimodule. As in Chapter 15, set $I(M) = Hom_{\mathbb{B}}(R, \mathbb{B}^M)$. For $r \in R$ and $\eta \in I(M)$ we define the function $r\eta: R \to \mathbb{B}^M$ by setting $r\eta: r' \mapsto (r'r)\eta$. It is straightforward to verify that $r\eta \in I(M)$ and that this definition turns $I(M)$ into a left R-semimodule, which is additively idempotrent since R is an additively-idempotent semiring.

We claim that $I(M)$ is in fact complete. Indeed, let $\{\eta_i \mid i \in \Omega\}$ be a family of elements of $I(M)$ and define $\sum_{i\in\Omega} \eta_i$ to be the function from R to $\mathbb{B}^M$ given by $(\sum_{i\in\Omega} \eta_i): r \mapsto \sum_{i\in\Omega}(r)\eta_i$. This function belongs to $I(M)$ since, if $r, r' \in R$ then

$$(r + r')\left(\sum_{i\in\Omega} \eta_i\right) = \sum_{i\in\Omega}(r + r')\eta_i = \sum_{i\in\Omega}[r\eta_i + r'\eta_i] = \sum_{i\in\Omega} r\eta_i + \sum_{i\in\Omega} r'\eta_i$$

while

$$(r')\left(r\sum_{i\in\Omega} \eta_i\right) = (r'r)\left(\sum_{i\in\Omega} \eta_i\right) = \sum_{i\in\Omega}(r'r)\eta_i = \sum_{i\in\Omega}(r')r\eta_i = (r')\left(r\sum_{i\in\Omega} \eta_i\right).$$

Clearly $\mathbb{B}$ is a finitary complete semimodule over itself, and so, as previously noted, $\mathbb{B}^M$ is also finitary and complete. Now suppose that $\{\eta_i \mid i \in \Omega\}$ is a family of elements of $I(M)$ and let $\eta' \in I(M)$ satisfy the condition that $\sum_{j\in\Lambda} \eta_j \leq \eta'$ for each finite subset Λ of Ω. Then for each such Λ and each $r \in R$ we have $(r)\left(\sum_{j\in\Lambda} \eta_j\right) \leq (r)\eta'$ in $\mathbb{B}^M$. Since $\mathbb{B}^M$ is finitary, this implies that $(r)\left(\sum_{i\in\Omega} \eta_i\right) \leq (r)\eta'$ for each $r \in R$ and so $\sum_{i\in\Omega} \eta_i \leq \eta'$. Thus $I(M)$ is a finitary complete left R-semimodule.

We are left to show that there exists a monic R-homomorphism $\alpha: M \to I(M)$ and, indeed, such a map is defined as follows: if $m \in M$ and $r \in R$ let $r(m\alpha) \in \mathbb{B}^M$ map $m'inM$ to 0 if $m' \geq rm$ and map it to 1 otherwise. The fact that α is an R-homomorphism is straightforward to verify. □

(23.4) PROPOSITION. *If R is an additively-idempotent semiring and M is a finitary [countably-] complete left R-semimodule then $CEnd_R(M)$ [resp. $CCEnd_R(M)$] can be made a finitary [countably-] complete additively-idempotent semiring.*

PROOF. It is easy to see that $CEnd_R(M)$ [resp. $CCEnd_R(M)$] is a subsemiring of $End_R(M)$ and so is additively idempotent. We claim that this semiring can be made [countably-] complete by an appropriate definition of infinite summation. Indeed, if Ω is a [countable] index set and if $\alpha_i \mid i \in \Omega\}$ is a family of elements of $CEnd_R(M)$ [resp. $CCEnd_R(M)$], we define $\sum_{i\in\Omega}\alpha_i$ by $(\sum_{i\in\Omega}\alpha_i) : m \mapsto \sum_{i\in\Omega} m\alpha_i$. If $m, m' \in M$ and $r \in R$ then we have

$$\begin{aligned}(m+m')\left(\sum_{i\in\Omega}\alpha_i\right) &= \sum_{i\in\Omega}(m+m')\alpha_i = \sum_{i\in\Omega}(m\alpha_i + m'\alpha_i)\\ &= \sum_{i\in\Omega} m\alpha_i + \sum_{i\in\Omega} m'\alpha_i = m\left(\sum_{i\in\Omega}\alpha_i\right) + m'\left(\sum_{i\in\Omega}\alpha_i\right)\end{aligned}$$

and

$$\begin{aligned}(rm)\left(\sum_{i\in\Omega}\alpha_i\right) &= \sum_{i\in\Omega}(rm)\alpha_i = \sum_{i\in\Omega} r(m\alpha_i)\\ &= r\left(\sum_{i\in\Omega} m\alpha_i\right) = r\left[m\left(\sum_{i\in\Omega}\alpha_i\right)\right].\end{aligned}$$

Therefore $\sum_{i\in\Omega}\alpha_i \in End_R(M)$. Moreover, if $\{m_j \mid j \in \Lambda\}$ is a [countable] family of elements of M then

$$\begin{aligned}\left(\sum_{j\in\Lambda} m_j\right)\left(\sum_{i\in\Omega}\alpha_i\right) &= \sum_{i\in\Omega}\left(\sum_{j\in\Lambda} m_j\right)\alpha_i = \sum_{i\in\Omega}\sum_{j\in\Lambda} m_j\alpha_i\\ &= \sum_{j\in\Lambda}\sum_{i\in\Omega} m_j\alpha_i = \sum_{j\in\Lambda} m_j\left(\sum_{i\in\Omega}\alpha_i\right).\end{aligned}$$

Therefore $\sum_{i\in\Omega}\alpha_i$ belongs to $CEnd_R(M)$ [resp. $CCEnd_R(M)$]. It is now straightforward to check that $CEnd_R(M)$ [resp. $CCEnd_R(M)$] is [countably] complete as a left and right semimodule over itself, and so is a [countably-] complete semiring.

We are left to show that this semiring is finitary. Let α be an endomorphism of M and let c be an infinite cardinal. Set $\alpha_c = (\alpha)\theta_c$ Then for each $m \in M$ we have $m\alpha_c = (m\alpha)\theta_c = m\alpha$ by Proposition 23.2. Therefore $\alpha = \alpha_c$ and so $CEnd_R(M)$ [resp. $CCEnd_R(M)$] is finitary. □

(23.5) PROPOSITION. *Every additively-idempotent semiring R can be embedded in a finitary complete semiring.*

PROOF. By Proposition 23.3, we know that R, considered as a left semimodule over itself, can be embedded in a finitary complete left R-semimodule M. By

Proposition 23.2, the semiring $S = CEnd_R(M)$ is finitary complete, as is the semiring $T = CEnd_S(M)$. But there is a canonical embedding of R into T, establishing the desired result. □

(23.6) PROPOSITION. *If R is a zerosumfree semiring then there exists a morphism of semirings from R to a complete additively-idempotent semiring.*

PROOF. Define a relation ρ on R by setting $r \ \rho \ r'$ if and only if there exist elements $x, y \in R$ and nonnegative integers n and m satisfying $r + x = nr'$ and $r' + y = mr$. This relation is clearly symmetric and reflexive. To show that it is transitive, assume that $r \ \rho \ r'$ and $r' \ \rho \ r''$. Then $r + x = mr'$ and $r' + y = nr''$ for some $x, y \in R$ and $m, n \in \mathbb{N}$. It then follows that

$$r + (x + my) = mr' + my = m(r' + y) = m(nr'') = (mn)r''.$$

Similarly, we can show that $r'' + z = kr$ for some $z \in R$ and $k \in \mathbb{N}$. Therefore $r \ \rho \ r''$, proving that ρ is an equivalence relation on R. Note that if $r \ \rho \ 0$ then $r + x = 0$ for some $x \in R$. Since R is assumed to be zerosumfree, this implies that $r = 0$.

We now prove that ρ is compatible with the operations on R. Indeed, suppose that $r_1 \ \rho \ s_1$ and that $r_2 \ \rho \ s_2$. Then we have $r_1 + x_1 = m_1 s_1$, $s_1 + y_1 = m_1 r_1$, $r_2 + x_2 = m_2 s_2$, and $s_2 + y_2 = m_2 r_2$, from which we deduce that

$$r_1 r_2 + (r_2 x_2 + x_1 r_2 + x_1 x_2) = (m_1 s_1)(m_2 s_2) = (m_1 m_2) s_1 s_2$$

and

$$s_1 s_2 + (s_1 y_2 + y_1 s_2 + y_1 y_2) = (n_1 r_1)(n_2 r_2) = (n_1 n_2) r_1 r_2$$

so $r_1 r_2 \ \rho \ s_1 s_2$. Also

$$r_1 + r_2 + [(m_1 - 1)r_1 + m_2 x_1 + (m_1 - 1)r_2 + m_1 x_2] = m_1 m_2 (s_1 + s_2)$$

and

$$s_1 + s_2 + [(n_2 - 1)s_1 + n_2 y_1 + (n_1 - 1)s_2 + n_1 y_2] = n_1 n_2 (r_1 + r_2)$$

so $(r_1 + r_2) \ \rho \ (s_1 + s_2)$. Thus we see that ρ is a congruence relation on R. Moreover, the semiring R/ρ is additively idempotent. By Proposition 23.5, this semiring can be embedded in a complete semiring and so this embedding, composed with the canonical morphism $R \to R/\rho$, is the morphism we seek. □

(23.7) PROPOSITION. *Let R be a positive additively-idempotent semiring and let M be a left R-semimodule which is a retract of a finitary [countably-] complete left R-semimodule N. Then M itself is finitary and [countably-] complete.*

PROOF. By hypothesis, there exist R-homomorphisms $\alpha: M \to N$ and $\beta: N \to M$ such that $\alpha\beta$ is the identity map on M. By Proposition 23.2 we know that N is a [countably-] complete lattice in which $\vee_{i \in \Omega} n_i = \sum_{i \in \Omega} n_i$ for any [countable] family $\{n_i \mid i \in \Omega\}$ of elements of N. If $n = \sum_{i \in \Omega} m_i \alpha$, then $m_h \alpha \leq n$ for all $h \in \Omega$ and so $m_h = m_h \alpha\beta \leq n\beta$. Therefore $n\beta$ is an upper bound for $\{m_i \mid i \in \Omega\}$ in M. Assume that $m \in M$ is another upper bound for this family of elements.

Then $m_h \leq m$ for all $h \in \Omega$ and so $m_h\alpha \leq m\alpha$, Therefore $n = \sum_{i\in\Omega} m_i\alpha \leq m\alpha$ so $n\beta \leq m\alpha\beta = m$. Thus $n\beta$ is the unique least upper bounded for $\{m_i \mid i \in \Omega\}$ in M, proving that M is a complete lattice. We now define $\sum_{i\in\Omega} m_i$ in M to be equal to $n\beta \in M$ and claim that this definition turns M into a complete R-semimodule. Indeed, if $r \in R$ then

$$\sum_{i\in\Omega} rm_i = \left(\sum_{i\in\Omega}(rm_i)\alpha\right)\beta = \left(\sum_{i\in\Omega} r(m_i\alpha)\right)\beta$$
$$= \left(r\left(\sum_{i\in\Omega} m_i\alpha\right)\right)\beta = \left(\left(\sum_{i\in\Omega} m_i\alpha\right)\beta\right) = r\left(\sum_{i\in\Omega} m_i\right).$$

Again, by Proposition 23.2, the semimodule M is finitary complete. □

(23.8) PROPOSITION. *If R is a positive additively-idempotent semiring then every injective left R-semimodule can be made finitary and complete.*

PROOF. Assume that M is an injective left R-semimodule. By Proposition 23.3, we know that M can be embedded in a finitary complete R-semimodule. Since M is injective, this embedding is a retraction and so, by Proposition 23.7, we see that M too is finitary and complete. □

Note that the converse of Proposition 23.8 is false. Indeed, any complete lattice is a finitary complete $\mathbb{B}$-semimodule, but injective left $\mathbb{B}$-semimodules must be frames.

(23.9) PROPOSITION. *Let R be a complete semiring and let $u: A \to B$ be a function between nonempty sets. Then:*

(1) *If $g \in R^B$ then there exists a function $g_1 \in R^B$ satisfying $h_u[h_u^{-1}[g]]+g = g$; and*

(2) *If $f \in R^A$ then $f + h_u^{-1}[h_u[f]] = h_u^{-1}[h_u[f]]$.*

PROOF. (1) If $b_0 \in B$

$$h_u[h_u^{-1}[g]]: b_0 \mapsto \sum_{a\in A} h_u^{-1}[g](a)h_u(a,b_0) = \sum_{a\in A}\sum_{b\in B} h_u(a,b)g(b)h_u(a,b_0).$$

But this sum equals 0_R except in the case $b = b_0 = u(a)$, in which case it equals $g(b_0)$. Thus, if $g_1 \in R^B$ is the function defined by

$$g_1(b) = \begin{cases} 0 & \text{if } b \in im(u) \\ g(b) & \text{otherwise.} \end{cases}$$

then $h_u[h_u^{-1}[g]] + g_1 = g$.

(2) If $a_0 \in A$ then

$$h_u^{-1}[h_u[f]]: a_0 \mapsto \sum_{b\in B} h_u(a_0,b)h_u[f](b)$$
$$= \sum_{b\in B}\sum_{a\in A} h_u(a_0,b)f(a)h_u(a,b)$$
$$= \sum_{f(a)=f(a_0)} f(a)$$

and so $f + h_u^{-1}[h_u[f]] = h_u^{-1}[h_u[f]]$. □

Note that Proposition 23.9 implies that if R is a complete semiring and if $u: A \to B$ is a function between nonempty sets then the function $g \mapsto h_u[h_u^{-1}[g]]$ is an interior operator on R^B and the function $f \mapsto h_u^{-1}[h_u[f]]$ is a closure operator on R^A.

24. CLO-SEMIRINGS

A lattice-ordered semiring R is a **complete-lattice-ordered semiring (CLO-semiring)** if and only if $(R, \vee, \wedge)$ is a complete lattice. A CLO-semiring is a **quantalic lattice-ordered semiring (QLO-semiring)** if and only if multiplication distributes over arbitrary joins from either side; it is a **frame-ordered semiring (FO-semiring)** if and only if it is a QLO-semiring and the underlying complete lattice is a frame. As an immediate extension of Proposition 21.12(1), we see that if a a is an element of a CLO-semiring R then $\vee(Ra) = a = \vee(aR)$.

The study of complete lattices equipped with an additional operation which distributes over arbitrary joins goes back to [Krull, 1924], [Dilworth, 1939], and [Ward & Dilworth, 1939]. CLO-semirings are considered in [Fuchs, 1954, 1963] under the name of **complete lattice-ordered semigroups**. Refer also to [Anderson, 1976]. The related notion of a **quantale** is studied in [Borceux & Van den Bossche, 1986] and is based on an attempt by Mulvey to provide a constructive foundation for quantum mechanics. Also see [Borceux, Rosický & Van den Bossche, 1989] and [Brown & Gurr, 1993a]. In a quantale Q, multiplication is associative but 1 is only a one-sided multiplicative identity: $a1 = a$ for all elements a of Q but $1a$ is not necessarily equal to a. An element a of Q is **two-sided** if $1a = a = a1$ and the collection Q^* of all two-sided elements of Q is a sublattice of Q which is in fact a QLO-semiring. Since the meet of any arbitrary family of two-sided elements of Q is again two-sided, we see that for any $a \in Q$ there exists a unique smallest element a^* among those elements b of Q^* satisfying $b \geq a$. In order to relate a quantale and its subsemiring of two-sided elements, the notion of a quantum frame was introduced in [Rosický, 1989, 1995]. In particular, we can consider Q as a Q^*-semimodule. A more general notion of a quantale given in [Niefield & Rosenthal, 1988] and [Román & Rumbos, 1991a]. For the application of quantales to process semantics of computer programs, refer to [Abramsky & Vickers, 1993]. Another related notion is that of a **net**, introduced in [Blikle; 1971, 1977], which differs from that of a CLO-semiring in that multiplication is assumed to distribute only over finite or countable joins. Such structures are also known as **σ-frames**. For an application of nets to the analysis of computer programs, refer to [Janicki, 1977].

Quantic lattice-ordered semirings provided a natural setting for fuzzy set theory, as proposed in [Goguen, 1967] and for a logic of inexact concepts. The notion of a logic over a QLO-semiring has been extensively expanded in [Pavelka, 1979a, 1979b,

1979c].

(24.1) EXAMPLE. If $(R, \vee, \wedge)$ is a complete lattice then R is surely a CLO-lattice; it is a frame precisely when this lattice is a QLO-semiring. Any frame is a frame-ordered semiring if we take multiplication to coincide with meet. Indeed, as an immediate consequence of Proposition 21.14 we see that frames are precisely the multiplicatively idempotent QLO-semirings. Thus, if R is a frame-ordered semiring in which multiplication and meet are not the same, R has two canonical structures of a complete semiring: $(R, \vee, \cdot)$ and $(R, \vee, \wedge)$.

(24.2) EXAMPLE. If R is a semiring then $(ideal(R), +, \cdot)$ is a QLO-semiring. Not every QLO-semiring is isomorphic to a subsemiring of a QLO-semiring of this type. See [Bogart, 1969b]. If $(S, *)$ is a monoid then $(ideal(S), \cup, *)$ is an FO-semiring, where $A * B = \{a * b \mid a \in A, b \in B\}$ for all ideals A and B of S.

Similarly, if $\mathcal{O}$ is a sheaf of commutative rings on a locale then the sheaf of ideals of $\mathcal{O}$ is a semiring under the operations of sum and sheaf product. See [Niefield & Rosenthal, 1990] for details.

(24.3) EXAMPLE. The semiring $(\mathbb{I}, max, \cdot)$ is frame ordered, where the frame operations are sup and inf.

(24.4) EXAMPLE. If R is a semiring then the dual lattice $(R-fil)^{du}$ of $R-fil$ is a CLO-semiring. However, multiplication in $(R-fil)^{du}$ distributes over arbitrary joins (i.e. intersections in $R-fil$)) from the left but not necessarily from the right. Therefore $(R-fil)^{du}$ is not a QLO-semiring. For the case of R a ring, this semiring is studied in detail in [Golan, 1986].

(24.5) EXAMPLE. If A is a nonempty set then a relation on A is a subset of $A \times A$. The family R of all relations on A is a frame under the operations of intersection and union. In addition, we can define the product of two elements of R by setting $BC = \{(a, a') \in A \times A \mid \text{ there exists an } a'' \in A \text{ with } (a, a'') \in B, (a'', a') \in C\}$. This turns R into a FO-semiring. For the extension of this notion to fuzzy relations refer, for example, to [Dubois & Prade, 1980].

This example can be generalized in several directions, one of the most important being that of [Chin & Tarski, 1951]. In these generalizations, as a rule, we end up with a QLO-semiring the underlying lattice of which is a complete atomic boolean algebra with, perhaps, some additional structure.

(24.6) EXAMPLE. If A is a nonempty set then the semirings $R = (sub(A^*), +, \cdot)$ and $R' = (sub(A^\infty), +, \cdot)$ of formal languages and formal ∞-languages on A, respectively defined in Example 1.11, are QLO-semirings. In the semiring R' we can also define countably-infinite products as follows: if $L_1, L_2, \ldots$ are elements of R'. define $L_1L_2L_3 \cdot \ldots$ to be the set of all words $w \in A^\infty$ of the form $w = a_1a_2a_3 \cdot \ldots$ where, for each i, we have $\square \neq a_i \in (L_i \cap A^*) \cup (L_i \cap A^*)^* \cdot (L_i \cap A^\infty)$.

(24.7) EXAMPLE. The semiring $R = (\mathbb{R} \cup \{-\infty\}, max, +)$ is certainly a CLO-semiring. If A is a nonempty set, then the set M of all bounded elements of R^A, namely the set of those functions $f \in R^A$ for which there exists an element $r \in R$ such that $f(a) \leq r$ for each $a \in A$, is an R-semimodule. For the importance of this semimodule in optimization theory built around idempotent measures, refer to [Gunawardena, 1998] or [Akian, Quadrat & Viot, 1998].

(24.8) PROPOSITION. *If a is a prime element of a CLO-semiring R then $A = \{a' \in R \mid a'$ prime and $a' \leq a\}$ has a minimal element.*

PROOF. The proof of this is a straightforward variant of the proof of Proposition 7.14. □

The notion of a compact element of a complete semiring holds in particular for complete lattices: an element a of a complete lattice R is compact if and only if for each nonempty set A of R satisfying $\vee A \geq a$ there exists a finite subset A' satisfying $\vee A' \geq a$. If a and b are compact elements of R then surely so is $a \vee b$.

(24.9) EXAMPLE. The semiring $(R - fil, \cap, \cdot)$ presented in Example 1.7 is a CLO-semiring. The compact elements of this semiring are studied in [Golan, 1987] (where they are called "ducompact"). A sufficient condition for $\kappa \in R - fil$ to be compact is that it have a cofinal subset of finitely-generated left ideals. This would always be true, of course, if R were left noetherian.

The product of two compact elements of R need not be compact. We say that R is **compactly generated** (or **algebraic**) if and only if every nonzero element of R is the join of compact elements. Compactly-generated CLO-semirings satisfying the condition that the product of compact elements is compact are studied in [Keimel, 1972]. The semiring $sub(A)$ of subsets of a nonempty set A is compactly generated.

(24.10) EXAMPLE. If R is a semiring then any finitely-generated ideal of R is a compact element of the CLO-semiring $ideal(R)$. Therefore $ideal(R)$ is compactly generated.

(24.11) PROPOSITION. *If R is a CLO-semiring and A is a nonempty set then R^A is compactly-generated if and only if R is.*

PROOF. For each $a \in A$ and $r \in R$, let $w_{a,r}$ be the function defined by $w_{a,r}(a) = r$ and $w_{a,r}(a') = 0$ for $a' \neq a$. We claim that an element r of R is compact if and only if $w_{a,r}$ is a compact element of R^A for all $a \in A$. Indeed, assume that r is a compact element of R and, for an element a of A, let U be a nonempty subset of R^A satisfying $w_{a,r} \leq \vee U$. Then $r = w_{a,r}(a) \leq \vee_{g \in U} g(a)$ and so there exists a finite subset U' of U such that $r \leq \vee_{g \in U'} g(a)$. Then $w_{a,r} \leq \vee U'$, proving that $w_{a,r}$ is compact. Conversely, assume that $w_{a,r}$ is compact for all $a \in A$ and let Y be a nonempty subset of R satisfying $r \leq \vee Y$. Then for any $a \in A$ we have $w_{a,r} \leq \vee_{s \in Y} w_{a,s}$. Since $w_{a,r}$ is compact, there is a finite subset Y' of Y such that $w_{a,r} \leq \vee_{s \in Y'} w_{a,s}$, whence $r = w_{a,r}(a) \leq \vee_{s \in Y'} w_{a,s}(a) = \vee Y'$.

For $r \in R$ we set $C(r) = \{s \leq r \mid s \text{ is compact}\}$. Similarly, for $f \in R^A$ we set $C(f) = \{g \leq f \mid g \text{ is compact}\}$. Assume that R^A is compactly-generated and let $0 \neq r \in R$. Then, by the above, we see that for $a \in A$ we have $w_{a,r} = \vee_{s \in C(r)} w_{a,s}$ and so $r = w_{a,r}(a) = \vee_{s \in C(r)} w_{a,s}(a) = \vee C(r)$. Thus R is compactly-generated.

Conversely, suppose that R is compactly-generated. If $0 \neq f \in R^A$ then $f = \vee_{a \in A} w_{a,f(a)} = \vee\{w_{a,s} \mid a \in A; s \in C(f(a))\}$ where each of the functions $w_{a,s}$ is compact. Thus R^A is compactly-generated. $\square$

(24.12) PROPOSITION. *If R is a CLO-semiring in which 1 is compact then, for each $1 \neq r \in R$, the set $B = \{r' \in R \mid r \leq r' < 1\}$ has a maximal element, which is prime.*

PROOF. If B' is a chain of elements of B then $\vee B' \in B$ since 1 is compact. Therefore, by Zorn's Lemma, B has a maximal element. Since R is simple, additively idempotent, and difference ordered we see by Proposition 20.47, taking the case of $A = \{1\}$, that such a maximal element is prime. $\square$

(24.13) PROPOSITION. *Let A be a nonempty set of compact elements of a CLO-semiring R and let $B = \{r \in R \mid a \leq r$ for all $a \in A\}$. Then for each $b \in B$ there exists a maximal element b' of B with $b \leq b'$.*

PROOF. Let B' be a maximal totally-ordered subset of B with $b \in B'$. Since the elements of A are compact, $b' = \vee B' \in B$. Clearly b' is a maximal element of B satisfying $b \leq b'$. $\square$

The following result generalizes Proposition 7.25.

(24.14) PROPOSITION. *In a compactly-generated CLO-semiring, every semiprime element is the meet of primes.*

PROOF. Let s be a semiprime element of a compactly-generated CLO-semiring R. It suffices to show that if $r > s$ in R there exists a prime element b of R satisfying $b \geq s$ and $b \not\geq r$. Indeed, let $r > s$. Define a sequence $A = \{a_1, a_2, \dots\}$ of compact elements of R in the following manner: Let a_1 be a compact element of R satisfying $a_1 \leq r$ and $a_1 \not\leq s$. Such an element exists since R is compactly generated. Now assume that we have found compact elements $a_2, \dots, a_n$ of R such that $a_i \not\leq s$ for all i and $a_i \leq (a_{i-1})^2$ for all $2 \leq i \leq n$. Then $a_n^2 \not\leq s$ since s is semiprime and hence there exists a compact element a_{n+1} of R such that $a_{n+1} \not\leq s$ and $a_{n+1} \leq a_n^2$.

By construction, we note that if $a_i, a_j \in A$ then there exists an element a_h of A satisfying $a_h \leq a_i a_j$. Since $a_i \not\leq s$ for all i, we see by Proposition 24.13 that there exists a maximal element b of $\{r' \in R \mid a_i \not\leq r'$ for all $i\}$ such that $s \leq b$. By Proposition 21.15, R is simple and positive so by 20.47, we conclude that b is prime. Clearly $r \not\leq b$. $\square$

(24.15) PROPOSITION. *For a compactly-generated CLO-semiring R the following conditions are equivalent:*

(1) *Every $1 \neq a \in R$ is the intersection of prime elements;*
(2) *Every $1 \neq a \in R$ is semiprime;*
(3) *R is multiplicatively idempotent;*
(4) *If $a, b \in R$ then $ab = a \wedge b$.*

PROOF. (1) $\Leftrightarrow$ (2): Assume (1). If $1 \neq a \in R$ then there exists a nonempty set A of prime elements of R satisfying $a = \wedge A$. If $r^2 \leq a$ in R then $r^2 \leq b$ for all $b \in A$ so $r \leq b$ for all such b. Hence $r \leq a$, proving that a is semiprime. Thus we have (2). The converse follows from Proposition 24.14.

(2) ⇔ (3): Clearly (3) implies (2). Conversely, assume (2). If $1 \neq a \in R$ then $a^2 \leq a^2$ and so $a \leq a^2$. Conversely, $a^2 \leq a \wedge a = a$ and so $a = a^2$. Thus we have (3).

(3) ⇔ (4). This follows from Proposition 21.14. □

(24.16) Example. [Keimel, 1972] Let $(M, *)$ be a semigroup. Then the set R of all semigroup-ideals of M is a compactly-generated CLO-semiring and so every semiprime element I of R is the intersection of prime elements of R. Moreover, the following conditions are equivalent:

(1) Every $I \in R$ is semiprime;
(2) Every $I \in R$ is idempotent;
(3) $IH = I \cap H$ for all $I, H \in R$.

If R is a CLO-semiring and if a and b are elements of R, then we define the **left residual** $ab^{\langle -1 \rangle} = \vee\{r \in R \mid rb \leq a\}$ and the **right residual** $b^{\langle -1 \rangle}a = \vee\{r \in R \mid br \leq a\}$. Clearly $b^{\langle -1 \rangle}a \wedge ab^{\langle -1 \rangle} \geq a$ for all $b \in R$. Note that if R is a QLO-semiring then $(ab^{\langle -1 \rangle})b = \vee\{rb \mid rb \leq a\} \leq a \wedge b$ and similarly $b(b^{\langle -1 \rangle}a) = \vee\{br \mid br \leq a\} \leq a \wedge b$. Note too that any CLO-semiring is simple and positive by Proposition 21.15 and so we see that $a0^{\langle -1 \rangle} = 1 = 0^{\langle -1 \rangle}a$ for any element a of R and, if R is entire, we also have $0b^{\langle -1 \rangle} = 0 = b^{\langle -1 \rangle}0$ for any nonzero element b of R. In general, if a is an element of a QLO-semiring R then $0a^{\langle -1 \rangle} = \vee\{r \in R \mid ra \leq 0\} = \vee\{r \in R \mid ra = 0\}$ and $(0a^{\langle -1 \rangle})a = 0$. Thus $0a^{\langle -1 \rangle}$ is the unique maximal left annihilator of a. Similarly, $a^{\langle -1 \rangle}0$ is the unique maximal right annihilator of a.

(24.17) Example. If R is the CLO-semiring $(\mathbb{I}, max, min)$ and $a, b \in R$ then $b^{\langle -1 \rangle}a = 1$ when $a \geq b$ while $b^{\langle -1 \rangle}a = a$ when $a < b$.

(24.18) Example. Recall the notion of a triangular norm on $\mathbb{I}$ as defined in Example 1.13. If $\sqcap$ is a triangular norm on $\mathbb{I}$ and if a and b are elements of $\mathbb{I}$ then $a \leq 1$ so $a \sqcap b \leq 1 \sqcap b = b$. Similarly, $a \sqcap b \leq a$ and so $a \sqcap b \leq min\{a, b\}$. Thus $(\mathbb{I}, max, \sqcap)$ is a lattice-ordered semiring. (Example 24.3 is a special case of this.) If $\sqcap$ is a triangular norm on $\mathbb{I}$ which is lower semicontinuous as a function from $\mathbb{I} \times \mathbb{I}$ to $\mathbb{I}$ then in fact $R = (\mathbb{I}, max, \sqcap)$ is a QLO-semiring and so we can define residuals (sidedness is unimportant here, because of the commutativity of multiplication) in R. Some of these are presented in detail in [Gottwald, 1984] for various triangular norms. Thus, for example, if we consider the fundamental triangular norms we have

(1) In the semiring $(\mathbb{I}, max, \sqcap_0)$ we see that $ab^{\langle -1 \rangle}$ equals 1 if $b \leq a$ and equals a otherwise;
(2) In the semiring $(\mathbb{I}, max, \sqcap_1)$ we see that $ab^{\langle -1 \rangle}$ equals 1 if $b = 0$ and equals $min\{1, b/a\}$ otherwise; and
(3) In the semiring $(\mathbb{I}, max, \sqcap_\infty)$ we see that $ab^{\langle -1 \rangle} = min\{1, 1 - a + b\}$. This result is due to J. Łukasiewicz in connection with his studies in logic.

Other examples of residuals in semirings of this type are given in [Weber, 1983]. Thus, for instance, if $0 < c \in \mathbb{R}$ one can define a triangular norm $\sqcap_{H(c)}$ on $\mathbb{I}$ by setting $a \sqcap_{H(c)} b = ab/[c + (1-c)(a+b-ab)]$. In the semiring $(\mathbb{I}, max, \sqcap_{H(c)})$ we have $ab^{\langle -1 \rangle} = [a + (c-1)a(1-b)]/[b + (c-1)a(1-b)]$ for $b > a$.

(24.19) EXAMPLE. Let R be a CLO-semiring and let A, B, and C be nonempty sets. If $g \in R^{A \times B}$, $h \in R^{B \times C}$, and $k \in R^{A \times C}$ then we can define

$$k \circ h^{\langle -1 \rangle} = \sum \{g' \in R^{A \times B} \mid g' \circ h \leq k\}$$

and

$$g^{\langle -1 \rangle} \circ k = \sum \{h' \in R^{B \times C} \mid g \circ h' \leq k\}.$$

In particular, if $h \in R^{A \times A}$ then

$$h \circ h^{\langle -1 \rangle} = \sum \{k \in R^{A \times A} \mid k \circ h \leq h\}$$

and

$$h^{\langle -1 \rangle} \circ h = \sum \{k \in R^{A \times A} \mid h \circ k \leq h\}.$$

These R-valued relations on $A \times A$ are called, respectively, the right and left **traces** of h and have been studied, for the special case of $R = \mathbb{I}$, in [Doignon et al., 1986], [Fodor, 1992], and [Sanchez, 1976].

(24.20) EXAMPLE. If R is a CLO-semiring and A, B, and C are nonempty sets then there are ways of defining compositions between relations $h \in R^{A \times B}$ and $k \in R^{B \times C}$ other than those given previously. Some of these, along with their applications, were considered in [De Baets & Kerre, 1993b] for the case of $(\mathbb{I}, \vee, \wedge)$. These can be easily extended further. For example, over an arbitrary CLO-semiring we can set

$$(h \lhd k): (a, c) \mapsto \left(\sum_{b \in B} h(a,b)\right) \left(\bigwedge_{b \in B} h(a,b)^{\langle -1 \rangle} k(b,c)\right) \left(\sum_{b \in B} k(b,c)\right).$$

The reasons for considering such compositions are detailed in [De Baets & Kerre, 1993b].

If r and s are elements of a CLO-semiring R then, as an immediate consequence of the definitions, we see that $r(r^{\langle -1 \rangle} s) \leq s \wedge r$. If we have equality for all $s \in R$, then the element r of R is **left weakly meet principal**. Similarly, $r^{\langle -1 \rangle}(rs) \geq s + r^{\langle -1 \rangle} 0$ for all $r, s \in R$. If we have equality for all $s \in R$, then r is **left weakly sum principal**. An element of r is **left weakly principal** if and only if it is both left weakly meet principal and left weakly sum principal. An element r of R is **left meet principal** if and only if $ra \wedge b = r[a \wedge r^{\langle -1 \rangle} b]$ for all elements a and b of R and is **left sum principal** if and only if $r^{\langle -1 \rangle}[ra + b] = a + r^{\langle -1 \rangle} b$ for all a and b in R. Clearly left meet principal elements are left weakly meet principal and left sum principal elements are left weakly sum principal. An element of R is **left principal** if and only if it is both left meet principal and left sum principal. Since any CLO-semiring is positive and simple, we note that 0 is always both left and right principal.

Note that if an element r of R is left weakly meet principal then an element s of R satisfies $s \leq r$ if and only if $s = rr'$ for some element r' of R.

(24.21) EXAMPLE. [Anderson, 1975] If S is a commutative semigroup with 0 and $R = ideal(S)$, then a principal ideal Ss_0 of S is always weakly meet principal. It is weakly sum principal if and only if $ss_0 = s's_0 \neq 0$ implies that $Ss = Ss'$ for all $s, s' \in S$.

(24.22) EXAMPLE. [Alarcón & Anderson, 1994a] Let $R = ideal(\mathbb{Q}^+[t])$, where t is an indeterminate. Then the principal ideal $(1+t)$ is a meet-principal element of R which is weakly sum principal but not sum principal.

(24.23) PROPOSITION. *If a is an element of a CLO-semiring R which is not a left zero divisor then $a^{\langle -1 \rangle}0 = 0$. If, in addition, a is weakly sum principal then $a^{\langle -1 \rangle}(ab) = b$ for all elements b of R.*

PROOF. If a is not a left zero divisor then $a^{\langle -1 \rangle}0 = \vee\{r \in R \mid ar = 0\} = 0$. If, in addition, a is weakly sum principal then $a^{\langle -1 \rangle}(ab) = b + a^{\langle -1 \rangle}0 = b$. □

(24.24) PROPOSITION. *A commutative QLO-semiring R is isomorphic to $ideal(S)$ for some commutative multiplicative monoid S with 0 if and only if R is an FO-semiring and there exists a subset S' of R satisfying the following conditions:*

(1) $0 \in S'$;
(2) *Every element of S' is weakly meet principal;*
(3) *Every element of R is of the form $\vee A$ for some nonempty subset A of S';*
(4) *S' is closed under products;*
(5) *If $s \in S'$ then there do not exist nonzero elements s' and s'' of S' satisfying $s = s' + s''$.*

PROOF. If S is a commutative multiplicative monoid with 0 then in Example 24.2 we already noted that $ideal(S)$ is an FO-semiring. Moreover, the set $S' = \{Sa \mid a \in S\}$ of principal ideals of S has the desired properties.

Conversely, let R be an FO-semiring having a subset S' satisfying (1) – (5). Then $(S', \cdot)$ is a commutative monoid with 0. We define a function $\gamma\colon ideal(S') \to R$ by $\gamma\colon H \mapsto \sum H$. Clearly $\gamma(S') = 1$ and $\gamma(\{0\}) = 0$. Moreover, $\gamma(H \cup K) = \gamma(H)+\gamma(K)$ and $\gamma(HK) = \gamma(H)\gamma(K)$ since R is additively idempotent. Thus γ is a morphism of semirings. For $r \in R$ there exists a nonempty subset A of S' satisfying $r = \vee A$. Then $r = \vee\{Sa \mid a \in A\} = \gamma(H)$, where $H = \cup\{Sa \mid a \in A\} \in ideal(S')$. Thus γ is surjective. To show that γ is injective, it suffices to show that $\gamma(H) \leq \gamma(K)$ implies that $H \subseteq K$. Indeed, let $h \in H$. Then $h \leq \gamma(H) \leq \gamma(K) = \vee K$ and so $h = h \wedge (\vee K) = \vee\{h \wedge k \mid k \in K\}$. Since each element k of K is weakly meet principal, we have $h = h \wedge k = k(k^{\langle -1 \rangle}h)$ for all $k \in K$. Moreover, $k^{\langle -1 \rangle}h = \vee A_k$ for some subset A_k of S' and so $h = \vee\{kA_k \mid k \in K\}$. By (5), this implies that $h = ks$ for some $s \in S'$, proving that $h \in S'k \subseteq K$. Thus $H \subseteq K$, as desired. □

(24.25) PROPOSITION. *For elements a, b, and c of a QLO-semiring R the following conditions hold:*

(1) $a \leq b \Leftarrow c^{\langle -1 \rangle}a \leq c^{\langle -1 \rangle}b$;
(2) $a \leq b \Leftarrow b^{\langle -1 \rangle}c \leq a^{\langle -1 \rangle}c$;
(3) $a \geq bc \Leftrightarrow b^{\langle -1 \rangle}a \geq c$;
(4) $c^{\langle -1 \rangle}(b^{\langle -1 \rangle}a) = (bc)^{\langle -1 \rangle}a$;

(5) $c^{\langle -1\rangle}(b^{\langle -1\rangle}a) \geq (c^{\langle -1\rangle}b)a$;
(6) $(c^{\langle -1\rangle}b)^{\langle -1\rangle}(c^{\langle -1\rangle}a) \geq c^{\langle -1\rangle}a$;
(7) $(c^{\langle -1\rangle}b)^{\langle -1\rangle}(c^{\langle -1\rangle}a) \geq b^{\langle -1\rangle}a$;
(8) $a \geq b \Leftrightarrow b^{\langle -1\rangle}a = 1$;
(9) $a = 1^{\langle -1\rangle}a$;
(10) $a + b = 1 \Leftarrow b^{\langle -1\rangle}a = a$;
(11) $a^{\langle -1\rangle}(b \wedge c) = a^{\langle -1\rangle}b \wedge a^{\langle -1\rangle}c$;
(12) $c^{\langle -1\rangle}(ba) \geq (c^{\langle -1\rangle}b)a$;
(13) $(c^{\langle -1\rangle}b)(b^{\langle -1\rangle}a) \leq c^{\langle -1\rangle}a$.

Similarly, the analogous conditions for right residuals are also true.

PROOF. (1) By definition, $c(c^{\langle -1\rangle}a) \leq a \leq b$ so $c^{\langle -1\rangle}a \leq c^{\langle -1\rangle}b$.

(2) $c \geq b(b^{\langle -1\rangle}c) \geq a(b^{\langle -1\rangle}c)$ and hence $a^{\langle -1\rangle}c \geq b^{\langle -1\rangle}c$.

(3) If $a \geq bc$ then $b^{\langle -1\rangle}a \leq c$ by definition. Conversely, if $b^{\langle -1\rangle}a \geq c$ then $a \geq b(b^{\langle -1\rangle}a) \geq bc$.

(4) This is an immediate consequence of the definitions.

(5) By definition, $a \leq b(b^{\langle -1\rangle}c)$ and $b^{\langle -1\rangle}a \geq c[c^{\langle -1\rangle}(b^{\langle -1\rangle}a)]$ and so $a \geq b(b^{\langle -1\rangle}a) \geq b(c[c^{\langle -1\rangle}(b^{\langle -1\rangle}a)]) = (bc)[c^{\langle -1\rangle}(b^{\langle -1\rangle}a)]$. This implies that $(bc)^{\langle -1\rangle}a \geq c^{\langle -1\rangle}(b^{\langle -1\rangle}a)$. Conversely, $a \geq (bc)[(bc)^{\langle -1\rangle}a] = b(c[(bc)^{\langle -1\rangle}a])$ and so $b^{\langle -1\rangle}a \geq c[(bc)^{\langle -1\rangle}a]$. Thus $c^{\langle -1\rangle}(b^{\langle -1\rangle}a) \geq (bc)^{\langle -1\rangle}a$, proving equality.

(6) By definition, $b \geq c(c^{\langle -1\rangle}b)$ and so $ba \geq [c(c^{\langle -1\rangle}b)]a = c[(c^{\langle -1\rangle}b)a]$. This implies that $c^{\langle -1\rangle}(ba) \geq (c^{\langle -1\rangle}b)a$.

(7) We know that $b \geq c(c^{\langle -1\rangle}b)$ so

$$b \geq b(b^{\langle -1\rangle}a) \geq [c(c^{\langle -1\rangle}b)](b^{\langle -1\rangle}a) = c[(c^{\langle -1\rangle}b)(b^{\langle -1\rangle}a)].$$

Then $c^{\langle -1\rangle}a \geq (c^{\langle -1\rangle}b)(b^{\langle -1\rangle}a)$ so we have (7).

(8), (9) These are immediate consequences of the definition.

(10) Since $a + b = 1$, we have $a = 1a = (a+b)a = a^2 + ba$. In particular, $a \geq ba$ so $a \leq b^{\langle -1\rangle}a$. Conversely, $a \geq a(b^{\langle -1\rangle}a)$ since $ab \leq a \wedge b$ and $a \geq b(b^{\langle -1\rangle}a)$ by definition. Therefore $a \geq a(b^{\langle -1\rangle}a) + b(b^{\langle -1\rangle}a) = (a+b)(b^{\langle -1\rangle}a) = b^{\langle -1\rangle}a$, proving equality.

(11) By definition, $b \wedge c \geq a[a^{\langle -1\rangle}(b \wedge c)]$ so $a^{\langle -1\rangle}b \wedge a^{\langle -1\rangle}c \geq a^{\langle -1\rangle}(b \wedge c)$. Conversely, $b \wedge c \geq a(a^{\langle -1\rangle}b) \wedge a(a^{\langle -1\rangle}c) \geq a(a^{\langle -1\rangle}b \wedge a^{\langle -1\rangle}c)$ and so $b \wedge c \geq a(a^{\langle -1\rangle}b \wedge a^{\langle -1\rangle}c)$. This suffices to prove the reverse inequality, and so we have equality.

(12) If $r \in R$ satisfies $cr \leq b$ then $cra \leq ba$ since R is positive and so $ra \leq c^{\langle -1\rangle}(ba)$. In particular, for $r = c^{\langle -1\rangle}b$ we have $(c^{\langle -1\rangle}b)a \leq c^{\langle -1\rangle}ba$.

(13) Since $b(b^{\langle -1\rangle}a) \leq a$ we have, by (12) and (1), that $(c^{\langle -1\rangle}b)(b^{\langle -1\rangle}a) \leq c^{\langle -1\rangle}[b(b^{\langle -1\rangle}a)] \leq c^{\langle -1\rangle}a$.

The analogous conditions for right residuals are proven in the same manner. □

Note that by Proposition 24.25(8) we have $b^{\langle -1\rangle}1 = 1 = 1b^{\langle -1\rangle}$ for all $b \in R$.

(24.26) PROPOSITION. *The following conditions on a QLO-semiring are equivalent:*

(1) *Every element of R is left meet principal;*
(2) *Every element of R is left weakly meet principal.*

PROOF. Clearly (1) implies (2). Conversely, assume (2). If $a, b, r \in R$ then, by (2) and Proposition 24.25(4), we have $ra \wedge b = ra[(ra)^{\langle -1 \rangle}b] = r(a[a^{\langle -1 \rangle}(r^{\langle -1 \rangle}b)]) = r(r^{\langle -1 \rangle}b \wedge a)$, proving (1). □

(24.27) PROPOSITION. *If R is a QLO-semiring then:*

(1) *The set of all meet-principal elements of R is a submonoid of $(R, \cdot)$;*
(2) *The set of all sum-principal elements of R is a submonoid of $(R, \cdot)$;*
(3) *The set of all principal elements of R is a submonoid of $(R, \cdot)$.*

PROOF. (1) By Proposition 24.25(9) we see that 1 is a meet-principal element of R. Assume that r and s are meet-principal. If $a, b \in R$ then, using Proposition 24.25(4), we note that $(rs)[(rs)^{\langle -1 \rangle}b \wedge a] = r(s[s^{\langle -1 \rangle}(r^{\langle -1 \rangle}b) \wedge a]) = r(r^{\langle -1 \rangle}b \wedge sa) = b \wedge (rs)a$ and so rs is also left meet-principal.

(2) Again, by Proposition 24.25(9) we see that 1 is a sum-principal element of R. Assume that r and s are sum-principal. If $a, b \in R$ then

$$\begin{aligned}(rs)^{\langle -1 \rangle}[(rs)a + b] &= s^{\langle -1 \rangle}(r^{\langle -1 \rangle}[r(sa) + b]) \\ &= s^{\langle -1 \rangle}(sa + r^{\langle -1 \rangle}b) \\ &= a + s^{\langle -1 \rangle}(r^{\langle -1 \rangle}b) \\ &= a + (rs)^{\langle -1 \rangle}b\end{aligned}$$

and so rs is also left sum-principal.

(3) This is a direct consequence of (1) and (2). □

In particular, if R is a commutative QLO-semiring and if A is the set of all principal elements of R which are not zero divisors then A is an Øre set and we can construct the classical semiring of fractions $A^{-1}R$ of R. Indeed, recall that $A^{-1}R$ is defined to be $(A \times R)/\sim$, where $\sim$ is the equivalence relation defined by $(a, r) \sim (a', r')$ if and only if there exist elements u and u' of R satisfying $ur = u'r'$ and $ua = u'a'$. If (a, r) and (a', r') are elements of $A \times R$ satisfying $ar' = a'r$ then, taking $u = a'$ and $u' = a$, we see that $(a, r) \sim (a', r')$. Conversely, if $(a, r) \sim (a', r')$ then $uu'ar' = uu'a'r$ and so, by Proposition 24.23, $ar' = (uu')^{\langle -1 \rangle}(uu'ar') = (uu')^{\langle -1 \rangle}(uu'a'r) = a'r$. Therefore this construction generalizes that given in [Burton, 1975]. Note that $A^{-1}R$ is a semiring which is partially-ordered by the relation $a^{-1}r \leq b^{-1}s$ if and only if $rb \leq as$ in R. Indeed, $A^{-1}R$ has the structure of a lattice with operations given by $a^{-1}r \vee b^{-1}s = (ab)^{-1}[rb + sa] = a^{-1}r + b^{-1}s$ and $a^{-1}r \wedge b^{-1}s = (ab)^{-1}[rb \wedge sa]$. Howeover, $A^{-1}R$ is not necessarily a lattice-ordered semiring since it does not follow from the definitions that $(a^{-1}r)(b^{-1}s) \leq a^{-1}r \wedge b^{-1}s$.

(24.28) PROPOSITION. *The following conditions on an element r of a QLO-semiring R are equivalent:*

(1) *r is left weakly principal;*
(2) *r is left principal.*

PROOF. Clearly (2) implies (1). Conversely, assume (1). Making use of (1) and

Proposition 24.25(1), we see that for all $a, b \in R$ we have

$$\begin{aligned} r^{\langle -1 \rangle}(a + rb) &= r^{\langle -1 \rangle}(a + rb) + r^{\langle -1 \rangle}0 = r^{\langle -1 \rangle}(r[r^{\langle -1 \rangle}(a + rb)]) \\ &= r^{\langle -1 \rangle}[r \wedge (a + rb)] = r^{\langle -1 \rangle}[(r \wedge a) + rb] \\ &= r^{\langle -1 \rangle}[r(r^{\langle -1 \rangle}a) + rb] = r^{\langle -1 \rangle}[r(r^{\langle -1 \rangle}a + b)] \\ &= r^{\langle -1 \rangle}a + b + r^{\langle -1 \rangle}0 = r^{\langle -1 \rangle}a + b. \end{aligned}$$

Furthermore, making use of Proposition 24.25(11) as well, we have

$$\begin{aligned} ra \wedge b &= ra \wedge r \wedge b = r[r^{\langle -1 \rangle}(ra \wedge b)] \\ &= r[r^{\langle -1 \rangle}(ra \wedge r \wedge b)] = r(r^{\langle -1 \rangle}[ra \wedge r(r^{\langle -1 \rangle}b)]) \\ &= r(r^{\langle -1 \rangle}[ra] \wedge r^{\langle -1 \rangle}[r(r^{\langle -1 \rangle}b)]) = r[r^{\langle -1 \rangle}(a + 0) \wedge (r^{\langle -1 \rangle}b + r^{\langle -1 \rangle}0)] \\ &= r[r^{\langle -1 \rangle}(a + 0) \wedge r^{\langle -1 \rangle}b] = r[r^{\langle -1 \rangle}(a \wedge b) + r^{\langle -1 \rangle}0] \\ &= r[rr^{\langle -1 \rangle}(a \wedge b)] + 0 = r[r^{\langle -1 \rangle}(a \wedge b)], \end{aligned}$$

proving (2). □

(24.29) Proposition. *If a, b, and c are elements of a QLO-semiring R then $a^{\langle -1 \rangle}(bc^{\langle -1 \rangle}) = (a^{\langle -1 \rangle}b)c^{\langle -1 \rangle}$.*

Proof. As a direct consequence of the definitions, it is easy to see that both of the mentioned expressions are equal to $\vee\{r \in R \mid arc \leq b\}$. □

(24.30) Proposition. *If a and b are elements of a QLO-semiring R then:*

(1) *$b = a^{\langle -1 \rangle}(ab)$ if and only if $b = a^{\langle -1 \rangle}c$ for some $c \in R$;*
(2) *$b = a(a^{\langle -1 \rangle}b)$ if and only if $b = ac$ for some $c \in R$;*
(3) *$b = (ba)a^{\langle -1 \rangle}$ if and only if $b = ca^{\langle -1 \rangle}$ for some $c \in R$;*
(4) *$b = (ba^{\langle -1 \rangle})a$ if and only if $b = ca$ for some $c \in R$.*

Proof. (1) If $b = a^{\langle -1 \rangle}c$ for some $c \in R$ then $a(a^{\langle -1 \rangle}c) \leq c$ and so, by Proposition 24.25, $a^{\langle -1 \rangle}[a(a^{\langle -1 \rangle}c)] \leq a^{\langle -1 \rangle}c$. Moreover, $a^{\langle -1 \rangle}c \leq a^{\langle -1 \rangle}[a(a^{\langle -1 \rangle}c)]$ and so $b = a^{\langle -1 \rangle}c = a^{\langle -1 \rangle}[a(a^{\langle -1 \rangle}c)] = a^{\langle -1 \rangle}(ab)$. The reverse implication is trivial.

(2) If $b = ac$ for some $c \in R$ then $a^{\langle -1 \rangle}(ac) \geq c$ by Proposition 24.25 and so $a[a^{\langle -1 \rangle}(ac)] \geq ac$. On the other hand, $ac \geq a[a^{\langle -1 \rangle}(ac)]$ by definition and so $b = ac = a[a^{\langle -1 \rangle}(ac)] = a(a^{\langle -1 \rangle}b)$. Again, the reverse implication is trivial.

The proofs of (3) and (4) are given in an analogous manner. □

(24.31) Proposition. *If a, b, and c are elements of a QLO-semiring R then:*

(1) *$(ab^{\langle -1 \rangle})^{\langle -1 \rangle}a \geq a + b$;*
(2) *$a(b^{\langle -1 \rangle}a)^{\langle -1 \rangle} \geq a + b$;*
(3) *$a[(ab^{\langle -1 \rangle})^{\langle -1 \rangle}a]^{\langle -1 \rangle} = ab^{\langle -1 \rangle}$;*
(4) *$[a(b^{\langle -1 \rangle}a)^{\langle -1 \rangle}]^{\langle -1 \rangle}a = b^{\langle -1 \rangle}a$.*

Proof. (1) From the definitions, we know that $(ab^{\langle -1 \rangle})^{\langle -1 \rangle}a \geq a$. To show that $(ab^{\langle -1 \rangle})^{\langle -1 \rangle}a \geq b$ we have to show that $a \geq (ab^{\langle -1 \rangle})b$ and this, indeed, is an immediate consequence of the definition.

(2) The proof of this is similar to the proof of (1).

(3) Since $a \geq (ab^{\langle -1\rangle})^{\langle -1\rangle}a \geq b$ we have $(ab^{\langle -1\rangle})^{\langle -1\rangle}a \geq b$ and so $ab^{\langle -1\rangle} \geq a[(ab^{\langle -1\rangle})^{\langle -1\rangle}a]^{\langle -1\rangle}$. The reverse containment follows by definition.

(4) The proof is similar to that of (3). □

(24.32) PROPOSITION. *For an element a of a QLO-semiring R which is not a unit the following conditions are equivalent:*

(1) *a is prime;*
(2) *$ab^{\langle -1\rangle} = a$ for all $b \in R$ satisfying $b \not\leq a$;*
(3) *$b^{\langle -1\rangle}a = a$ for all $b \in R$ satisfying $b \not\leq a$.*

PROOF. Assume (1). If $r \in R$ satisfies $rb \leq a$ then $r \leq a$ since $b \not\leq a$ and a is prime. Therefore $ab^{\langle -1\rangle} \leq a$. The reverse inequality is trivial and so we have equality. Conversely, assume (2) and let $b, c \in R$ satisfy $b, c \not\leq a$ and $bc \leq a$. Then $b \leq ac^{\langle -1\rangle} = a$, a contradiction. Thus a is prime, proving the equivalence of (1) and (2). The equivalence of (1) and (3) is proven similarly. □

(24.33) COROLLARY. *If b is a prime element of a QLO-semiring R then $a^{\langle -1\rangle}b$ and $ba^{\langle -1\rangle}$ are also prime for all $a \not\leq b$ in R.*

PROOF. Assume that $c \not\leq a^{\langle -1\rangle}b$ in R. By Proposition 24.25(4), we have $c^{\langle -1\rangle}(a^{\langle -1\rangle}b) = (ac)^{\langle -1\rangle}b$. Since $c \not\leq a^{\langle -1\rangle}b$, $ac \not\leq b$ and so, by Proposition 24.32, $c^{\langle -1\rangle}(a^{\langle -1\rangle}b) = b$. But $b \leq a^{\langle -1\rangle}b \leq c^{\langle -1\rangle}(a^{\langle -1\rangle}b)$ so we have $c^{\langle -1\rangle}(a^{\langle -1\rangle}b) = a^{\langle -1\rangle}b$, proving that $a^{\langle -1\rangle}b$ is prime by Proposition 24.32. Similarly, $ba^{\langle -1\rangle}$ is prime. □

(24.34) PROPOSITION. *Let a be an element of a QLO-semiring R satisfying the condition that $\{r \in R \mid r > a\}$ has a unique minimal element c. Then $ac^{\langle -1\rangle}$ is prime.*

PROOF. If r and r' are elements of R satisfying $rr' \leq ac^{\langle -1\rangle}$ then $rr'c \leq a$. Assume that $r' \not\leq ac^{\langle -1\rangle}$. Then $r'c \not\leq a$ and so $a < a + r'c \leq c$. By the minimality of c, this implies that $a + r'c = c$ and so $rc = r(a + r'c) = ra + rr'c \leq a$. Thus $r \leq ac^{\langle -1\rangle}$, proving that $ac^{\langle -1\rangle}$ is prime. □

(24.35) PROPOSITION. *Let a be an element of a QLO-semiring R and let U be a nonempty subset of R. Then:*

(1) $(\vee U)^{\langle -1\rangle}a = \wedge\{b^{\langle -1\rangle}a \mid b \in U\}$;
(2) $a(\vee U)^{\langle -1\rangle} = \wedge\{ab^{\langle -1\rangle} \mid b \in U\}$;
(3) $a^{\langle -1\rangle}(\wedge U) = \wedge\{a^{\langle -1\rangle}b \mid b \in U\}$;
(4) $(\wedge U)a^{\langle -1\rangle} = \wedge\{ba^{\langle -1\rangle} \mid b \in U\}$.

PROOF. (1) If $b \in U$ then $\vee U \geq b$ implies that $b^{\langle -1\rangle}a \geq (\vee U)^{\langle -1\rangle}a$ and so $c = \wedge\{b^{\langle -1\rangle}a \mid b \in U\} \geq (\vee U)^{\langle -1\rangle}a$. Conversely, $b^{\langle -1\rangle}a \geq c$ for all b in U and so $(\vee U)c = \vee\{bc \mid b \in U\} \leq a$. Therefore $c \leq (\vee U)^{\langle -1\rangle}a$, proving equality.

(2) This is proven similarly.

(3) Note that $a[a^{\langle -1\rangle}(\wedge U)] \leq \wedge U \leq b$ for all $b \in U$ and so $a^{\langle -1\rangle}(\wedge U) \leq a^{\langle -1\rangle}b$ for all b in U. Thus $a^{\langle -1\rangle}(\wedge U) \leq d = \wedge\{a^{\langle -1\rangle}b \mid b \in U\}$. Conversely, $d \leq a^{\langle -1\rangle}b$ and so $ad \leq b$ for all such b. Thus $ad \leq \wedge U$ and so $d \leq a^{\langle -1\rangle}(\wedge U)$, proving equality.

(4) Note that $[(\wedge U)a^{\langle -1\rangle}]a \leq \wedge U \leq b$ for each $b \in U$ and so $(\wedge U)a^{\langle -1\rangle} \leq ba^{\langle -1\rangle}$ for all such b. Therefore $(\wedge U)a^{\langle -1\rangle} \leq c = \wedge\{ba^{\langle -1\rangle} \mid b \in U\}$. Conversely, if $b \in U$ then $ba^{\langle -1\rangle} \leq c$ and so $b \geq (ba^{\langle -1\rangle})a \geq ca$. Therefore $\wedge U \geq ca$ and so $(\wedge U)a^{\langle -1\rangle} \geq c$, proving the desired equality. □

(24.36) PROPOSITION. *Let R be an entire QLO-semiring. For any subset A of R and any nonzero left principal element b of R we have $b^{\langle -1\rangle}(\vee A) = \vee\{b^{\langle -1\rangle}a \mid a \in A\}$.*

PROOF. Set $d = \vee A$. By Proposition 24.23 and the fact that b is weakly left meet principal, we have $b^{\langle -1\rangle}d = b^{\langle -1\rangle}(b(b^{\langle -1\rangle}d)) = b^{\langle -1\rangle}(d \wedge b)$. Since R is a QLO-semiring we then have

$$\begin{aligned} b^{\langle -1\rangle}d &= b^{\langle -1\rangle}(d \wedge b) = b^{\langle -1\rangle}(\vee\{a \wedge b \mid a \in A\}) \\ &= b^{\langle -1\rangle}(\vee\{b(b^{\langle -1\rangle}a) \mid a \in A\}) = b^{\langle -1\rangle}b[\vee\{b^{\langle -1\rangle}a \mid a \in A\}] \\ &= \vee\{b^{\langle -1\rangle}a \mid a \in A\}. \end{aligned}$$

□

(24.37) COROLLARY. *If R is an entire QLO-semiring in which 1 is compact then every left principal element of R is compact.*

PROOF. If b is a left principal element of R and A is a nonempty subset of R satisfying $b \vee A$ then, by Proposition 24.36, $1 = b^{\langle -1\rangle}(\vee A) = \vee\{b^{\langle -1\rangle}a \mid a \in A\}$. Since 1 is compact, there exists a finite subset A' of A such that $1 = \vee\{b^{\langle -1\rangle}a \mid a \in A'\} = b^{\langle -1\rangle}(\vee A')$ and so $b \leq \vee A'$, proving that b is compact. □

Let b be a nonzero left principal element of an entire QLO-semiring R. By Proposition 24.25(11) and Proposition 24.36 we see that b defines a congruence relation $\equiv_b$ on R by setting $r \equiv_b s$ if and only if $b^{\langle -1\rangle}r = b^{\langle -1\rangle}s$.

(24.38) PROPOSITION. *The following conditions on an element a of a QLO-semiring R are equivalent:*

(1) *$ab^{\langle -1\rangle} = b^{\langle -1\rangle}a$ for all $b \in R$;*
(2) *If $b_1 \cdot \ldots \cdot b_n \leq a$ then $b_2 \cdot \ldots \cdot b_n b_1 \leq a$ for all $n \geq 2$ and all $b_1, \ldots, b_n \in R$.*

PROOF. (1) ⇒ (2): Set $c = b_2 \cdot \ldots \cdot b_n$. Then $b_1 c \leq a$ and so $b_1 \leq ac^{\langle -1\rangle} = c^{\langle -1\rangle}a$, proving that $b_2 \cdot \ldots \cdot b_n b_1 = cb_1 \leq a$.

(2) ⇒ (1): For any $r \in R$, if $rb \leq a$ then, by (2), $br \leq a$. Then (1) follows from the definitions. □

(24.39) PROPOSITION. *Let R be a CLO-semiring and let A be a nonempty set. Then $S = R^{A\times A}$ with operations $\oplus$ and $\odot$ defined by*

(1) $(s \oplus s')(a, a') = s(a, a') + s'(a, a')$;
(2) $(s \odot s')(a, a') = \vee\{s(a, b)s'(b, a') \mid b \in A\}$;

is a semiring. Furthermore, if D is a nonempty set then $(R^{D\times A}, \wedge)$ is a left S-semimodule, with scalar multiplication defined by

$$(sf)(d, a) = \wedge\{f(d, b)s(a, b)^{\langle -1\rangle} \mid b \in A\}.$$

PROOF. The proof that S is a semiring is straightforward, as is the fact that $(R^A, \wedge)$ is an abelian monoid.

If $s, s' \in S$, $f \in R^{D\times A}$, and $(d,a) \in D \times A$ then, by Proposition 24.35 and Proposition 24.25(4), we have

$$\begin{aligned}[(s \odot s')f](d,a) &= \wedge\{f(d,b)[s \odot s')(a,b)^{\langle -1\rangle} \mid b \in A\} \\ &= \wedge\{f(d,b)[\vee\{s(a,c)s'(c,b) \mid c \in A\}]^{\langle -1\rangle} \mid b \in A\} \\ &= \wedge\{f(d,b)[s(a,c)s'(c,b)]^{\langle -1\rangle} \mid b, c \in A\} \\ &= \wedge\{[f(d,b)s'(c,b)^{\langle -1\rangle}]s(a,c)^{\langle -1\rangle} \mid b, c \in A\} \\ &= [s(s'f)](d,a)\end{aligned}$$

and so $(s \odot s')f = s(s'f)$.

If $s \in S$ and $f, g \in R^{D\times A}$ then, by Proposition 24.25(11),

$$\begin{aligned}[s(f \wedge g)](d,a) &= \wedge\{(f \wedge g)(d,b)s(a,b)^{\langle -1\rangle} \mid b \in A\} \\ &= \wedge\{f(d,b)s(a,b)^{\langle -1\rangle} \wedge g(d,b)s(a,b)^{\langle -1\rangle} \mid b \in A\} \\ &= (sf)(d,a) \wedge (sg)(d,a)\end{aligned}$$

for all $(d,a) \in D \times A$ and so $s(f \wedge g) = sf \wedge sg$.

If $s, s' \in S$ and $f \in R^{D\times A}$ then for each $(d,a) \in D \times A$ we have, by Proposition 24.35,

$$\begin{aligned}[(s + s')f](d,a) &= \wedge\{f(d,b)(s \vee s')(a,b)^{\langle -1\rangle} \mid b \in A\} \\ &= \wedge\{f(d,b)s(a,b)^{\langle -1\rangle} \wedge f(d,b)s'(a,b)^{\langle -1\rangle} \mid b \in A\} \\ &= [sf \wedge s'f](d,a)\end{aligned}$$

and so $(s + s')f = sf \wedge s'f$.

If $f \in R^{D\times A}$ and $(d,a) \in D \times A$ then

$$(1_S f)(d,a) = \wedge\{f(d,b)0^{\langle -1\rangle} \mid a \neq b \in A\} \wedge f(d,a)1^{\langle -1\rangle} = f(d,a)$$

and so $1_S f = f$.

Finally, if $e \in R^{D\times A}$ is defined by $e(d,a) = 1$ for all $(d,a) \in D \times A$ then $e \wedge f = f \wedge e = e$ for all $f \in R^{D\times A}$. Moreover, if $s \in S$ and $(d,a) \in D \times A$ then, by Proposition 24.25(8), we have $se(d,a) = \wedge\{1s(a,b)^{\langle -1\rangle} \mid b \in A\} = 1 = e(d,a)$ and so $se = e$. If $f \in R^{D\times A}$ and $(d,a) \in D \times A$ then $(0f)(d,a) = \wedge\{f(d,b)0^{\langle -1\rangle} \mid b \in A\} = 1$ and so $0f = e$. □

Note that we can take the particular case of D being a singleton and thus see that if R is a CLO-semiring and A is a nonempty set then $(R^A, \wedge)$ is a left S-semimodule, where $S = R^{A\times A}$ is the semiring defined in Proposition 24.39. Also, note that, in this case, if we take $R = \mathbb{B}$ we come back to the semiring S constructed in Example 22.8.

Furthermore, we remark in passing that it is possible to consider residuation in more general situations. For example, if R is a CLO-semiring and M is a partially-ordered left R-semimodule then for each pair m, m' of elements of M we can define

$m'm^{\langle -1 \rangle}$ to be the element $\vee\{r \in R \mid rm \leq m'\}$ of R. See, for example, [Johnson & Johnson, 1970]. However, we will not pursue this matter further here.

In Proposition 6.48 we showed that if R is a simple semiring, and hence, in particular, if R is a CLO-semiring, then for each $1 \neq d \in R$ the set I_d of all d-small elements of R is an ideal of R. It is also immediate that the set I_1 of all 1-small elements of R is all of R. If $d \leq d'$ are elements of a CLO-semiring R and if $a \in I_d$ then $a + b = 1$ implies that $d + b = 1$ and so $d' + b = d + d' + b = d' + 1 = 1$. Therefore $a \in I_{d'}$. Thus $I_d \subseteq I_{d'}$.

(24.40) PROPOSITION. *If R is a CLO-semiring in which 1 is compact and if $d \in R$ then $I_d = (a]$ for some $a \in I_d$.*

PROOF. The result is clearly true for $d = 1$ and so we can assume that $d \neq 1$. Set $a = \vee I_d$. We claim that $a \in I_d$. Indeed, assume that $a + b = 1$. Since 1 is compact, there exists a finite subset U of I_d such that $\vee U + b = 1$. Since I_d is an ideal, $\vee U \in I_d$ and so this implies that $d + b = 1$. Hence $a \in I_d$. This implies that $I_d \subseteq (a]$. Conversely, if $a' \in (a]$ then $a' + a = a \in I_d$ and, since the ideal I_d is strong by Proposition 6.48, we see that $a' \in I_d$, proving equality. □

In Chapter 18 we considered nuclei on partially-ordered semirings. We now turn to consider them in the special case of CLO-semirings and QLO-semirings.

(24.41) PROPOSITION. *If R is a CLO-semiring in which 1 is compact then the function $\sigma: R \to R$ defined by $d \mapsto \wedge I_d$ is a nucleus.*

PROOF. If $d \leq d'$ are elements of R then, as we have noted above, $I_d \subseteq I_{d'}$ and so $\sigma(d) \leq \sigma(d')$. Moreover, we know that if $d \in R$ then $d \in I_d$ and so $d \leq \sigma(d)$ and similarly $\sigma(d) \leq \sigma^2(d)$. Suppose that a is a $\sigma(d)$-small element of R. If $a + b = 1$ then $\sigma(d) + b = 1$ and so, by the compactness of 1, there is a finite subset U of I_d satisfying $\vee U + b = 1$. But I_d is an ideal of R so $\vee U \in I_d$ and hence $d + b = 1$. Thus a is d-small. This shows that $I_d = I_{\sigma(d)}$ and so $\sigma(d) = \sigma^2(d)$.

Finally, assume that d and d' are elements of R and that $\sigma(d)\sigma(d') + b = 1$. Then $\sigma(d) + b = 1$ and $\sigma(d') + b = 1$ and so, since $\sigma(d)$ is d-small and $\sigma(d')$ is d'-small, we have $d + b = 1 = d' + b$. Thus $dd' + bd' = d'$ so $1 = dd' + bd' + b = dd' + b$, proving that $\sigma(d)\sigma(d')$ is dd'-small or, in other words, that $\sigma(d)\sigma(d') \leq \sigma(dd')$. □

(24.42) EXAMPLE. [Banaschewski & Harting, 1985] If R is the semiring of open subsets of a topological space X and $d \in R$ then $a \in R$ is d-small if and only if every closed subset of X contained in a is contained in d. In particular, if X is a T_1-space then σ is the identity map on R. The converse is false.

(24.43) PROPOSITION. *If R is a QLO-semiring then a function $\nu: R \to R$ is a nucleus if and only if $\nu(b)\nu(a)^{\langle -1 \rangle} = \nu(b)a^{\langle -1 \rangle}$ and $\nu(a)^{\langle -1 \rangle}\nu(b) = a^{\langle -1 \rangle}\nu(b)$ for all $a, b \in R$.*

PROOF. Assume that ν is a nucleus and let a, b be elements of R. Since $a \leq \nu(a)$, we have $\nu(b)\nu(a)^{\langle -1 \rangle} \leq \nu(b)a^{\langle -1 \rangle}$ by Proposition 24.25(2). On the other hand, $[\nu(b)a^{\langle -1 \rangle}]\nu(a) \leq \sigma\nu[\nu(b)a^{\langle -1 \rangle}]\nu(a) \leq \nu([\nu(b)a^{\langle -1 \rangle}]a) \leq \nu(\nu(b)) = \nu(b)$ so $\nu(a)a^{\langle -1 \rangle} \leq \nu(b)\nu(a)^{\langle -1 \rangle}$, proving the first equality. The second equality is proven similarly.

Now assume that both equalities hold for all elements a and b of R. By Proposition 24.25, we see that $1 = \nu(a)^{\langle -1\rangle}\nu(a) = a^{\langle -1\rangle}\nu(a)$ and so $a = a1 \leq \nu(a)$ for each $a \in R$. Moreover, if $a \leq b$ in R then $a \leq \nu(b)$ and so, by Proposition 24.25(8), we have $1 = a^{\langle -1\rangle}\nu(b) = \nu(a)^{\langle -1\rangle}\nu(b)$. Thus $\nu(a) = \nu(a)\cdot 1 \leq \nu(b)$. If $a \in R$ then $1 = \nu(a)\nu(a)^{\langle -1\rangle} = \nu(a)\nu^2(a)^{\langle -1\rangle}$ and so, as before, $\nu^2(a) \leq \nu(a) \leq \nu^2(a)$, proving that $\nu(a) = \nu^2(a)$. Finally, if $a, b \in R$ then $ab \leq \nu(ab)$ implies that $a \leq \nu(ab)b^{\langle -1\rangle} = \nu(ab)\nu(b)^{\langle -1\rangle}$ and so $a\nu(b) \leq \nu(ab)$. This in turn means that $\nu(b) \leq a^{\langle -1\rangle}\nu(ab) = \nu(a)^{\langle -1\rangle}\nu(ab)$ so $\nu(a)\nu(b) \leq \nu(ab)$. □

(24.44) Proposition. *If R is a QLO-semiring and if $\nu: R \to R$ is a nucleus then $\nu(ba^{\langle -1\rangle}) \leq \nu(b)a^{\langle -1\rangle}$ and $\nu(a^{\langle -1\rangle}b) \leq a^{\langle -1\rangle}\nu(b)$ for all $a, b \in R$.*

Proof. If $a, b \in R$ then $\nu(ba^{\langle -1\rangle})\nu(a) \leq \nu([ba^{\langle -1\rangle}]a) \leq \nu(b)$ so $\nu(ba^{\langle -1\rangle}) \leq \nu(b)\nu(a)^{\langle -1\rangle} = \nu(b)a^{\langle -1\rangle}$. The other inequality is proven similarly. □

If R is a QLO-semiring, set $Y(R) = \{a \in R \mid ab^{\langle -1\rangle} = b^{\langle -1\rangle}a \text{ for all } b \in R\}$. This set is nonempty since $1 \in Y(R)$. Moreover, as a direct consequence of Proposition 24.35 we see that $Y(R)$ is closed under taking arbitrary meets. If R is commutative then surely $Y(R) = R$. For each $d \in Y(R)$ we can define a function $\nu_d: R \to R$ by $\nu_d: a \mapsto d(da^{\langle -1\rangle})^{\langle -1\rangle}$. Note that $\nu_1: a \mapsto 1$ for all $a \in R$.

(24.45) Proposition. *If R is a QLO-semiring and $d \in Y(R)$ then ν_d is a nucleus on R.*

Proof. By Proposition 24.25(2), we see that $a \leq b$ implies that $da^{\langle -1\rangle} \geq db^{\langle -1\rangle}$ and so $\nu_d(a) \leq n_d(b)$. By Proposition 24.31(1) we have $\nu_d(a) \geq a+d \geq a$. Moreover, by Proposition 24.31(4), we have

$$\nu_d^2(a) = d(d(d(da^{\langle -1\rangle})^{\langle -1\rangle})^{\langle -1\rangle})^{\langle -1\rangle} = d(da^{\langle -1\rangle})^{\langle -1\rangle} = \nu_d(a).$$

If $a, b \in R$ then $[d(da^{\langle -1\rangle})^{\langle -1\rangle}][da^{\langle -1\rangle}] \leq d$. By Proposition 24.25(12), we have $[d(db^{\langle -1\rangle})^{\langle -1\rangle}][db^{\langle -1\rangle})a^{\langle -1\rangle}] \leq da^{\langle -1\rangle}$ and so

$$d(da^{\langle -1\rangle})^{\langle -1\rangle}][d(db^{\langle -1\rangle})^{\langle -1\rangle}][(db^{\langle -1\rangle})a^{\langle -1\rangle}] \leq d.$$

That is to say, $[\nu_d(a)\nu_d(b)][(db^{\langle -1\rangle})a^{\langle -1\rangle}] \leq d$. By Proposition 24.25(4), this says that $[\nu_d(a)\nu_d(b)][d(ab)^{\langle -1\rangle}] \leq d$ and so $\nu_d(a)\nu_d(b) \leq \nu_d(ab)$. □

A QLO-semiring R is a **Girard semiring** if and only if there exists an element d of $Y(R)$ such that ν_d is the identity map on R. In such semirings the function from R to R defined by $a \mapsto da^{\langle -1\rangle}$ is called the **linear negation** map. See [Yetter, 1989] for details. Complete boolean algebras are certainly Girard semirings.

(24.46) Application. In Example 1.10 we saw that if $(M, *)$ is a monoid then $(sub(M), \cup, \cdot)$ is a semiring, which is in fact easily seen to be a QLO-semiring. Note that if $A, B \in sub(M)$ then

$$AB^{\langle -1\rangle} = \cup\{C \in sub(M) \mid BC \subseteq A\} = \{m \in M \mid b * m \in A \text{ for all } b \in B\}.$$

An element A of $sub(M)$ satisfies $AB^{\langle -1\rangle} = B^{\langle -1\rangle}A$ for all $B \in sub(M)$ if and only if $m * m' \in A \Leftrightarrow m' * m \in A$ for all $m, m' \in M$. Under the assumption that M

is commutative, this situation was studied in [Girard, 1987] in connection with the semantics of **linear logic**, developed as a suitable logic for the study of parallelism in computer systems. For a development of proof theory using linear logic, see [Girard, 1989]; for a sequent calculus for noncommutiative intuitionistic linear logic based on QLO-semirings, see [Brown & Gurr, 1993b].

Given an element D of $sub(M)$, let $S_D = \{A \in sub(M) \mid \nu_D(A) = A\}$. Then $(S, \vee_D, \cdot_D)$ is a Girard semiring with operations defined by $\vee_D A_i = \nu_D(\cup A_i)$ and $A \cdot_D B = \nu_D(AB)$. Rosenthal [1990] shows that in fact any Girard semiring is isomorphic to a semiring obtained in this manner.

(24.47) PROPOSITION. *If R is a QLO-semiring then there exists a Girard semiring S and a surjective morphism of semirings $S \to R$.*

PROOF. Let R be a QLO-semiring and set $S = R \times R$. Define lattice operations $\vee$ and $\wedge$ on S by setting $\vee(a_i, b_i) = (\vee A_i, \wedge b_i)$ and $\wedge(a_i, b_i) = (\wedge a_i, \vee b_i)$. Then $(S, \vee, \wedge)$ is a complete lattice. Define operations $+$ and $\cdot$ on S by setting $(a,b) + (c,d) = (a \vee c, b \wedge d)$ and $(a,b) \cdot (c,d) = (ac, da^{\langle -1 \rangle} \wedge c^{\langle -1 \rangle} b)$ for all $a, b, c, d \in R$.

Clearly $(S, +)$ is a commutative monoid with identity element $(0, 1)$. As a consequence of Propositions 24.25, 24.29, and 24.35, we can verify that $(S, \cdot)$ is a monoid with identity $(1, 1)$ and that multiplication distributes over addition from either side. Furthermore, if $(a, b) \in S$ then $(0,1) \cdot (a,b) = (0,1) = (a,b) \cdot (0,1)$. Thus $(S, +, \cdot)$ is a semiring. As a consequence of Proposition 24.35, we also see that multiplication distributes over arbitrary joins in S and so S is in fact a QLO-semiring.

In S we see that

$$\begin{aligned}(a,b) \cdot (c,d) \leq (e,f) &\Leftrightarrow ac \leq e \text{ and } f \leq da^{\langle -1 \rangle} \wedge c^{\langle -1 \rangle} b \\ &\Leftrightarrow a \leq ec^{\langle -1 \rangle}, fa \leq d, \text{ and } cf \leq b \\ &\Leftrightarrow a \leq ec^{\langle -1 \rangle}, a \leq f^{\langle -1 \rangle} d, \text{ and } cf \leq b \\ &\Leftrightarrow (a,b) \leq (ec^{\langle -1 \rangle} \wedge f^{\langle -1 \rangle} d, cf)\end{aligned}$$

and so we see that, in S, $(e,f)(c,d)^{\langle -1 \rangle} = (ec^{\langle -1 \rangle} \wedge f^{\langle -1 \rangle} d, cf)$. Similarly,

$$(c,d)^{\langle -1 \rangle}(e,f) = (c^{\langle -1 \rangle} e \wedge df^{\langle -1 \rangle}, fc)$$

in S. Thus, $(1,1) \in Y(S)$ and, indeed, for each element (a,b) of S we have $(1,1)[(1,1)(a,b)^{\langle -1 \rangle}]^{\langle -1 \rangle} = (a,b)$ and so $\nu_{(1,1)}$ is the identity map on S. This shows that S is a Girard semiring.

Finally, we note that we have a surjective morphism of semirings $\gamma: S \to R$ defined by $\gamma: (a,b) \mapsto a$. □

In frame-ordered semirings we can define not only infinite sums but infinite products as well, using a construction based on that in [Levitzki, 1946]. Let R be a frame-ordered semiring and let $\theta: \Omega \to R$. Without loss of generality we can assume that there exists an ordinal h such that Ω is the set of all ordinals less than h. Then we can define the element $a = \prod^r \theta(\Omega)$ inductively as follows:

(1) If $h = 0$ then $a = 1$;
(2) If $h = k + 1 > 0$ is not a limit ordinal and if $\Omega' = \Omega \setminus \{k\}$ then $a = [\prod^r \theta(\Omega')]\theta(k)$; (3) If $h > 0$ is a limit ordinal then $a = \vee\{\prod^r \theta(\Omega') \mid \Omega' \subset \Omega\}$.

Note that, since R is lattice-ordered, we have $\prod^r \theta(\Omega') \subseteq \prod^r \theta(\Omega)$ whenever $\Omega' \subseteq \Omega$.

We can similarly define $a = \prod^l \theta(\Omega)$ by changing the definition in (2) to be $a = \theta(k)[\prod^l \theta(\Omega')]$.

(24.48) EXAMPLE. For infinite applications of triangular norms and conorms, see [González, 1999].

25. FIXED POINTS OF AFFINE MAPS

Let R be a semiring and let M be a left R-semimodule. If $a \in R$ and $m \in M$ then the R-**affine map** from M to itself defined by a and m is the function $\lambda_{a,m}: M \to M$ given by $\lambda_{a,m}: m' \mapsto am' + m$. We will denote the set of all R-affine maps from M to itself by $Aff(M)$. Note that affine maps are written as acting on the side opposite scalar multiplication, in this case on the right. We can define affine maps of right R-semimodules in a similar fashion: if M is a right R-semimodule, if $a \in R$, and if $m \in M$ then we have an R-affine map $\rho_{a,m}: M \to M$ given by $\rho_{a,m}: m' \mapsto m + m'a$. These maps will be written as acting on the left.

(25.1) EXAMPLE. Let R be a semiring and let M the the set of all functions from $\mathbb{N}$ to R. This is clearly an (R,R)-bisemimodule. If S is the semiring of all R-endomorphisms of M as a right R-semimodule, then M is also an (S,R)-bisemimodule. Indeed, we have a morphism of semirings $\gamma: R \to S$ defined by $\gamma(a): f \mapsto af$ for all $a \in R$ and $f \in M$. One of the elements of S not in the image of γ is the **right shift** R-endomorphism α defined in Example 14.34. We also have an R-homomorphism of right R-semimodules $\theta: R \to M$ defined by $(\theta a)(i) = a$ if $i = 0$ and $(\theta a)(i) = 0$ if $i > 0$. Combining these two, we see that for each $a \in R$ we have an S-affine map $\sigma_a = \lambda_{a,\theta a}$ from M to itself defined by $\sigma_a: f \mapsto \alpha f + \theta a$ for all $f \in M$. Maps of this form are called **affine right shifts** of M.

We define an operation $+$ on $Aff(M)$ as follows: if $\lambda_{a,m}$ and $\lambda_{b,n}$ are elements of $Aff(M)$ then $\lambda_{a,m} + \lambda_{b,n} = \lambda_{a+b,m+n}$. It is easily verified that $(Aff(M),+)$ is a commutative monoid the identity of which is the map $\lambda_{0,0}$ which sends every element m' of M to 0_M. If $\lambda_{a,m} \in Aff(M)$ and $r \in R$, set $r\lambda_{a,m}$ equal to $\lambda_{ra,rm}$. This turns $Aff(M)$ into a left R-semimodule. Moreover, we have an R-monomorphism from M to $Aff(M)$ defined by $m \mapsto \lambda_{0,m}$. Similarly, we have an R-epimorphism from $R \times M$ to $Aff(M)$ defined by $(r,m) \mapsto \lambda_{r,m}$.

We can also define the product of two elements of $Aff(M)$ to be their composition: $\lambda_{a,m}\lambda_{b,n} = \lambda_{ba,bm+n}$. Again, it is clear that $(Aff(M),\cdot)$ is a monoid with identity element $\lambda_{1,0}$ which takes every element of M to itself. However, $Aff(M)$ is not a semiring with respect to these operations. Indeed, while it is true that if $\lambda_{a,m}$, $\lambda_{b,n}$, and $\lambda_{c,p}$ are elements of $Aff(M)$ then $\lambda_{c,p}(\lambda_{a,m} + \lambda_{b,n}) =$

$\lambda_{c,p}\lambda_{a,m} + \lambda_{c,p}\lambda_{b,n}$, distributivity from the other side does not necessarily hold. Similarly, while $\lambda_{a,m}\lambda_{0,0} = \lambda_{0,0}$, the same is not necessarily true for multiplication by $\lambda_{0,0}$ on the left. (The structure of $Aff(M)$ is known as a **seminearring**, an interesting generalization of the notion of a semiring which we will not go into since it is beyond the scope of this book.) Note, in particular, that if $a \in R$ and $m \in M$ then for each $k > 0$ we have $\lambda_{a,m}^k = \lambda_{a^k,n}$, where $n = (\sum_{i=0}^{k-1} a^i)m$. In particular, if a is nilpotent with index of nilpotency k then $m'\lambda_{a,m}^h = (\sum_{i=0}^{k-1} a^i)m$ for all $h \geq k$.

An affine map in $Aff(M)$ of the form $\lambda_{1,m}$ for some $m \in M$ is a **transformation** of M. The set $Trans(M)$ of all transformations of M is a submonoid of $(Aff(M), \cdot)$. The function $M \to Trans(M)$ defined by $m \mapsto \lambda_{1,m}$ is an isomorphism from the additive monoid $(M, +)$ to the multiplicative monoid $(Trans(M), \cdot)$.

If X is any set and φ is a function from X to itself then a **fixed point** of φ is an element x of X satisfying $\varphi(x) = x$. In particular, let R be a semiring and let M be a left R-semimodule. For $a \in R$ and $m \in M$ we will denote the set of all fixed points of the R-affine map $\lambda_{a,m}$ by $\mathcal{L}(a, m)$. Similarly, if M is a right R-semimodule then we denote the set of all fixed points of the R-affine map $\rho_{a,m}$ by $\mathcal{R}(a, m)$. If M is an (R, R)-bisemimodule, and in particular in the case $M = R$, then we denote $\mathcal{L}(a, m) \cap \mathcal{R}(a, m)$ by $\mathcal{T}(a, m)$. Thus $\mathcal{L}(a, m) = \{m' \in M \mid m' = am' + m\}$, $\mathcal{R}(a, m) = \{m' \in R \mid m' = m'a + m\}$, and $\mathcal{T}(a, m) = \{m' \in M \mid m' = am' + m = m'a + m\}$. These sets may be empty. Thus, for example, if $R = M = \mathbb{N}$ then $\mathcal{L}(1, 1) = \mathcal{R}(1, 1) = \mathcal{T}(1, 1) = \varnothing$. If $\mathcal{L}(a, m)$ [resp. $\mathcal{R}(a, m)$] is nonempty then we say that the equation $X = aX + m$ [resp. the equation $X = Xa + m$] is **solvable** in M. If this set is a singleton, then the corresponding equation is said to be **uniquely solvable**.

In what follows, when we will state results for sets of the form $\mathcal{L}(a, m)$, the corresponding results for sets of the form $\mathcal{R}(a, m)$ will also be implied.

If R is a semiring and M is a left R-semimodule, and if A is a finite or countably-infinite set, then the system of linear equations

$$\{X_i = a_iX_i + m_i \mid i \in A; a_i \in R, m_i \in M\}$$

can be represented as one linear equation of the form $X = fX + g$, where $f \in \mathcal{M}_{A,rc}(R)$ and $g \in M^A$ are given by $g(i) = m_i$ for all $i \in A$ and $f(i, j) = a_i$ if $i = j$ while $f(i, j) = 0$ otherwise.

(25.2) EXAMPLE. If M is a left R-semimodule then the set $\mathcal{L}(a, 0_M)$ is never empty for any element a of R since $0_M \in \mathcal{L}(a, 0_M)$. Indeed, $\mathcal{L}(1, 0_M) = M$ and $\mathcal{L}(a, 0_M)$ is a right ideal of R for all $1 \neq a \in R$.

(25.3) EXAMPLE. If R is a commutative semiring then clearly $\mathcal{L}(a, b)$, $\mathcal{R}(a, b)$, and $\mathcal{T}(a, b)$ are equal for all $a, b \in R$.

(25.4) EXAMPLE. It is clear that $\mathcal{L}(1, b) = \mathcal{R}(1, b) = \mathcal{T}(1, b)$ for all elements b of R. Moreover, $Z(R) = \{b \in R \mid \mathcal{T}(1, b) \neq \varnothing\}$.

(25.5) EXAMPLE. If a is an infinite element of R, then $a \in \mathcal{T}(1,r)$ for every element r of R. In particular, if R is a simple semiring then $1 \in \mathcal{T}(1,r)$ for all r in R.

Let a be a nilpotent element of a semiring R having positive index of nilpotency n and let $b = 1 + a + \cdots + a^{n-1}$. Then $b \in \mathcal{T}(a,1)$. More generally, if a is an element of a semiring R for which there exists a nonnegative integer n such that $1 + a + \cdots + a^n = 1 + a + \cdots + a^{n+1}$ then $b = 1 + a + \cdots + a^n$ belongs to $\mathcal{T}(a,1)$. Such an element a of R is said to be **stable** and the least nonnegative integer n such that $1+a+\cdots+a^n = 1+a+\cdots+a^{n+1}$ is the **stability index** of a. Note that if a and b are elements of R satisfying the condition that ab is stable with stability index n, then ba is stable with stability index at most $n+1$. Indeed,

$$\begin{aligned} 1 + ba + \cdots + (ba)^{n+1} &= 1 + b[1 + \cdots + (ab)^n a \\ &= 1 + b[1 + \cdots + (ab)^{n+1}]a \\ &= 1 + ba + \cdots + (ba)^{n+2}. \end{aligned}$$

Note that if a is a stable element of an additively-idempotent semiring having stability index n then $1 + a + \cdots + a^n \in I^{\times}(R)$.

The semiring R is simple if and only if every element is stable with stability index 0. Some conditions for the transition matrix of a graph with values in a semiring R to be stable are given in [Wongseelashote, 1979]. Thus, if R is a bounded distributive lattice and if n is a positive integer, any element of $\mathcal{M}_n(R)$ is stable [Give'on, 1964].

By Proposition 20.37, we note that a semiring R is difference ordered if and only if $\mathcal{L}(1, b+c) \subseteq \mathcal{L}(1,b)$ for all $b, c \in R$. Also, we note that an element a of a simple semiring R is small if and only if $1 \notin \mathcal{L}(a,b)$ and $1 \notin \mathcal{L}(b,a)$ for any element b of R.

(25.6) APPLICATION. One of the major motivations for the study of linear equations in semirings comes from graph theory. Let Γ be a directed graph on a finite set V and let U be the set of edges of Γ. Without loss of generality, we can assume that $V = \{1, \ldots, n\}$ for some positive integer n. A **path** from vertex v to vertex v' in the graph is a finite sequence of arcs of the form $\langle (v, i_2), (i_2, i_3), \ldots, (i_{t-1}, v') \rangle$. We also assume that we have a function $len \colon U \to \mathbb{R}^+$ which assigns to each arc in U a value called the **length** of the arc. For the sake of convenience, we extend len to a function from $V \times V$ to $\mathbb{R}^+ \cup \{\infty\}$ by setting $len(i,j) = \infty$ whenever $(i,j) \notin U$. The length of a path $P = \langle (v, i_2), \ldots, (i_{t-1}, v') \rangle$ is then defined to be $len(P) = \sum \{len(j,k) \mid (j,k) \in P\}$. The **Shortest Path Problem** in Γ is that of finding, for a given set v and v' of vertices, a path from v to v' having minimal length. There is no loss of generality in assuming that the vertices are numbered such that $v = 1$.

The Shortest Path Problem is, to quote [Mahr, 1981], the "most famous and important problem in combinatorial optimization"; it is extensively studied there and in [Gondran & Minoux, 1984a], in which several solution algorithms are presented. Refer also to [Lawler, 1976] for a good introduction to this problem. For an analysis of the algebraic complexity of path problems see [Mahr, 1982], and for additional algorithms and explicit computer routines to solve this problem see [Brucker, 1972], [Gallo & Pallottino, 1986, 1988], [Gondran, 1975], [Kolokol'tsov,

1992], [Mehlhorn, 1984], [Minoux, 1979], [Moffat & Takaoka, 1987], [Shier, 1973], [Tarjan, 1981a, 1981b], [Tong & Lam, 1996], and the other papers listed in [A. R. Pierce, 1975]. Moreover, it is shown there that the problem can be considered in the following setting: let R be the semiring defined in Example 1.22 and let $A = [a_{ij}]$ be the matrix in $\mathcal{M}_n(R)$ (where n is the number of vertices in V) defined by $a_{ij} = len(i,j)$. Then solutions to the Shortest Path Problem are obtained by finding elements of $\mathcal{L}(A, 1)$. Refer to [Backhouse & Carré, 1975], [Carré, 1971, 1979], and [Zimmermann, 1981]. This application motivated Litvinov, Maslov & Rodionov [1998] to refer to the problem of finding solutions of equations of the for $X = AX + B$, where $A, B \in \mathcal{M}_n(R)$ for some additively-idempotent semiring R, as the "generalized stationary Bellman equation". For solutions to the Shortest Path Problem in networks with fuzzy lengths, refer to [Chanas, 1987].

Martelli [1974, 1976] used similar reasoning to solve a related problem. If Γ is a directed graph as above with vertex set V and if A is a set of arcs of Γ, let Γ^{-A} be the graph having the same set of vertices and defined by setting

$$\Gamma^{-A}(v, v') = \begin{cases} \Gamma(v, v') & \text{if } (v, v') \notin A \\ 0 & \text{otherwise} \end{cases}.$$

If $v \neq v'$ are distinct vertices of Γ then a (v, v')-**cut set** for Γ is a set A A of arcs of Γ such that in the graph Γ^{-A} there is no path from v to v'. The family of all (v, v')-cut sets for Γ is partially ordered by inclusion. Making use of the semiring R defined in Example 1.20, he considers elements of $\mathcal{L}(A, 1)$ for certain matrices A in $\mathcal{M}_n(R)$, and in this manner identifies all minimal (v, v')-cut sets for any given pair (v, v') of vertices in V. A variant on this problem, in which one wants to find the k shortest paths for some $k > 1$, is discussed in [Minieka & Shier, 1973], [Shier, 1976], and [Wongseelashote, 1976]. The analysis of this problem is done in the additively-idempotent commutative semiring $(R, \oplus, \odot)$, where R is the set of all k-tuples $(a_1, \ldots, a_k)$ of elements of $\mathbb{R} \cup \{\infty\}$ satisfying $a_1 < a_2 < \cdots < a_k$ (where we take $\infty < \infty$) and the operations are defined by

(1) $(a_1, \ldots, a_k) \oplus (b_1, \ldots, b_k) = (c_1, \ldots, c_k)$ with c_j being the jth-smallest distinct element of $\{a_1, \ldots, a_k\} \cup \{b_1, \ldots b_k\}$;

(2) $(a_1, \ldots, a_k) \odot (b_1, \ldots, b_k) = (d_1, \ldots, d_k)$ with d_j being the jth-smallest distinct element of $\{a_i + b_h \mid 1 \leq i, h \leq k\}$.

The additive identity of R is $(\infty, \ldots, \infty)$ and the multiplicative identity of R is $(0, \infty, \ldots, \infty)$.

Certain variants of the Shortest Path Problem in which the length (or cost) of an edge in a path depends on the number of edges already passed through are considered in [Minoux, 1976] and are solved using endomorphism semirings of $\mathbb{N}$-semimodules.

(25.7) APPLICATION. For an example of the use of semirings to solve graph-theoretic problems associated with the design of VLSI chips, see [Iwano, 1987]. For the application of semirings in designing reconfigurable-architecture hardware systems using dynamic computational structures which can be used to solve graph-theoretic problems, see [Babb et al., 1998].

(25.8) APPLICATION. A **Petri net** Γ is a finite directed graph the vertices of which are partitioned into two disjoint classes, the set $P = P(\Gamma)$ of **places** of the net and the set $T = T(\Gamma)$ of **transitions** of the net, satisfying the condition that every edge of the graph either connects a place to a transition or a transition to a place. A **marking** on a Petri net Γ is a function from $P(\Gamma)$ to $\mathbb{N}$. Each transition $t \in T$ defines two important markings on Γ: the marking u_t which assigns to each $p \in P$ the number of edges from p to t and the marking v_t which assigns to each $p \in P$ the number of edges from t to p. If f is a marking on Γ and $t \in T$ satisfies the condition that $f(p) \geq u_t(p)$ for all $p \in P$ then the net can be **fired** at t to obtain a new marking f' on Γ defined by $f'(p) = f(p) - u_t(p) + v_t(p)$ for each $p \in P$. This process is denoted by $f[t\rangle f'$. If a marking f' is obtained from a marking f by the successive firing of (not necessarily distinct) transitions $t_1, \ldots, t_k$ we write f $[t_1, \ldots, t_k\rangle f'$. If such a succession of firings is possible, we say that f' is **reachable** from f.

Petri nets were initially developed by Carl Adam Petri in 1962 and received considerable impetus through their use in MIT's Project MAC; they are of great importance in modeling the behavior of multi-node asynchronous systems, such as large computer networks or industrial processes. See [Peterson, 1981] or [Reisig, 1985] for an introduction to the theory and application of Petri nets. For the use of Petri nets in the theory of discrete event dynamical systems, refer to [Baccelli et al., 1992].

If Γ is a Petri net satisfying $P(\Gamma) = \{p(1), \ldots, p(n)\}$ and $T(\Gamma) = \{t(1), \ldots, t(m)\}$ then the structure of Γ can be characterized by two $m \times n$ matrices, $U = [u_{ij}]$ and $V = [v_{ij}]$, defined by $u_{ij} = u_{t(j)}(p(i))$ and $v_{ij} = v_{t(j)}(p(i))$ for all $1 \leq i \leq n$ and $1 \leq j \leq m$. Any marking f on Γ can be considered as an element $a_f = (a_1, \ldots, a_n)$ of $\mathbb{N}^n$ defined by $a_i = f(p(i))$. It is clear that a necessary condition for a marking f' to be reachable from a marking f on Γ is that there exist a solution to the system of linear equations $a_{f'} = a_f + xU$.

Several variants on the theme of Petri nets have been extensively studied, one of the most useful being **timed Petri nets**. For an analysis of timed Petri nets using semirings, refer to [Cohen, Gaubert & Quadrat, 1998].

A framework for consideration of Petri nets with values in a difference-ordered semiring was introduced in [Golan, 1997] and extensively studied in [Wu, 1998]. Let R be such a semiring. Given a nonempty set P (not necessarily finite), the elements of which will be called **places**, and a nonempty set T (also not necessarily finite) disjoint from P, the elements of which will be called **transitions**, an R-**net** on the pair $\Gamma = (P, T)$ is a pair of functions $\mu \in R^{(P \times T)}$ and $\nu \in R^{(T \times P)}$. A **marking** with values in R is a function $f \in R^{(P)}$. Thus, every transition $t \in T$ defines two markings with values in R: $\mu_t: p \mapsto \mu(p, t)$ and $\nu_t: p \mapsto \nu(t, p)$. Similarly, a **guard** (sometimes also called a **threshold**) with values in R is a function $g \in R^{(T)}$. Thus, every place $p \in P$ defines two guards with values in R: $\mu_p: t \mapsto \mu(p, t)$ and $\nu_p: t \mapsto \nu(t, p)$. If $t \in T$ then the **source** of t is $\{p \in P \mid \mu_p(t) \neq 0_R\}$ and the **target** of t is $\{p \in P \mid \nu_p(t) \neq 0_R\}$. Similarly, if $p \in P$ then the **inset** of p is $\{t \in T \mid \nu_t(p) \neq 0_R\}$ and the **outset** of p is $\{t \in T \mid \mu_t(p) \neq 0_R\}$.

If $f \in R^{(P)}$ is a marking with values in R and if $t \in T$ then we say that the net (μ, ν) can be **fired** at t if and only if $f \geq \mu_t$. In that case, there exists a marking $f'' \in R^{(P)}$ satisfying $f = f'' + \mu_t$ and so the marking $f' = f'' + \nu_t$ satisfies

$f' + \mu_t = f + \nu_t$. In this case we write $f\ [R|t\rangle\ f'$ and say that f' is the marking **obtained** from f as a result of firing the net at t. One problem we encounter when working over an arbitrary zerosumfree semiring which is not encountered over $\mathbb{N}$ is that the marking f'' above, and hence the marking f', is not necessarily unique. Thus, in order for our notation to be well-defined we have to have a method of designating a canonical element of $S(f) = \{f' \in R^{(P)} \mid f\ [R|t\rangle\ f'\}$ to select in case this set is not a singleton. This depends on the properties of the semiring R and is the central and most difficult problem in applying the model given here. If R is cancellative then it is easy to see that $S(f)$ is either empty or a singleton so there is no problem. If R is additively-idempotent and complete then $S(f)$ has a unique maximal element and we can choose that to be the canonical one. (Recall tha by Proposition 23.5, every additively-idempotent semiring can be embedded in a complete such semiring.) Similarly, if $g \in R^{(T)}$ is a guard with values in R and if $p \in P$ then we say that the net (μ, ν) can be **activated** at p to obtain a new guard g' if and only if $g \geq \nu_p$ and we write $g\ [R|p\rangle\ g'$, where g' is a guard satisfying $g' + \nu_p = g + \mu_p$. Again, since $S(g) = \{g' \in R^{(T)} \mid g\ [R|p\rangle g'\}$, when nonempty, may not be a singleton, we need a method of designating a canonical element of this set and that method will depend on the semiring of which we are working.

If a marking f' is obtained from a marking f with values in R by successive firings of a sequence $w = t_1 \ldots t_n$ of (not necessarily distinct) transitions, we write $f\ [R|w\rangle\ f'$. Thus, every marking $f \in R^{(P)}$ defines a subset $L(f)$ of the free monoid T^* of all finite sequences $t_1 \ldots t_n$ of elements of T, defined by the condition that $w \in L(f)$ if and only if there exists a marking $f' \in R^{(P)}$ such that $f\ [R|w\rangle\ f'$. The set $L(f)$ is the **formal language** defined by f. Similarly, if a guard g' is obtained from a guard g with values in R by successive firings of a sequence $y = p_1 \ldots p_n$ of (not necessarily distinct) places, we write $g\ [R|y\rangle\ g'$. Thus, every guard $g \in R^{(T)}$ defines a subset $M(g)$ of the free monoid P^* of all finite sequences $p_1 \ldots p_n$ of elements of P, defined by the condition that $y \in M(g)$ if and only if there exists a guard $g' \in R^{(T)}$ such that $g\ [R|y\rangle\ g'$. The set $M(g)$ is the **formal language** defined by g.

If we choose $R = \mathbb{N}$ we obtain the usual Petri nets defined above; if we choose $R = \mathbb{I}$ we obtain **fuzzy Petri nets**.

(25.9) APPLICATION. For an use of fixed points of affine maps in the analysis of the semantics of programming languages and in the definition of abstract data types, see [Manes & Arbib, 1986]. While the explanation there is often presented in categorical language, it is clear how the same statements can often be made in the context of semirings and semimodules over them. The use of fixed-points of affine maps in the study of programming languages and data types harks back to the work of Dana Scott.

(25.10) APPLICATION. Countably complete semirings are an appropriate framework for studying recursion in database systems. Indeed, the family R of all linear relational operators on an arbitrary database is such a semiring. The effect of several mutual recursion operators on a database can then be evaluated as a fixed point of a suitable affine map on a matrix semiring over R. See [Ioannidis & Wong, 1991] and [Du & Ishii, 1995] for further details.

(25.11) EXAMPLE. Finding fixed points of an affine map is a special case of solving the general **affine equation problem**: given a left R-semimodule M and affine functions $\lambda_{a,m}, \lambda_{b,n}: M \to M$, find $K = \{m' \in M \mid \lambda_{a,m}(m') = \lambda_{b,n}(m')\}$. If R is the schedule algebra $(\mathbb{R} \cup \{-\infty\}, max, +)$ and $M = R$, the solution to this problem is the following:

(1) If both $b < a$ and $n < m$ or if both $a < b$ and $m < n$, then

$$K = \{max\{m, n\} - max\{a, b\}\};$$

(2) If (1) does not hold and if neither $a = b$ nor $m = n$, then $K = \varnothing$;

(3) If $a = b$ and $m \neq n$, then

$$K = \{x \in R \mid x \geq max\{m, m'\} - a\};$$

(4) If $a \neq b$ and $m = n$, then

$$K = \{x \in R \mid x \leq b - max\{a, a'\}\};$$

(5) If $a = b$ and $m = n$, then $K = R$.

See [Baccelli et al., 1992] for details. Refer to the same source for consideration of the affine equation problem also for the semiring S defined in Example 8.3t.

(25.12) PROPOSITION. *Let R be a semiring and let M be a left R-semimodule. For $a, b \in R$ and $m \in M$ we then have:*

(1) *If $d \in \mathcal{L}(a, 1)$ then $dm \in \mathcal{L}(a, m)$;*

(2) *If $d \in \mathcal{L}(a, a)$ then $(1 + d)m \in \mathcal{L}(a, m)$;*

(3) *If $m' \in \mathcal{L}(a, m)$ and if $m'' \in \mathcal{L}(a, 0_M)$ then $m' + m'' \in \mathcal{L}(a, m)$;*

(4) *If $c \in \mathcal{L}(a, 1)$ and $d \in \mathcal{L}(bc, 1)$ then $cd \in \mathcal{L}(a + b, 1)$;*

(5) *$\mathcal{L}(a, m) \subseteq \mathcal{L}(a^{n+1}, \sum_{i=0}^{n} a^i m)$ for all $n \geq 0$.*

PROOF. (1) - (3): These are immediate consequences of the distributive laws in R.

(4) This follows since $(a + b)cd + 1 = acd + bcd + 1 = acd + d = (ac + 1)d = cd$.

(5) If $m' \in \mathcal{L}(a, m)$ then, by repeated substitution, we have $m' = am' + m = a^2m' + am + m = \dots$. □

(25.13) PROPOSITION. *Let R be a semiring and let M be an additively idempotent left R-semimodule. If $a \in R$ and $m \in M$ then $\mathcal{L}(a, m)$ is closed under addition Moreover, $\mathcal{L}(a, m) \subseteq \mathcal{L}(a + 1, m)$ for all $a \in R$ and all $m \in M$.*

PROOF. If $m', m'' \in \mathcal{L}(a, m)$ then $m' + m'' = am' + m + am'' + m = a(m' + m'') + m$ so $m' + m'' \in \mathcal{L}(a, m)$. The second part is surely true if $\mathcal{L}(a, m) = \varnothing$. Otherwise, if $m' \in \mathcal{L}(a, m)$ then $(a + 1)m' + m = am' + m' + m = m' + m' = m'$ and so $m' \in \mathcal{L}(a + 1, m)$. □

(25.14) PROPOSITION. *Let R be a semiring and let $\alpha: M \to N$ be an R-homomorphism of left R-semimodules. Let $a \in R$ and $m \in M$. If $m' \in \mathcal{L}(a, m)$ then $m'\alpha \in \mathcal{L}(a, m\alpha)$.*

PROOF. This is immediate since we have $m' = am' + m$ and so $m'\alpha = a(m'\alpha) + m\alpha$. □

An element a of a semiring R is **right quasiregular** if and only if $\mathcal{L}(a,a) \neq \varnothing$, it is **left quasiregular** if and only if $\mathcal{R}(a,a) \neq \varnothing$, and it is **quasiregular** if and only if $\mathcal{T}(a,a) \neq \varnothing$. If a is a quasiregular element of R then an element of $\mathcal{T}(a,a)$ is a **quasi-inverse** of a. A subsemiring R is **rationally closed** if and only if it contains the quasi-inverses of each of its quasiregular elements. If $\{S_i \mid i \in \Omega\}$ is a family of rationally-closed subsemirings of a semiring R then $\cap_{i\in\Omega} S_i$ is also rationally closed. Thus each subsemiring R' of R is contained in a minimal rationally-closed subsemiring, called the **rational closure** of R' in R.

(25.15) EXAMPLE. If a is an element of an additively-idempotent semiring R then $1 + a + \cdots + a^n = (1+a)^n$. Moreover, $a^n = a^{n+1}$ if and only if $(1+a)^n = (1+a)^{n+1}$. If there indeed exists a positive integer n for which this is true, then a is quasiregular and $a^* = (1+a)^n$.

(25.16) EXAMPLE. If A is a nonempty set and R is a semiring then we had called an element f of $R\langle\langle A\rangle\rangle$ quasiregular if $f(\square) = 0$. If f is such an element then $g = lim_{n\to\infty} \sum_{k=1}^{n} f^k$ exists, where the limit is taken with respect to the topology defined in Chapter 2, and is the unique member of $\mathcal{T}(f,f)$. Thus it is also quasiregular in the sense defined here.

The rational closure of $R\langle A\rangle$ in $R\langle\langle A\rangle\rangle$ is called the semiring of **rational series** in A over R. A basic result in algebraic automata theory, known as Kleene's Theorem, states that these are precisely the formal power series which are the behaviors of finite $R\langle A\rangle$-automata (in the sense of [Kuich & Salomaa, 1986]). For a generalization of this theorem to finite automata having more than one initial state see [Kuich, 1987]: if A' is a subset of a complete semiring with necessary summation which contains 0 and 1 then the rational closure of the subsemiring of R generated by A' consists of the behaviors of finite automata the transition matrices of which have entries in A'. A consequence of this result, known as the Schützenberger Representation Theorem, is that an element f of $R\langle\langle A\rangle\rangle$ is rational if and only if, for some $n \geq 1$, there exist a morphism of semigroups $\mu: A^* \to \mathcal{M}_n(R)$, a $1 \times n$ vector U, and an $n \times 1$ vector V such that, for each $w \in A^*$, we have $f(w) = U\mu(w)V$. For a detailed study of this semiring and its application to formal language theory, see [Berstel & Reutenauer 1988], [Choffrut, 1992], [Salomaa & Soittola, 1978] and [Kuich & Salomaa, 1986].

One should note that dealing with rational series can be very difficult, even if R is very nice. Indeed, if $R = \mathbb{Z}$ it is undecidable as to whether a given rational series in $R\langle\langle A\rangle\rangle$

(1) has a zero coefficient;
(2) has infinitely-many zero coefficients;
(3) has a positive coefficient;
(4) has infinitely-many positive coefficients;
(5) has its coefficients ultimately nonnegative;
(6) has two equal coefficients.

See Proposition 9.15 of [Kuich & Salomaa, 1986] for a proof of this, based on the undecidability of Hilbert's Tenth Problem. It is, however, decidable as to whether a given rational series is equal to 0 or equal to a polynomial, and whether two given rational series are equal.

If the semiring R is finite and commutative then we have another means of characterizing the rational series in A over R. If $f \in R\langle\langle A\rangle\rangle$ then we define the **Hankel matrix** of f to be the function $H(f) \in R^{A^* \times A^*}$ defined by $H(f): (w, w') \mapsto f(ww')$. Then f is a rational series if and only if the number of distinct columns in the $H(f)$ is finite. If R is a field then f is rational if and only if the rank of the matrix $H(f)$ is finite.

Because of the importance of quasi-inverses in the solution of various problems in applied mathematics and computer science, many algorithms have been constructed to compute the quasi-inverse of a quasiregular element in various semirings, especially in semirings of matrices over simpler semirings. Refer, for example, to [Gondran & Minoux, 1984a, 1984b].

(25.17) PROPOSITION. *Let R be a semiring and let M be a left R-semimodule. Then the following conditions on an element a of R are equivalent:*

(1) *a is right quasiregular;*
(2) *$\mathcal{L}(a, m) \neq \varnothing$ for all $m \in M$;*
(3) *$\mathcal{L}(a, 1) \neq \varnothing$.*

PROOF. (1) $\Rightarrow$ (3): If $r \in \mathcal{L}(a, a)$ then one easily sees that $r + 1 \in \mathcal{L}(a, 1)$.
(3) $\Rightarrow$ (2): If $m \in M$ and $r \in \mathcal{L}(a, 1)$ then $rm \in \mathcal{L}(a, m)$.
(2) $\Rightarrow$ (1): This is immediate. □

(25.18) PROPOSITION. *Let $\gamma: R \to S$ be a morphism of semirings. If a is a [right, left] quasiregular element of R then $\gamma(a)$ is a [right, left] quasiregular element of S.*

PROOF. If a is a right quasiregular element of R then there exists an element r of R satisfying $r = ar + a$ and so $\gamma(r) = \gamma(a)\gamma(r) + \gamma(a)$. Thus $\gamma(a)$ is a right quasiregular element of S. The proof for left quasiregular elements is similar. □

(25.19) COROLLARY. *If $\gamma: R \to S$ is a morphism of semirings and S' is a rationally-closed subsemiring of S then $R' = \gamma^{-1}(S')$ is a rationally-closed subsemiring of R.*

PROOF. This is an immediate consequence of Proposition 25.18. □

If R is a semiring then a left R-semimodule M is **totally [uniquely] solvable** if and only if $\mathcal{L}(a, m)$ is nonempty [resp. is a singleton] for each a in R and each m in M. It is clear that a necessary and sufficient condition for M to be totally solvable is that every element of M be contained in a totally solvable submodule of M. If M is a totally solvable left R-semimodule and if $\alpha: M \to N$ is a surjective R-homomorphism of left R-semimodules then, by Proposition 25.14, we see that N is also totally solvable.

A semiring R is **right totally [uniquely] solvable** if it is totally uniquely solvable as a *left* R-semimodule. That is to say, R is right totally solvable if and only if it satisfies the equivalent conditions of Proposition 25.17. Similarly, R is **left totally [uniquely] solvable** if and only if $\mathcal{R}(a, b)$ is nonempty [resp. is a singleton] for all a and b in R. A semiring R is **totally [uniquely] solvable** if and only if $\mathcal{T}(a, b)$ is nonempty [resp. is a singleton] for all $a, b \in R$. By Proposition 25.12(1) we see that if R is right totally solvable then every left R-semimodule is totally solvable. A **Lehmann semiring** is a semiring R satisfying the condition that $\mathcal{T}(a,1) \neq \emptyset$ for every element a of R. This condition says that for every element a of R there exists an element a^* of R such that $a^* = aa^* + 1 = a^*a + 1$. Note that, as an immediate consequence of this definition, we see that $aa^* = aa^*a + a = a^*a$ for each element a of R and that $0^* = 1$. A Lehmann semiring is a **Conway semiring** if we can choose the elements a^* such that, for all a and b in R we have $(ab)^* = 1 + a(ba)^*b$ and $(a+b)^* = (a^*b)^*a^*$.

(25.20) EXAMPLE. From [Conway, 1971] and [Krob, 1992] we know that there exist additively-idempotent semirings R_0 and R_1, each equipped with an endofunction $a \mapsto a^*$, such that

(1) $(a+b)^* = (a^*b)^*a^*$ and $a^* = (a^n)^*\left(\sum_{i=0}^{n-1} a^i\right)$ for all $a, b \in R_0$ and all $n > 1$, but there exist elements $a', b' \in R_0$ for which $(a'b')^* \neq 1 + a'(b'a')^*b'$.

(2) $(ab)^* = 1 + a(ba)^*b$ and $a^* = (a^n)^*\left(\sum_{i=0}^{n-1} a^i\right)$ for all $a, b \in R_1$ and all $n > 1$, but there exist elements $a', b' \in R_1$ for which $(a'+b')^* \neq (a'^*b')^*a'^*$.

Moreover, for each prime p there exists a Conway semiring R_p such that $a^* = (a^q)^*\left(\sum_{i=0}^{q-1} a^i\right)$ for all prime integers $q \neq p$ but there exist elements $a', b' \in R_p$ such that $a'^* \neq (a'^p)^*\left(\sum_{i=0}^{p-1} a'^i\right)$.

For the use of such $*$-operators in process algebras, see [Bergstra & Ponse, 1995].

(25.21) EXAMPLE. Clearly every totally solvable semiring R is a Lehmann semiring. If every element of R is stable then R is a Conway semiring [Backhouse & Carré, 1975].

(25.22) EXAMPLE. In Example 1.10 we noted that if $(M, *)$ is a monoid then $R = (sub(M), +, \cdot)$ is a semiring under the operations $A + B = A \cup B$ and $AB = \{a * b \mid a \in A, b \in B\}$. For each element A of R, let A^* be the smallest submonoid of M containing A. Then $A^* \in \mathcal{T}(A, 1_R)$ and so R is a Lehmann semiring.

(25.23) EXAMPLE. As already noted, the semiring $\mathbb{N}$ of nonnegative integers is neither right totally solvable nor left totally solvable.

(25.24) EXAMPLE. A field F is not right totally solvable since $\mathcal{L}(1,1) = \emptyset$.

(25.25) EXAMPLE. [Lehmann, 1977] If R is an entire zerosumfree semiring and ∞ is an element not in R then, $R\{\infty\}$ is totally solvable and so is a Lehmann semiring. Indeed, if a and b are elements of $R\{\infty\}$ then, by definition, $\infty \in \mathcal{T}(a,b)$. For any element a of R there may be several ways of closing the element a^*. For example, let $R = (\mathbb{R}^+\{\infty\}, +, \cdot)$. Then:

(1) We can take $0^* = 1$ and $a^* = \infty$ for all $0 \neq a \in R$; or
(2) We can take $1^* = \infty$ and $a^* = (1-a)^{-1}$ for $1 \neq a \in R$.

(25.26) EXAMPLE. If R is a simple semiring then R is totally solvable and hence a Lehmann semiring. Indeed, if $a, b \in R$ then, by Proposition 4.3, we have $b \in \mathcal{T}(a, b)$. In particular, the boolean semiring $\mathbb{B}$ is a Lehmann semiring. Since $\mathbb{B}$ is complete by Example 22.1, we see that if A is a nonempty set then the semiring of formal power series $\mathbb{B}\langle\langle A\rangle\rangle$ is also complete by Example 22.9 and so, by (2), it is also a Lehmann semiring in which, for each $f \in \mathbb{B}\langle\langle A\rangle\rangle$ we define f^* by $f^*(w) = \sum_{n\geq 0} f^n(w)$ for each $w \in A^*$, where the the infinite sum in $\mathbb{B}$ is defined as in Example 22.1.

(25.27) EXAMPLE. [Lehmann, 1977] Let $R = \mathbb{R} \cup \{-\infty, \infty\}$ and extend the natural order and addition on $\mathbb{R}$ to R by setting

(1) $-\infty < r < \infty$ for all $r \in \mathbb{R}$,
(2) $-\infty + r = -\infty$ for all $r \in \mathbb{R}$,
(3) $r + \infty = \infty$ for all $r \in \mathbb{R}$, and
(4) $-\infty + \infty = \infty$.

Then $(R, min, +)$ is a semiring. If we define the operator $*$ on R by $a^* = 0$ for all $a \geq 0$ while $a^* = -\infty$ for all $a < 0$, then it is in fact a Lehmann semiring.

(25.28) EXAMPLE. If R is an additively-idempotent semiring and if $a \mapsto a^*$ and $a \mapsto a^\diamond$ are functions from R to R satisfying the condition that $a^*, a^\diamond \in \mathcal{T}(a, 1)$ for all $a \in R$ then, by Proposition 25.13, the function $a \mapsto a^* + a^\diamond$ also satisfies this condition.

(25.29) EXAMPLE. Let $(sub(A^*), +, \cdot)$ be the semiring of all formal languages on a nonempty set A, as defined in Example 1.11, and for each language L in $sub(A^*)$ let $L^* = \{\square\} \cup [\cup_{i=1}^{\infty} L^i]$. Then $L^* \in \mathcal{T}(L, A^*)$ and so $sub(A^*)$ is a Lehmann semiring. Similarly, if S is the semiring $sub(A^\infty) \cup \{-\infty\}$ defined in Example 3.21, then S is a Lehmann semiring. An explicit algorithm for the computation of L^* on the Instruction Systolic Array (ISA), together bounds on the implementation time for computation, is given in [Lang, 1987].

(25.30) APPLICATION. The semiring $S = (R^{A\times A}, \oplus, \cdot)$ defined in Proposition 24.39 is a Lehmann semiring where, for each $s \in S$ we define $s^* \in S$ by

$$s^*(a, a') = \vee_{n\geq 1} s^n(a, a')$$

for all $a, a' \in A$. Basing himself on the work of Goguen [1967, 1969], Wechler [1986b] uses this semiring to construct a proof system for the partial correctness of nondeterministic computer programs. To do this, he notes that the left S-semimodule R^A is difference-ordered by the componentwise order induced from the difference order on R and satisfies the condition that if $f, g \in R^A$ and $s \in S$ satisfy $f \geq s^n g$ for all $n \in \mathbb{P}$ then $f \geq s^* g$. The elements of S are just R-valued relations on the set A which can be considered as nondeterministic "programs" on A. The elements of R^A can be considered as "assertions" pertaining to these programs.

(25.31) EXAMPLE. [Kuich & Salomaa, 1986] In Application 3.19 we defined the behavior $||\mathcal{A}||$ of an $R\langle\langle A\rangle\rangle$-automaton $\mathcal{A} = (S, M, s_0, P)$ and noted that this behavior may not always exist. If M is totally uniquely solvable in $R' = \mathcal{M}_{S,r}(R\langle\langle A\rangle\rangle)$ and $\mathcal{T}(M, 1'_R) = \{M^*\}$ then $||\mathcal{A}||$ always exists and equals e_0M^*P, where $e_0 \in (R\langle\langle A\rangle\rangle)^S$ is defined by $e_0(s_0) = 1$ and $e_0(t) = 0$ for $t \neq s_0$.

(25.32) PROPOSITION. *Every countably-complete semiring is left and right totally solvable.*

PROOF. Let R be a countably complete semiring and let $a \in R$. Set $a^* = \sum_{i=0}^{\infty} a^i$. Then $aa^* + 1 = a^* = a * a + 1$ and so $a^* \in \mathcal{T}(a, 1)$. If $b \in R$ then $a^*b \in \mathcal{L}(a, b)$ and $ba^* \in \mathcal{R}(a, b)$. Thus R is left and right totally solvable. □

Every countably-complete semiring R is a Lehmann semiring by Proposition 25.32. In this case, we have $a^* = \sum_{i=0}^{\infty} a^i$. If R is a countably-complete semiring then, as a direct consequence of Proposition 4.1, we see that for all elements a and b in R we have $(a + b)^* = \sum_{i=0}^{\infty}\sum_{j=0}^{\infty} a^{[i]}b^{[j]}$ and $\sum_{j=0}^{\infty} a^{[j]}b^{[n]} = (a^*b)^n a^*$ and so $(a+b)^* = (a^*b)^*a^*$ for all $a, b \in R$. See [Kuich, 1987] and [Hebisch, 1990] for details. Thus, for example, if D is a nonempty set and if R is the semiring of relations on D as defined in Example 22.8, then for each $r \in R$, the element $r^* = \sum_{i=0}^{\infty} r^i$ is just the transitive and reflexive closure of r.

(25.33) COROLLARY. *Any additively-idempotent semiring can be embedded in a semiring which is both left and right totally solvable.*

PROOF. This is a direct consequence of Proposition 25.32 and Proposition 23.5. □

A **Kleene semiring** is a subsemiring S of a countably-complete semiring R satisfying the condition that if a is in S then $a^* = \sum_{i=0}^{\infty} a^i$ is also in S. Such semirings are Lehmann semirings since we clearly have $a^* \in \mathcal{T}(a, 1)$ for each $a \in S$ and in fact they are Conway semirings [Conway, 1971]. The consideration of such operations has its roots in the work of Schröder [1895] and of Dedekind. However, unlike Kozen [1990], we will not assume that S is necessarily additively idempotent.

(25.34) PROPOSITION. *A semiring R is a Lehmann semiring if and only if every element of R is quasiregular.*

PROOF. Assume that R is a Lehmann semiring. Let $a \in R$ and let $b = aa^* = a^*a$. Then $b = ab + a = ba + a$ and so a is quasiregular. The converse follows from Proposition 25.17. □

(25.35) PROPOSITION. *Let $\gamma: R \to S$ be a surjective morphism of semirings. If R is a Lehmann semiring then so is S.*

PROOF. Let s be an element of S and let a be an element of R satisfying $\gamma(a) = s$. If $s^* = \gamma(a^*)$, where $a^* \in \mathcal{T}(a, 1_R)$, then $s^* = \gamma(aa^* + 1_R) = \gamma(a)\gamma(a^*) + 1_S = ss^* + 1_S$ and similarly $s^* = s^*s + 1_S$. Thus $s^* \in \mathcal{T}(s, 1_S)$. □

(25.36) PROPOSITION. *If R is a right totally solvable semiring then so is $\mathcal{M}_n(R)$ for each $n \geq 1$.*

PROOF. By Proposition 25.17, it suffices to show that if $A = [a_{ij}]$ is an element of $\mathcal{M}_n(R)$ then $\mathcal{L}(A, A)$ is nonempty. Indeed, let us define matrices $A_0, \ldots, A_n$ in $\mathcal{M}_n(R)$ inductively as follows:

(1) $A_0 = A$;
(2) $A_k = [a_{k;ij}]$, where $a_{k;ij} = a_{k-1;ij} + a_{k-1;ik}b_{k-1}a_{k-1;kj}$, for b_{k-1} an element of $\mathcal{L}(a_{k-1;kk}, 1)$.

We claim that $A_n \in \mathcal{L}(A, A)$. Clearly this holds if and only if, for all $1 \le i, j \le n$, we have $a_{n;ij} = a_{ij} + \sum_{h=1}^{n} a_{ih}a_{n;hj}$. To prove this, we will prove, by induction on k, that

(*) $a_{k;ij} = a_{ij} + \sum_{h=1}^{k} a_{ih}a_{k;hj}$

for all $1 \le i, j \le n$. If $k = 0$ this is trivial, since the sum over an empty set of elements of R is 0. Assume, therefore, that $k > 0$ and that (*) has been established for $k - 1$. Then

$$\begin{aligned}
a_{ij} + \sum_{h-1}^{k} a_{ih}a_{k;hj} &= a_{ij} + \sum_{h=1}^{k} a_{ih}\left[a_{k-1;hj} + a_{k-1;hk}b_{k-1}a_{k-1;kj}\right] \\
&= a_{ij} + \left[\sum_{h=1}^{k-1} a_{ih}\left(a_{k-1;hj} + a_{k-1;hk}b_{k-1}a_{k-1;kj}\right)\right] \\
&\quad + \left[a_{ik}a_{k-1;kj} + a_{ik}a_{k-1;kk}b_{k-1}a_{k-1;kj}\right] \\
&= a_{ij} + \left[\sum_{h=1}^{k-1} a_{ih}a_{k-1;hj}\right] + \left[\sum_{h=1}^{k-1} a_{ih}a_{k-1;hk}b_{k-1}a_{k-1;kj}\right] \\
&\quad + \left[a_{ik}b_{k-1}a_{k-1;kj}\right] \\
&= \left[a_{ij} + \sum_{h=1}^{k-1} a_{ih}a_{k-1;hj}\right] + \left[a_{ik} + \sum_{h=1}^{k-1} a_{ih}a_{k-1;hk}\right] b_{k-1}a_{k-1;kj}.
\end{aligned}$$

By the induction hypothesis, this equals $a_{k-1;ij} + a_{k-1;ik}b_{k-1}a_{k-1;kj}$, which is just $a_{k;ij}$, as desired. □

(25.37) COROLLARY. *The following conditions on a semiring R are equivalent:*

(1) *R is right totally solvable;*
(2) *$\mathcal{M}_n(R)$ is right totally solvable for all $n \ge 1$;*
(3) *There exists a positive integer k such that $\mathcal{M}_k(R)$ is right totally solvable.*

PROOF. By Proposition 25.36 we see that (1) implies (2), and clearly (2) implies (3). Now assume (3). Let a be an element of R and let $A = [a_{ij}]$ be the element of $\mathcal{M}_k(R)$ defined by $a_{11} = a$ and $a_{ij} = 0$ if $(i, j) \ne (1, 1)$. By (3), there exists a matrix $B = [b_{ij}] \in \mathcal{L}(A, 1)$. Then, in particular,

$$b_{11} = 1 + \sum_{h=1}^{k} a_{1h}b_{h1} = 1 + ab_{11}$$

and so $b_{11} \in \mathcal{L}(a, 1)$. By Proposition 25.17, this suffices to prove (1). □

(25.38) COROLLARY. *If R is a Lehmann semiring so is the semiring $\mathcal{M}_n(R)$ of for each positive integer n.*

PROOF. Let R be a Lehmann semiring. An easy adaptation of the proof of Proposition 25.36 shows that if $A \in \mathcal{M}_n(R)$ then there exists an element $A^* \in \mathcal{T}(A, A)$. If I is the multiplicative identity of $\mathcal{M}_n(R)$, we then see that $A + I \in \mathcal{T}(A, I)$. □

Several algorithms for the computation of A^* for a square matrix A over a Lehmann semiring R, including parallel computation models, are given in [Abdali, 1994] and [Abdali & Saunders, 1985].

(25.39) PROPOSITION. *The following conditions on a semiring R are equivalent:*

(1) *R is a Lehmann semiring;*
(2) *$\mathcal{M}_n(R)$ is a Lehmann semiring for all $n \geq 1$;*
(3) *There exists a positive integer k such that $\mathcal{M}_k(R)$ is a Lehmann semiring.*

PROOF. The proof is the same as the proof of Corollary 25.37. □

(25.40) COROLLARY. *Let R be a semiring and let A be a countably-infinite set. If $S = \mathcal{M}_{A,rc}(R)$ is a Lehmann semiring then so is R.*

PROOF. Assume S is a Lehmann semiring and let $a \in R$. Let $B = [b_{ij}]$ be the element of S defined by $b_1 1 = a$ and $b_{ij} = 0$ for $(i, j) \neq (1, 1)$. Then there exists a matrix $B^* = [c_{ij}]$ in S satisfying $BB^* + I = B^* = B^*B + I$, where I is the multiplicative identity of S. Multiplying out, we see that this means that $c_{11} \in \mathcal{T}(a, 1)$. Thus R is a Lehmann semiring. □

(25.41) PROPOSITION. *If R is a simple semiring and $A \in \mathcal{M}_n(R)$ for some $n \geq 1$, then we can select $A^* = I + A + \cdots + A^{n-1}$, where I is the multiplicative identity of $\mathcal{M}_n(R)$.*

PROOF. This is an immediate consequence of the construction in Proposition 25.36, beginning with the choice of $a^* = 1$ for all $a \in R$. □

Let R be a semiring and let A be a set which is either finite or countably infinite. Set $S = \mathcal{M}_A(R)$ if A is finite or $S = \mathcal{M}_{A,rc}(R)$ if A is countably infinite. If S is a Lehmann semiring then, by Corollary 25.39 and Corollary 25.40, we see that $\mathcal{M}_{A'}(R)$ is a Lehmann semiring for each subset A' of A. Let $M \in S$ and assume that M can be written in block form as $\begin{bmatrix} B & C \\ D & E \end{bmatrix}$, where B and E are square matrices. Then, by assumption, there exists a matrix $B^* \in \mathcal{T}(B, I)$, where I is the identity matrix of the appropriate size. Set $U = E + DB^*C$. Again, there exists a matrix $U^* \in \mathcal{T}(U, I)$. A straightforward calculation then shows that $M^* = \begin{bmatrix} F & G \\ H & K \end{bmatrix}$ where $F = B^* + B^*CU^*DB^*$, $G = B^*CU^*$, $H = U^*DB^*$, and $K = U^*$. If S is a Conway semiring, we can also take $F = (B + CED)^*$, $K = (E + DB^*C)^*$, $G = B^*CK$, and $H = E^*DF$. See [Conway, 1971] for details. Indeed, a straightforward computation shows that if M^* is always given by these conditions then R must be a Conway semiring.

(25.42) PROPOSITION. *Let R be a Conway semiring satisfying the condition that the semiring $S = \mathcal{M}_{\omega,rc}(R)$ is also a Conway semiring. Then any polynomial equation of the form $X = \sum_{i=0}^{n} a_i X^i$ $(a_i \in R)$ has a solution in R.*

PROOF. Let $M = [m_{ij}] \in S$ be the matrix defined as follows:

(1) $m_{1j} = a_j$ for $1 \leq j \leq n$ and $m_{1j} = 0$ for $j > n$;

(2) $m_{21} = a_0$ and $m_{i1} = 0$ for $i > 2$;

(3) $m_{ij} = m_{i-1,j-1}$ for $i > 1$ and $j > 1$.

Then we can write $M = \begin{bmatrix} B & C \\ D & M \end{bmatrix}$, where $B = [a_1]$, C is a row matrix, and D is a column matrix. By hypothesis, there exists a matrix $M^* = [v_{ij}]$ in S satisfying $M^* = MM^* + I = M^*M + I$, where I is the multiplicative identity of S. By the above remark, we know that $M^* = \begin{bmatrix} F & G \\ H & K \end{bmatrix}$ where $F = (B + CM^*D)^*$, $K = (M + DB^*C)^*$, $G = B^*CK$, and $H = M^*DF$. In our particular case, we obtain:

(*) $v_{11} = \left[a_1 + \sum_{i=1}^{n-1} a_{i+1} v_{i1} a_0\right]^*$

and

(**) $v_{i1} = v_{i-1,1} a_0 v_{11}$ for all $i \leq n$.

By (*), we have

(***) $v_{11} = 1 + \left[a_1 + \sum_{i=1}^{n-1} a_{i+1} v_{i1} a_0\right] v_{11}$.

By (**), we see that $v_{i1} a_0 = (v_{11} a_0)^i$ for all $1 < i \leq n$. Multiplying (***) on the right by a_0 and substituting these values for the v_{i1}, we obtain

$$v_{11} a_0 = a_0 + a_1 (v_1 1 a_0) + \sum_{h=1}^{n} a_h (v_{11} a_0)^h$$

and so we see that $v_{11} a_0$ is a solution of the given polynomial equation. □

Given a semiring R and an indeterminate t, any polynomial $f = \sum a_i t^i \in R[t]$ defines a **polynomial function** $\hat{f}: R \to R$ given $r \mapsto \sum a_i r^i$. This function is not, in general, monic.

Proposition 25.42 shows that if R is a Conway semiring then any polynomial function in R^R has a fixed point. Baccelli et al. [1992] point out that the function $f \mapsto \hat{f}$ is closely related to the Fenchel transform in convexity theory [Fenchel, 1949]. They also study polynomial functions over the schedule algebra, and their applications, in detail.

(25.43) PROPOSITION. *Every entire zerosumfree semiring can be embedded in a totally solvable semiring.*

PROOF. This is a direct consequence of Example 25.25. □

If U and V are nonempty subsets of a semiring R we set $\mathcal{L}(U, V) = \cup\{\mathcal{L}(a, b) \mid a \in U, b \in V\}$. Similarly, we define $\mathcal{R}(U, V)$ and $\mathcal{T}(U, V)$.

(25.44) PROPOSITION. *If H is a right ideal of a semiring R then $\{a \in R \mid \mathcal{L}(a, H) \neq \varnothing\}$ is either R or a right ideal of R.*

PROOF. If $r, r' \in \mathcal{L}(a, H)$ then there exist elements b and b' in H such that $r = ar + b$ and $r' = ar' + b'$. Therefore $r + r' = a(r + r') + (b + b')$ so $r + r' \in \mathcal{L}(a, H)$.

If $r'' \in R$ then $rr'' = a(rr'') + br''$ so $rr' \in \mathcal{L}(a, H)$. Hence, if $\mathcal{L}(a, H)$ is not all of R, it must be a right ideal of R. □

We now turn our attention to partially-ordered semirings. In this situation, we can not only consider the solution sets $\mathcal{L}(a, b)$ and $\mathcal{R}(a, b)$ but also see if they have minimal elements.

(25.45) PROPOSITION. *Let a and b be elements of a difference-ordered semiring R and let $r \in \mathcal{L}(a, b)$. Then:*

(1) *$a^n b \leq r$ for all $n \geq 0$;*
(2) *$(1 + a + \cdots + a^n)b \leq r$ for all $n \geq 0$; and*
(3) *If a is stable with index of stability n then $c = (1 + a + \cdots + a^n)b$ is the minimal element of $\mathcal{L}(a, b)$.*

PROOF. (1) We will proceed by induction on n. If $n = 0$ then $r = ar + b$ implies that $r \geq b = a^0 b$. Now assume that $n > 0$ and that $a^{n-1}b \leq r$. Then $r = ar + b$ implies that $r \geq ar \geq a(a^{n-1}b) = a^n b$.

(2) Again, we proceed by induction on n. If $n = 0$ the result follows from (1). If $n = 1$ we have $(1 + a)b = b + ab \leq b + ar = r$. Now suppose that $n > 1$ and that $cb \leq r$, where $c = 1 + a + \cdots + a^{n-1}$. Then $(1 + a + \cdots + a^n)b = b + (ac)b = b + a(cb) \leq b + ar = r$.

(3) If a is stable with index of stability n then surely $c \in \mathcal{L}(a, b)$. It is the minimal element of $\mathcal{L}(a, b)$ by (2). □

Recall that every additively-idempotent semiring is canonically difference ordered.

(25.46) PROPOSITION. *Let a be an element of an additively-idempotent semiring R.*

(1) *If $a^n = a^{n+1}$ for some nonnegative integer n then $b = 1 + a + \cdots + a^n$ is the minimal element of $\mathcal{L}(a, 1)$.*
(2) *We have $a \leq 1$ if and only if 1 is an element of $\mathcal{L}(a, 1)$. Moreover, in this case 1 is the minimal element of $\mathcal{L}(a, 1)$.*

PROOF. (1) This is an immediate consequence of Proposition 25.45(3).

(2) If $a \leq 1$ then, by Example 20.26, $1 = a + 1 = a1 + 1$ and so $1 \in \mathcal{L}(a, 1)$. Moreover, if $r \in \mathcal{L}(a, 1)$ then $r = ar + 1$ so $1 \leq r$. Thus 1 is the minimal element of $\mathcal{L}(a, 1)$. Conversely, if 1 is an element of $\mathcal{L}(a, 1)$ then, in particular, $1 = 1 + a$ so $a \leq 1$. □

Recall that simple semirings are additively idempotent.

(25.47) PROPOSITION. *If a and b are elements of a simple semiring R then b is the unique minimal element of $\mathcal{L}(a, b)$.*

PROOF. Since R is simple we have $1 = a + 1$ and so $b = (a + 1)b = ab + b$, proving that $b \in \mathcal{L}(a, b)$. If $r \in \mathcal{L}(a, b)$ then $r = ar + b$ so $b \leq r$. Thus b is the unique minimal element of $\mathcal{L}(a, b)$. □

(25.48) Proposition. *If a and b are elements of a lattice-ordered semiring R then $db = b$ for all $d \in \mathcal{L}(a,1)$.*

Proof. By Proposition 25.12 we know that $db \in \mathcal{L}(a,b)$ and so $b \leq db$ by Proposition 25.47 since, by Proposition 21.15, R is simple. Since R is lattice ordered, we have $db \leq b$ by definition, and thus we have equality. □

In Chapter 18 we noted that extremal semirings are also additively idempotent.

(25.49) Proposition. *Let R be an extremal semiring and let a be an element of R satisfying the condition that $\mathcal{L}(a,0) = \{0\}$. Then $\mathcal{L}(a,1) = \{1\}$ whenever $a \leq 1$, and $\mathcal{L}(a,1) = \varnothing$ otherwise.*

Proof. Assume that $a \leq 1$. Then $1 = 1 + a$ and so $1 \in \mathcal{L}(a,1)$. Conversely, if $r \in \mathcal{L}(a,1)$ then $r = 1 + ar$ so $1 \leq r$. By Proposition 25.46(2), we have $r \leq 1$ and so $r = 1$. Thus $\mathcal{L}(a,1) = \{1\}$. Now assume that $a \not\leq 1$. If $r \in \mathcal{L}(a,1)$ then $r = ar + 1$. Since R is extremal, this means that either $r = 1$ or $r = ar$. If $r = 1$ then $a \leq 1$ by Proposition 25.46(2), counter to our assumption. Therefore $r = ar$ and so $r \in \mathcal{L}(a,0)$. Thus, by hypothesis, $r = 0$ and so $0 = a0 + 1 = 1$, which is impossible. Thus we conclude that if $a \not\leq 1$ then $\mathcal{L}(a,1) = \varnothing$. □

In QLO-semirings the result is even simpler.

(25.50) Proposition. *If a and b are elements of a QLO-semiring R then $\mathcal{L}(a,b)$ has a unique maximal element and a unique minimal element.*

Proof. By Proposition 25.47 we know, since QLO-semirings are simple, that $\mathcal{L}(a,b)$ is nonempty and has a unique minimal element b. If $c = \vee\mathcal{L}(a,b)$ then $ac + b = \vee\{ar \mid r \in \mathcal{L}(a,b) + b\} = \vee\{ar + b \mid r \in \mathcal{L}(a,b) = c\}$ and so c is the unique maximal element of $\mathcal{L}(a,b)$. □

(25.51) Example. [Park, 1981] If S is the semiring $sub(A^{\infty}) \cup \{-\infty\}$ defined in Example 3.21, we can define $L^* = \{\square\} \cup [\cup_{i=1}^{\infty} L^i]$ for each $-\infty \neq L \in S$. For each such L, set $L^{\omega} = A^*$ if $L = \{\square\}$ and

$$L^{\omega} = \{w_0 w_1 \cdot \cdots \mid w_i \in L\} \cup [L^* \cap A^*]$$

otherwise. For each $-\infty \neq L \in S$, set $L^{\diamond} = L^* \cup L^{\omega}$. Then for each $-\infty \neq L, L' \in S$ one can check that L^*L' is the minimal element of $\mathcal{L}(L,L')$ and $L^{\diamond}L'$ is the maximal element of $\mathcal{L}(L,L')$. Note that if $L \in S$ then $\cap_{i \geq 0} L^i A^{\infty} \subseteq L^{\omega}$, but we do not necessarily have equality.

Note that it is also possible to consider fixed points of nonlinear maps from a semiring to itself, though this study is much less developed. For an example of such a problem over the schedule algebra which arises from optimization theory, see [K. Zimmermann, 1982]. Similarly, certain such problems over semirings of the form $\mathcal{M}_n(R\langle\langle A\rangle\rangle)$ arise from the study of context-free languages. These are considered in detail in [Kuich & Salomaa, 1986]; see also [Manes & Arbib, 1986].

Another example is found in [Gondran, 1979]; there one considers a difference-ordered semiring R and an element r of R for which there exists a natural number p satisfying $1 + r + \cdots + r^p = 1 + r + \cdots + r^{p+1}$. For each $k \geq 0$, let $c_k =$

$(2k)!/(k+1)!k \in \mathbb{N}$ be the kth **Catalan number** and let $t_k(r) = \sum_{i=0}^{k} c_i r^i$. Then $t_p(r) = t_{p+1}(r) = \ldots$ and $t_p(r)$ is the unique minimal fixed point of the function φ from R to itself defined by $\varphi: x \mapsto rx^2 + 1$.

A different approach to such problems is described in [Wechler, 1984]. An equation of the form $x = a + bx^2$ can be transformed into a system of countably-many linear equations in countably-many unknowns: $x_{n+1} = ax_n + bx_{n+2}$. This leads to the consideration of the linear equation $X = AX + C$, where X is the column vector consisting of the variables x_n, where $A = [a_{ij}]$ is the matrix defined by

$$a_{i,j} = \begin{cases} a & \text{for all } j = i-1 \\ b & \text{for all } j = i+1 \\ 0 & \text{otherwise} \end{cases},$$

and C is the column vector having a as its top entry and 0 elsewhere. If $X^* \in \mathcal{L}(A, C)$ then the first component of X^*C is a fixed point of the map $x \mapsto a + bx$.

Solutions of infinite systems of linear equations over semirings are studied in detail in [Kuich & Urbanek, 1983].

One method of guaranteeing the existence of fixed points of affine maps is introducing by an iterative method of some sort. In order to do this, we have to introduce the notion of convergence of sequences of elements of a semiring.

Let R be a semiring and define addition and multiplication on $R^{\mathbb{N}}$ componentwise. We will denote the multiplicative identity of $R^{\mathbb{N}}$ by f_1. Then $R^{\mathbb{N}}$ is also an (R, R)-bisemimodule. Among the maps from $R^{\mathbb{N}}$ to itself which we have already noted are the affine right shifts σ_a defined in Example 25.1: for each $a \in A$ and $f \in R^{\mathbb{N}}$, the function $\sigma_a(f)$ is defined by

$$[\sigma_a(f)](i) = \begin{cases} a & \text{if } i = 0 \\ f(i-1) & \text{if } i > 0 \end{cases}.$$

Let R be a semiring. A nonempty subset D of $R^{\mathbb{N}}$ satisfying the conditions

(1) $f_1 \in D$;
(2) D is an (R, R)-subbisemimodule of $R^{\mathbb{N}}$;
(3) $\sigma_a(D) \subseteq D$ for all $a \in R$.

is called a **convergence domain**. A function $f \in D$ is said to be D-**convergent**. If D is a convergence domain, a homomorphism of bisemimodules $lim_D: D \to R$ is a **limit function** provided that $lim_D(f_1) = 1$ and $lim_D(\sigma_a(f)) = lim_D(f)$ for all $a \in R$ and $f \in D$. The element $lim_D(f)$ of R is the D-**limit** of f.

(25.52) EXAMPLE. [Kuich & Salomaa, 1986] A function $f \in R^{\mathbb{N}}$ is **eventually constant** if and only if there exist a natural number $k(f)$ and an element a of R such that $f(i) = a$ for all $i \geq k(f)$. Note that if f is eventually constant then $f = \sigma_{f(0)}\sigma_{f(1)} \cdot \ldots \cdot \sigma_{f(k(f)-1)}(af_1)$ and so the set E of all eventually-constant functions is contained in every convergence domain in $R^{\mathbb{N}}$. Moreover, we have a limit function lim_E defined on E by $lim_E(f) = f(k(f))$. Indeed, it is easy to see that this is the only limit function definable on E.

(25.53) EXAMPLE. [Kuich, 1987; Karner, 1992] If R is a complete semiring it does not necessarily follow that there exists a convergence domain D in $R^{\mathbb{N}}$ and a limit function $lim_D: D \rightarrow R$ given by $lim_D(f) = \sum_{i \in \mathbb{N}} g(i)$, where $g(i) = \sum_{j=0}^{i} f(j)$ for each $i \geq 0$. To see this, consider the semiring R defined in Example 22.17 together with the second of the definitions of $\sum$ given there. Under this definition of $\sum$, we note that $\sum_{i \in \mathbb{N}} f_1(i) = \infty$, and so it can define no limit function.

Let A be a nonempty set, let R be a semiring, and let $S = R\langle\langle A \rangle\rangle$. If $s \neq t$ are distinct elements of S, set $m(s,t) = min\{|w| \mid w \in A^* \text{and } s(w) \neq t(w)\}$. Pick a real number c satisfying $0 < c < 1$ and define a function $d: S \times S \rightarrow \mathbb{R}^+$ by setting $d(s,s) = 0$ for all $s \in S$ and $d(s,t) = c^{m(s,t)}$ for $s \neq t$ in S. It is straightforward to verify that d is a complete ultrametric on S (i.e. $d(s,s') \leq max\{d(s,s''), d(s',s'')\}$ for all $s, s', s'' \in S$) and that the functions $(s,t) \mapsto s+t$ and $(s,t) \mapsto st$ from $S \times S$ to S are continuous with respect to the topology defined by this ultrametric. This in turn defines a limit function lim on a convergence domain D in $S^{\mathbb{N}}$ as follows: if $f \in S^{\mathbb{N}}$ then $f \in D$ and $lim(f) = s \in S$ if and only if for each $k \geq 0$ there exists an $m \geq 0$ such that $w \in A^*$ and $|w| \leq k$ imply that $f(j)(w) = s(w)$ for all $j \geq m$.

(25.54) PROPOSITION. *Let R be a semiring and let lim_D be a limit function defined on a convergence domain D in $R^{\mathbb{N}}$. Let A be a nonempty set and let E be the set of all those functions f in $R\langle\langle A \rangle\rangle^{\mathbb{N}}$ satisfying the condition that, for each $w \in A^*$, the function $f_w: \mathbb{N} \rightarrow R$ given by $f_w: n \mapsto f(n)(w)$ belongs to D. Then:*

(1) *E is a convergence domain in $R\langle\langle A \rangle\rangle^{\mathbb{N}}$; and*
(2) *The function $lim_E: E \rightarrow R\langle\langle A \rangle\rangle$ defined by $[lim_E(f)]: w \mapsto lim_D(f_w)$ for all $w \in A^*$ is a limit function on E.*

PROOF. (1) Set $S = R\langle\langle A \rangle\rangle$. If $f, g \in E$ and $w \in A^*$ then

$$(f+g)_w(n) = (f+g)(n)(w) = f(n)(w) + g(n)(w) = f_w(n) + g_w(n)$$

and so $(f+g)_w = f_w + g_w$. Hence $f+g \in E$. If $s \in S$ then

$$\begin{aligned}(sf)_w(n) = (sf)(n)(w) &= \sum\{s(w')f(n)(w'') \mid w'w'' = w\} \\ &= \sum\{s(w')f_{w''}(n) \mid w'w'' = w\} \\ &= \left[\sum\{s(w')f_{w''} \mid w'w'' = w\}\right](n).\end{aligned}$$

Since $s(w')f_{w''} \in D$ for all $w', w'' \in A^*$, it follows that $(sf)_w \in D$ for each $w \in A^*$ and so $sf \in E$. Similarly, $fs \in E$ and so E is an (S,S)-subbisemimodule of $S^{\mathbb{N}}$.

Let $s \in S$ and $f \in E$. If $w \in A^*$ then $[\sigma_s(f)]_w(0) = [\sigma_s(f)](0)(w) = s(w)$ and $[\sigma_s(f)]_w(i+1) = [\sigma_s(f)](i+1)(w) = f(i)(w) = f_w(i)$. Therefore $[\sigma_s(f)]_w = \sigma_s(w)(f_w) \in D$ for all $w \in A^*$, proving that $\sigma_s(f) \in E$. Thus E is a convergence domain.

(2) This is a straightforward consequence of the definition. □

Every element a of a semiring R defines the **power sequence** p_a in $R^{\mathbb{N}}$ given by $p_a: i \mapsto a^i$ and the **canonical sequence** g_a in $R^{\mathbb{N}}$ given by $g_a: k \mapsto \sum_{i=1}^{k} a^i$. Note that if R is additively idempotent then $g_a(k) = p_{a+1}(k)$ for all $a \in R$ and all $k \in \mathbb{N}$.

If there exists a convergence domain D and a limit function $lim_D\colon D \to R$ such that $g_a \in D$ then we will denote $lim_D(g_a)$ by $a^{+(D)}$ or simply by a^+ if there is no room for confusion concerning which limit function we are using. If $f_1 + g_a \in D$ then we denote $lim_D(f_1 + g_a)$ by $a^{*(D)}$ or simply by a^* if there is no room for confusion.

(25.55) PROPOSITION. *Let a be an element of a semiring R and let lim_D be a limit function defined on a convergence domain D in $R^{\mathbb{N}}$. Then g_a is D-convergent if and only if $f_1 + g_a$ is and, in this case,*

(1) $a^* = 1 + a^+$;
(2) $aa^* = a^*a = a^+$;
(3) *For all $n \geq 0$ we have* $a^* = \sum_{j=0}^{n} a^j + a^{n+1}a^* = \sum_{j=0}^{n} a^j + a^*a^{n+1}$.

PROOF. Since f_1 is eventually constant we know by Example 25.52 that $f_1 \in D$. Therefore $g_a \in D$ implies that $f_1 + g_a \in D$. Conversely, if $f_1 + g_a \in D$ then $g_a = a[f_1 + g_a] \in D$.

(1) Assume that a^+ exists. Then

$$1 + a^+ = lim_D(f_1) + lim_D(g_a) = lim_D(f_1 + g_a) = a^*$$

so a^* exists. Conversely, if a^* exists then $aa^* = a[lim_D(f_1 + g_a)] = lim_D(g_a) = a^+$ so a^+ exists.

(2) and (3) follow immediately from (1). □

(25.56) PROPOSITION. *Let a and b be elements of a semiring R and let lim_D be a limit function defined on a convergence domain D in $R^{\mathbb{N}}$. Then g_{ab} is D-convergent if and only if g_{ba} is. Moreover, in this case, $(ab)^*a = a(ba)^*$.*

PROOF. To prove the first assertion, it suffices to show that $g_{ab} \in D$ implies that $g_{ba} \in D$. Indeed, if $g_{ab} \in D$ then $h = b[f_1 + g_{ab}]a \in D$ and so $\sigma_0(h) + f_1 \in D$. But, for each $n \geq 0$, $h(n) = b[\sum_{i=0}^{n}(ab)^i]a = \sum_{i=1}^{n+1}(ba)^i$ and so $[\sigma_0(h) + f_1](n) = \sum_{i=0}^{n-1}(ba)^i = g_{ba}(n)$. Thus $g_ba \in D$. As for the second assertion, we note that $(ab)^*a = [lim_D(f_1 + g_{ab})]a = lim_D(f_1a) + lim_D(g_{ab}a) = lim_D(af_1) + lim_D(ag_{ba}) = a(ba)^*$. □

If R is a semiring and if M is a left R-semimodule then the set of all functions from $\mathbb{N}$ to M is a left R-semimodule with addition and scalar multiplication defined componentwise, denotes by $M^{\mathbb{N}}$. If $f \in R^{\mathbb{N}}$ and if $m \in M$ then fm is an element of $M^{\mathbb{N}}$ defined by $fm\colon i \mapsto f(i)m$ for all $i \in \mathbb{N}$. In particular, f_1m is the constant function $i \mapsto m$ for all $i \in \mathbb{N}$.

If D is a convergence domain in $R^{\mathbb{N}}$ and if $lim_D\colon D \to R$ is a limit function defined on D then a left R-semimodule M is **compatible** with lim_D if and only if, for each finite set $\{g_1, \ldots, g_n\}$ of elements of D and finite set $\{m_1, \ldots, m_n, m\}$ of elements of M satisfying $\sum_{i=1}^{n} g_im_i = f_1m$ we have $\sum_{i=1}^{n}[lim_D(g_i)]m_i = m$. Clearly the left R-semimodule R is compatible with any such limit function.

(25.57) PROPOSITION. *Let R be a semiring, let $lim_D\colon D \to R$ be a limit function defined on a convergence domain D in $R^{\mathbb{N}}$, and let M be a left R-semimodule compatible with D. Let a be an element of R satisfying the conditions that*

$p_a, g_a \in D$ and $lim_D(p_a) = 0$. Then the equation $X = aX + m'$ is uniquely solvable for each element m' of M.

PROOF. Since $g_a \in D$, we know by Proposition 25.55 and Proposition 25.12 that a^* exists and that $a * m' \in \mathcal{L}(a, m')$ for each element m' of M. All we are left to show is uniqueness. Indeed, assume that $m \in \mathcal{L}(a, m')$. By Proposition 25.12(5) we see that $m = a^{n+1}m + \sum_{i=0}^{n} a^i m'$ for all $n \geq 0$ and so $f_1 = (ap_a)m + (g_a)m'$ in $M^{\mathbb{N}}$. By compatibility, we therefore have $m = [lim_D(ap_a)]m + [lim_D(g_a)]m' = 0_M + a^*m' = a^*m'$. □

We now turn to some results due to Karner [1992, 1994].

(25.58) EXAMPLE. Let R be a semiring and let $S = R\langle\langle A\rangle\rangle$ for some nonempty set A. It is easy to see that if $s \in S$ is quasiregular then the sequence $p_s \in S^{\mathbb{N}}$ converges to 0_S. Therefore, by Proposition 25.57, we see that the equation $X = sX + m'$ is uniquely solvable for each quasiregular $s \in S$ and each element m' of a left S-semimodule M.

(25.59) PROPOSITION. *For a complete semiring R the following conditions are equivalent:*

(1) *If $\theta: \mathbb{N} \to R$ and if there exist a natural number n_0 and an element a of R such that*
$$\sum_{i=0}^{n} \theta(i) = a$$
for all $n \geq n_0$ then $\sum \theta = a$;

(2) *If $\theta: \mathbb{N} \to R$ and if there exists an element a of R such that $\theta(i) \in \mathcal{L}(1, a)$ for all $i \in \mathbb{N}$ then $\sum \theta \in \mathcal{L}(1, a)$;*

(3) *If $\theta: \Omega \to R$, where Ω is countable, satisfies the condition that $\sum_{i\in\Gamma} \theta(i) = \sum_{i\in\Lambda} \theta(i)$ for some finite subset Λ of Ω and all finite subsets Γ of Ω containing Λ, then $\sum \theta = \sum_{i\in\Lambda} \theta(i)$;*

(4) *Given $a \in R$, if $\theta: \Omega \to R$, where Ω is countable, satisfies the condition that for each finite subset Λ of Ω there is a finite subset $\psi(\Lambda)$ of Ω containing Λ satisfying $\sum_{i\in\psi(\Lambda)} \theta(i) = a$, then $\sum \theta = a$.*

PROOF. (1) $\Rightarrow$ (4): Without loss of generality we can assume that $\Omega = \mathbb{N}$. Define a sequence $\{\Lambda_n \mid n \in \mathbb{N}\}$ of finite sets inductively, by setting $\Lambda_0 = \psi(\{0\})$ and $\Lambda_{n+1} = \psi(\Lambda_n \cup \{min(\mathbb{N} \setminus \Lambda_n)\}$ for each $n \in \mathbb{N}$. Then clearly $\Lambda_n \subset \Lambda_{n+1}$ for all n and $\cup_{n\in\mathbb{N}}\Lambda_n = \mathbb{N}$. Now define the function $\theta': \mathbb{N} \to R$ inductively as follows:

(1) $\theta'(0) = \sum\{\theta(i) \mid i \in \Lambda_0\}$;

(2) $\theta'(n+1) = \sum\{\theta(i) \mid i \in \Lambda_{n+1} \setminus \Lambda_n\}$.

then $\sum_{i\leq n} \theta'(i) = \sum_{i\in\Lambda_n} \theta(i) = a$ and so, by (1), $\sum \theta = \sum \theta' = a$.

(4) $\Rightarrow$ (3): This is immediate.

(3) $\Rightarrow$ (2): Suppose $\theta(i) \in \mathcal{L}(1, a)$ for all $i \in \mathbb{N}$ Define $\theta': \mathbb{N} \to R$ by setting $\theta'(0) = a$ and $\theta'(i+1) = \theta(i)$ for all $i \in \mathbb{N}$. Let $\Lambda = \{0\}$. Then for all finite subsets Γ of $\mathbb{N}$ containing Λ we have $\sum_{i\in\Gamma} \theta'(i) = \sum_{i\in\Lambda} \theta'(i)$ and so
$$a + \sum \theta = \sum \theta' = \sum_{i\in\Lambda} \theta'(i) = a.$$

(2) ⇒ (1): Assume that $\sum_{i=0}^{n} \theta(i) = a$ for all $n \geq n_0$. Define a function $\theta' \colon \mathbb{N} \to R$ by setting $\theta'(i) = \theta(n_0 + i + 1)$ for all $i \in \mathbb{N}$. Then

$$a + \theta'(i) = \sum_{j=0}^{n_0+i} \theta(j) + \theta(n_0 + i + 1) = \sum_{j=0}^{n_0+i+1} \theta(j)\theta(j) = a$$

for all $i \in \mathbb{N}$ and so

$$\sum \theta = \sum_{i=0}^{n_0} \theta(i) + \sum_{i>n_0} \theta(i) = \sum \theta' = a$$

and we are done. □

(25.60) Proposition. *For a complete semiring R the following conditions are equivalent:*

(1) *If $\theta \colon \Omega \to R$ is a function and if $a \in R$ satisfies the condition that $im(\theta) \subseteq \mathcal{L}(1, a)$ then $\sum \theta \in \mathcal{L}(1, a)$;*
(2) *If $\theta \colon \Omega \to R$ is a function satisfying the condition that for some fixed finite $\Lambda_0 \subseteq \Omega$ and all finite $\Lambda_0 \subseteq \Lambda \subseteq \Omega$ we have $\sum_{i \in \Lambda} \theta(i) = a$ then $\sum \theta = a$;*
(3) *If $\theta \colon \Omega \to R$ is a function satisfying the condition that for some fixed $a \in R$ and for all finite $\Lambda \subseteq \Omega$ there is a finite subset $\psi(\Lambda)$ of Ω containing Λ and satisfying $\sum_{i \in \psi(\Lambda)} \theta(i) = a$, then $\sum \theta = a$.*

Proof. Note that R is partially-ordered by the relation $a \leq b$ iff there exists a $c \in R$ satisfying $a + c = b$.

(1) ⇒ (3): Let $\theta \colon \Omega \to R$ be a function and let a be an element of R such that the conditions in (3) are satisfied. Set $\Gamma = \psi(\varnothing)$ and $a = \sum_{i \in \Lambda} \theta(i)$. Then for each $j \in \Omega \setminus \Lambda$ we have

$$a \leq a + \theta(j) = \sum_{i \in \Lambda \cup \{j\}} \theta(i) \leq \sum_{i \in \psi(\Lambda \cup \{j\})} \theta(i) = a$$

and so $theta(i) \in \mathcal{L}(1, a)$ for each $i \in \Omega$. Therefore, by (1),

$$\sum \theta = a + \sum_{i \in \Omega \setminus \Lambda} \theta(i) = a.$$

(3) ⇒ (2): This is immediate.

(2) ⇒ (1): The proof is the same as the corresponding part of the proof of Proposition 25.59. □

REFERENCES

S. KAMAL ABDALI

[1994] Parallel computations in $*$-semirings, in Klaus G. Fischer et al. (eds.): *Computational Algebra*, Marcel Dekker, New York, 1994.

S. KAMAL ABDALI & B. DAVID SAUNDERS

[1985] Transitive closure and related semiring properties via eliminants, Theor. Comp. Sci. 40 (1985), 257 - 274.

SAMSON ABRAMSKY & STEVEN VICKERS

[1993] Quantales, observational logic, and process semantics, Math. Structures Comput. Sci. 3 (1993), 161 - 227.

ACHILLE ACHACHE

[1985/6] Condition pour que la puissance I^X du treillis complet I soit algébrique, Portugal. Math. 43 (1985/6), 317- 319.

S. K. ACHARYYA, K. C. CHATTOPADHYAY, & G. G. RAY

[1993] Hemirings, congruences and the Stone-Čech compactification, Simon Stevin 67 (1993) Supplement, 21 -35.

M. R. ADHIKARI & M. K. DAS

[1994] Structure spaces for semirings, Bull. Calcutta Math. Soc. 86 (1994), 313 - 317.

M. R. ADHIKARI, JONATHAN S. GOLAN & M. K. SEN

[1994] A note concerning congruence relations on austere commutative semirings, Indian Science Cruiser 8 (1994) 25 - 26.

ALFRED V. AHO, JOHN E. HOPCROFT & JEFFERY D. ULLMAN

[1974] *The Design and Analysis of Computer Algorithms*, Addison-Wesley, New York, 1974.

JAVED AHSAN

[1993] Fully idempotent semirings, Proc. Japan Acad. 69, Ser. A (1993), 185 - 188.

MARIANNE AKIAN

[1995a] Densities of idempotent measures and large deviations, preprint, 1995.

[1995b] Theory of cost measures: convergence of decision variables, preprint, 1995.

MARIANNE AKIAN, RAVINDRA BAPAT & STÉPHANE GAUBERT

[1998] Asymptotics of the Perron eigenvalue and eigenvector using max-algebra, C. R. Acad. Sci. Paris (I) 327 (1998), 927 - 932.

MARIANNE AKIAN, JEAN-PIERRE QUADRAT & MICHEL VIOT

[1994] Bellman processes, in G. Cohen & J. P. Quadrat (eds): *Eleventh International Conference on Analysis and Optimization of Systems: Discrete Event Systems*, Lecture Notes in Control and Information Sciences #199, Springer-Verlag, Berlin, 1994.

[1998] Duality between probability and optimization, in J. Gunawardena (ed.): *Idempotency (Bristol, 1994)*, Publ. Newton Inst. #11, Cambridge Univ. Press, Cambridge, 1998, 331 - 353.

FRANCISCO E. ALARCÓN & DANIEL D. ANDERSON

[1994a] Commutative semirings and their lattices of ideals, Houston J. Math. 20 (1994), 571 -590.

[1994b] The lattice of ideals of a polynomial semiring, Algebra Universalis 31 (1994), 147 - 149.

GOTZ ALEFELD & JURGEN HERZBERGER

[1983] *Introduction to Interval Computations*, Academic Press, New York, 1983.

PAUL J. ALLEN

[1969] A fundamental theorem of homomorphisms for semirings, Proc. Amer. Math. Soc. 21 (1969), 412 - 416.

[1970] Cohen's theorem for a class of Noetherian semirings, Publ. Math. Debrecen 17 (1970), 169 - 171.

PAUL J. ALLEN & LOUIS DALE

[1975] Ideal theory in the semiring $\mathbb{Z}^+$, Publ. Math. Debrecen, 22 (1975), 219 - 224.

PAUL J. ALLEN & WILLIAM R. WINDHAM

[1973] Operator algebras with applications to semirings, Publ. Math. Debrecen, 20 (1973), 161 - 175.

ERIC ALLENDER, JIA JIAO, MEENA MAHAJAN & V. VINAY

[1996] Non-commutative arithmetic circuits: depth reduction and size limits, DIMACS Technical Report 96-03, 1996.

ELISABETTA ALLEVI

[1987] Prodotti sottodiretti di $(+,\cdot)$-bande commutative e di quasianelli distributivi, Ist. Lombardo (Rend. Sc.) A 121 (1987), 41 - 53.

J. P. ALLOUCE, E. CATELAND, H.-O. PEITGEN, G. SKORDEV, & J. SHALLIT

[1995] Automatic maps on a semiring with digits, Fractals 3 (1995), 663 - 677.

ANTONIO ALMEIDA COSTA

[1963] Sur la théorie générale des semi-anneaux, Publ. Math. Debrecen 10 (1963), 14 – 29.

[1968] Sur les μ-demi-anneaux, Math. Z. 108 (1968), 10 – 14.

[1974] *Cours d'Algèbre Générale*, vol. III, Fundacao Calouste Gulbenkian, Lisbon, 1974.

ANTONIO ALMEIDA COSTA & MARGARITA RAMALHO

[1978] Semirings, ideals and congruences, in *Estudos de Algebra, Geometria et Análise*, No. II, Centenário da Academia das Ciencias de Lisboa, Lisbon, 1978.

HUDA MOHAMMED J. AL-THANI

[1995] A note on projective semimodules, Kobe J. Math. 12 (1995), 89 – 94.

[1996] k-projective semimodules, Kobe J. Math. 13 (1996), 49 – 59.

DANIEL D. ANDERSON

[1975] Distributive noether lattices, Michigan Math. J. 22 (1975), 109 – 115.

[1976] Abstract commutative ideal theory without chain conditions, Algebra Univ. 6 (1976), 131 – 145.

[1977] Fake rings, fake modules, and duality, J. Algebra 47 (1977), 425 – 432.

DANIEL D. ANDERSON & DAVID F. ANDERSON

[1991] Examples of star operations on integral domains, Comm. Alg. 18 (1991), 1621 – 1643.

JACK M. ANDERSON

[1993] The operational calculus of Mikusiński, preprint, 1993.

CHARLES ANDRE

[1989] An algebra for "SENS", Technical Report LASSY-13S, Laboratoire de Signaux et Systèmes, Université de Nice-Sophia Antipolis, 1989.

A. J. ANTUNES MONTEIRO

[1978] Congruences, ideals and homomorphisms in semirings, in *Estudos de Algebra, Geometria et Análise*, No. II, Centenário da Academia das Ciencias de Lisboa, Lisbon, 1978.

MICHAEL A. ARBIB, A. J. KFOURY & ROBERT N. MOLL

[1981] *A Basis for Theoretical Computer Science*, Springer-Verlag, Berlin, 1981.

B. H. ARNOLD

[1951] Distributive lattices with a third operation defined, Pacific J. Math. 1 (1951), 33 – 41.

JEAN-PIERRE AUBIN

[1993] *Optima and Equilibria*, Springer-Verlag, Berlin, 1993.

ROBERT J. AUMANN & L. S. SHAPLEY

[1974] *Values of Non-Atomic Games*, Princeton University Press, Princeton, 1974.

JONATHAN BABB, MATTHEW FRANK & ANANT AGARWAL

[1998] Solving graph problems with dynamic computation structures, preprint, 1998.

FRANÇOIS LUIS BACCELLI, GUY COHEN, GEERT JAN OLSDER & JEAN-PIERRE QUADRAT

[1992] *Synchronization and Linearity*, John Wiley & Sons, New York, 1992.

FRANÇOIS LUIS BACCELLI & JEAN MAIRESSE

[1998] Ergodic theorems for stochastic operators and discrete event networks, in J. Gunawardena (ed.): *Idempotency (Bristol, 1994)*, Publ. Newton Inst. #11, Cambridge Univ. Press, Cambridge, 1998, 171 -208.

ACHIM BACHEM

[1978] The theorem of Minkowski for polyhedral monoids and aggregated linear Diophantine equations, in R. Henn et al. (eds.): *Optimization and Operations Research*, Lecture Notes in Economics and Mathematical Systems #157, Springer-Verlag, Berlin, 1978.

R. C. BACKHOUSE & BERNARD A. CARRÉ

[1975] Regular algebra applied to path-finding problems, J. Inst. Maths. Applics. 15 (1975), 161 - 186.

J. C. M. BAETEN, J. A. BERGSTRA & J. W. KLOP

[1988] An operational semantics for process algebra, in G. Mirkowska & H. Rasiowa (eds.): *Mathematical Problems in Computation Theory*, Banach Center Publications #21, PWN, Warsaw, 1988.

L. B. BALL

[1986] The arithmetic of differentiation, Math. Mag. 59 (1986), 275 - 282.

BERNHARD BANASCHEWSKI & ROSWITHA HARTING

[1985] Lattice aspects of radical ideals and choice principles, Proc. London Math. Soc. (3) 50 (1985), 385 - 404.

HANS-JURGEN BANDELT

[1983] Free objects in the variety generated by rings and distributive lattices, in K. H. Hofmann et al. (eds.): *Recent Developments in the Algebraic, Analytical, and Topological Theory of Semigroups*, Lecture Notes in Mathematics #998, Springer-Verlag, Berlin, 1983.

HANS-JURGEN BANDELT & MARIO PETRICH

[1982] Subdirect products of rings and distributive lattices, Proc. Edinburgh Math. Soc. 25 (1982), 155 - 171.

EROL BARBUT

[1967] *On Certain Semirings with Ascending Chain Conditions*, PhD. Thesis, University of California, Riverside, 1967.

LeROY B. BEASLEY & NORMAN J. PULLMAN

[1988a] Operators that preserve semiring matrix functions, Lin. Alg. Appl. 99 (1988), 199 - 216.

[1988b] Semiring rank versus column rank, Lin. Alg. Appl. 101 (1988), 33 - 48.

V. P. BELAVKIN

[1987] Ordered $*$-semirings and generating functionals of quantum statistics, Doklady Akademii Nauk SSSR 293 (1987), 18 - 21. [Russian]

RICHARD E. BELLMAN & MAGNUS GIERTZ

[1973] On the analytic formalism of the theory of fuzzy sets, Inf. Sc. 5 (1973), 149 - 156.

JEAN BENABOU

[1959] Trellis locaux et paratopologies, Séminaire C. Ehresmann 1957/58, Fac. des Sciences de Paris, 1959.

J. A. BERGSTRA & J. W. KLOP

[1983] Process algebra for communication and mutual exclusion, Report IW 218/83, Mathematisch Centrum, Amsterdam, 1983.

[1984] The algebra of recursively defined processes and the algebra of regular processes, in J. Paredaens (ed.): *Automata, Languages and Programming*, Lecture Notes in Computer Science #172, Springer-Verlag, Berlin, 1984.

[1986] Process algebra: specification and verification in bisimulation semantics, in M. Hazewinkel et al. (eds): *Mathematics and Computer Science II*, CWI Monograph #4, North Holland, Amsterdam, 1986.

[1989] ACP_τ, a universal axiom system for process specificiation, in M. Wirsing & J. A. Bergstra (eds.): *Algebraic Methods: Theory, Tools and Applications*, Lecture Notes in Computer Science #394, Springer-Verlag, Berlin, 1989.

J. A. BERGSTRA & ALBAN PONSE

(1) Frame-based process logics, in Alban Ponse et al. (eds): *Modal Logic and Process Algebra*, CSDLI Lecture Notes # 53, CSLI Publications, Stanford, 1995.

JOEL BERMAN & MASAO MUKAIDONO

[1984] Enumerating fuzzy switching functions and free Kleene algebras, Comp. and Maths. with Appls. 10 (1984), 25 - 35.

JEAN BERSTEL & DOMINIQUE PERRIN

[1985] *Theory of Codes*, Academic Press, New York, 1985.

JEAN BERSTEL & CHRISTOPHE REUTENAUER

[1982] Recognizable formal power series on trees, Theor. Comp. Sci. 18 (1982), 115 - 148.

[1988] *Rational Series and Their Languages*, Springer-Verlag, Berlin, 1988.

PIERRE BERTHIAUME

[1967] The injective envelope of S – sets, Canad. Math. Bull. 10 (1967), 261 – 273.

MARTHA O. BERTMAN

[1978] Bicyclic semirings, J. Austral. Math. Soc. (Ser. A) 26 (1978), 419 – 441.

ANDREJ BIALYNICKI – BIRULA

[1958] On the spaces of ideals of semirings, Fund. Math. 45 (1958), 247 – 253.

ALAIN BIGARD, KLAUS KEIMEL & SAMUEL WOLFENSTEIN

[1977] *Groupes et Anneaux Réticulés*, Lecture Notes in Mathematics #608, Springer-Verlag, Berlin, 1977.

GARY F. BIRKENMEIER

[1989] Seminearrings and nearrings induced by the circle operation, pre – print, 1989.

GARY F. BIRKENMEIER & HENRY HEATHERLY

[1987] Operation inducing systems, Algebra Universalis 24 (1987), 137 – 148.

[1989] Embeddings of strongly right bounded rings and algebras, Comm. Algebra 17 (1989), 573 – 586.

G. D. BIRKHOFF & GARRETT BIRKHOFF

[1946] Distributive postulates for systems like boolean algebras, Trans. Amer. Math. Soc. 60 (1946), 3 – 11.

GARRETT BIRKHOFF

[1971] The role of modern algebra in computing, in G. Birkhoff & M. Hall, Jr. (eds.): *Computers in Algebra and Number Theory*, SIAM=AMS Proceedings #4, American Mathematical Society, Providence, R. I., 1971.

[1973] *Lattice Theory*, 3rd Edition, Colloquium Publication #25, American Mathematical Society, Providence, R.I., 1973.

STEFANO BISTARELLI, UGO MONTANARI & FRANCESCA ROSSI

[1997a] Constraint solving over semirings, in *Proceedings of IJCAI'95*, Morgan Kaufmann, 1995.

[1997b] Semiring-based constraint solving and optimization, J. Assoc. Comp. Mach. 44 (1997), 201 – 236.

R. L. BLAIR

[1953] Ideal lattice and the structure of rings, Trans. Amer. Math. Soc. 75 (1953), 136 – 153.

MICHAEL N. BLEICHER & SAMUEL BOURNE

[1965] On the embeddability of partially ordered halfrings, J. Math. Mech. 14 (1965), 109 – 116.

ANDRZEJ BLIKLE

[1971] Nets complete lattices with a composition, Bull. Acad. Polon. Sci., Ser sci. math. astr. phys. 19 (1971), 1123 - 1127.

[1977] Analysis of programs by algebraic means, in *Mathematical Foundations of Computer Science*, Banach Center Publications #2, PWN, Warsaw, 1977.

WAYNE BLIZARD

[1989] Multiset theory, Notre Dame J. Formal Logic 30 (1989), 36 - 66.

P. A. BLONIARZ, HARRY B. HUNT III & DANIEL J. ROSENKRANTZ

[1984] Algebraic structures with hard equivalence and minimization problems, J. Assoc. Comp. Mach. 31 (1984), 879 - 904.

STEPHEN L. BLOOM & ZOLTÁN ÉSIK

[1993] *Iteration Theories*, Springer-Verlag, Berlin, 1993.

STEPHEN L. BLOOM, N. SABADINI & R. F. C. WALTERS

[1996] Matrices, machines and behaviors, Appl. Categ. Struct. 4 (1996), 343 - 360.

THOMAS S. BLYTH

[1964] Matrices over ordered algebraic structures, J. London Math. Soc. 39 (1964), 427 - 432.

[1965] The general form of residuated algebraic structures, Bull. Soc. Math. France 93 (1965), 109 - 127.

THOMAS S. BLYTH & MELVIN F. JANOWITZ

[1972] *Residuation Theory*, Pergamon Press, Oxford, 1972.

KENNETH P. BOGART

[1969a] Distributive local noether lattices, Michigan Math. J. 16 (1969), 215 - 223.

[1969b] Nonimbeddable noether lattices, Proc. Amer. Math. Soc. 22 (1969), 129 - 133.

FRANCIS BORCEUX, J. ROSICKÝ & GILBERTE VAN DEN BOSSCHE

[1989] Quantales and C^*-algebras, J. London Math. Soc. (2) 40 (1989), 398 - 404.

FRANCIS BORCEUX & GILBERTE VAN DEN BOSSCHE

[1986] Quantales and their sheaves, Order 3 (1986), 61 - 87.

MARIA MARGARITA BOTERO DE MEZA & HANNS J. WEINERT

[1971] Erweiterung topologischer Halbringe durch Quotienten und Differentzenbildung, Jber. Deutsch Math.-Verein. 73 (1971), 60 - 85.

SAMUEL BOURNE

[1951] The Jacobson radical of a semiring, Proc. Nat. Acad. Sci. U.S. A. 37 (1951), 163 - 170.

[1952] On the homomorphism theorem for semirings, Proc. Nat. Acad. Sci. U.S.A. 38 (1952), 118 - 119.

[1956] On multiplicative idempotents of a potent semiring, Proc. Nat. Acad. Sci. U.S.A. 42 (1956), 632 – 638.

[1961/2a] On normed semialgebras, Studia Math. 21 (1961/2), 45 – 54.

[1961/2b] On Banach *-semialgebras, Studia Math. 21 (1961/2), 207 – 214.

[1962] On locally compact positive halffields, Math. Ann. 146 (1962), 423 – 426.

SAMUEL BOURNE & HANS J. ZASSENHAUS

[1958] On the semiradical of a semiring, Proc. Nat. Acad. Sci. USA 44 (1958), 907 – 914.

J. G. BRAKER & GEERT JAN OLSDER

[1993] The power algorithm in max algebra, Linear Alg. Appl. 182 (1993), 67 – 89.

J. G. BRAKER & J. A. C. RESING

[1993] On a generalized asymptoticity problem in max algebra, *Discrete Event Systems: Modeling and Control (Prague, 1992)*, Progr. Systems Control Theory, #13, Birkhäuser, Basel, 1993, 125–139.

CAROLYN BROWN & DOUG GURR

[1993a] A representation theorem for quantales, J. Pure Appl. Alg. 85 (1993), 27 – 42.

[1993b] Relations and non-commutative linear logic, preprint, 1993.

PETER BRUCKER

[1972] R-Netzwerke und Matrix Algorithmen, Computing 10 (1972), 271 – 283.

PETER BRUCKER, W. PAPENJOHANNI & UWE ZIMMERMANN

[1984] A dual optimality criterion for algebraic linear programs, Ann. Discrete Math. 19 (1984), 35 – 40.

GREGORY W. BRUNFIEL

[1979] *Partially Ordered Rings and Semi-Algebraic Geometry*, London Math. Soc. Lecture Note Series #37, Cambridge University Press, Cambridge, 1979.

G. BRUNGS & H. LAKSER

[1975] Injective hulls of semilattices, Canad. Math. Bull. 13 (1975), 297 – 298.

JANUSZ A. BRZOZOWSKI

[1964] Derivatives of regular events, J. Assoc. Comp. Mach. 11 (1964), 481 – 494.

M. PAZ BUJANDA JAUREGUI & MIGUEL L. LAPLAZA GRACIA

[1967] Una generalizacion del concepto de anillo local (conclusión), Rev. Mat. Hispano-Americana 27 (1967), 63 – 101.

RICHARD G. BURTON

[1975] Fractional elements in multiplicative lattices, Pacific J. Math. 56 (1975), 35 – 49.

PETER BUTKOVIC

[1985] Necessary solvability conditions of systems of linear extremal equations, Discrete Applied Math. 10 (1985), 19 – 26.

DAN BUTNARIU & ERICH PETER KLEMENT

[1993] *Triangular Norm-Based Measures and Games with Fuzzy Coalitions*, Kluwer, Dordrecht, 1993.

DAN BUTNARIU, ERICH PETER KLEMENT & SAMY ZAFRANY

[1995] On triangular norm-based propositional fuzzy logics, Fuzzy Sets and Systems 69 (1995), 241 – 255.

CAO ZHIQIANG

[1984] Comparison between two kinds of semilattice-semigroups, Acta Math. Scientia 4 (1984), 311 – 317.

[1993] An algebraic system generalizing the fuzzy subsets of a set, in P. Wang (ed.): *Advances in Fuzzy Sets, Possibility Theory, and Applications*, Plenum, New York, 1993, 71 – 80.

CAO ZHIQIANG, KI HANG KIM & FRED W. ROUSH

[1984] *Incline Algebra and Applications*, Ellis Horwood, Chichester, 1984.

BERNARD A. CARRÉ

[1971] An algebra for network routing problems, J. Inst. Math. Applics. 7 (1971), 273 – 294.

[1979] *Graphs and Networks*, Oxford Univ. Press, Oxford, 1979.

E. D. CASHWELL & C. J. EVERETT

[1959] The ring of number-theoretic functions, Pacific J. Math. 9 (1959), 975 – 985.

GIANPIERO CATTANEO & ALESSANDRO MANIA

[1974] Abstract orthogonality and orthocomplementation, Proc. Cambridge Phil. Soc. 76 (1974), 115 – 132.

GIANPIERO CATTANEO & GIUSEPPE MARINO

[1988] Non-usual orthocomplementations on partially ordered sets and fuzziness, Fuzzy Sets and Systems 25 (1988), 107 – 123.

KATARINA CECHLÁROVÁ & JÁN PLÁVKA

[1996] Linear independence in bottleneck algebras, Fuzzy Sets and Systems 77 (1996), 337 – 348.

STEFAN CHANAS

[1987] Fuzzy optimization in networks, in J. Kacprzyk & S. A. Orlovski (eds.): *Optimization Models Using Fuzzy Sets and Possibility Theory*, Reidel, Dordrecht, 1987.

RITA CHAUDHURI & A. MUKHERJEA

[1980] Idempotent boolean matrices, Semigroup Forum 21 (1980), 273 – 282.

SHIOW-WEN WENDY JIANG CHEN

[1987] *Decision Support in Manufacturing Systems Using Activity Network*, PhD. Thesis, University of Iowa, Ames, 1987.

LUDMILA A. CHERKASOVA

[1988] On models and algebras for concurrent processes, in M. P. Chytil et al. (eds.): *Mathematical Foundations for Computer Science 1988*, Lecture Notes in Computer Science #324, Springer-Verlag, Berlin, 1988.

L. H. CHIN & ALFRED TARSKI

[1951] Distributive and modular laws in the arithmetic of relation algebras, Univ. Calif. Publ. Math. 1 (1951), 341 - 384.

HAN HYUK CHO

[1992] On the regular fuzzy matrices, in Myung-Hwan Kim (ed.): *Prospects of Modern Algebra*, Proceedings of Workshops in Pure Mathematics #12, Pure Mathematics Research Assn., The Korean Academic Council, 1992.

CHRISTIAN CHOFFRUT

[1992] Rational relations and rational series, Theor. Comp. Sci. 98 (1992), 5 - 13.

CHRISTIAN CHOFFRUT & GIOVANNI PIGHIZZINI

[1997] Distances between languages and reflexivity of relations, in I. Prívara & P. Ružička (eds.): *Mathematical Foundations of Computer Science 1997*, Lecture Notes in Computer Science #1295, Springer-Verlag, Berlin, 1997.

NOAM CHOMSKY AND MARCEL PAUL SCHÜTZENBERGER

[1963] The algebraic theory of context-free languages, in P. Brafford & D. Hirschberg (eds.): *Computer Programming and Formal Systems*, North Holland, Amsterdam, 1963.

A. C. CHOUDHURY

[1957] The doubly distributive m-lattice, Bull. Calcutta Math. Soc. 49 (1957), 71 - 74.

R. PROCESI CIAMPI & R. ROTA

[1987] The hyperring spectrum, Rivis. Mat. Pura ed Appl. 1 (1987), 71 - 80.

ALFRED H. CLIFFORD

[1958] Totally-ordered commutative semigroups, Bull. Amer. Math. Soc. 64 (1958), 305 - 316.

ALFRED H. CLIFFORD & G. B. PRESTON

[1961] *The Algebraic Theory of Semigroups*, vol. I, Mathematical Surveys #7, American Mathematical Society, Providence, 1961.

JEAN COCHET-TERRASSON, GUY COHEN, STÉPHANE GAUBERT, MICHAEL McGETTRICK & JEAN-PIERRE QUADRAT

[1998] Numerical computation of spectral elements in max-plus algebra, to appear, Proceedings of the IFAC Workshop on Systems, Structure and Control, Nantes, 1998.

GUY COHEN, STÉPHANE GAUBERT & JEAN-PIERRE QUADRAT

[1996] Kernels, images and projections in Dioids, preprint, 1996.

[1997] Linear projectors in the max-plus algebra, preprint, 1997.

[1998] Algebraic system analysis of timed Petri nets, in J. Gunawardena (ed.): *Idempotency (Bristol 1994)*, Publ. Newton Inst. #11, Cambridge Univ. Press, Cambridge, 1998, 145 – 170.

PAUL M. COHN

[1965] *Universal Algebra*, Harper & Row, New York, 1965.

[1975] Algebra and language theory, Bull. London Math. Soc. 7 (1975), 1 – 29.

JOHN H. CONWAY

[1971] *Regular Algebra and Finite Machines*, Chapman and Hall, London, 1971.

WILLIAM H. CORNISH

[1971] Direct summands in semirings, Math. Japon. 16 (1971), 13 – 19.

MARIE-FRANCOISE COSTE-ROY, MICHEL COSTE & LOUIS MAHÉ

[1980] Contribution to the study of the natural number object in elementary topoi, J. Pure Appl. Algebra 17 (1980), 36 – 68.

R. C. COURTER

[1969] Rings all of whose factor rings are semi-prime, Canad. Math. Bull. 12 (1969), 417 – 426.

THOMAS C. CRAVEN

[1991] Orderings on semirings, Semigroup Forum 43 (1991), 45 – 52.

ROGER CRUON & PHILIPPE F. HERVÉ

[1965] Quelques résultats relatifs a une structure algebrique et a son application au probleme central de l'ordonnancement, Rev. Franc. Rech. Oper. 34 (1965), 3 – 19.

RAYMOND A. CUNINGHAME-GREEN

[1962] Describing industrial processes with interference and approximating their steady-state behavior, Op. Res. Quart. 13 (1962), 95 – 100.

[1976] Projections in minimax algebra, Math. Programming 10 (1976), 111 – 123.

[1979] *Minimax algebra*, Lecture Notes in Economics and Mathematical Systems #166, Springer-Verlag, Berlin, 1979.

[1983] The characteristic maxpolynomial of a matrix, J. Math. Anal. Appl. 95 (1983), 110 – 116.

[1984] Using fields for semiring computations, Ann. Discrete Math. 19 (1984), 55 – 74.

[1991] Minimax algebra and its applications, Fuzzy Sets and Systems 41 (1991), 251 – 267.

RAYMOND A. CUNINGHAME-GREEN & W. F. BORAWITZ

[1984] Scheduling by non-commutative algebra, Ann. Discrete Math. 19 (1984), 75 – 82.

RAYMOND A. CUNINGHAME-GREEN & RAINER E. BURKARD

[1984] Eigenfunctions and optimal orbits, J. Comp. Appl. Math. 11 (1984), 83 – 94.

RAYMOND A. CUNINGHAME-GREEN & F. HUISMAN

[1982] Convergence problems in minimax algebra, J. Math. Anal. & Appl. 88 (1982), 196 – 203.

RAYMOND A. CUNINGHAME-GREEN & P. F. J. MEIJER

[1980] An algebra for piecewise-linear minimax problems, Discrete Appl. Math. 2 (1980), 267 – 294.

LOUIS DALE

[1976a] Monic and monic free ideals in a polynomial semiring, Proc. Amer. Math. Soc. 56 (1976), 45 – 50.

[1976b] Monic and monic free ideals in a polynomial semiring in several variables, Proc. Amer. Math. Soc. 61 (1976), 209 – 216.

[1977a] The k-closure of monic and monic free ideals in a polynomial semiring, Proc. Amer. Math. Soc. 64 (1977), 219 – 226.

[1977b] Direct sums of semirings and the Krull-Schmidt theorem, Kyungpook Math. J. 17 (1977), 135 – 141.

[1980] A note on the structure of ideals in a Euclidean ring, Kyungpook Math. J. 20 (1980), 279.

[1981] Standard halfrings and standard ideals, Kyungpook Math. J. 21 (1981), 281 – 288.

[1982] The structure of monic ideals in a noncommutative polynomial semiring, Acta Math. Acad. Sci. Hungar. 39 (1982), 163 – 168.

[1983] Extending certain semiring homomorphisms to ring homomorphisms, Kyungpook Math. J. 23 (1983), 13 – 18.

LOUIS DALE & PAUL J. ALLEN

[1976] Ideal theory in polynomial semirings, Publ. Math. Debrecen 23 (1976), 183 – 190.

LOUIS DALE & DOROTHY L. HANSON

[1977] The structure of ideals in a Euclidean semiring, Kyungpook Math. J. 17 (1977), 21 – 29.

LOUIS DALE & JANET D. PITTS

[1978] Euclidean and gaussian semirings, Kyungpook Math. J. 18 (1978), 17 – 22.

B. DE BAETS & E. E. KERRE

[1993a] The generalized modus ponens and triangular fuzzy data model, Fuzzy Sets and Systems 59, (1993), 305 - 317.

[1993b] Fuzzy relational compositions, Fuzzy Sets and Systems 60 (1993) 109 - 120.

RICHARD DEDEKIND

[1894] Über die Theorie der ganzen algebraischen Zahlen, Supplement XI to P. G. Lejeune Dirichlet: *Vorlesungen über Zahlentheorie*, 4 Aufl., Druck und Verlag, Braunschweig, 1894.

S. DELSARTE

[1948] Fonctions de Möbius sur les groupes Abeliens finis, Annals of Math. 49 (1948), 600 - 609.

A. DE LUCA & S. TERMINI

[1972] Algebraic properties of fuzzy sets, J. Math. Anal. Appl. 40 (1972), 373 - 386.

PIERRE DEL MORAL

[1998] Maslov optimization theory, topological aspects, in J. Gunawardena (ed.): *Idempotency (Bristol 1994)*, Publ. Newton Inst. #11, Cambridge Univ. Press, Cambridge, 1998, 354 - 382.

PIERRE DEL MORAL & M. DOISY

[1999a] Maslov idempotent probability calculus, preprint, 1999.

[1999b] On the applications of Maslov optimization theory, preprint, 1999.

PIERRE DEL MORAL & GÉRARD SALUT

[1998] Random particle methods in $(max, +)$ optimization problems, in J. Gunawardena (ed.): *Idempotency (Bristol 1994)*, Publ. Newton Inst. #11, Cambridge Univ. Press, Cambridge, 1998, 383 -391.

M.-M. DEZA & I. G. ROSENBERG

[1986] General convolutions motivated by designs, Acta Univ. Carolinae Math. et Phys. 27 (1986), 49 - 66.

ANTONIO DI NOLA

[1985] Relational equations in totally ordered lattices and their complete solution, J. Math. Anal. Appl. 107 (1985), 148 - 155.

[1990] On solving relational equations in Brouwerian lattices, Fuzzy Sets and Systems 34 (1990), 365 - 376.

ANTONIO DI NOLA & ADA LETTIERI

[1989] Relation equations in residuated lattices, Rend. Circ. Mat. Palermo 3 (1989), 246 - 256.

TAMMO tom DIECK

[1979] *Transformation Groups and Representation Theory*, Lecture Notes in Mathematics #766, Springer-Verlag, Berlin, 1979.

ROBERT P. DILWORTH

[1939] Non-commutative residuated lattices, Trans. Amer. Math. Soc. 46 (1939), 426 - 444.

[1962] Abstract commutative ideal theory, Pacific J. Math. 12 (1962), 481 - 498.

HANS DOBBERTIN

[1979] Note on associative Newman algebras, Algebra Univ. 9 (1979), 396 - 397.

[1982] On Vaught's criterion for isomorphisms of countable Boolean algebras, Algebra Univ. 15 (1982), 95 - 114.

J.-P. DOIGNON, P. MONJARET, M. ROUBENS & Ph. VINCKE

[1986] Biorders families, valued relations and preference modelling, J. Math. Psych. 30 (1986), 435 - 480.

I. L. DORROH

[1932] Concerning adjunctions to algebra, Bull. Amer. Math. Soc. 38 (1932), 85 - 88.

RONALD EUGENE DOVER & HOMER E. STONE

[1965] On semisubtractive halfrings, Bull Austral. Math. Soc. 12 (1965), 371 - 378.

CLIFFORD HUGH DOWKER & DONNA STRAUSS

[1974] Separation axioms for frames, in *Topics in Topology, Keszthely (Hungary) 1972*, Colloq. Math. Soc. János Bolyai #8, North Holland, Amsterdam, 1974.

XIAOYONG DU & NAOHIRO ISHII

[1995] An algebraic rewriting theorem of multiple linear recursions and its applications, in N. Revell & A Min Tjoa (eds.): *Sixth Conference on Database and Expert Systems Applications*, Lecture Notes in Computer Science #978, Springer-Verlag, Berlin, 1995.

R. DUARTE RAMOS

[1978] Des questions d'absorption en demi-anneaux, in *Estudos de Algebra, Geometria et Análise*, No. II, Centenário da Academia das Ciencias de Lisboa, Lisbon, 1978.

DIDIER DUBOIS & HENRI PRADE

[1980] *Fuzzy Sets and Systems, Theory and Applications*, Academic Press, New York, 1980.

MARIE-LOUISE DUBREIL-JACOTIN, LEONCE LESIEUR & ROBERT CROISOT

[1953] *Leçons sur la Théorie des Treillis, des Structures Algébriques Ordonnées et Treillis Géométriques*, Gauthier-Villars, Paris, 1953.

G. DUCHAMP & J. Y. THIBON

[1988] Théorèmes de transfert pour les polynômes partiellement commutatifs, Theoretical Comp. Sci. 57 (1988), 239 - 249.

P. I. DUDNIKOV

[1992] Endomorphisms of the submodule of bounded functions, in V. P. Maslov & S. N. Sambourskiĭ (eds.): *Idempotent Analysis*, Advances in Soviet Mathematics #13, American Mathematical Society, Providence, 1992.

P. I. DUDNIKOV & S. N. SAMBOURSKIĬ

[1989] Spectra of endomorphisms of semimodules over semirings with an idempotent operation, Dokl. Acad. Nauk SSR 308 (1989), 780 - 783. [Russian]

[1991] Endomorphisms of semimodules over semirings with an idempotent opration, Izv. Akad. Nauk SSSR Ser. Math. 55 (1991), 93 - 109. [Russian]

[1992] Endomorphisms of finitely generated free semimodules, in V. P. Maslov & S. N. Sambourskiĭ (eds.): *Idempotent Analysis*, Advances in Soviet Mathematics #13, American Mathematical Society, Providence, 1992.

BILL J. DULIN & JAMES R. MOSHER

[1972] The Dedekind property for semirings, J. Austral. Math. Soc. 14 (1972), 82 - 90.

ARYEH DVORETZKY, ABRAHAM WALD & J. WOLFOWITZ

[1951] Relations among certain ranges of vector measures, Pacific J. Math. 1 (1951), 59 - 74.

CHARLES EHRESMANN

[1957] Gattungen von lokalen Strukturen, Jber. Deutsch. Math.-Verein. 60 (1957/8), 59 - 77.

SAMUEL EILENBERG

[1974] *Automata, Languages, and Machines*, vol. A, Academic Press, New York, 1974.

SAMUEL EILENBERG & MARCEL PAUL SCHÜTZENBERGER

[1969] Rational sets in commutative monoids, J. Algebra 13 (1969), 173 - 191.

R. EILHAUER

[1968] Zur Theorie der Halbkörper. I, Acta Math. Acad. Sci. Hungar. 19 (1968), 23 - 45.

CALVIN C. ELGOT

[1975] Monadic computation and iterative algebraic theories, in H. E. Rose & J. C. Shepherdson (eds.): *Logic Colloquium '73*, North Holland, Amsterdam, 1975.

[1976] Matricial theories, J. Algebra 42 (1976), 391 - 422.

D. D. ERMOLAEVA

[1989] Automorphisms of the max-matrix system, Vest. Moskov. Univ., Matematika 44 (1989), 3 - 8. [Russian]

FANG JIN-XUAN

[1993] Studies in fuzzy relations using triangular norms, Info Sci. 67 (1993), 127 - 135.

SHALOM FEIGELSTOCK

[1980] Radicals of the semiring of abelian groups, Publ. Math. Debrecen 27 (1980), 89 - 92.

W. FENCHEL

[1949] On the conjugate convex functions, Canad. J. Math. 1 (1949), 73 - 77.

A. V. FINKELSTEIN & M. A. ROYTBERG

[1993] Computation of biopolymers: a general approach to different problems, BioSystems 30 (1993), 1 - 20.

ISIDORE FLEISCHER

[1995] Abstract ideal theory, Normal 43 (1995), 120 - 135.

JOHN G. FLETCHER

[1980] A more general algorithm for computing closed semiring costs between vertices of a directed graph, Comm. Assoc. Comp. Mach. 23 (1980), 350 - 351.

MICHEL FLIESS

[1971] Sur la définition des séries formelles algébriques, C. R. Acad. Sci. Paris 273, A284 - A287.

[1974] Matrices de Hankel, J. Math. Pures et Appl. 53 (1974), 197 - 224.

JANÓS FODOR

[1992] Traces of fuzzy binary relations, Fuzzy Sets and Systems 50 (1992), 331 - 341.

TATYANA S. FOFANOVA

[1971] Polygons over distributive lattices, Sibir. Mat. Ž. 12 (1971), 1195 - 1199. [Russian]

[1982] Polygons over distributive lattices, in B. Csákány et al. (eds.): *Universal Algebra*, Colloq. Math. Soc. János Bolyai #29, North Holland, Amsterdam, 1982.

M. J. FRANK

[1979] On the simultaneous associativity of $F(x,y)$ and $x+y-F(x,y)$, Aeq. Math. 19 (1979), 194 - 226.

D. H. FREMLIN

[1974] *Topological Riesz Spaces and Measure Theory*, Cambridge University Press, Cambridge, 1974.

HANS FREUDENTHAL

[1936] Teilweise geordnete Moduln, Proc. Akad. Wet. Amsterdam 39 (1936), 641 - 651.

PETER FREYD

[1966] Algebra valued functors in general and tensor products in particular, Colloq. Math. 14 (1966), 89 - 106.

VÉRONIQUER FROIDURE & JEAN-ÉRIC PIN

[1998] Algorithms for computing finite semigroups, preprint, 1998.

LASZLO FUCHS

[1954] A lattice-theoretic discussion of some problems in additive ideal theory, Acta Math. Acad. Sc. Hung. 5 (1954), 299 - 313.

[1963] *Partially Ordered Algebraic Systems*, Addison-Wesley, Reading, Mass., 1963.

LASZLO FUCHS & OTTO STEINFELD

[1963] Principal components and prime factorization in partially ordered semigroups, Ann. Univ. Sci. Budapest. Sec. Math. 6 (1963), 103 - 111.

R. FULLER & H.-J. ZIMMERMANN

[1992] On computation of the compositional rule of inference under triangular norms, Fuzzy Sets and Systems 51 (1992), 267 - 275.

GIORGIO GALLO & STEFANO PALLOTTINO

[1986] Shortest path methods: a unifying approach, in G. Gallo & C. Sandi (eds.): *Netflow at Pisa*, Mathematical Programming Study #20, North Holland, Amsterdam, 1986.

[1988] Shortest path algorithms, Ann. Oper. Res. 13 (1988), 3 - 79.

F. R. GANTMACHER

[1959] *Applications of the Theory of Matrices*, Interscience, New York, 1959.

BARRY J. GARDNER

[1993] Some abstract algebra from the elementary calculus course, Int. J. Math. Educ. Sci. Technol. 24 (1993), 781 - 789.

STÉPHANE GAUBERT

[1995] Performance evaluation of $(max, +)$ automata, IEEE Transactions on Automatic Control 40 (1995), 2014 - 2025.

[1996a] *Systèmes Dynamiques à Événements Discrets* (Notes de cours commun, EN-SMP, Option Automatique & DEA ATS Orsay), INRIA Rocquencourt, Le Chesnay, 1996.

[1996b] On the Burnside problem for semigroups of matrices in the $(max, +)$ algebra, Semigroup Forum 52 (1996), 271 - 292.

STÉPHANE GAUBERT & JEAN MAIRESSE

[1998] Task resource models and $(max, +)$ automata, in J. Gunawardena (ed.): *Idempotency (Bristol 1994)*, Publ. Newton Inst. #11, Cambridge Univ. Press, Cambridge, 1998, 133 - 144.

[1999] Modeling and analysis of timed Petri nets using heaps of pieces, to appear, IEEE Transactions on Automatic Control, 1999.

STÉPHANE GAUBERT & *MAX PLUS*

[1997] Methods and applications of $(max, +)$ linear algebra, in R. Reischuk & M. Morvan (eds.): *STACS 97*, Lecture Notes in Computer Scvience #1200, Springer-Verlag, Berlin, 1997.

BRUNO GAUJAL & ALAIN JEAN-MARIE

[1998] Computational issues in recursive stochastic systems, in J. Gunawardena (ed.): *Idempotency (Bristol, 1994)*, Publ. Newton Inst. #11, Cambridge Univ. Press, Cambridge, 1998, 209 –230.

F. GÉCSEG & I. PEÁK

[1972] *Algebraic Theory of Automata*, Akadémai Kaidó, Budapest, 1972.

IZRAIL' GEL'FAND

[1941] Normierte Ringe, Mat. Sbornik 9 (1941), 3 – 23.

GEORGE GEORGESCU

[1988] Some sheaf constructions for distributive lattices, Bull. Math. Soc. Sci. Math. R. S. Roum. 32 (1988), 299 – 303.

YAN GEORGET & PHILIPPE CODOGNET

[1998] Compiling semiring-based constraints with `clp(FD,S)`, preprint, 1998.

GERHARD GIERZ et al.

[1980] *A Compendium of Continuous Lattices*, Springer-Verlag, Berlin, 1980.

B. GIFFLER

[1963] Scheduling general production system using schedule algebra, Nav. Res. Log. Quart. 10 (1963), 237 – 255.

[1968] Schedule algebra: a progress report, Nav. Res. Log. Quart. 15 (1968), 255 – 280.

ROBERT GILMER

[1972] *Multiplicative Ideal Theory*, Marcel Dekker, New York, 1972.

E. GIRALDES

[1978a] Sur les demi-anneaux à division à droite, in *Estudos de Algebra, Geometria et Análise*, No. II, Centenário da Academia das Ciencias de Lisboa, Lisbon, 1978.

[1978b] Sue les demi-anneaux inverses, in *Estudos de Algebra, Geometria et Análise*, No. II, Centenário da Academia das Ciencias de Lisboa, Lisbon, 1978.

JEAN-YVES GIRARD

[1987] Linear Logic, J. Theoret. Comp. Sci. 50 (1987), 1 – 102.

[1989] Towards a geometry of interaction, in J. W. Gray & A. Scedrov (eds.): *Categories in Computer Science and Logic*, Contemporary Mathematics #92, American Mathematical Society, Providence, 1989.

YEHOSHAFAT GIVE'ON

[1964] Lattice matrices, Inform. and Control 7 (1964), 477 – 484.

KAZIMIERZ GLAZEK

[1968] On certain characterizations of distributive lattices, Colloq. Math. 19 (1968), 195 – 198.

[1985] *A Short Guide through the Literature on Semirings*, Mathematical Institute, University of Wroclaw, Poland, 1985.

JOSEPH A. GOGUEN

[1967] L-fuzzy sets, J. Math. Anal. Appl. 18 (1967), 145 - 174.

[1969] The logic of inexact concepts, Synthese 19 (1968/69), 325 - 373.

JONATHAN S. GOLAN

[1986] *Torsion Theories*, Longman Scientific and Technical, Harlow, 1986.

[1987] *Linear Topologies on a Ring: an Overview*, Longman Scientific and Technical, Harlow, 1987.

[1991] Information semimodules and absorbing subsemimodules, J. Math. (Hanoi) 19 (1991), 1 - 21.

[1992] *The Theory of Semirings, with Applications in Mathematics and Theoretical Computer Science*, Longman Scientific and Technical, Harlow, 1992.

[1996a] Norms, semirings, and power algebras, in S. Parvathi et al. (eds): *Proceedings of the Fourth Ramanujan Symposium on Algebra and its Applications*, Publications of the Ramanujan Institute #5, University of Madras, Madras, 1996.

[1996b] Semirings of formal series over hypermonoids: some interesting cases, Kyungpook Math. J. 36 (1996), 107 - 111.

[1997] A framework for consideration of fuzzy Petri nets, in *Proceedings of FUZZY'97*, Israel Ministry of Science, 1997.

JONATHAN S. GOLAN, ALEXANDRU MATEESCU & DRAGOS VAIDA

[1996] Semirings and parallel composition of processes, J. Automata, Languages and Combinatorics 1 (1996), 199-217.

JONATHAN S. GOLAN & HAROLD SIMMONS

[1988] *Derivatives, Nuclei and Dimensions on the Frame of Torsion Theories*, Longman Scientific and Technical. Harlow, 1988.

JONATHAN S. GOLAN & WANG HAUXIONG

[1996] On embedding in complete semirings, Comm. Alg. 24 (1996), 2945 - 2962.

MARTIN GOLDSTERN

[1985] *Vervollständigung von Halbringen*, Diplomarbeit, Technische Universität Wien, 1985.

MICHEL GONDRAN

[1975] Path algebra and algorithms, in B. Roy (ed.): *Combinatorial Programming: Methods and Applications*, D. Reidel, Dordrecht, 1975.

[1979] Les elements p-reguliers dans les dioïdes, Discrete Math. 25 (1979), 33 - 39.

MICHEL GONDRAN & MICHEL MINOUX

[1978] L'indépendance linéaire dans les dioïdes, Bulletin de la Direction Etudes et Recherches, EDF, Sér. C, 1 (1978), 67 - 90.

[1984a] *Graphs and Algorithms*, Wiley-Interscience, New York, 1984.

[1984b] Linear algebra in dioids: a survey of recent results, Ann. Discrete Math. 19 (1984), 147 - 164.

LEÓN GONZÁLEZ

[1999] A note on the infinitary action of triangular norms and conorms, Fuzzy Sets and Systems 101 (1999), 177 - 180.

SIEGFRIED GOTTWALD

[1984] T-Normen und φ-Operatoren als Wahrheitswertfunktionen mehrwertiger Junktoren, in G. Wechsung (ed.): *Frege Conference, 1984*, Akademie-Verlag, Berlin, 1984.

CURTIS GREENE

[1982] The Möbius function of a partially ordered set, in I. Rival (ed.): *Ordered Sets*, D. Reidel, Dordrecht, 1982.

JEREMY GUNAWARDENA

[1994] Min-Max functions, Discrete Event Dynamical Systems 4 (1994), 377 - 406.

[1998] An introduction to idempotency, in J. Gunawardaea (ed.): *Idempotency (Bristol 1994)*, Publ. Newton Inst. #11, Cambridge Univ. Press, Cambridge, 1998, 1 - 49.

GUO SI-ZHENG, WANG PEI-ZHUANG, ANTONIO DI NOLA & S. SESSA

[1988] Further contributions to the study of finite fuzzy relation equations, Fuzzy Sets and Systems 26 (1988), 93 - 104.

M. GUPTA & T. YAMAKAWA

[1988] *Fuzzy Computing: Theory, Hardware, and Applications*, Elsevier, Amsterdam, 1988.

VINEET GUPTA

[1994] *Chu Spaces: A Model of Concurrency*, PhD. Thesis, Department of Computer Science, Stanford University, 1994.

VERENA GURUSWAMI

[1976] *Torsion Theories and Localizations for M-sets*, PhD. Thesis, McGill University, Montreal, 1976.

FERNANDO GUZMÁN

[1992] The variety of Boolean semirings, J. Pure and Appl. Alg. 78 (1992), 253 - 270.

DORTE HAFTENDORN

[1978] Additiv kommutative und idempotente Halbringe mit Faktorbedingung. I, Publ. Math. Debrecen 25 (1978), 107 - 116.

[1979] Additiv kommutative und idempotente Halbringe mit Faktorbedingung, II, Publ. Math. Debrecen 26 (1979), 5 - 12.

M. HALL & S. PIANSKOOL

[1996] Injectivity for cancellative semimodules, Southeast Asian Bull. Math. 20, no. 4 (1996), 85 - 93.

FRANZ HALTER-KOCH

[1998] *Ideal Systems, an Introduction to Multiplicative Ideal Theory*, Marcel Dekker, New York, 1998.

ELDON HANSEN

[1992] *Global Optimization Using Interval Analysis*, Marcel Dekker, New York, 1992.

J. HANUMANTHACHARI, K. VENU RAJU & HANNS J. WEINERT

[1986] Some results on partially ordered semirings and semigroups, in S. Wolfenstein (ed): *Algebra and Order, Proceedings of the First International Symposium on Ordered Algebraic Structures*, Heldermann-Verlag, Berlin, 1986.

KOSABURO HASHIGUCHI

[1982] Limitness theorem for finite automata with distance functions, J. Comp. Syst. Sci. 24 (1982), 233 - 244.

HIROSHI HASHIMOTO

[1984] Subinverses of fuzzy matrices, Fuzzy Sets and Systems 12 (1984), 155 - 168.

[1985] Transitivity of generalized fuzzy matrices, Fuzzy Sets and Systems 17 (1985), 83 - 90.

HENRY E. HEATHERLY

[1974] Semiring multiplications on commutative monoids, Publ. Math. Debrecen 21 (1974), 119 - 123.

UDO HEBISCH

[1990] The Kleene theorem in countably complete semirings, Bayreuther Math. Schrift. 31 (1990), 55 - 66.

[1992] Eine algebraische Theorie unendlicher Summen mit Anwendungen auf Halbgruppen und Halbringe, Bayreuther. Math. Schrift. 40 (1992), 21 - 152.

UDO HEBISCH & HANNS J. WEINERT

[1987] On Euclidean semirings, Kyungpook Math. J. 27 (1987), 61 - 88.

[1993] *Halbringe*, Teubner, Stuttgart, 1993.

[1996] Semirings and semifields, in M. Hazewinkel (ed.): *Handbook of Algebra, vol. I*, North-Holland, Amsterdam, 1996, 425 - 462.

REINHOLD HECKMANN

[1993] Observable modules and power domain constructions, Algebra, Logic & Applications 5 (1993), 159 - 187.

H. HELLENDOORN

[1990] Closure properties of the compositional rule of inference , Fuzzy Sets and Systems 35 (1990), 163 - 183.

MATTHEW HENNESSY

[1988] *Algebraic Theory of Processes*, MIT Press, Cambridge, Mass., 1988.

MELVIN HENRIKSEN

[1958] Ideals in semirings with commutative addition, Notices Amer. Math. Soc. 5 (1958), 321.

[1958/9] The $a^{n(a)} = a$ theorem for semirings, Math. Japon. 5 (1958/9), 21 – 24.

WOLFGANG HERFORT & WERNER KUICH

[1984] Semitopological semirings and pushdown automata, Math. Systems Theory 17 (1984), 279 – 291.

WIM H. HESSELINK

[1990] Command algebras, recursion, and program transformation, Formal Aspects of Computing 2 (1990), 60 – 104.

DENIS HIGGS

[1980] Axiomatic infinite sums – an algebraic approach to integration theory, in W. Graves (ed.): *Proceedings of the Conference on Integration, Topology, and Geometry in Linear Spaces*, Contemporary Mathematics #2, American Mathematical Society, Providence, 1980.

DAVID HILBERT

[1890] Über die Theorie der algebraischen Formen, Math. Ann. 36 (1890), 473 – 534.

[1899] Über den Zahlbegriff, Jber. Deutsch. Math.-Verein. 8 (1899), 180 – 184.

PETER J. HILTON

[1967] Some remarks concerning the semiring of polyhedra, Bull. Soc. Math. Belg. 19 (1967), 277 – 288.

EVAN HOUSTON & MUHAMMAD ZAFRULLAH

[1988] Integral domains in which each t-ideal is divisorial, Michigan Math. J. 35 (1988), 291 – 300.

JOHN M. HOWIE

[1976] *An Introduction to Semigroup Theory*, Academic Press, New York, 1976.

SUNG C. HU

[1975] A ternary algebra for probability computation of digital circuits, in *Proceedings of the 1975 International Symposium on Multiple-Valued Logic*, IEEE Computer Society, Long Beach, 1975.

ROBERT A. HUMMEL & MICHAEL S. LANDY

[1987] A statistical viewpoint on the theory of evidence, IEEE Trans. Pattern Analysis and Machine Intel. 10 (1988), 235 – 247.

HARRY B. HUNT III

[1983] Finite languages and the computational complexity of algebraic structures, in L. J. Cummings (ed.): *Combinatorics on Words, Progress and Perspectives*, Academic Press, New York, 1983.

EDWARD V. HUNTINGTON

[1902] Complete sets of postulates for the theories of positive integral and positive rational numbers, Trans. Amer. Math. Soc. 3 (1902), 280 - 284.

SYED A. HUQ

[1973] The semiring of quotients of commutative semirings, Colloq. Math. 28 (1973), 185 - 188.

HARRY C. HUTCHINS & HANNS J. WEINERT

[1990] Homomorphims and kernels of semifields, Period. Math. Hung. 21 (1990), 113 - 152.

KENZO IIZUKA

[1959] On the Jacobson radical of a semiring, Tohoku Math. J. (2) 11 (1959), 409 - 421.

YANNIS E. IOANNIDIS & EUGENE WONG

[1991] Towards an algebraic theory of recursion, J. Assoc. Comp. Mach., 38 (1991) 329 - 381.

KIYOSHI ISÉKI

[1956a] Notes on topological spaces V. On structure spaces of semiring, Proc. Japan. Acad. 32 (1956), 426 - 429.

[1956b] Ideal theory of semiring, Proc. Japan. Acad. 32 (1956), 554 - 559.

[1958a] Ideals in semirings, Proc. Japan. Acad. 34 (1958), 29 - 31.

[1958b] On ideals in semirings, Proc. Japan. Acad. 34 (1958), 507 - 509.

KIYOSHI ISÉKI & YASUE MIYANAGA

[1956a] Notes on topological spaces III. On space of maximal ideals of a semiring, Proc. Japan. Acad. 32 (1956), 325 - 328.

[1956b] Notes on topological spaces IV. Function semiring on topological spaces, Proc. Japan. Acad. 32 (1956), 392 - 395.

KAZUO IWANO

[1987] *Two-dimensional Dynamic Graphs and their VLSI Applications*, PhD. Thesis, Princeton University, 1987.

HIROYUKI IZUMI, YASUYOSHI INAGAKI & NAMIO HONDA

[1984] A complete axiom system for algebra of closed-regular expressions, in J. Paredaens (ed.): *Automata, Languages and Programming*, Lecture Notes in Computer Science #172, Springer-Verlag, Berlin, 1984.

J. JACAS & L. VALVERDE

[1996] On fuzzy relations, metrics, and cluster analysis, preprint, 1996.

GERARD JACOB

[1975] Sur un Théorème de Shamir, Inform. and Control 27 (1975), 218 - 226.

BERNARD JACOBSON & ROBERT J. WISNER

[1966] Matrix number theory, I: factorization of 2×2 unimodular matrices, Pub. Math. Debrecen 13 (1966), 67 - 72.

NATHAN JACOBSON

[1980] *Basic Algebra II*, Freeman, San Francisco, 1980.

[1985] *Basic Algebra I* (2nd edition), Freeman, San Francisco, 1985.

RYSZARD JANICKI

[1977] Vectors of coroutines over Blikle nets, in M. Karpiński (ed.): *Fundamentals of Computation Theory*, Lecture Notes in Computer Science #56, Springer-Verlag, Berlin, 1977.

MELVIN F. JANOWITZ

[1970] Principal multiplicative lattices, Pacific J. Math. 33 (1970), 653 - 656.

[1976] A note on Rickart rings and semi-Boolean algebras, Algebra Univ. 6 (1976), 9 - 12.

MARK JERRUM & MARC SNIR

[1982] Some exact complexity results for straight-line computations over semirings, J. Assoc. Comp. Mach. 29 (1982), 874 - 897.

JAROSLAV JEZEK & TOMAS KEPKA

[1983] Simple semimodules over commutative semirings, Acta. Sci. Math. (Szeged) 46 (1983), 17 - 27.

E. W. JOHNSON & JOHNNY A. JOHNSON

[1970] Lattice modules over semi-local Noether lattices, Fund. Math. 68 (1970), 187 - 201.

JAMES S. JOHNSON & ERNEST G. MANES

[1970] On modules over a semiring, J. Algebra 15 (1970), 57 - 67.

PETER T. JOHNSTONE

[1982] *Stone Spaces*, Cambridge University Press, Cambridge, 1982.

[1984] Wallman compactification of locales, Houston J. Math. 10 (1984), 201 - 206.

PASCUAL JORDAN

[1949] Über nichtcommutative Verbande, Arch. Math. 2 (1949), 56 - 59.

ANDRE JOYAL & MYLES TIERNEY

[1984] *An Extension of the Galois Theory of Grothendieck*, A. M. S. Memoir #309, American Mathematical Society, Providence, 1984.

M. A. KAASHOEK & T. T. WEST

[1974] *Locally Compact Semi-Algebras*, North Holland, Amsterdam, 1974.

R. E. KALMAN, P. L. FALB & MICHAEL A. ARBIB

[1969] *Topics in Mathematical System Theory*, McGraw-Hill, New York, 1969.

GEORG KARNER

[1992] On limits in complete semirings, Semigroup Forum 45 (1992), 148 - 165.

[1994] A topology for complete semirings, in P. Enjalbert et al. (eds.), *Proceedings of STACS'94*, Lecture Notes in Computer Science #775, Springer-Verlag, Berlin, 1994, 389 - 400.

GEORG KARNER & WERNER KUICH

[1997] A characterization of abstract families of algebraic power series, in I. Prívara & P. Ružička (eds.): *Mathematical Foundations of Computer Science 1997*, Lecture Notes in Computer Science #1295, Springer-Verlag, Berlin, 1997.

PAUL H. KARVELLAS

[1974] Inversive semirings, J. Austral. Math. Soc. 18 (1974), 277 - 288.

[1975] Extension of a semigroup embedding theorem to semirings, Canad. Math. Bull. 18 (1975), 297 - 298.

YEFIM KATSOV

[1997] Tensor products and injective envelopes of semimodules over additive regular semirings, Algebra Colloq. 4 (1997), 121 - 131.

ARNOLD KAUFMANN

[1975] *Introduction to the Theory of Fuzzy Subsets*, vol. I, Academic Press, New York, 1975.

ARNOLD KAUFMANN & MADAN M. GUPTA

[1985] *Introduction to Fuzzy Arithmetic*, Van Nostrand Reinhold, New York, 1985.

YASUO KAWAHARA & HITOSHI FURUSAWA

[1999] An algebraic formulation of fuzzy relations, Fuzzy Sets and Systems 101 (1999), 125 - 135.

K. A. KEARNES

[1995] Semilattice modes I: the associated semiring, Algebra Universalis 34 (1995), 220 - 272.

KLAUS KEIMEL

[1972] A unified theory of minimal prime ideals, Acta Math. Acad. Sci. Hungar. 23 (1972), 51 - 69.

JIN BAI KIM

[1984] Inverses of Boolean matrices, Bull. Inst. Math. Acad. Sinica 12 (1984), 125 - 128.

JIN BAI KIM, ALPHONSE BAARTMANS & NOR SHAKILA SAHADIN

[1989] Determinant theory for fuzzy matrices, Fuzzy Sets and Systems 29 (1989), 349 - 356.

KI HANG KIM

[1982] *Boolean Matrix Theory and Applications*, Marcel Dekker, New York, 1982.

KI HANG KIM, MASAHARU MIZUMOTO, JUNICHI TOYODA & KOKICHI TANAKA

[1975] L-fuzzy grammars, Inform. Sci. 8 (1975), 123 - 140.

KI HANG KIM & FRED W. ROUSH

[1980] Generalized fuzzy matrices, Fuzzy Sets and Systems 4 (1980), 293 - 315.

[1995] Inclines of algebraic structures, Fuzzy Sets and Systems 72 (1995), 189 - 196.

DAVID KIRBY

[1969] Closure operations on ideals and submodules, J. London Math. Soc. 44 (1969), 283 - 291.

STEPHEN C. KLEENE

[1956] Representation of events in nerve nets and finite automata, in C. E. Shannon & J. McCarthy (eds.): *Automata Studies*, Princeton University Press, Princeton, 1956.

ERICH PETER KLEMENT

[1982] On the relationship between different notions of fuzzy measures, in R. Yager (ed.): *Fuzzy Set and Possibility Theory*, Pergamon Press, New York, 1982.

DONALD E. KNUTH

[1969] *The Art of Computer Programming, vol. 2: Seminumerical Algorithms*, Addison-Wesley, Reading, Mass., 1969.

VASSILI N. KOLOKOL'TSOV

[1992] On linear, additive, and homogeneous operators in idempotent analysis, in V. P. Maslov & S. N. Sambourskiĭ (eds.): *Idempotent Analysis*, Advances in Soviet Mathematics #13, American Mathematical Society, Providence, 1992.

VASSILI N. KOLOKOL'TSOV & VICTOR P. MASLOV

[1988] The general form of the endomorphisms in the space of continuous functions with values in a numerical commutative semiring (with the operation $\oplus = max$), Soviet Math. Dokl. 36 (1988), 55 - 59.

[1989a] Idempotent analysis as a technique of the theory of control and optimal synthesis I, Funktsional. Anal. i Prilozhen. 23 (1989), 1 - 14. [Russian]

[1989b] Idempotent analysis as a technique of the theory of control and optimal synthesis II, Funktsional. Anal. i Prilozhen. 23 (1989), 53 - 62, 96. [Russian]

[1997] *Idempotent Analysis and Applications*, Kluwer, Dordrecht, 1997.

[1998] New differential equation for the dynamics of Pareto sets, in J. Gunawardena (ed.): *Idempotency (Bristol, 1994)*, Publ. Newton Inst. #11, Cambridge Univ. Press, Cambridge, 1998, 322 - 330.

MAURICE KOSKAS

[1969] Forme algébrique des théories de Stone-Gel'fand, Sém. P. Dubreil, M. L. Dubreil-Jacotin, L. Lesieur et C. Pisot 21 (1967/68), Fasc. 2, Secrétariat Mathématique, Paris, 1969.

DEXTER KOZEN

[1990] On Kleene algebras and closed semirings, in B. Rovan (ed.): *Mathematical Foundations of Computer Science, 1990*, Lecture Notes in Computer Science #452, Springer-Verlag, Berlin, 1990.

DANIEL KROB

[1987] Monoids et semi-anneaux complets, Semigroup Forum 36 (1987), 323 - 339.

[1988] *Equations K-Rationnelles*, thesis, Université Paris VII, 1988.

[1992] Models of a K-rational identity system, J. Comp. System Sci. 45 (1992), 396 - 434.

[1998] Some automata-theoretic aspects of min-max-plus semirings, in J. Gunawardena (ed.): *Idempotency (Bristol, 1994)*, Publ. Newton Inst. #11, Cambridge Univ. Press, Cambridge, 1998, 70 - 79.

WOLFGANG KRULL

[1924] Axiomatische Begründung der Algemeinen Idealtheory, SitZ. phys.-med. Soc. Erlangen 56 (1924), 47 - 63.

WERNER KUICH

[1987] The Kleene and the Parikh theorem in complete semirings, in T. Ottmann (ed.): *Automata, Languages and Programming*, Lecture Notes in Computer Science #267, Springer-Verlag, Berlin, 1987.

[1991] Automata and languages generalized to ω-continuous semirings, Theor. Comp. Sci. 79 (1991), 137 - 150.

[1995] Representations and complete semiring morphisms, Info. Processing Letters 56 (1995), 293 - 298.

WERNER KUICH & ARTO SALOMAA

[1986] *Semirings, Automata, Languages*, EATCS Monographs on Theoretical Computer Science no. 5, Springer-Verlag, Berlin, 1986.

WERNER KUICH & F. J. URBANEK

[1983] Infinite linear systems and one counter languages, Theor. Comp. Sci. 22 (1983), 95 - 126.

J. KUNTZMANN

[1972] *Théorie des Réseaux (Graphes)*, Dunod, Paris, 1972.

SUSAN LaGRASSA

[1995] *Semirings: Ideals and Polynomials*, PhD. Thesis, University of Iowa, 1995.

JOHN LAKE

[1976] Sets, fuzzy sets, multisets and functions, J. London Math. Soc. (2) 12 (1976), 323 - 326.

GERARD LALLEMENT

[1979] *Semigroups and Combinatorial Applications*, Wiley, New York, 1979.

JOACHIM LAMBEK

[1966] *Lectures on Rings and Modules*, Blaisdell, Waltham, Mass., 1966.

MAX D. LARSEN & PAUL J. McCARTHY

[1971] *Multiplicative Theory of Ideals*, Academic Press, New York, 1971.

DONALD R. LaTORRE

[1965] On h-ideals and k-ideals in hemirings, Publ. Math. Debrecen 12 (1965), 219 - 226.

[1967a] A note on the Jacobson radical of a hemiring, Publ. Math. Debrecen 14 (1967), 9 - 13.

[1967b] The Brown-McCoy radicals of a hemiring, Publ. Math. Debrecen 14 (1967), 15 - 28.

[1970] A note on quotient semirings, Proc. Amer. Math. Soc. 24 (1970), 463 - 465.

EUGENE L. LAWLER

[1976] *Combinatorial Optimization: Networks and Matroids*, Holt, Rinehart and Winston, New York, 1976.

F. WILLIAM LAWVERE

[1964] An elementary theory of the category of sets, Proc. Nat. Acad. Sci. 52 (1964), 1506 - 1511.

[1973] Metric spaces, generalized logic and closed categories, Rend. Sem. Mat. e Fis. Milano (1973), 135 - 166.

REBECCA LEE

[1979] Covers and associated primes in noetherian lattice modules, Houston J. Math. 5 (1979), 219 - 239.

LEE SIN-MIN

[1971] Axiomatic characterization of Σ-semirings, Acta Sci. Math. (Szeged) 32 (1971), 337 - 343.

DANIEL J. LEHMANN

[1977] Algebraic structures for transitive closure, Theor. Comp. Sci. 4 (1977), 59 - 76.

S. A. LESIN & S. N. SAMBOURSKIĬ

[1992] Spectra of compact endomorphisms, in V. P. Maslov & S. N. Sambourskiĭ (eds.): *Idempotent Analysis*, Advances in Soviet Mathematics #13, American Mathematical Society, Providence, 1992.

HING LEUNG

[1988] On the topological structure of a finitely-generated semigroup of matrices, Semigroup Forum 37 (1988), 273 - 287.

YAAKOV LEVITZKI

[1946] On powers with transfinite exponents, Riveon LeMathematika 1 (1946), 8 - 13. [Hebrew]

LING CHO-HSIN

[1966] Representation of associative functions, Publ. Math. Debrecen 12 (1966), 189 - 212.

GRIGORI L. LITVINOV & VICTOR P. MASLOV

[1998] The correspondence principle for idempotent calculus and some computer applications, in J. Gunawardena (ed.): *Idempotency (Bristol, 1994)*, Publ. Newton Inst. #11, Cambridge Univ. Press, Cambridge, 1998, 420 - 443.

GRIGORI L. LITVINOV, VICTOR P. MASLOV & A. Ya. RODIONOV

[1998] Unifying approach to software and hardware design for scientific calculations, preprint, 1998.

GRIGORI L. LITVINOV, VICTOR P. MASLOV & A. N. SOBOLEVSKIĬ

[1998] Idempotent mathematics and interval analysis, preprint, 1998.

LIU WANG-JIN

[1982] Fuzzy invariant subgroups and fuzzy ideals, Fuzzy Sets and Systems 8 (1982), 133 - 139.

DANIEL LOEB

[1992] Sets with a negative number of elements, Advances in Math 91 (1992), 64 -74.

HOOI-TONG LOH

[1967] Notes on semirings, Math. Mag. 40 (1967), 150 - 152.

HOOI-TONG LOH & H. H. TEH

[1966/7] Some theorems on groupoid semirings, Nanta Math. 1 (1966/7), 33 - 37.

P. LORENZEN

[1939] Abstrakte Begründung der multiplikativen Idealtheorie, Math. Z. 45 (1939), 533 - 553.

PAOLA LORETI & MARCO PEDICINI

[1998] Idempotent Analogue of Resolvent Kernels for a Deterministic Optimal Control Problem, preprint, 1998.

M. LOTHAIRE

[1983] *Combinatorics on Words*, Addison-Wesley, Reading, Mass., 1983.

JOHN K. LUEDEMAN

[1983] Torsion theories and semigroups of quotients, in K. H. Hofmann et al. (eds.): *Recent Developments in the Algebraic, Analytical, and Topological Theory of Semigroups*, Lecture Notes in Mathematics #998, Springer-Verlag, Berlin, 1983.

HERBERT LUGOWSKI

[1962] Über die Vervollständigung geordneter Halbringe, Publ. Math. Debrecen 9 (1962), 213 - 222.

[1964a] Über gewisse geordnete Halbmoduln mit negativen Elementen, Publ. Math. Debrecen 11 (1964), 23 - 31.

[1964b] Über die Vervollst"andigung gewisser geordneter Halbmoduln mit negativen Elementen, Publ. Math. Debrecen 11 (1964), 135 - 138.

W. A. J. LUXEMBURG & ADRIAAN C. ZAANEN

[1971] *Riesz Spaces*, vol. 1, North Holland, Amsterdam, 1971.

ROGER C. LYNDON

[1961] Relation algebras and projective geometries, Michigan Math. J. 8 (1961), 21 - 28.

SHUICHIRO MAEDA

[1960] On a ring whose principal right ideals generated by idempotents form a lattice, J. Science Hiroshima Univ. Sec. A 24 (1960), 509 - 525.

F. S. MACAULAY

[1916] *Algebraic Theory of Modular Systems*, Cambridge University Press, Cambridge, 1916.

BERND MAHR

[1981] A bird's-eye view of path problems, in H. Noltemeier (ed.): *Graph-theoretic Concepts in Computer Science*, Lecture Notes in Computer Science #100, Springer-Verlag, Berlin, 1981.

[1982] Algebraic complexity of path problems, RAIRO Inform. Theor. 16 (1982), 263 - 292.

[1984] Iteration and summability in semirings, Ann. Discrete Math. 19 (1984), 229 - 256.

MICHAEL G. MAIN & DAVID B. BENSION

[1985] Free semiring-representations and nondeterminism, J. Comp. Sci. Sys. 30 (1985), 318 - 328.

MICHAEL G. MAIN & DAVID L. BLACK

[1993] Semantic models for total correctness and fairness, Theor. Comp. Sci. 107 (1993), 305 - 332.

JEAN MAIRESSE

[1985] A graphical approach to the spectral theory in the $(max, +)$ algebra, IEEE Transactions on Automatic Control 20 (1985), 1-6.

[1997] Products of irreducilbe random matrices in the $(max, +)$ algebra, Adv. in Appl. Probab. 29 (1997), 444 - 477.

KING-TIM MAK & KERMIT SIGMON

[1988] Standard threads and distributivity, Aeq. Math. 36 (1988), 252 - 267.

A. MANDEL & IMRE SIMON

[1977] On finite semigroups of matrices, Theor. Comp. Sci. 5 (1977), 101 - 111.

ERNEST G. MANES

[1976] *Algebraic Theories*, Springer-Verlag, Berlin, 1976.

ERNEST G. MANES & MICHAEL A. ARBIB

[1986] *Algebraic Approaches to Program Semantics*, Springer-Verlag, Berlin, 1986.

ERNEST G. MANES & D. B. BENSON

[1985] The inverse semigroup of a sum-ordered semiring, Semigroup Forum 31 (1985), 129 - 152.

ALBERTO MARTELLI

[1974] An application of regular algebra to the enumeration of cut sets in a graph, in J. Rosenfeldt (ed.): *Information Processing 74*, North Holland, Amsterdam, 1974.

[1976] A gaussian elimination algorithm for the enumeration of cut sets in a graph, J. Assoc. Comp. Mach. 23 (1976), 58 - 73.

OLIVIER MARTIN, ANDREW M. ODLYZKO & STEPHEN WOLFRAM

[1984] Algebraic properties of cellular automata, Comm. Math. Phys. 93 (1984), 219 - 258.

JEAN-PAUL MASCLE

[1986] Torsion matrix semigroups and recognizable transductions, in L. Kott (ed.): *Automata, Languages and Programming*, Lecture Notes in Computer Science #226, Springer-Verlag, Berlin, 1986.

VICTOR P. MASLOV

[1987] *Méthodes Opératorielles*, Mir, Moscow, 1987.

VICTOR P. MASLOV & S. N. SAMBOURSKIĬ

[1992a] *Idempotent Analysis*, Advances in Soviet Mathematics #13, American Mathematical Soc., Providence, R.I., 1992.

[1992b] Stationary Hamilton-Jacobi and Bellman equations (existence and uniqueness of solutions) in V. P. Maslov & S. N. Sambourskiĭ (eds.): *Idempotent Analysis*, Advances in Soviet Mathematics #13, American Mathematical Society, Providence, 1992.

MAX-PLUS WORKING GROUP

[1995] Max-plus algebra and applications to system theory and optimal control, in *Proceedings of the International Congress of Mathematicians*, Vol. 1, 2 (Zürich, 1994), 1511–1522, Birkäuser, Basel, 1995.

RALPH N. McKENZIE, GEORGE F. McNULTY & WALTER F. TAYLOR

[1987] *Algebras, Lattices, Varieties*, vol. I, Wadsworth & Brooks/Cole, Monterey, 1987.

KURT MEHLHORN

[1984] *Data Structures and Algorithms, vol. 2: Graph Algorithms and NP-Completeness*, Springer-Verlag, Berlin, 1984.

KARL MENGER

[1942] Statistical metrics, Proc. Nat. Acad. Sci. USA 28 (1942), 535 - 537.

RADKO MESIAR & ENDRE PAP

[1998] Different interpretations of triangular norms and related operations, Fuzzy Sets & Systems 96 (1998), 183 - 189.

HENRYK MINC

[1978] *Permanents*, Addison-Wesley, New York, 1978.

[1988] *Nonnegative matrices*, Wiley, New York, 1988.

E. MINIEKA & D. R. SHIER

[1973] A note on an algebra for the k best routes in a network, J. Inst. Maths. Applics. 11 (1973), 145 – 149.

MICHEL MINOUX

[1976] Structures algébriques généralisées des problèmes de cheminement dans les graphes: théorèmes, algorithmes et applications, RAIRO Rech. Oper. 10 (1976), 33 – 62.

[1979] Generalized path algebras, in A. Prékopa (ed.): *Survey of Mathematical Programming*, vol. 2, North Holland, Amsterdam, 1979.

SIDNEY S. MITCHELL & PAUL B. FENOGLIO

[1988] Congruence-free commutative semirings, Semigroup Forum 37 (1988), 79 – 91.

SIDNEY S. MITCHELL & PORNTIP SINUTOKE

[1982] The theory of semifields, Kyungpook Math. J. 22 (1982), 325 – 347.

MASAHARU MIZUMOTO

[1989] Pictorial representations of fuzzy connectives. Part I: Cases of t-norms, t-conorms, and averaging operators, Fuzzy Sets and Systems 31 (1989), 217 – 242.

MASAHARU MIZUMOTO & KOKICHI TANAKA

[1981] Fuzzy sets and their operations, Inform. and Control 48 (1981), 30 – 48.

MASAHARU MIZUMOTO, J. TOYODA & KOKICHI TANAKA

[1975] Various kinds of automata with weights, J. Computer System Sci. 10 (1975), 219 – 236.

JIRI MOCKOR

[1977] A realization of d-groups, Czech. Math. J. 27 (1977), 296 – 312.

[1983] *Groups of Divisibility*, D. Reidel, Dordrecht, 1983.

A. MOFFAT & T. TAKAOKA

[1987] An all pairs shortest path algorithm with expected time $O(n^2 log\ n)$, SIAM J. Computing 16 (1987), 1023 – 1031.

ITALICO MOLINARO

[1960] Demi-groupes résiduifs, J. Math. Pures et Appl. (Ser. 9) 39 (1960), 319 – 356.

RAMON E. MOORE

[1966] *Interval Analysis*, Prentice-Hall, Englewood Cliffs, N. J., 1966.

[1979] *Methods and Applications of Interval Analysis*, SIA Studies in Applied Math. #2, Soc. Indust. Appl. Math., Philadelphia, 1979.

JAMES R. MOSHER

[1970] Generalized quotients of hemirings, Compos. Math. 22 (1970), 275 - 281.

[1971] Semirings with descending chain condition and without nilpotent elements, Compos. Math. 23 (1971), 79 – 85.

P. MUKHOPADHYAY

[1996] Characterization of regular semirings, Math. Vesnik 48 (1996), 83 - 85.

ALBERT A. MULLIN

[1975] On the algebraic structure of pre-rings, Notices Amer. Math. Soc. 22 (1975), A703.

V. MURALI

[1989] Fuzzy equivalence relations, Fuzzy equivalence relations 30 (1989), 155 - 163.

KENTARO MURATA

[1950] On the quotient semi-group of a nonkommutative semi-group, Osaka Math. J. 2 (1950), 1 – 5.

P. V. RAMANA MURTY

[1974] Ideal topology on a distributive lattice, J. Austral. Math. Soc. 18 (1974), 503 – 508.

J. C. MUZIO & T. C. WESSELKAMPER

[1986] *Multiple-valued Switching Theory*, Adam Hilger Ltd., Bristol, 1986.

HIDEGORO NAKANO

[1966] *Linear Lattices*, Wayne State University Press, Detroit, 1966.

TAKEO NAKANO

[1967] Rings and partly ordered systems, Math. Z. 99 (1967), 355 – 376.

G. NAUDÉ & C. NOLTE

[1982] A survey of the realization and duality theories of linear systems over rings, Quaest. Math. 5 (1982), 135 – 164.

M. H. A. NEWMAN

[1941] A characterization of Boolean lattices and rings, J. London Math. Soc. 16 (1941), 256 – 272.

SUSAN B. NIEFIELD & KIMMO I. ROSENTHAL

[1988] Constructing locales from quantales, Math. Proc. Camb. Phil. Soc. 104 (1988), 215 – 234.

[1990] Sheaves of semiprime ideals, Cahiers Topologie Gèom. Diffèrentielle Catègoriques 31 (1990), 213 – 228.

MAURICE NIVAT

[1968] Transductions des langages de Chomsky, Ann. Inst. Fourier, Grenoble 18 (1968), 339 – 456.

EMMY NOETHER

[1927] Abstrakter Aufbau der Idealtheorie in algebraischen Zahl und Funktionenkörpern, Math. Ann. 96 (1927), 26 - 61.

M. L. NORONHA GALVÃO

[1959] On the theory of residuals, Univ. Lisboa Rev. Fac. Ci. A (2) 7 (1959), 283 - 300.

[1960a] Sur les idéaux réguliers d'un semi-anneaux, Univ. Lisboa Rev. Fac. Ci. A (2) 8 (1960), 169 - 172.

[1960b] Sobre a teoria de Noether-Krull em seimianeis, Univ. Lisboa Rev. Fac. Ci. A (2) 8 (1960), 175 - 256.

[1978a] Les relations de congruence sur le demi-anneau N des nobres naturels, Period. Math. Hungar. 9 (1978), 231 - 235.

[1978b] Ideals in the semiring N, Portugal. Math. 37 (1978), 113 - 117.

[1978c] Sur les demi-anneaux à division, in *Estudos de Algebra, Geometria et Análise*, No. II, Centenário da Academia das Ciencias de Lisboa, Lisbon, 1978.

SAKIKO OHASHI

[1968] On axiom systems of commutative rings, Proc. Japan. Acad. 44 (1968), 915 - 924.

GEERT JAN OLSDER

[1991] Eigenvalues of dynamic max-min systems, Discrete Event Dynamical Systems: Theory and Applications 1 (1991), 177-207.

[1992] About difference equations, algebras and discrete events, in A. van der Burgh & J. Simonis (eds.): *Topics in Engineering Mathematics*, Kluwer, Dordrecht, 1992, 121 - 150.

GEERT JAN OLSDER & C. ROOS

[1988] Cramér and Cayley-Hamilton in the max algebra, Lin. Alg. Appl. 101 (1988), 87–108.

D. M. OLSON

[1978] A note on the homomorphism theorem for hemirings, Internat. J. Math. & Math. Sci. 1 (1978), 439 - 445.

D. M. OLSON & T. L. JENKINS

[1983] Radical theory for hemirings, J. Nat. Sci. Math. 23 (1983), 23 - 32.

SERGEI V. OVCHINNIKOV

[1981] Structure of fuzzy binary relations, Fuzzy Sets and Systems 6 (1981), 169 - 175.

[1993] On some fuzzy binary relations related to implication functions, in R. Lowen and M. Roubens (eds.): *Fuzzy Logic, State of the Art*, Kluwer, Dordrecht, 1993.

R. PADMANABHAN & H. SUBRAMANIAN

[1968] Ideals in semirings, Math. Japon. 13 (1968), 123 - 128.

S. N. N. PANDIT

[1961] A new matrix calculus, SIAM J. Appl. Math. 9 (1961), 632 - 639.

SEYMOUR PAPERT

[1959] Which distributive lattices are lattices of closed sets?, Proc. Camb. Phil. Soc. 55 (1959), 172 - 176.

[1964] An abstract theory of topological subspaces, Proc. Camb. Phil. Soc. 60 (1964), 197 - 203.

DAVID PARK

[1981] Concurrency and automata on infinite sequences, in P. Duessen (ed.): *Theoretical Computer Science*, Lecture Notes in Computer Science #104, Springer-Verlag, Berlin, 1981.

JAN PASEKA

[1986] Regular and normal quantales, Archiv. Math. (Brno) 22 (1986), 203 - 210.

JAN PAVELKA

[1979a] On fuzzy logic I, Z. Math. Logik Grund. Math. 25 (1979), 45 - 52.

[1979b] On fuzzy logic II, Z. Math. Logik Grund. Math. 25 (1979), 119 - 134.

[1979c] On fuzzy logic III, Z. Math. Logik Grund. Math. 25 (1979), 447 - 464.

K. R. PEARSON

[1966] Interval semirings on R_1 with ordinary multiplication, J. Austral. Math. Soc. 6 (1966), 273 - 288.

[1983a] Certain topological semirings in R_1, J. Austral. Math. Soc. 8 (1968), 171 - 182.

[1983b] Embedding semirings in semirings with multiplicative unit, J. Austral. Math. Soc. 8 (1968), 183 - 191.

KETY PEEVA

[1983] On algebraic structures for matrices, relations and graphs over a semiring, Mat. Bul. 7-8 (1983/4), 36 - 48.

[1991] Equivalence, reduction and minimization of finite automata over semirings, Theor. Comp. Sci. 88 (1991), 269 - 285.

DIONISIO PEREZ ESTEBAN

[1988] Semi-rings and spectral spaces, in J. L. Bueso et al. (eds.): *Ring Theory Proceedings, Granada, 1986*, Lecture Notes in Mathematics #1328, Springer-Verlag, Berlin, 1988.

VASILE PETEANU

[1967] An algebra of the optimal path in networks, Mathematica (Cluj) 9 (1967), 335 - 342.

[1969] Optimal paths in networks and generalizations (I), Mathematica (Cluj) 11 (1969), 311 - 327.

[1970] Optimal paths in networks and generalizations (II), Mathematica (Cluj) 12 (1970), 159 - 186.

JAMES L. PETERSON

[1981] *Petri Net Theory and the Modeling of Systems*, Pretice-Hall, Englewood Cliffs, N.J., 1981.

MARIO PETRICH

[1965] Associative polynomial multiplications over an infinite integral domain, Math. Nachr. 29 (1965), 67 – 75.

A. R. PIERCE

[1975] Bibliography on algorithms for shortest path, shortest spanning tree, and related circuit routing problems (1956-1974), Networks 5 (1975), 129 – 149.

RICHARD S. PIERCE

[1972] *Compact Zero-Dimensional Metric Spaces of Finite Type*, A. M. S. Memoir #130, American Mathematical Society, Providence, 1972.

[1983] Tensor products of boolean algebras, in R. S. Freese & O. C. Garcia (eds.): *Universal Algebra and Lattice Theory*, Lecture Notes in Mathematics #1004, Springer-Verlag, Berlin, 1983.

[1989] Countable boolean algebras, in J. Donald Monk (ed.): *Handbook of Boolean Algebras*, vol. 3, North Holland, Amsterdam, 1989.

JEAN-ÉRIC PIN

[1998] Tropical semirings, in J. Gunawardena (ed.): *Idempotency (Bristol, 1994)*, Publ. Newton Inst. #11, Cambridge Univ. Press, Cambridge, 1998, 50 – 69.

B. PIOCHI

[1988] Congruences on inversive hemirings, Stud. Sci. Math. Hungar. 23 (1988), 251 – 255.

MIREILLE O. POINSIGNON GRILLET

[1970] Subdivision rings of a semiring, Fund. Math. 67 (1970), 67 – 74.

S. V. POLIN

[1974] Simple skew semifields and semifields, Sibir. Math. Ž. 15 (1974), 90 – 101. [Russian]

FRANCISCO POYATOS SUAREZ

[1967] Descomposiciones irreducibles en suma directa interna de ciertas estructuras algebraicas, Rev. Mat. Hispano-Americana (4) 27 (1967), 151 – 170.

[1969] Sobre ciertas descomposiciones de semimódulos en suma directa, Rev. Mat. Hispano-Americana (4) 29 (1969), 50 – 58.

[1971] Modulos de complecion de A-semimódulos, Rev. Mat. Hispano-Americana (4) 31 (1971), 123 – 156.

[1972] El teorema de Jordan-Hölder para A-semimódulos, Rev. Mat. Hispano-Americana (4) 32 (1972), 251 – 260.

[1973a] El teorema de Jordan-Hölder para A-semimódulos (continuación), Rev. Mat. Hispano-Americana (4) 33 (1973), 36 – 48.

[1973b] El teorema de Jordan-Hölder para A-semimódulos (conclusión), Rev. Mat. Hispano-Americana (4) 33 (1973), 122 - 132.

[1974] Sucesiones principales y series de composicion de A-semimódulos, Rev. Mat. Hispano-Americana (4) 34 (1974), 60 - 66.

[1976] Series principales y series de composicion de semianillos, Rev. Mat. Hispano-Americana (4) 36 (1976), 63 - 76.

[1977] Sobre ciertas propiedades de los bi-ideales de un semianillo, Rev. Mat. Hispano-Americana (4) 37 (1977), 107 - 113.

[1980] The Jordan-Hölder theorem for semirings, Rev. Mat. Hispano-Americana (4) 40 (1980), 49 - 65.

RAQUEL PRATA DOS SANTOS

[1981] Pseudocomplementation in semirings and Ω-algebras, in *Actas VIII Jornadas Luso Espanholas Matemática*, vol. I, Universidade Coimbra, 1981.

VAUGHN PRATT

[1986] Modeling concurrency with partial orders, Int. J. Parallel Processing 15 (1986), 33 - 71.

[1993] The second calculus of binary relations, in *MFCS '93*, Lecture Notes in Computer Science #711, Springer-Verlag, Berlin, 1993, 142 - 155.

[1994] Chu spaces, complementarity and uncertainty in rational mechanics, preprint, 1994.

[1995a] The Stone gamut: a coordinatization of mathematics, in *Proc. 10th Annual Symposium on Logic in Computer Science*, IEEE Computer Society Press, 1995, 444 - 454.

[1995b] Chu spaces and their interpretation as concurrent objects, in *Computer Science Today*, Lecture Notes in Computer Science #1000, Springer-Verlag, Berlin, 1995, 392 - 405.

[1996] Reconciling event structures and higher dimensional automata, preprint, 1996.

[1997] Chu spaces from the representational viewpoint, preprint, 1997.

WALTER PRENOWITZ & JAMES JANTOSCIAK

[1979] *Join Geometries*, Springer-Verlag, Berlin, 1979.

KENDALL PRESTON, Jr. & MICHAEL J. B. DUFF

[1984] *Modern Cellular Automata, Theory and Applications*, Plenum, New York, 1984.

HILARY ANN PRIESTLEY

[1970] Representation of distributive lattices by means of ordered Stone spaces, Bull. London Math. Soc. 2 (1970), 186 - 190.

DANUTA PRZEWORSKA-ROLEWICZ

[1988] *Algebraic Analysis*, Kluwer, Dordrecht, 1988.

[1998] *Logarithms and Antilogarithms*, Kluwer, Dordrecht, 1998.

MARGARITA RAMALHO

[1974] Demi-anneaux: morphismes et idéaux II, in *Estudos de Matemática* em *Homenagem ao Prof. A. Almeida Costa*, Lisbon, 1974.

[1976] A characterization of μ-semirings, Proc. Amer. Math. Soc. 60 (1976), 49 - 52.

P. RANGA RAO

[1981] Lattice ordered semirings, Math. Sem. Notes Kobe Univ. 9 (1981), 119 - 149.

HELENA RASIOWA & ROMAN SIKORSKI

[1963] *The Mathematics of Metamathematics*, PWN, Warsaw, 1963.

LASZLO RÉDEI

[1952] Die Verallgemeinerung der Schreierschen Erweiterungstheorie, Acta Sci. Math. (Szeged) 14 (1952), 252 - 273.

[1967] *Algebra*, vol. I, Akadémiai Kaidó, Budapest, 1967.

HORST REICHEL

[1987] *Initial Computability, Algebraic Specifications, and Partial Algebras*, Oxford University Press, Oxford, 1987.

WOLFGANG REISIG

[1985] *Petri Nets*, Springer-Verlag, Berlin, 1985.

CHRISTOPHE REUTENAUER

[1978] Sur les series rationnelles en variables non commutatives, in G. Ausiello and C. Böhm (eds.):, *Automata, Languages and Programming*, Lecture Notes in Computer Science #62, Springer-Verlag, Berlin, 1978.

CHRISTOPHE REUTENAUER & HOWARD STRAUBING

[1984] Inversion of matrices over a commutative semiring, J. Algebra 88 (1984), 350 - 360.

PIERRE ROBERT

[1971] An algorithm for finding the essential sets of arcs of certain graphs, J. Comb. Theory 10 (1971), 288 - 298.

GAETANO RODRIQUEZ

[1977] Bande di semianelli monoidali, Boll. Un. Mat. Ital. (5) 14B (1977), 569 - 591.

[1980] Decomposizione di un semianello in un semireticolo di semianelli, Boll. Un. Mat. Ital. Suppl. 2 (1980), 53 - 67.

LEOPOLDO ROMÁN

[1989] Cartesian categories with natural numbers object, J. Pure Appl. Algebra 58 (1989), 267 - 278.

LEOPOLDO ROMÁN & BEATRIZ RUMBOS

[1991a] Quantic lattices, Internat. J. Theoret. Phys. 30 (1991), 1555–1563.

[1991b] A characterization of nuclei and implication in orthomodular lattices, J. Pure Appl. Algebra 73 (1991), 155 - 163.

ANNA B. ROMANOWSKA & JONATHAN D. H. SMITH

[1985] *Modal Theory, An Algebraic Approach to Order, Geometry, and Convexity*, Heldermann Verlag, Berlin, 1985.

IVO ROSENBERG

[1988] Algebraic properties of a general convolution, in M.-M. Deza et al. (eds.): *Algebraic, Extremal and Metric Combinatorics, 1986*, London Math. Soc. Lecture Note Series #131, Cambridge University Press, Cambridge, 1988.

DANIEL J. ROSENKRANTZ & HARRY B. HUNT, III

[1988] Matrix multiplication for finite algebraic systems, Inform. Proc. Letters 28 (1988), 189 - 192.

KIMMO I. ROSENTHAL

[1990] A note on Girard quantales, Cahiers Topologie Gèom. Diffèrentielle Catègoriques 31 (1990), 3 - 11.

[1992] A general approach to Gabriel filters on quantales, Comm. Algebra 20 (1992), 3393 - 3409.

JIRI ROSICKÝ

[1987] Multiplicative lattices and frames, Acta Math. Hung. 49 (1987), 391 - 395.

[1989] Multiplicative lattices and C^* - algebras, Cahiers Topologie Gèom. Diffèrentielle Catègoriques 30 (1989), 99 - 105.

[1995] Characterizing spatial quantales, Algebra Universalis 34 (1995), 175 - 178.

JIRI ROSICKÝ & S. ŠMARDA

[1985] T_1-Locales, Math. Proc. Cambridge Phil. Soc. 98 (1985), 81 - 86.

LOUIS H. ROWEN

[1988a] *Ring Theory*, Vol. I, Academic Press, San Diego, 1988.

[1988b] *Ring Theory*, Vol. II, Academic Press, San Diego, 1988.

KAMALARANJAN ROY

[1960] Newmannian geometry I, Bull. Calcutta Math. Soc. 52 (1960), 187 - 194.

GRZEGORZ ROZENBERG & ARTO SALOMAA

[1980] *The Mathematical Theory of L Systems*, Academic Press, New York, 1980.

D. E. RUTHERFORD

[1963/4] The Cayley-Hamilton theorem for semi-rings, Proc. Roy. Soc. Edinburgh A66 (1963/4), 211 - 215.

JACQUES SAKAROVITCH

[1987] Kleene's theorem tevisited, in A. Kelemenová & J. Kelemen (eds.): *Trends, Techniques and Problems in Theoretical Computer Science*, Lecture Notes in Computer Science #281, Springer-Verlag, Berlin, 1987.

ARTO SALOMAA

[1966] Two complete axiom systems for the algebra of regular events, J. Assoc. Comp. Mach. 13 (1966), 158 – 169.

ARTO SALOMAA & MATTI SOITTOLA

[1978] *Automata-theoretic Aspects of Formal Power Series*, Springer-Verlag, Berlin, 1978.

S. N. SAMBOURSKIĬ & A. A. TARASHCHAN

[1992] The Fourier transform and semirings of Pareto sets, in V. P. Maslov & S. N. Sambourskiĭ (eds.): *Idempotent Analysis*, Advances in Soviet Mathematics #13, American Mathematical Society, Providence, 1992.

ELIE SANCHEZ

[1976] Resolution of composite fuzzy relation equations, Inform. and Control 30 (1976), 38 – 48.

JUAN B. SANCHO DE SALAS & TERESA SANCHO DE SALAS

[1989] Dimension of dense subalgebras of C(X), Proc. Amer. Math. Soc. 105 (1989), 491 – 499.

TERESA SANCHO DE SALAS

[1987] *Methods of Commutative Algebra for Topology*, Dpto. de Matemáticas, Universidad de Salamanca, Salamanca, 1987.

BORIS M. SCHEIN

[1976] Regular elements of the semigroup of all binary relations, Semigroup Forum 13 (1976), 95 – 102.

JURG SCHMID

[1983] Distributive lattices and rings of quotients, in A. P. Huhn & E. T. Schmidt (eds.): *Contributions to Lattice Theory*, Colloq. Math. Soc. Jànos Bolyai #33, North Holland, Amsterdam, 1983.

G. SCHMIDT & T. STROHLEIN

[1985] Relation algebras: concept of points and representability, Discrete Math. 54 (1985), 83 – 92.

K. D. SCHMIDT

[1982] A general Jordan decomposition, Arch. Math. 38 (1982), 556 – 564.

E. SCHRÖDER

[1895] *Volresung über die Algebra der Logic (Exakte Logik). Dritter Band: Algebra und Logik der Relative*, B. G.Teubner, Leipzig, 1895.

B. SCHWEIZER & A. SKLAR

[1961] Associative functions and statistical triangle inequalities, Publ. Math. Debrecen 8 (1961), 169 – 186.

[1963] Associative functions and abstract semigroups, Publ. Math. Debrecen 10 (1963), 69 – 81.

DANIEL J. SCULLY

[1991] Maximal rank-one spaces of matrices over chain semirings I: u-spaces, Lin. Alg. Appl. 150 (1991), 67 – 80.

[1993] Maximal rank-one spaces of matrices over chain semirings II: (u,i)-spaces, Lin. Alg. Appl. 181 (1993), 29 – 43.

M. K. SEN & M. R. ADHIKARI

[1992] On k-ideals of semirings, Internat. J. Math. & Math. Sci. 15 (1992), 347 – 350.

[1993] On maximal k-ideals of semirings, Proc. Amer. Math. Soc. 119 (1993), 699 –703.

B. SHAFER

[1976] *A Mathematical Theory of Evidence*, Princeton University Press, Princeton N. J., 1976.

D. R. SHIER

[1973] A decomposition algorithm for optimality problems in tree-structured networks, Discrete Math. 6 (1973), 175 – 189.

[1976] Iterative methods for determining the k shortest paths in a network, Networks 6 (1976), 205 – 229.

M. A. SHUBIN

[1992] Algebraic remarks on idempotent semirings and the kernel theory in spaces of bounded functions, in V. P. Maslov & S. N. Sambourskiĭ (eds.): *Idempotent Analysis*, Advances in Soviet Mathematics #13, American Mathematical Society, Providence, 1992.

H. J. SHYR

[1979] *Free Monoids and Languages*, Soochow University, Taipei, Taiwan, 1979.

IMRE SIMON

[1988] Recognizable sets with multiplicities in the tropical semiring, in M. P. Chytil et al. (eds.): *Mathematical Foundations for Computer Science 1988*, Lecture Notes in Computer Science #324, Springer-Verlag, Berlin, 1988.

[1994] On semigroups of matrices over the tropical semiring, RAIRO Inform. Theor. Appl. 28 (1994), 277–294.‘

PRAHLAD SINGH & N. S. YADAV

[1981] A type of semiring, Nepali Math. Sc. Rep. 6 (1981), 45 – 48.

R. SIVARAMAKRISHNAN

[1989] *Classical Theory of Arithmetic Functions*, Marcel Dekker Inc., New York, 1989.

LEV A. SKORNJAKOV

[1986] Invertible matrices over distributive lattices, Sib. Mat. Ž. 27 (1986), 182 – 185. [Russian]

W. SŁOWIKOWSKI & WACLAW ZAWADOWSKI

[1955] A generalization of maximal ideals method of Stone and Gelfand, Fund. Math. 42 (1955), 215 – 231.

DAVID A. SMITH

[1966] On semigroups, semirings, and rings of quotients, J. Sci. Hiroshima Univ. 30 (1966), 123 – 130.

F. A. SMITH

[1966] A structure theory for a class of lattice ordered semirings, Fund. Math. 59 (1966), 49 – 64.

ALEXANDRU SOLIAN & T. M. VISWANATHAN

[1988] Hochster's theorem, coherent locales, and lattices of radical ideals, Comm. Algebra 16 (1988), 2625 – 2648.

EDUARDO D. SONTAG

[1976] On linear systems and noncommutative rings, Math. Systems Theory 9 (1976), 327 – 344.

DONALD F. STANAT

[1972] Approximation of weighted type 0 languages by formal power series, Inform. and Control 21 (1972), 344 – 381.

OTTO STEINFELD

[1959] Über die Struktursätze der Semiringe, Acta Math. Acad. Sci. Hungar. 10 (1959), 149 – 155.

[1963] Über Semiringe mit multiplikativer Kürzungsregel, Acta Sci. Math (Szeged) 24 (1963), 190 – 195.

[1964] Über die Operatorendomorphismen gewisser Operatorhalbgruppen, Acta. Math. Sci. Hungar. 15 (1964), 123 – 131.

BO STENSTRÖM

[1969] Radicals and socles of lattices, Arch. Math. 20 (1969), 258 – 261.

HOMER E. STONE

[1972] Ideals in halfrings, Proc. Amer. Math. Soc. 33 (1972), 8 – 14.

[1977] Matrix representation of simple halfrings, Trans. Amer. Math. Soc. 233 (1977), 339 – 353.

MARSHALL STONE

[1937] Topological representation of distributive lattices and Brouwerian logics, Casopis Pest. Mat. Fys. 67 (1937), 1 – 25.

HOWARD STRAUBING

[1983a] A combinatorial proof of the Cayley-Hamilton Theorem, Discrete Math. 43 (1983), 273 – 279.

[1983b] The Burnside problem for semigroups of matrices, in L. J. Cummings (ed.): *Combinatorics on Words, Progress and Perspectives*, Academic Press, New York, 1983, 279 – 295.

TEO STURM

[1986] On an algebraization of measure theory, abstract semirings, Quaest. Math. 9 (1986), 393 – 441.

H. SUBRAMANIAN

[1970] Von Neumann regularity in semirings, Math. Nachr. 45 (1970), 73 – 79.

M. SUGENO

[1979] *Theory of Fuzzy Integrals and Applications*, PhD. Thesis, Tokyo Institute of Technology, Tokyo, 1979.

PHILIPP SÜNDERHAUF

[1997] Tensor products and powerspaces in quantitive domain theory, Electronic Notes in Theoretical Computer Science 6 (1997).

MICHIHIRO TAKAHASHI

[1981] On the bordism categories. II, Math. Sem. Notes Kobe Univ. 9 (1981), 495 – 530.

[1982a] On the bordism categories. III, Math. Sem. Notes Kobe Univ. 10 (1982), 211 – 236.

[1982b] Completeness and c-completeness of the category of semimodules, Math. Sem. Notes Kobe Univ. 10 (1982), 551 – 562.

[1982c] Extensions of semimodules. I, Math. Sem. Notes Kobe Univ. 10 (1982), 563 – 592.

[1983] Extensions of semimodules. II, Math. Sem. Notes Kobe Univ. 11 (1983), 83 – 118.

[1984a] On semimodules. I, Kobe J. Math. 1 (1984), 67 – 97.

[1984b] On semimodules. II, Kobe J. Math. 1 (1984), 177 – 190.

[1985] On semimodules. III, Kobe J. Math. 2 (1985), 131 – 141.

[1987] Structures of semimodules, Kobe J. Math 4. (1987), 79 – 101.

[1996a] Isomorphism theorems in semimodule theory, preprint, 1996.

MICHIHIRO TAKAHASHI & WANG HUAXIONG

[1993] Injective semimodule over a 2-semiring, Kobe J. Math. 10 (1993), 59 – 70.

BIT-SHUN TAM

[1981] On the semiring of cone preserving maps, Lin. Alg. and its Appl. 35 (1981), 79 – 108.

TAKAYUKI TAMURA

[1981] Notes on semirings whose multiplicative semigroups are groups, in K. Numakura (ed): *Proceedings of the 5th Symposium on Semigroups*, Josai University, Sakado, 1981.

G. S. TANG & H. H. TEH

[1966/7] A remark on the relation between groupoid semirings and matrix semirings, Nanta Math. 1 (1966/7), 38 – 39.

ROBERT E. TARJAN

[1981a] A unified approach to path problems, J. Assoc. Comp. Mach. 28 (1981), 577 – 593.

[1981b] Fast algorithms for solving path problems, J. Assoc. Comp. Mach. 28 (1981), 594 – 614.

M. K. THAKARE & C. S. MANJAREKAR

[1984] Abstract spectral theory: multiplicative lattices in which every character is contained in a unique maximal character, in H. L. Manocha & J. B. Srivastava (eds): *Algebra and its Applications*, Marcel Dekker, New York, 1984.

M. K. THAKARE, C. S. MANJAREKAR & S. MAEDA

[1988] Abstract spectral theory II: Minimal characters and minimal spectrums of multiplicative lattices, Acta Sci. Math. 52 (1988), 53 - 67.

MELVIN C. THORNTON

[1972] Semirings of functions determine finite T_0 topologies, Proc. Amer. Math. Soc. 34 (1972), 307 - 310.

W. J. THRON

[1962] Lattice-equivalence of topological spaces, Duke Math. J. 29 (1962), 671 - 680.

YUPAPORN TIRASUPA

[1979] Factorizable transformation semigroups, Semigroup Forum 18 (1979), 15 - 19.

C. W. TONG & K. P. LAM

[1996] Closed semiring optimization circuits using a connectionist approach, IEEE Trans. Circuits and Systems I: Fundamental Theory and Applications 43 (1996), 478 - 482.

A. S. TROELSTRA

[1992] *Lectures on Linear Logic*, CSLI Lecture Notes #29, Center for the Study of Language and Information, Stanford, Calif., 1992.

ASKAR A. TUGANBAEV

[1998] *Semidistributive Modules and Rings*, Kluwer, Dordrecht, 1998

L. VALVERDE

[1985] On the structure of F-indistinguishability operators, Fuzzy Sets and Systems 17 (1985), 313 - 328.

ROBERT M. VANCKO

[1974] The class of algebras in which weak independence is equivalent to direct sums independence, Colloq. Math. 30 (1974), 187 - 191.

H. S. VANDIVER

[1934] Note on a simple type of algebra in which cancellation law of addition does not hold, Bull. Amer. Math. Soc. 40 (1934), 914 - 920.

[1939] On some simple types of semi-rings, Amer. Math. Monthly 46 (1939), 22 - 26.

[1940] On the embedding of one semi-group in another, with application to semi-rings, Amer. J. Math. 62 (1940), 72 - 78.

H. S. VANDIVER & M. H. WEAVER

[1956] A development of associative algebra and an algebraic theory of numbers. IV, Math. Mag. 30 (1956), 1 - 8.

JULES C. VARLET

[1969] Fermetures multiplicatives, Bull. Soc. Roy. Sci. Liège 38 (1969), 101 - 115.

JOHN VON NEUMANN

[1966] *Theory of Self-Reproducing Automata*, Univ. of Illinois Press, Urbana, 1966.

N. N. VOROBJEV

[1963] The extremal matrix algebra, Dokl. Akad. Nauk SSSR 152 (1963), 24 - 27. [Russian]

EDOUARD WAGNEUR

[1991] Moduloïds and pseudomodules I. Dimension theory, Discrete Math. 98 (1991), 57 - 73.

[1998] The geometry of finite dimensional pseudomodules, in J. Gunawardena (ed.): *Idempotency (Bristol, 1994)*, Publ. Newton Inst. #11, Cambridge Univ. Press, Cambridge, 1998, 392 - 405.

ABRAHAM WALD

[1943] On a statistical generalization of metric spaces, Proc. Nat. Acad. Sci. USA 29 (1943), 196 - 197.

WANG HAUXIONG

[1988] A note on endomorphism semirings of semimodules, Kobe J. Math. 5 (1988), 155 - 160.

[1994] Injective hulls of semimodules over additively-idempotent semirings, Semigroup Forum 48 (1994), 377 - 379.

[1996] *Some Semiring-theoretic Aspects of Rational Events*, PhD. Thesis, University of Haifa, 1996.

[1997] On characters of semirings, Houston J. Math. 23 (1997), 391 - 405.

[1998] On rational series and rational languages, Theor. Comp. Sci. 205 (1998), 329 - 336.

WANG HAUXIONG & MICHIHIRO TAKAHASHI

[1989] On epimorphisms of semimodules, Kobe J. Math. 6 (1989), 297 - 298.

MORGAN WARD

[1937] Residuation in structures over which a multiplication is defined, Duke Math. J. 3 (1937), 627 - 636.

[1940] Residuated distributive lattices, Duke Math. J. 6 (1940), 641 - 651.

MORGAN WARD & ROBERT P. DILWORTH

[1939] Residuated lattices, Trans. Amer. Math. Soc. 45 (1939), 335 - 354.

SIEGFRIED WEBER

[1983] A general concept of fuzzy connectives, negations, and implications based on t-norms and t-conorms, Fuzzy Sets and Systems 11 (1983), 115 - 134.

WOLFGANG WECHLER

[1975] R-fuzzy grammars, in J. Becvár (ed.): *Mathematical Foundations of Computer Science, 1975*, Lecture Notes in Computer Science #32, Springer-Verlag, Berlin, 1975.

[1977] Families of R-fuzzy languages, in M. Karpiński (ed.): *Fundamentals of Computation Theory*, Lecture Notes in Computer Science #56, Springer-Verlag, Berlin, 1977.

[1984] Linearization of systems of algebraic equations. Part I: Dyck matrices, in G. Wechsung (ed.): *Frege Conference, 1984*, Akademie-Verlag, Berlin, 1984.

[1986a] R-fuzzy computation, J. Math. Anal. Appl. 115 (1986), 225 - 232.

[1986b] Hoare algebras versus dynamic algebras, in J. Demetrovics et al. (eds.): *Algebra, Combinatorics and Logic in Computer Science*, Colloq. Math. Soc. János Bolyai #42, North Holland, Amsterdam, 1986.

[1988] Iterative nondeterministic algebras, in G. Mirkowska & H. Rasiowa (eds.): *Mathematical Problems in Computation Theory*, Banach Center Publications #21, PWN, Warsaw, 1988.

HANNS J. WEINERT

[1962] Über Halbringe und Halbkörper. I, Acta Math. Acad. Sci. Hungar. 13 (1962), 363 - 378.

[1963] Über Halbringe und Halbkörper. II, Acta. Math. Acad. Sci. Hungar. 14 (1963), 209 - 227.

[1964] Über Halbringe und Halbkörper. III, Acta. Math. Acad. Sci. Hungar. 15 (1964), 177 - 194.

[1980] Multiplicative cancellativity of semirings and semigroups, Acta. Math. Acad. Sci. Hungar. 35 (1980), 335 - 338.

[1982] *Theory of Semigroups and Ring-like Algebras*, Notes of Lectures at the Universidad de los Andes, Bogotá, Columbia, 1982.

[1986] Partially ordered semirings and semigroups, in S. Wolfenstein (ed.): *Algebra and Order, Proceedings of the First International Symposium on Ordered Algebraic Structures*, Heldermann-Verlag, Berlin, 1986.

[1988] Generalized semialgebras over semirings, in H. Jürgensen et al. (eds.): *Semigroups, Theory and Applications*, Lecture Notes in Mathematics #1320, Springer-Verlag, Berlin, 1988.

RICHARD WIEGANDT

[1959] On the general theory of Möbius inversion formula and Möbius product, Acta Sci. Math. (Szeged) 20 (1959), 164 - 180.

[1962] Über die Struktursätze der Halbringe, Ann. Univ. Sci. Budapestinensis Sec. Math. 5 (1962), 51 - 68.

AHNONT WONGSEELASHOTE

[1976] An algebra for determining all path-values in a network with application to k-shortest paths problems, Networks 6 (1976). 307 – 334.

[1979] Semirings and path spaces, Discrete Math. 26 (1979), 55 – 78.

WU FUMING

[1998] *A Framework for Dynamic Modelling of Information Systems: Petri Nets on Difference-Ordered Hemirings*, PhD. Thesis, University of Haifa, 1998.

DAVID N. YETTER

[1989] Quantales and (non-commutative) linear logic, preprint, 1989.

YING XIE

[1991] *ω-Complete Semirings and Matrix Iteration Theories*, PhD. Thesis, Stevens Institute of Technology, Hoboken, 1991.

MICHAEL YOELI

[1961] A note on a generalization of boolean matrix theory, Amer. Math. Monthly 68 (1961), 552 – 557.

[1965] Lattice-ordered semigroups, graphs, and automata, J. Soc. Indust. Appl. Math. 13 (1965), 411 – 422.

S. M. YUSUF & M. SHABIR

[1988] Radical classes and semisimple classes for hemirings, Stud. Sci. Math. Hungar. 23 (1988), 231 – 235.

ADRIAAN C. ZAANEN

[1983] *Riesz Spaces II*, North Holland, Amsterdam, 1983.

LOFTI A. ZADEH

[1965] Fuzzy Sets, Inform. and Control 8 (1965), 338 – 353.

HANS J. ZASSENHAUS

[1958] *The Theory of Groups*, 2nd ed., Chelsea, New York, 1958.

JOHN ZELEZNIKOW

[1981] Regular semirings, Semigroup Forum 23 (1981), 119 – 136.

ZHANG KUN-LUN

[1994] Determinant theory for D_{01}-lattice matrices, Fuzzy Sets and Systems 62 (1994), 347 – 353.

ZHAO CUI-KUI

[1987] On matrix equations in a class of complete and completely distributive lattices, Fuzzy Sets and Systems 22 (1987), 303 – 320.

[1990] Inverses of L-fuzzy matrices, Fuzzy Sets and Systems 34 (1990), 103 – 116.

KAREL ZIMMERMANN

[1982] The solution of some nonlinear optimization problems in a minimax algebra, Z. Angew. Math. Mech. 62 (1982), T391 – T392.

UWE ZIMMERMANN

[1981] *Linear and Combinatorial Optimization in Ordered Algebraic Structures*, North Holland, Amsterdam, 1981.

INDEX OF APPLICATIONS

Semiring theory has been developed not as an exercise in generalization for generalization's sake but because of its value as a tool in many significant applications in mathematics, computer science, and other fields. A sample of these applications is given in the text, though, since our primary emphasis is on the mathematics of semirings, most are only mentioned *en passant*. References are provided, however, to allow the interested reader to pursue these on his/her own.

INDEX OF TERMINOLOGY

A

■ B

■ C

■ D

■ E

F

■ G

H

I

■ J

■ K

■ L

■ M

■ N

■ O

P

■ Q

■ R

■ S

■ T

Y

Z

GPSR Compliance
The European Union's (EU) General Product Safety Regulation (GPSR) is a set of rules that requires consumer products to be safe and our obligations to ensure this.

If you have any concerns about our products, you can contact us on

ProductSafety@springernature.com

In case Publisher is established outside the EU, the EU authorized representative is:

Springer Nature Customer Service Center GmbH
Europaplatz 3
69115 Heidelberg, Germany

www.ingramcontent.com/pod-product-compliance
Ingram Content Group UK Ltd.
Pitfield, Milton Keynes, MK11 3LW, UK
UKHW021504200726
13877UKWH00027B/489

* 9 7 8 9 4 0 1 5 9 3 3 4 2 *